UNCONVENTIONAL METHODS IN EXPLORATION FOR PETROLEUM AND NATURAL GAS IV

UNCONVENTIONAL METHODS IN EXPLORATION FOR PETROLEUM AND NATURAL GAS IV

A SYMPOSIUM UNDER THE AUSPICES OF THE
INSTITUTE FOR THE STUDY OF
EARTH AND MAN
AT
SOUTHERN METHODIST UNIVERSITY

Edited by
MARTIN J. DAVIDSON

INSTITUTE FOR THE STUDY OF EARTH AND MAN
SOUTHERN METHODIST UNIVERSITY PRESS
DALLAS

UNCONVENTIONAL METHODS IN EXPLORATION
FOR PETROLEUM AND NATURAL GAS IV

The paper in this book meets the standards for permanence and durability established by the Committee on Production Guidelines for Book Longevity of the Council on Library Resources.

Typesetting and production by AnthroGraphics,
Southern Methodist University, Dallas, Texas 75275.

Distributed by SMU Press
Box 415
Southern Methodist University
Dallas, Texas 75275

Library of Congress Cataloging-in-Publication Data

Unconventional methods in exploration for petroleum and natural gas IV.

1. Prospecting—Congresses. 2. Petroleum—Congresses.
3. Gas, Natural—Congresses.

I. Davidson, Martin J., 1924-
II. Southern Methodist University. Institute for the Study of Earth and Man.

TN271.P4U534 1986 622'.1828 86-13962
ISBN 0-87074-221-3

INSTITUTE FOR THE STUDY OF EARTH AND MAN

"Nature and history, the earth and man, stand in the closest relations to each other, and form only one grand harmony."

—*Arnold Guyot, 1849*

The Institute, under whose auspices this symposium was arranged, is organized as an affiliate of Southern Methodist University "to encourage, foster, promote and advance research and development and education in the field of the anthropological and geological sciences and other related fields of scholarship at Southern Methodist University."

Incorporated in Texas on May 11, 1966

The Institute notes the great loss to innovative exploration in the deaths, since Symposium III, of Dr. Sylvain Pirson and Dr. Leo Horvitz. Each was a pioneer in the most complete and authentic sense of the term. Each, to his own great honor, never ceased the pursuit of new science.

CONTENTS

FOREWORD

The Institute's fourth symposium on Unconventional Methods in Exploration for Petroleum and Natural Gas was held on May 1 and 2, 1985. The planning of the symposium was under the general supervision of the Institute's symposium committee which was chaired by Benjamin M. Gottlieb and included Claude C. Albritton, Jr., Marvin Bloomquist, Martin J. Davidson, Charles F. Dodge, William B. Heroy, Jr., W. T. Lehman, Earl R. Price, Dietmar Schumacher, Richard C. Weart, and William Webber. The detailed planning was done by a subcommittee consisting of Martin J. Davidson, Program Chairman, Earl R. Price, and Richard C. Weart. Members of the ISEM staff also played key and continuous roles in staging the symposium; especially to be noted are Netta G. Blanchard and Pamela Smith.

The symposium attendance of 135 contributed extensively during discussion periods. The technical program covering a variety of topics received favorable comment from the participants. Of special interest was the keynote luncheon address which was given by Dr. Juan Chacin Guzman, Vice President of Petroleos de Venezuela. His topic was "An Overview of the Petroleum Industry in Latin America." The address was given to a large invited audience including members of the Dallas Council on World Affairs.

Sincere appreciation is due Martin J. Davidson for extensive and significant contributions in the technical editing of this volume.

James E. Brooks
Dallas, Texas

PREFACE

Publication of the fourth volume of this series provides the occasion to weigh the success of William B. Heroy's mid-1960s vision of establishing a forum in exploration that would overcome the barriers to innovation characteristic of all conventional wisdom. Those barriers, in petroleum exploration, have hardly been overcome but they have been decidedly breached and, to that extent, Bill Heroy's vision has become a reality.

The early emphasis of this series stressed case histories. The present emphasis is to recognize, and especially to encourage, theoretical analysis. What is needed now is general acquiescence to a theoretical framework, not necessarily unchallengeable, that would explain the reported empirical findings.

Even without convincing explanatory mechanisms, empirical research has value in itself. In most fields, technology preceded science, which is to say that "know-how" preceded "know-why." Beginning technology, which was essentially empirical, invariably paved the way for investment in "science research" that not only refined the existing technology but forced the creation of newer technologies. Society is painfully perceiving that human welfare is a function of the management of science, in the private as well as in the public sector.

Moving from the general to the specific, and in particular to those exploration methods based on vertical migration (with which so much of this series has been concerned), we have reached a watershed. Empirical data abound and have stimulated the use of vertical migration methods, but that is not enough. The continuing and presently critical task is to accommodate that empirical data to a theoretical structure that has yet to be convincingly demonstrated.

In the absence of expanded theoretical research (independently replicated and made public) vertical migration methods will deteriorate into the application of inexpensive approximations of valid techniques. The end result may well be technological regression. While it is true that the best is often the enemy of the good, those who perform and pay for research, and especially those who perform and pay for exploration, must come to understand that the "poor-boy" approach is not only too expensive, but its failures will destroy the momentum that is now apparent in the increased use of a number of different vertical migration methods.

Leigh Price's insistence (this volume, p. 245) on the need for greater and more intensive theoretical research is certainly on point. That such a keen awareness of *the* critical issue should emanate from the U.S. Geological Survey is high irony when one considers the conscious and purposeful benign neglect of vertical migration methods by the Survey over most of the last fifty years.

Production of a volume of this nature is always a more complex task than initially envisaged. This book owes its very existence to the indomitable constancy of support by Charlotte Whaley as copy editor, by Chris Christopher, producer of the camera-ready copy, and by Liebe Purnell who coordinated production with the staff of the Institute and with SMU Press.

Martin J. Davidson
Dallas, Texas
July 1986

UNCONVENTIONAL METHODS IN EXPLORATION FOR PETROLEUM AND NATURAL GAS

IV

WHY NOT LIMIT GEOPHYSICAL EXPLORATION TO THE SEISMIC METHOD?

T. R. LaFehr

EDCON, INC.

INTRODUCTION

The overwhelming dominance of geophysical exploration by the seismic reflection method is a reality well appreciated by everyone at this fourth symposium on unconventional methods in exploration for petroleum and natural gas. The underlying question in the title, "Why not limit geophysical exploration to the seismic method?", is not intended to be an affront to the serious efforts of unconventional hydrocarbon explorationists. Similarly, when addressing our seismic colleagues, acknowledgement is taken of the superb advances which that discipline has enjoyed over the last fifty or more years. It may prove useful to those developing nonseismic techniques to try to understand this exploration marketplace.

We may wish to question the logic of those many exploration managers who do choose not to use any geophysical methods other than seismic reflection. Why have they not been persuaded over the years that some of the more serious problems inherent in the seismic technique can be helped, if not solved, by bringing into their program some of the technology described in this kind of symposium? Even when our methods are employed, we seldom enjoy comparable priority to the seismic interpretation. If at all possible, geologic disagreements are resolved in favor of the seismic interpretation even in the presence of known acoustic pitfalls. Obviously, more than science is involved in such decisions.

You may question the logic of exploration decision makers, as I have done, and you may also be concerned about the very fabric of our market. The petroleum industry has been well known for its "flocking tendency," and it still is. Professor Martin Davidson said in a previous meeting like this one, "Explorationists tend to huddle in areas where subsurface control exists and producing zones are familiar. Even seismic is at its best close to where velocities are known." We can all recall some of the more outstanding dogmas-gone-wrong which at one time guided a whole industry. It was often lonely, expensive, and frustrating to challenge the prevailing wisdom; but occasionally the payoff for such contrariness was very large indeed.

Our marketplace is not a simple one and cannot be described easily in a few words. There may be almost as many opinions as there are people propounding them. Nonetheless, for purposes relevant to this meeting, we can analyze geophysical exploration in terms of three philosophic schools: (1) the Reconnaissance School, (2) the Seismic-only School, and (3) the School of Coordinated Geophysical Exploration. Each of these schools has a substantial following. Each has experienced its exploration and economic successes. No doubt defectors exist—those who previously belonged exclusively to one school and are now firmly entrenched in another. Other schools also exist and some explorationists may belong to one or more. With all due respect, however, the other schools are not (at least not yet) viable forces in the exploration marketplace.

RECONNAISSANCE EXPLORATION: A MYTH?

For quite some time now, the reconnaissance theory of exploration has enjoyed a certain level of respectability, partly because it makes sense and partly because our leading textbooks on geophysical exploration tell us that it makes sense. The reconnaissance theory may be stated as follows: "Nonseismic methods are cheap and fast; it is only prudent and cost-effective to let these methods guide the more expensive seismic surveys."

Dr. Sigmund Hammer stated a similar view at Symposium III and elsewhere, citing that the 1981

Components of this opening address of Symposium IV were first developed in a panel discussion on exploration in mature areas at the SEG Midwestern Meeting in San Antonio, Texas, on March 17, 1985.

geophysical activity report (SEG) indicates about $500 million (estimated at half the total on land) was spent for seismic reconnaissance in the continental United States. Dr. Hammer goes on to say that trade journals are loaded with display ads offering nonexclusive seismic surveys with line spacings of 10–20 mi or more. He asks: What is the chance that such open seismic reconnaissance will find or, even more basically, fail to find an oil field in the area? His answer is that, while the seismic interests call for more and more seismic effort at enormous increases in cost, what the petroleum industry *really* needs is effective geophysical land reconnaissance which yields 100 times the rate of coverage at less than 1/10 of the cost.

Although gravity and magnetics have experienced, and may continue to experience, success in the role of lead reconnaissance tools, two problems exist with Dr. Hammer's thesis. The first is the implication that seismic exploration is too expensive to invade substantially turf previously believed to be the rightful domain of gravity and magnetics. The reason that so much money is spent on reconnaissance seismic activity is not the perversion of "seismic interests" (they, too, are subject to exploration marketplace whims; witness the current SEG Crew Count), but the unique geologic value usually resulting from such surveys. Reflection seismology is our *best reconnaissance* method. Seismic records can yield multiple intervals and interval velocities, information which is invaluable at the reconnaissance stage as well as at later stages of exploration.

The second problem with Dr. Hammer's thesis is that it tends to perpetuate the textbook mythology of the reconnaissance role for gravity and magnetics. Gravity and magnetics do play a role in reconnaissance exploration, sometimes a decisive one. There is, however, a real danger in stereotyping these methods as reconnaissance only. Like all good mythologies, this one is based on some past truth, perhaps enough to do more harm than good. Unfortunately, the "reconnaissance only" argument obscures a far more important future for all nonseismic techniques: aiding the critical decisions regarding detailed prospect evaluation even in mature exploration provinces.

For economic reasons most offshore work is conducted as joint operations. Seismic is the primary tool, and gravity and magnetics are often added because the incremental cost is not large. Such joint operations preclude the use of previously obtained gravity and magnetic data as a basis for the seismic program. Onshore, the locations of basins are generally well established and rather detailed gravity and magnetic surveys have already been acquired. Reconnaissance is a term whose meaning has expanded with the use of detailed gravity and magnetics, if only onshore. There are exceptions to these conclusions, but we must be cognizant of the way the world actually works while we are working to move it toward accepting new technology.

These are times when more accuracy, more resolution, more detail, and more closely spaced data are needed; we will not be accepted in this seismically dominated exploration world if we cannot or will not help explorationists to solve their problems. The reconnaissance mode for nonseismic methods, while occasionally important, is clearly secondary to the far more important consideration of participating in detailed prospect evaluation.

THE SEISMIC-ONLY SCHOOL

We wonder why companies ignore our technology in the face of so many cases which clearly demonstrate its exploration utility. Carl Savit, a former SEG president, said:

> About the only way that we really have to explore under the earth is with sound. We've tried all kinds of other things over the years, but the only thing that will get us any kind of penetration and detail is sound. We've tried electrical and radio waves, and everything that people have come up with, including psychics who've come around and told us what there is underground, but the reliable one is sound, the seismic waves that we normally use.

A widespread view prevails that only seismic data contain the characteristics required for modern petroleum geophysical exploration. A parallel exists between this dogma and the reconnaissance one. Their simplistic themes do not provide optimal solutions when confronted with even normally complex geologic environments. *Only* sound gives us "*any kind* of penetration and detail?"

Let us give credit where credit is due and reject the view that it is "seismic interests" that are responsible for the dominance of seismic methods. We are not going to make much progress by ignoring the real reasons for the fact that our market share is miniscule. The seismic method works amazingly well a respectable fraction of the time that it is put to use; it is one of mankind's finest scientific and engineering accomplishments. It is so useful and important in deciphering the subsurface geology that few explorationists would consider drilling their wells without including it.

It is rather important for those of us engaged in nonseismic exploration to understand another reason for the existence of the seismic-only school. Many in our marketplace use the seismic method as their exclusive geophysical tool because a seismic cross section *looks* like a geologic cross section. When imploring management to risk their company's money to drill a well, most geophysicists are far more at home with their seismic sections that look like geologic sections than in trying to explain how milligals or gammas (nanoTeslas?) relate to structure, depth, or lithology. You and I know that direct geologic interpretation of seismic data is not so simple that we can safely accept such an analogy.

COORDINATED GEOPHYSICAL EXPLORATION

The recognition that the seismic method plays an essential and important role, and that nonseismic methods offer more than reconnaissance, leads to another exploration philosophy: coordinated geophysics. Geophysical measurements respond to variations in rock properties, which we hope are diagnostic of the economic potential of the prospect. We can, therefore, logically ask: "Can the geologic results we seek somehow be extracted from the geophysical data?"

For all extant geophysical methods, we might discuss this question in terms of four possible outcomes: (1) yes, (2) no, (3) we don't know the answer, and (4) partially. The following discussion of these four possibilities is excerpted from *The Leading Edge* President's Page, June, 1984.

> Fortunately, though not with the frequency we would like, the answer is sometimes "yes." This is, of course, precisely why we are in business. A "yes" response can run the gamut from low resolution, weak signal cases to unmistakable reflections, events, or anomalies. The strength of the signal is not what's at issue here, though obviously we require some minimal or threshold response. It is the *existence* of the signal that matters, and we'll leave for another discussion the ever-present problem of identifying the signal in what may be a noisy data set.
>
> Some believe that the signal is always somewhere in the data but that at times we just can't seem to extract it. This is, unfortunately, not always true because a geologic boundary may not be a geophysical boundary. Salt domes or layers, for example, may exist adjacent to sands or shales having the same density, so that no matter how large the structure it will produce no gravity anomaly from the zone of geologic contrast. Similarly, some geologic boundaries are transparent to the seismograph because of the absence of any acoustic impedance contrast. Some basement rocks we know to be nonmagnetic. Therefore, regrettably, sometimes the answer is "no"; the answer we seek is not contained in a particular data set.
>
> But I dare say the unequivocal "no" is unusual. More often, we simply do not know the answer. We have to acquire the geophysical data just to answer the question, and even then we may still be ambivalent in a difficult geologic area. And even within a given area, the answer may grade from "no" to "yes," with changing depths, porosity, lithology, or structure.
>
> Much of the time, a majority in my own experience, the answer is "partially." For example, salt overhang may exist in a zone of no density contrast, but the overhang may be connected to large positive and negative contrasts above and below it. Or a strong reflection on one side of a fault seemingly is lost altogether on the other side. Or a nonmagnetic basement terrane is interrupted by strongly magnetic rocks. In such cases and in many more, we need access to as much geophysical information as possible.

The "Coordinated" philosophy of geophysical exploration is not new, but I see it enjoying a renewed emphasis as we admit that pitfalls exist in each method. Sometimes the weakness of one method is minimized or even avoided by the introduction of another, sister, geophysical method.

CONCLUSIONS

L. L. Nettleton in *Gravity and Magnetics in Oil Prospecting* wrote:

> There is a natural tendency, by those concerned primarily with any one of the geophysical methods, to become myopic in their study of the physical or mathematical aspects . . . The large fraction of geophysicists employed in the oil industry will do well to keep in mind that the *objective of their jobs is to find oil.*

Let those of us who specialize in nonseismic technology recognize the extraordinary importance of seismic exploration. Let us also be clear about the limitations of the nonseismic techniques. And let us emphasize our strengths in two ways: (1) what our specialties can add toward improved understanding of exploration targets in a stand-alone mode, and (2) how we can help get more information out of seismic data.

This last, in my view, is the fundamental reason why we should not limit geophysical exploration to the seismic method. It makes sense to the exploration manager, who juggles far more than geophysical data, that no one specialty can flourish in a vacuum. Physicist Erwin Schrödinger said it well: "Specialization is not a virtue but an unavoidable evil. The isolated knowledge obtained by a group of specialists in a narrow field has no value whatsoever but only in its synthesis with all the rest of knowledge."

THE APPLICATION OF ELECTRICAL TECHNIQUES IN MAPPING HYDROCARBON-RELATED ALTERATION

Larry J. Hughes, Kenneth L. Zonge, and Norman R. Carlson

ZONGE ENGINEERING AND RESEARCH ORGANIZATION, INC.

ABSTRACT

An extensive amount of evidence is available to demonstrate that light hydrocarbons leak from their traps, rise upward through the overlying sediments, and generate various forms of alteration and mineralization in those sediments. Some phases of this migration/alteration process may be detectable by surface-based electrical measurements, at least in favorable geologic settings, and a high correlation between measured electrical anomalies and oil and gas fields have been reported by a number of researchers. However, an incomplete evaluation of geologic, topographic, and cultural influences on the data makes these results highly questionable in many cases, and a case study which evaluates all three of these effects has yet to be published.

Case histories over Whitney Canyon and Ryckman Creek fields in the Wyoming Overthrust are presented to illustrate this problem. The fields show strong conductive anomalies which are very well correlated with the fields, but a detailed analysis shows that the anomalies may be related to geology and culture, and not necessarily to alteration. A third case history over Ashland Field in Oklahoma's Arkoma Basin shows a shallow, resistive feature and a deeper, conductive feature. While these are roughly correlated with the limits of the field, important questions are left unanswered by a single line of data. This stresses the importance of grid-mapping fields rather than of running single-line test profiles, as is the normal mode of operation.

The results of the work to date clearly suggest that a radically different approach must be taken in proving whether or not electrochemical alteration is measurable with electrical surveys. Extremely detailed, grid-type electrical mapping with extensive off-field "background" coverage is essential. An analysis of the results of over 1,200 line-miles (1,930 line-km) of electrical work over forty-four fields and numerous prospects, as well as geologic consideration, suggests criteria for the types of fields to be studied. The data must be routinely correlated with geology and must be analyzed for geologic, topographic, and cultural effects in order to be useful. The results should be confirmed with surface geochemistry, shallow electric logging, and shallow coring. Lesser efforts will almost certainly lead to further ambiguous results.

INTRODUCTION

An extensive body of evidence now exists to demonstrate that hydrocarbons escape from their traps, migrate upwards, and alter the geochemistry of the overlying sediments. A very strong, direct argument for the migration of hydrocarbons is the worldwide occurrence of surface oil seeps over many petroleum deposits. Seeps have been observed throughout recorded history, and are responsible for the discovery of a significant portion of the world's oil reserves. A second argument for migration involves the results of numerous surface geochemical investigations undertaken since the 1930s. This work has shown anomalously high concentrations of surface soil gases over a great number of fields worldwide, often in a "halo" pattern surrounding the plan-view extent of the hydrocarbons.

Strong circumstantial evidence of tertiary migration can be found in the correlation of mineral assemblages with underlying commercial petroleum deposits. High concentrations of pyrite have been noted in the near-surface sandstones over many fields (Donovan, 1974; Ferguson, 1979; Oehler and Sternberg, 1984), as has calcite cement (Donovan et al., 1981; Sassen, 1980; Oehler and Sternberg, 1984). Sulfur is often found in petroliferous environments, and it is a common occurrence in many areas, such as the sediments overlying salt domes in the Gulf Coast area of Texas (Davis, 1979; Sassen, 1980) and in Sicily (Neglia, 1979). Magnetite has been noted in large quantities over Cement Field (Donovan et al., 1979) and low amplitude anomalies noted in airborne magnetic surveys have been attributed to magnetite (Henderson et al., 1984; Foote, 1984). Uranium has been linked with petroleum by numerous authors (Olmsted, 1975; Pirson, 1979; Curiale et al., 1983), and it is believed to be the source of widely noted radiometric anomalies (Pringle et al., 1953; Armstrong and Heemstra, 1973; Roberts et al., 1978). Petroleum-related copper, lead, zinc, and manganese have also been noted over hydrocarbons.

Other evidence for upward migration involves helium and radon anomalies in surface soils (Roberts, 1979; Fleischer and Turner, 1984), high iodine and occasional high bromine concentrations in surface soils (Gallagher, 1984), tonal anomalies due to bleaching of surface soils (Ferguson, 1979; Donovan et al., 1981), biochemical anomalies in plants over oil and gas fields (Richers et al., 1982;

Dalziel and Donovan, 1980, 1984), bacteria concentrations over petroleum traps (Davis, 1979), and thermal anomalies (Roberts, 1981).

The above features can be loosely connected in a single migration alteration hypothesis, illustrated in a very diagrammatic sense in Figure 1. Lighter hydrocarbons, primarily methane, leak from a reservoir and move either by diffusion or along faults and fractures through the overlying sediments. This may be accompanied by migration of saline formation waters which carry high concentrations of metallic salts and other elements. As the hydrocarbons rise upward via buoyant influences, they create a reduced plume or chimney. Toward the surface, the plume undergoes a number of geology-dependent changes. An assemblage of minerals is deposited as a result of exsolution in oxidizing meteoric waters or of geochemical interactions with the host rocks in the presence of anaerobic bacteria. In some cases mineralization and soil gases assume a halo-shaped feature around the periphery of the reducing plume. The reducing conditions may also provide an excess of free electrons in the plume or disrupt the cation-exchange equilibria or geometry of clays. The combination of all these effects would be to produce an anomalously conductive and/or resistive, relatively polarizable feature, sometimes with considerable lateral and depth extent. Such a feature could represent an excellent target for exploration by ground-based electrical surveys.

Unfortunately, surveys since the mid-1970s have produced ambiguous results. A large percentage of conductive and polarizable anomalies have been reported by several research groups, but in most cases the interpretation of the data as an alteration indicator is only a sufficient, not a necessary interpretation. Most data have been severely underevaluated in terms of geologic, topographic, and cultural influences, making conclusions regarding the source of the anomalies invalid. Additional problems involving low sample density and an incomplete understanding of the geologic complexities involved in alteration mechanisms have added further ambiguity.

The purpose of this paper is threefold. First, it is designed to review alteration mechanisms which might have electrical consequences, and to provide a basic understanding of the conditions under which they might be detected by an electrical survey. Second, the paper explores some of the specific reasons why many electrical results have been ambiguous. Several case histories are presented to illustrate this problem. Third, some concrete suggestions are offered for fundamentally changing the way the alteration problem is approached in future work.

ELECTRICALLY RESPONSIVE ALTERATION EFFECTS

While a number of migration/alteration effects have been discussed in the literature, only a few show significant potential of being detectable with electrical techniques. Five such phenomena are considered here:

1. Pyrite mineralization
2. Calcite cementation
3. Clay alteration
4. Brine effects
5. Redox potential cell

Many mineral assemblages associated with sediments overlying petroleum are believed to be intimately linked to the reaction of bacteria colonies which metabolize hydrocarbons. Two classes of these bacteria are involved: aerobic and anaerobic. Aerobic species exist in oxidizing environments characteristic of meteoric waters. Harwood (1973) indicates that biodegradation occurs in waters with a dissolved oxygen content of about 8 mg/l, and at temperatures between 20°C and 50°C. He also notes that hydrogen sulfide poisons aerobes, and therefore biodegradation is confined to oils which contain little or no hydrogen sulfide. Davis (1979) notes that aerobic bacteria are widespread and act on both naturally occurring and petroleum-derived methane. There are also species which act on ethane, propane, and higher-chain hydrocarbons, which normally have only a petroleum source. Oehler and Sternberg (1984) describe the action of aerobic microbes on methane as follows:

$$CH_4 + 2O_2 \rightarrow CO_2 + 2H_2O$$

Anaerobic bacteria exist in reducing environments, deriving their oxygen from sulfate ions in the groundwater. Phillipi (1977) indicates that biodegradation occurs at temperatures less than about 66°C. Davis

(1979) gives the following reaction under the influence of Desulfovibrio desulfuricans:

$$CH_4 + CaSO_4 \rightarrow H_2S + CaCO_3 + H_2O$$

Jobson, Cook, and Westlake (1979) suggest, however, that anaerobics grow on the residues of aerobic degradation under conditions of an increasingly reduced environment. Oehler and Sternberg (1984) summarize this as an action on organic carbon molecules, previously derived from aerobic microbe degradation of methane:

$$2CH_2O + SO_4^{=} \rightarrow 2HCO_3^- + H_2S$$

$$2CH_2O + SO_4^{=} \rightarrow HCO_3^- + HS^- + H_2O + CO_2$$

In this manner the organic carbon is oxidized to the bicarbonate ion, and sulfate ions in the groundwater are reduced to the reduced sulfur species HS^- and hydrogen sulfide (H_2S).

Pyrite Mineralization

Pyrite may be precipitated in a reducing environment, given a source of sulfur and iron. Sulfur may be obtained in petroleum provinces from hydrogen sulfide, which can be derived from the petroleum itself, from anaerobic bacterial action, or from oxidation of petroleum near the earth's surface. Iron may be derived from iron oxide grain coatings in sandstones, from pore-filling clays like chlorite, from rock fragment inclusions, or from deeper meteoric waters. The reaction of hydrogen sulfide and iron precipitates pyrite or marcasite:

$$Fe_2O_3 + 4H_2S \rightarrow 2FeO_2 + 3H_2O + 2H^+ + 2e^-$$

The existence of any pyritic zonation should be dependent upon the sulfur content of the oils, the geology and groundwater geochemistry of the sedimentary overburden, and the particular nature of bacterial degradation. For example, if the oils are high in sulfur content and the groundwater is high in iron, pyrite could be precipitated at any depth within the migration plume where there is sufficient porosity, such as in faults, fractures, and permeable formations. If oils are free from sulfur, the sulfur required for the reaction would be derived from bacterial degradation, in which case pyrite would occur within the upper sediments as a result of environmental restrictions on bacterial colonies. Since near-surface waters rarely contain significant amounts of iron (Berner, 1971), pyrite at shallow depths should be found primarily in sandstones with some interstitial iron content. Thus, zonation is intimately tied to local conditions. In all cases, however, seepage-related pyrite would be concentrated in the reduced plume of hydrocarbon migration.

Since disseminated pyrite leads to a high interfacial capacitance, and since pyrite occurs quite commonly in the earth, it is a common source of induced polarization anomalies. Numerous researchers have attributed IP anomalies measured in the field to pyrite (Snyder et al., 1981; Klein, 1983), and Oehler and Sternberg (1984) have directly tied the two causally over Ashland Field in Oklahoma. It is possibly the source of anomalies measured by the Elflex and Electraflex groups (Powell, 1981), which incorrectly had ascribed their results to "reflections" from oil-bearing strata at great depths.

Pyrite as an independent exploration parameter, however, has been shown to be unreliable due to its direct dependence upon favorable geologic circumstances. Ferguson (1979) shows no anomalous pyrite concentration over Altus Field and has attributed this observation to decreased porosity in surface sandstones and to subsurface structural relationships. In addition, Oehler and Sternberg (1984) showed a "false" pyrite anomaly over a nonproductive prospect in Texas, in which the mineralization was associated with fine-grained, organic-rich mudstones and was unrelated to petroleum migration. These results may explain the observation by Hughes (1983) that only 24% of oilfields show a correlatable polarization anomaly.

Carbonate Cementation

Calcite cements may be precipitated by three processes. The first involves the action of anaerobic bacteria, noted earlier, in which methane is reduced in the presence of sulfatic waters to yield calcite and hydrogen sulfide. The second process involves oxidation of petroleum without bacterial action, summarized by Kartsev et al. (1959):

$$C_nH_m + 3CaSO_4 \longrightarrow$$

$$3CaCO_3 + CO_2 + H_2O + 3H_2S$$

This may be a common source of secondary carbonate cements noted widely. The third calcite-producing process involves micropore filtration, in which negatively charged anions at the surfaces of clays repel carbonate ions, which readily combine with available calcium cations to precipitate a cement. This frequently occurs at the interface of a shale and a sandstone, and is not necessarily tied to influences from migrating hydrocarbons.

As was observed for pyrite, calcite zonation should be heavily dependent on the geology and groundwater geochemistry of the sedimentary overburden and the particular process causing the cementation. Presumably the bulk of the petroleum-caused reactions will be above the biodegradation zone, where bicarbonate ions are produced. Precipitation can be enhanced by a rise in pore-water pH due to the presence of bicarbonates, and can be diminished by pore-blocking materials such as chlorite. The most significant concentration should occur in the more porous sandstones and in fault-related porous zones, while little cementation should occur in tightly compacted shales and carbonates.

Small amounts of calcite can displace a large amount of pore fluid. Oehler and Sternberg (1984), for example, calculate that 4% calcite reduces the porsity of a typical porous sandstone by 35%. Hence, calcite-cemented zones should be detectable as high resistivity features in electrical surveys. To date only the work by Oehler and Sternberg over Ashland Field has provided a direct causal link between calcite cements and high resistivity anomalies. This work was supported by CSAMT measurements at Ashland by Hughes et al. (1984), which are summarized later in this paper.

Clay Alteration

Several influences on clay dynamics in the sedimentary overburden can be envisioned as a result of an induced reducing plume over a hydrocarbon trap. The first is diagenetic weathering of feldspars to produce clays. This can occur under the reducing, slightly acidic conditions created by significant amounts of migrating hydrogen sulfide and carbon dioxide, which are derived either from the trap or from the action of anaerobic bacteria. Clays formed in this manner will be chemically stable unless the environment is changed. Clay diagenesis has been documented at Cement Field (Lilburn and Al-Shaieb, 1983), where kaolinite and illite-smectite clays of late diagenetic origin are found in the surface sediments overlying the petroleum accumulation. Segal et al. (1984) report that a bleached portion of the Wingate Sandstone directly overlying Lisbon Field in Utah contains primarily kaolinite clays, while unbleached portions of the sandstone off the oilfield contain fresh plagioclase and muscovite. They show that the change over the field is diagenetic, not detrital.

A second influence on clay response involves possible disruption of cation-exchange characteristics of the clays due to the presence of organic matter and anomalous concentrations of cations in the reduced pore waters in the plume. This could possibly result in a third influence involving a change in clay geometry resulting from the absorption of cations of different size. At this time, the specifics of these influences under the reducing conditions over oilfields are not well understood.

Brine Effects

The upward migration of brines was postulated by Hughes (1983) to explain numerous conductive anomalies measured over known fields. Donovan et al. (1981) made a similar suggestion in their geologic analysis of Velma Field in Oklahoma. Roberts (1980) proposes an interesting hydrodynamic model which attempts to explain the dynamics of fluid migration through a trap. Roberts sees the trap as a "forced-draft" system, in which organic-rich waters flow upward through the trap. Abrupt changes in pressure, temperature, and salinity at the trap cause precipitation of the hydrocarbons, while waters are allowed to pass through. Leaking brines would then move upward because of hydraulic and/or hydrodynamic forces. If this is the case, one should be able to observe brine-plume effects directly in electric logs. Although the authors are aware of informal studies by several oil companies which show such effects, a rigorous analysis of the association of high salinity in sediments overlying petroleum accumulations has yet to be published. Hence, the brine plume mechanism is at best highly speculative. In addition, a number of unanswered questions exist in regard to flow rates, brine dissipation by laterally moving acquifer waters, permeability of brines in normally encountered sediments, interactions of brines and hydrocarbons at various temperatures, pressures, and salinities, etc. As a result, brine invasion is less preferred at this time as a potential source of electrical anomalies.

Redox Potential Cell

The generation of a reducing plume or chimney above a hydrocarbon trap produces an excess of free electrons within the plume. This is thought to produce an electrically conductive zone within the plume. No quantitative studies of this effect are known to the authors, however, and its expected magnitude is a mystery at this time. Pirson's "electrotelluric" currents are a variation on this idea (Pirson, 1976).

ELECTRICAL SURVEY RESULTS

Previous Studies

According to estimates by the authors, alteration-related expenditures for petroleum exploration worldwide were between \$20 and \$30 million during the period of 1976 to 1984. However, surprisingly few results of alteration mapping with electrical techniques have been published in the literature (see Appendix A for a review of these papers). The earliest mention known to the authors of alteration mapping refers to induced polarization (IP) work by Colfax Surveys, Ltd. in the mid-1970s (*Oilweek*, 1976, 1978). Some of this work was based on the electrochemical redox "fuel cell" popularized by Pirson (1971, 1976).

A serious investigation of alteration began in the late 1970s when Educational Data Consultants, Inc. (EDCON), a geophysical contractor, released the results of an induced polarization survey over Lambert Field in northern Texas (Snyder et al., 1981). The study showed a significant polarization anomaly which was attributed to micron-sized pyrite precipitated in the reduced hydrocarbon-migration zone. However, Holladay and West (1984) produced modeling results which suggested that well-casing effects were entirely sufficient to explain the Lambert Field data. Concurrent with EDCON's work, Zonge Engineering and Research Organization, Inc. (ZERO) produced evidence that frequent conductive anomalies and occasional polarizable anomalies were seen over known oil and gas fields in several basins (Zonge and Hughes, 1981; Carlson et al., 1981). Following the Holladay and West paper, an attempt was made to show the inability of well-casing effects to explain the measured anomalies at Cowboy Field in Utah (Hughes et al., 1982), and the results of an eight-field IP speculative survey included a complete treatment of well-casing and topographic effects (Hughes, 1983). Recent work by ZERO has focused on detection of conductive anomalies with a controlled source audio-frequency magnetotelluric (CSAMT) system (Ostrander et al., 1983; Hughes et al., 1984). A third major contractor, Phoenix Geophysics, Ltd., conducted an extensive IP speculative survey in 1983 and 1984, integrating electrical data with an analysis of well-casing effects and some core-drilling results. Klein (1983) presented the results of work over David Field in Alberta, where a conductive anomaly and a weak polarization anomaly were observed, a result which is representative of the overall spec project (Phoenix promotional literature, 1983). The David Field results are thought to be at least partly contaminated by well-casing effects in the polarization parameter (Klein, 1983; Sill, 1983), but the apparent resistivity data are thought to represent a conductive plume due to migration/alteration effects.

Probably the most interesting work on this subject presented so far, and one of the only three papers actually published in the technical journals, is work done by Oehler and Sternberg (1984) over Ashland Field in Oklahoma, in which IP surveys show a near-surface resistive, polarizable zone over the field. Head-space gas analysis and core cuttings on and off the field were used to link the resistive anomaly to calcite cementation and the polarizable anomaly to pyrite, both of which were attributed to migration/alteration effects. On the other hand, anomalies measured over a prospect were attributed to geologic effects unrelated to migration/alteration influences, suggesting that great care must be taken to integrate geologic and geophysical data for a proper interpretation.

Results of other IP and self-potential (SP) investigations have been published by Haren (1984), who noted conductive and polarizable anomalies over a number of fields.

In an effort to obtain more quantitative information on the results of alteration work to date, the authors sent out a questionnaire to a number of geophysical contractors and research groups. Of these, sixteen were returned by early 1984. Thirteen of the respondents were involved in alteration detection and mapping with IP or CSAMT. These respondents characterized their measurement of conductive anomalies over known fields as follows: frequent, 36%; occasional, 45%; rare, 19%. They characterized resistive anomalies as follows: frequent, 0%; occasional, 45%; rare 55%. They characterized polarizable anomalies as follows: frequent, 45%; occasional, 55%; rare, 0%. The authors have seen promotional literature from a number of contracting organizations claiming anomaly/petroleum correlations ranging from 60% to 100%. For example, the authors themselves examined a large amount of IP data obtained over twenty-eight oil and gas fields in North America between 1977 and 1982. The results (Table 1) show a high statistical correlation between conductive anomalies and field limits, and a slight correlation between polarizable anomalies and fields.

While these kinds of tabulations seem impressive at first, the authors believe that a great number of surveys done over oil and gas fields to date are heavily influenced by the contaminating effects of culture, especially well casings and pipelines, and a number are influenced by geologic and topographic effects. In general, these problems are not rigorously evaluated in production surveys because of economic limitations. In the absence of such a rigorous analysis, however, it is simply not possible to claim a link between electrical anomalies and an actual electrical alteration plume. Since no detailed study of all these effects has yet been published, we thought that it would be interesting to analyze several case histories to demonstrate some of the problems that can arise in these types of surveys.

Case History: Wyoming Overthrust IP Survey

An induced polarization survey was run across two of the western Wyoming Overthrust's larger producers, Whitney Canyon Field and Ryckman Creek Field. Whitney Canyon is classified as a gas giant, containing an estimated 3.1 trillion cu ft of sour natural gas and 66 million barrels of condensate. The trap is a high-relief anticline on the hanging wall of the Absaroka Thrust Plate. Production is from six Mesozoic and Paleozoic formations ranging in depth from 9,000 ft (2,740 m) to 15,000 ft (4,570 m). Ryckman Creek is also a thrust-faulted anticlinal trap and contains some 150 billion cu ft of sweet gas and 50 million barrels of 47° API gravity oil. Production is from the Triassic Nugget Sandstone and Thaynes Limestone at depths ranging from 7,800 ft (2,380 m) to 9,810 ft (2,990 m). Shurtliff Field, a one-well producer of oil and sweet gas from a thrust-faulted, Ankareh Formation trap, lies just south of the survey line.

The IP survey was performed in 1979 and 1980 using portable, digital receivers configured for six-channel, roll-along, dipole-dipole data acquisition.

TABLE 1

Correlation of Electrical Anomalies with Petroleum[1]

	Apparent Resistivity (Low Values)	Apparent Polarization (High Values)	REM (Negative Values)	At Least One Parameter
Excellent correlation	53%	27%	54%	61%
Fair correlation	12	18	19	22
Poor to no correlation	35	55	27	17
Negative correlation	0	0	0	0

[1]Based upon surveys of over twenty-eight oil and gas fields between 1977 and 1982.

The work utilized frequency-domain magnitude and phase angle data acquisition at 0.125, 0.25, 0.5, 1.0, 2.0, and 4.0 Hertz. An a-spacing of 1,700 ft (518 m) and a coverage of n=1 to n=6 were used, providing an effective penetration depth of approximately 3,400 ft (1,040 m). The line was run perpendicular to the prevailing structural trends in the area. The field work was completed prior to extensive step-out development at Whitney Canyon, which had been shut in awaiting completion of a gas processing plant. The Ryckman Creek data collection was done after the majority of the wells had been completed at that field, but before the Shurtliff discovery was drilled.

Figure 2 shows the topography along the survey line, the geology within the penetration depth limit of the survey (Lamerson, 1982), the cultural features existent at the time of the survey work, and the survey data. The stratigraphy of the Tertiary and Cretaceous overburden consists of shales, siltstones, sandstones, and mudstones. Several points should be noted about the geology. First, note the existence of an undisturbed Tertiary and Cretaceous overburden between station 23 and the east end of the line. The overburden is composed of sandy and silty shales with some conglomeratic units. As such, these units may favor clay diagenesis or alteration and perhaps some calcite cementation, but probably not a great deal of pyrite, due to the minimum number of clean sandstone beds. Hence, the apparent resistivity parameter would be expected to be more diagnostic than the apparent polarization parameter, should any alteration exist. On the east half of the line, structure beneath the overburden is at or below the penetration threshold of the data. Hence, the data in this region, especially east of station 16, respond to a layered, relatively unstructured geology. The west half of the line, however, is quite complex, with various Tertiary and Cretaceous sediments thrust to the surface via numerous high-angle thrust faults. Hence, one would expect the data to show complex patterns in this region.

The electrical data presented in Figure 2 include apparent resistivity at 0.125 Hz, decoupled phase angle at 0.125 Hz using a proprietary curve-matching algorithm (Wynn and Zonge, 1975), and an electromagnetic coupling term called residual electromagnetic (REM). As described by Zonge and Hughes (1981) and Carlson et al. (1981), the REM term consists of the total earth coupling with the calculated homogeneous earth coupling term removed by simple superposition. As such, it represents the electromagnetic coupling effects resulting from inhomogeneities such as layers or discrete bodies. The REM data presented here are the quadrature or imaginary component at 0.125 Hz. Negative numbers indicate a change to a more conductive environment.

The data over Ryckman Creek and Whitney Canyon fields show a striking correlation between conductive anomalies and the lateral extent of the fields. A strong conductor is also seen on the far west end of the line. The polarization data show a very subtle response over Whitney Canyon and a mildly polarizable response over Ryckman Creek. The REM data show very strong negative or conductive values over the fields and on the west end of the line, much as in the apparent resistivity data.

These results are quite typical of the results shown in Table 1—strong conductive anomalies, but rather weak to nonexistent polarization anomalies. In order to determine the source of these anomalies, it is necessary to examine the spurious effects of culture, geology, and topography.

Well-casing effects were evaluated with the integral-equation solution of Holladay and West (1984). This algorithm calculates the DC effects of semi-infinite conductive cylinders in a homogeneous half-space. We have used this algorithm extensively and have found it to be very satisfactory for general purposes. It is useful, however, to keep in mind three significant limitations of the model. First, it assumes all casings have an identical response, which is definitely not the case according to several studies we have done for private clients. Second, it does not account for layering effects, which can affect a well-casing response in a complex manner. Third, the model does not account for the interconnecting electrically continuous network of supply pipelines and casings present in most oilfields. Other limitations, such as the infinite pipe length assumption and the inclusion of only DC galvanic effects, also influence the calculations.

With the above limitations in mind, one can at least obtain a qualitative model of well-casing effects in the present data set. All casings within three dipole spacings of the line were modeled. Wells being drilled at the time of the survey were considered to be cased, as drillstem is expected to behave similarly to well casings. Dry holes were assumed to be uncased unless specifically known to be cased. A pipe conductivity of 2.2×10^{-7} ohm-meters and an outside pipe diameter of 8-3/8 in (34.0 cm) were used, based upon standard casing parameters. Surface impedance was adjusted so as to force a best fit with the polarization data. The final value used was 2 + 2i ohm-meter^2.

The modeling results and the residual data (apparent resistivity data with well-casing effect removed) for the apparent resistivity parameter are presented in Figure 3. The Ryckman Creek anomaly is reproduced quite well by the calculated well-casing effect, although the residual indicates that the model has overcorrected slightly. As expected, the Whitney Canyon anomaly and the anomaly to the far west are virtually unaffected, since the area was sparsely drilled at the time of the survey.

The modeling results and residual data for the apparent polarization parameter are presented in Figure 4. The residual data (measured polarization minus calculated well-casing effect) over Ryckman Creek show a relatively shallow effect over the field, with "pantsleg" type geometric effects dropping diagonally away from the affected zone. The simplest explanation of this zone involves a combined effect of the three pipelines crossing the field. Pipelines are a frequent, though unpredictable source of polarization anomalies. In this case, they do not result in an apparent resistivity effect. The polarization high may be enhanced by a slightly polarizable layer near the base of the Wasatch/Green River section. An alternative explanation is that the polarization high over Ryckman Creek is caused by pyrite or clay effects resulting from hydrocarbon migration. The fact that Whitney Canyon exhibits a minimal polarization response, while the surface geology is similar to that at Ryckman Creek, tends to make the alteration explanation the less preferred one.

The remaining responses on the line should be considered in terms of geology and topography. The Wasatch Formation is clearly resistive, while the underlying Evanston Formation is more conductive. The thrust faults near stations 24–25 and 23 appear to be strongly conductive. The proper combinations of conductive and resistive lithologic blocks, along with conductive faults, are able to account for the majority of the conductive response at Whitney Canyon and on the west end of the line. Topographic effects appear to be present in the data as well; for example, the surface high resistivities beneath stations 18 to 20 are the kind of effect one expects from a hill.

In conclusion, the Ryckman Creek anomalies can be explained as cultural features, and the Whitney Canyon apparent resistivity anomaly can be related to structure. It should be stressed that this is a sufficient, not a necessary explanation of the anomalies measured on the line. Modeling is at best imprecise, and at worst downright misleading. It is possible that conductive alteration effects are present in this data set, but are difficult to discern among the much larger apparent effects of well casings and structure. A possible solution to this problem is to use a technique which is less sensitive to culture, and is more easily interpreted for topographic and geologic effects. The controlled source audio-frequency magnetotelluric (CSAMT) method lends itself well to these objectives.

Case History: Arkoma Basin CSAMT Survey

The CSAMT technique was first used for petroleum exploration in 1982 (Ostrander et al., 1983) as a means of minimizing well-casing effects like those observed at Ryckman Creek. Although the technique provides only resistivity information, it is less affected by culture, especially off-line features such as well casings, and the effects are more easily interpretable.

A line of CSAMT was run over Ashland Field, located in the Arkoma Basin of south-central Oklahoma. The field was the site of the earlier investigations by Conoco (Oehler and Sternberg, 1984). Ashland produces dry gas from the Hartshorne Sandstone, a Pennsylvanian barrier reef. The trap is primarily stratigraphic with a thrust-fault seal on the far southern end of the field. The overlying stratigraphy is nearly horizontal and consists of Pennsylvanian sandstones, shales, and thin beds of coal and limestone. No major structure is evident in the units from the surface to the productive horizon at 3,000 ft (914 m).

Figure 5 shows the CSAMT data for the survey line, which was oriented in a north-south direction roughly perpendicular to the trend of the field. The apparent resistivity data show a very shallow zone of high resistivities overlying the field. The shallow CSAMT resistivity values agree quite well with the shallow Schlumberger soundings obtained by Conoco, which also showed a near-surface high-resistivity zone directly over the field. The maximum penetration of the CSAMT data is, however, approximately four times greater than that of the Schlumberger data, providing a more complete examination of the electrical response from the surface to depth. The CSAMT data show a deeper, conductive zone beneath the resistive surface and above the trap. The southern edge of the conductive anomaly is moderately well correlated with the southern edge of production, while the northern edge appears to be rather poorly correlated with the northern limits of the field. Phase difference data, which show the difference between the phase angles of the electric field and the magnetic field, are shown in Figure 5b. The data show a consistently high-over-low resistivity layering picture (numbers greater than 785 milliradians), with a slightly more pronounced effect over the general vicinity of the field. A third parameter, known as second derivative, is shown in Figure 5c. This parameter represents the second derivative of resistivity, and essentially shows the sign and degree of curvature of the inflection points on the resistivity-versus-frequency curve. Negative numbers indicate inflection points at which the curve is concave upward, changing from a low resistivity to a high resistivity with depth, and positive numbers reflect a concave downward curve, indicating high-over-low resistivity layering. The data in Figure 5c show a well-defined zone of positive second-derivative numbers which is well correlated with the extent of the field. This trend toward decreasing resistivities in the deep data can be seen in individual-station plots of resistivity-versus-frequency. The trend is subtle, however, and while it is consistent, it is difficult to interpret since it is in the so-called "transition zone" where assumptions governing resistivity calculations begin to break down. Hence, the results are slightly encouraging regarding conductive alteration directly over the field, but they are by no means conclusive.

It is worthwhile to note that several logistical changes could produce more conclusive results at Ashland Field. First, slightly lower frequencies should be obtained on a follow-up survey in order to ascertain the significance of deep second-derivative data. Second, the field should be gridded with CSAMT in order to determine the statistical correlation between the anomaly and the field in plan view. Third, a detailed, interactive, geological/geophysical interpretation should be made to derive the maximum amount of geophysical information on the project. This kind of work would require a considerable research commitment, but the results could be tied directly to alteration/migration processes demonstrated earlier by Oehler and Sternberg.

CONCLUSIONS

This paper has noted a number of observations showing that hydrocarbons leak from their traps and modify the overlying sediments through a number of mechanisms. Some of these effects should have electrical characteristics measurable by electrical techniques. However, no definite case histories, accompanied by detailed geologic correlation and properly interpreted for topographic and cultural effects, have been produced to prove that electrochemical alteration is measurable in sediments overlying petroleum accumulations. This paper illustrates this problem by examining electrical data over Whitney Canyon and Ryckman Creek fields. Strong electrical anomalies in the near-surface sediments are well correlated with the occurrence of petroleum at depth. A detailed examination of these anomalies shows, however, that cultural, geologic, and topographic influences can explain the anomalies quite satisfactorily. In a third example, data over Ashland Field show slightly anomalous results, which show only a fair correlation with the field limits. While these examples do not preclude the existence of alteration, they also do not prove it.

The fact that a measurable phenomenon exists to be measured, and has not yet been proven distinctly in field data, suggests a major rethinking of the way in which we approach the measurements. In the past, most demonstration surveys have been done with minimal background coverage, only a few lines at a time, and with an insufficient dedication to avoidance of or accounting for numerous complicating effects as a result of geology and culture. Proof or disproof of the alteration theory as a viable exploration technique will require the correcting of all of these deficiencies.

We recommend focusing future work on two goals, both designed to produce proof of an unambiguous electrical anomaly with a petroleum-derived alteration source. The first goal would be to measure alteration over a very large, untested area and to confirm the results with drilling. The second goal would be to measure alteration over a known field but in a manner which would not compromise the data as a result of cultural or other nonalteration effects.

Based upon experience with over 1,200 line-mi (1,930 line-km) of petroleum data, and upon geologic restrictions of alteration considered earlier, we suggest that the following criteria be considered in surveys over known fields:

1. Low well-casing and pipeline density.
2. Simple, horizontally layered stratigraphy and no lateral structure within the penetration limits of the survey, but with fault communication between the reservoir and the layered, unfaulted surface units.
3. Flat topography.
4. Geology favors alteration, i.e., no thick anhydrites to limit migration, sandstones available near the surface to host pyrite and calcite precipitation, shales available in the section to enhance clay effects, etc.
5. Reservoir is undepleted, has good production, has good reserves, and preferably is hydrostatically pressured or overpressured.
6. Reservoir geochemistry is normal, i.e., good supply of methane, normal biodegradation, average brine and mineral content, etc.
7. Field is representative of other traps in its basin.
8. Minimum number of possible pay zones.
9. Good well control on and off the field.

Work in untested areas should consider many of the above criteria as well.

The unsuccessful approach of single-line studies should be abandoned in favor of very tight grids. The work should include extensive "background" coverage off the field to establish regional variability. Thus, any local or regional trends not related to the field itself should be identifiable on a plan map of the data. The data should be rigorously evaluated for effects due to culture and geology. Interpretation should be done interactively using downhole electric logs and other geologic information for a "best fit" solution. If a case for electrical alteration is established during the course of this work, the electrical study should be followed up with a multidisciplinary investigation designed to link the anomalies causally to petroleum migration from traps at depth. Surface geochemistry and IP logging of tightly spaced shallow holes would be good follow-up possibilities in this effort. If a case for alteration is not established, care should be taken to obtain several negative case histories each in at least half a dozen, geologically diverse basins before the alteration approach is abandoned.

Once a dozen basins have been evaluated, one can answer several questions. First, does alteration mapping work in any basin? Second, if the results are mixed, are they attributable to varied geology or other factors? Third, if alteration mapping shows positive results in at least some areas, what is the most cost-efficient method of data acquisition to obtain the required amount of information, and what are the constraints upon the work? Finally, how do the results best complement ongoing seismic and geologic work, and how can these disparate data sets be used interactively for improved overall prospect evaluation?

The occurrence of geochemical alteration due to upward hydrocarbon migration is well established. In order to see whether or not this alteration produces measurable and consistent electrical responses, it will be necessary to increase dramatically the research commitment to this intriguing problem. The potential contribution of a "hydrocarbon indicator" technique to modern exploration makes such a commitment well worth considering.

REFERENCES

Armstrong, F. E., and Heemstra, R. J., 1973, Radiation halos and hydrocarbon reservoirs: A review: United States Bureau of Mines Information Circular 8579.

Berner, R. A., 1971, Principles of chemical sedimentology: New York, McGraw-Hill.

Carlson, N. R., Hughes, L. J., and Zonge, K. L., 1981, Hydrocarbon exploration using induced polarization, apparent resistivity and electromagnetic scattering: 51st Annual International Society of Exploration Geophysicists Meeting Technical Papers, v. 3, p. 1339–1358.

Curiale, J. A., Bloch, S., Rafalska-Bloch, J., and Harrison, W. E., 1983, Petroleum-related origin for uraniferous organic-rich nodules of southwestern Oklahoma: American Association of Petroleum Geologists Bulletin, v. 67, p. 588–608.

Dalziel, M. C., and Donovan, T. J., 1980, Biochemical evidence for subsurface hydrocarbon occurrence, Recluse Oil Field, Wyoming: United States Geological Survey Circular 837.

———, 1984, Correlations of suspected petroleum-generated biochemical and aeromagnetic anomalies, Bell Creek Oil Field, Montana, *in* Davidson, M. J., and Gottlieb, B. M., eds., Unconventional methods in exploration for petroleum and natural gas III: Dallas, Southern Methodist University Press, p. 59–69.

Davis, J. B., 1969, Microbiology in petroleum exploration, *in* Heroy, W. B., ed., Unconventional methods in exploration for petroleum and natural gas: Dallas, Southern Methodist University Press, p. 139–157.

Donovan, T. J., 1974, Petroleum microseepage at Cement, Oklahoma—evidence and mechanism: American Association of Petroleum Geologists Bulletin, v. 58, p. 429–446.

———, Forgey, R. L., and Roberts, A. A., 1979, Aeromagnetic detection of diagenetic magnetite over oil fields: American Association of Petroleum Geologists Bulletin, v. 63, p. 245–248.

———, Roberts, A. A., and Dalziel, M. C., 1981, Epigenetic zoning in surface and near-surface rocks resulting from seepage-induced, redox gradients, Velma Oil Field, Oklahoma: A synopsis: The Shale Shaker, v. 32, p. 1–7.

Ferguson, J. D., 1979, The subsurface alteration and mineralization of Permian red beds overlying several oil fields in southern Oklahoma: The Shale Shaker, v. 29, p. 172–178 and 200–208.

Fleischer, R. L., and Turner, L. G., 1984, Correlations of radon and carbon isotopic measurements with petroleum and natural gas at Cement, Oklahoma: Geophysics, v. 49, p. 810–817.

Foote, R. S., 1984, Significance of near-surface magnetic anomalies, *in* Davidson, M. J., and Gottlieb, B. M., eds., Unconventional methods in exploration for petroleum and natural gas III: Dallas, Southern Methodist University Press, p. 12–24.

Gallagher, A. V., 1984, Iodine: A pathfinder for petroleum deposits, *in* Davidson, M. J., and Gottlieb, B. M., eds., Unconventional methods in exploration for petroleum and natural gas III: Dallas, Southern Methodist University Press, p. 148–159.

Haren, R. J., 1984, An evaluation of electrical methods applied to petroleum in Australia: Exploration Geophysics, v. 15, p. 197–212.

Harwood, R. J., 1973, Biodegradation of oil, *in* The geology of fluids and organic matter in sediments: National Conference on Earth Science, Banff, Alberta, p. 149–156.

Henderson, R., Miyazaki, Y., and Wold, R., 1984, Direct indication of hydrocarbons from airborne magnetics: Exploration Geophysics, v. 15, p. 213–219.

Holladay, J. S., and West, G. F., 1984, Effects of well casings on surface electrical surveys: Geophysics, v. 49, p. 177–188.

Hughes, L. J., Nosal, E. A., Carlson, N. R., and Zonge, K. L., 1982, Distinguishing well casing from structural effects in electrical anomalies measured over hydrocarbons, a case history: Technical Program and Biographies, 52nd Annual Society of Exploration Geophysicists International Meeting and Exposition, p. 443.

Hughes, L. J., 1983, Case histories of an electromagnetic method for petroleum exploration: Tucson, Zonge Engineering and Research Organization, Inc.

———, Carlson, N. R., and Ostrander, A. G., 1984, Applications of CSAMT in mapping structure and alteration associated with petroleum: Technical Program and Biographies, 54th Annual International Society of Exploration Geophysicists Meeting and Exposition, p. 102–104.

Jobson, A. M., Cook, F. D., and Westlake, D. W. S., 1979, Interaction of aerobic and anaerobic bacteria in petroleum biodegradation: Chemical Geology, v. 24, p. 355–365.

Kartsev, A. A., Tabarsaranskii, Z. A., Subbota, M. I., and Mogilevskii, G. A., 1959, Geochemical methods of prospecting and exploration for petroleum and natural gas: Berkeley-Los Angeles, University of California Press.

Klein, J. D., 1983, Spectral induced polarization survey, David Field, Alberta: presented at the 36th Annual Meeting of the Midwest Society of Exploration Geophysicists, Mar. 6–9, Denver, Colorado.

Lamerson, P. R., 1982, The Fossil Basin area and its relationship to the Absaroka Thrust Fault system, *in* Powers, R. B., ed., Geologic Studies of the Cordilleran thrust belt, v. 1: Rocky Mountain Association of Geologists, p. 279–340.

Lilburn, R. A., and Al-Shaieb, Z., 1983, Geochemistry and isotopic composition of hydrocarbon-induced diagenetic aureole (HIDA), Cement Field, Oklahoma (Pt. 1): The Shale Shaker, v. 34, p. 40–56.

Neglia, S., 1979, Migration of fluids in sedimentary basins: American Association of Petroleum Geologists Bulletin, v. 63, p. 573–597.

Oehler, D. Z., and Sternberg, B. K., 1984, Seepage-induced anomalies, "false" anomalies, and implications for electrical prospecting: American Association of Petroleum Geologists Bulletin, v. 68, p. 1121–1145.

Oilweek, 1976, Principle of induced polarization applied to geochemical exploration: Sept. 27, p. 31–32.

Oilweek, 1978, Induced polarization explained: May 8, p. 21–23.

Olmsted, R. W., 1975, Geochemical studies of uranium in south-central Oklahoma: Master's thesis, Oklahoma State University, p. 116.

Ostrander, A. G., Carlson, N. R., and Zonge, K. L., 1983, Further evidence of electrical anomalies over hydrocarbon accumulations using CSAMT: Expanded abstracts and biographies, 53rd Annual International Society of Exploration Geophysicists Meeting and Exposition, p. 60–63.

Phillipi, G. T., 1977, On the depth, time, and mechanism of origin of the heavy to medium-gravity naphthenic crude oils: Geochimica Cosmochimica Acta, v. 41, p. 33–52.

Pirson, S. J., 1971, New electric technique can locate gas and oil: World Oil, Apr., p. 69–72, May, p. 72–74.

———, 1976, Prediction of hydrocarbons in place by magneto-electrotelluric exploration: Oil & Gas Journal, May 32, p. 82–86.

———, 1969, Geological, geophysical, and chemical modifications of sediments in the environment of oil fields, *in* Heroy, W. B., ed., Unconventional methods in exploration for petroleum and natural gas: Dallas, Southern Methodist University Press, p. 159–186.

Powell, J. A., 1981, Electric transient surveys for hydrocarbon: Technical papers, 50th Annual International Society of Exploration Geophysicists Meeting and Exposition, v. 3, p. 1773–1794.

Pringle, R. W., Roulston, K. I., Brownell, G. W., and Lundberg, H. T. F., 1953, The scintillation counter in the search for oil: Mining Engineering, Dec., p. 1251–1261.

Richers, D. M., Reed, R. J., Horstmann, K. C., Michels, G. D., Baker, R. N., Lundell, L., and Marrs, R. W., 1982, Landsat and soil-gas geochemical study of Patrick Draw Oil Field, Sweetwater County, Wyoming: American Association of Petroleum Geologists Bulletin, v. 66, p. 903–922.

Roberts, A. A., Donovan, T. J., Dalziel, M. C., and Forgey, R. L., 1978, Application of helium surveys to petroleum exploration (abstract): United States Geological Survey Professional Paper, no. 1100.

Roberts, A. A., 1981, Helium emanometry in exploring for hydrocarbons: Pt. II, *in* Gottlieb, B. M., ed., Unconventional methods in exploration for petroleum and natural gas II: Dallas, Southern Methodist University Press, p. 136–149.

Roberts, W. H., III, 1980, Design and function of oil and gas traps, *in* Roberts, W. H., III, and Cordell, R. J., eds., Problems of petroleum migration: American Association of Petroleum Geologists Studies in Geology, no. 10, p. 217–240.

———, 1981, Some uses of temperature data in petroleum exploration, *in* Gottlieb, B. M., ed., Unconventional methods in exploration for petroleum and natural gas II: Dallas, Southern Methodist University Press, p. 8–49.

Sassen, R., 1980, Biodegradation of crude oil and mineral deposition in a shallow Gulf Coast salt dome: Organic Geochemistry, v. 2, p. 153–166.

Segal, D. B., Ruth, M. D., Merin, I. S., Wantanabe, H., Soda, K., Takano, O., and Sano, M., 1984, Spectral remote sensing investigation of Lisbon Valley, Utah: presented at the International Symposium on Remote Sensing of Environment, Third Thematic Conference, Remote Sensing for Exploration Geology, Colorado Springs, Colorado, April 16–19.

Sill, W. R., 1983, Cultural effects in induced polarization data, *in* Electrical methods in oil and gas exploration, v. 2: Salt Lake City, Earth Science Laboratory, University of Utah Research Institute.

Snyder, D. D., Kolvoord, R. W., Frangos, W., Bajwa, Y., Fleming, D. B., and Tasci, M. T., 1981, Exploration for petroleum using complex resistivity measurements, *in* Advances in induced polarization and complex resistivity: Tucson, University of Arizona Press, p. 209–253.

Wynn, J. C., and Zonge, K. L., 1975, EM coupling, its intrinsic value, its removal and the cultural coupling problem: Geophysics, v. 40, p. 831–850.

Zonge, K. L., and Hughes, L. J., 1981, The complex resistivity method, *in* Advances in induced polarization and complex resistivity: Tucson, University of Arizona Press, p. 163–208.

APPENDIX

Selected Bibliography on Electrochemical Alteration Mapping

There are a limited number of published discussions on electrical detection of alteration over petroleum. Many of these are scattered in various journals and are not widely known. We have made an attempt to assemble all the papers which address this subject in one place for the convenience of the reader. A brief description regarding the material covered by the paper is offered as an aid to determining its content. This list is updated through December, 1985, and should be relatively complete.

Bertin, J., Farabolini, W., Gole, R., and Roudot, M., 1983, The transient method applied to hydrocarbon exploration: A statistical study of its results (abstract): Australian Society of Exploration Geophysicists Bulletin, v. 14, p. 189–190. Reviews the results of 59 time-domain IP surveys in which 50% to 85% of oil and gas occurrences show IP anomalies.

Carlson, N. R., and Hughes, L. J., 1984, Electrical profiling: American Society of Photogrammetry/American Congress on Surveying and Mapping Fall Convention, San Antonio, September 9–14. See Hughes and Carlson (1984).

———, Hughes, L. J., and Zonge, K. L., 1981, Hydrocarbon exploration using induced polarization, apparent resistivity and electromagnetic scattering: Technical papers, 51st Annual International Society of Exploration Geophysicists Meeting and Exposition, v. 3, p. 1339–1358. Sediments over oilfields show frequent conductive effects in apparent resistivity and REM (an EM coupling term) calculations, and occasional polarizable effects; IP case history over Lisbon Field (Utah) illustrates these effects; other examples show cultural contamination and usefulness of the REM parameter.

———, 1985, Delineation of oil and gas fields using a high resolution electrical technique: National Science Foundation/SBIR Grant no. EAR-8460562.

———, Hughes, L. J., Zonge, K. L., and Aiken, C. L. V., 1982, New applications of electrical methods in hydrocarbon exploration (abstract): American Association of Petroleum Geologists Bulletin, v. 66, p. 243. See Carlson, Hughes, and Zonge (1981).

Haren, R. J., 1984, An evaluation of electrical methods applied to petroleum in Australia: Exploration Geophysics, v. 15, p. 197–212. IP and SP case histories over several Queensland fields show resistivity and occasional high polarization anomalies are associated with the fields; these are attributed to alteration resulting from a redox zone in the path of hydrocarbon seepage from the traps; an electrochemical mechanism is suggested involving redox currents and pyrite precipitation.

———, and Reeves, J. A., 1983, Electrical petroliferous exploration in Australia: Australian Society of Exploration Geophysicists Bulletin, v. 14, p. 185–189. Field research shows low resistivity and low SP anomalies over several fields in Queensland; results are attributed to alteration of sediments overlying traps (see Haren, 1984, for a more complete discussion).

Holladay, J. S., and West, G. F., 1982, Effect of well casings on surface electrical surveys: Technical papers, 51st Annual International Society of Exploration Geophysicists Meeting and Exposition, v. 2, p. 815–838. A condensed and early version of Holladay and West (1984).

Holladay, J. S., and West, G. F., 1984, Effects of well casings on surface electrical surveys: Geophysics, v. 49, p. 177–188. Develops an integral-equation of solution to the resistivity and polarization effects of hollow, finite-length cylindrical conductors imbedded in a homogeneous half-space; presents modeling results for the anomalies found at Lambert Field in Texas by Snyder et al. (1981), and suggests that a large portion of the resistivity and IP anomalies measured there may have been well-casing effects; suggests the inclusion of pipeline-interconnection effects and earth layering effects for better modeling capabilities.

Hughes, L. J., 1983, Cast Histories of an Electromagnetic Method for Petroleum Exploration: Tucson, Zonge Engineering and Research Organization, Inc. Extensive analysis of IP interpretation, well-casing effects, topographic effects, anomaly mechanisms, mathematical theory, history of electrical exploration for petroleum; IP case histories over Ryckman Creek/Whitney Canyon fields (Wyoming), Desert Springs/Playa-Lewis/Desert Springs West fields (Wyoming), Lisbon Field (Utah), Trap Spring Field (Nevada), Little Buck Creek (Wyoming), Garza Field (Texas); case histories show mixed results which may correlate with field characteristics.

———, and Carlson, N. R., 1984, A review of electrical methods in a multidisciplinary exploration program: American Association of Petroleum Geologists Annual Convention Abstracts, San Antonio, Texas, May 20–23. Review of various electrical techniques for alteration detection, structure mapping, and direct detection; CSAMT case history at Trap Spring Field (Nevada) maps structure well; CSAMT case history at Ashland Field (Oklahoma) shows conductive anomaly at depth and resistivity anomaly at the surface, but anomalies not completely confined to field limits.

———, Carlson, N. R., and Ostrander, A. G., 1984, Applications of CSAMT in mapping structure and alteration associated with petroleum: Technical program and biographies, 54th Annual International Society of Exploration Geophysicists Meeting and Exposition, p. 102–104. CSAMT case history at Trap Spring Field (Nevada) maps structure well; CSAMT case history at Ashland Field (Oklahoma) shows resistive anomaly at surface, correlating with work by Oehler and Sternberg, and conductive anomaly at depth; Ashland anomalies not completely confined to field limits, for several possible reasons.

———, Carlson, N. R., and Zonge, K. L., 1985, Applications of electrical techniques in an integrated petroleum exploration program: presented at the China Petroleum Society/Society of Exploration Geophysicists Joint Meeting on Seismic Exploration, Beijing, PRC, April 16, 1985. More generalized version of Hughes, Carlson, and Ostrander (1984).

———, Zonge, K. L., and Carlson, N. R., 1985, The application of electrical techniques in mapping alteration due to hydrocarbon accumulations, *in* Davidson, M. J., ed., Unconventional Methods in Exploration for Petroleum and Natural Gas IV: Dallas, Southern Methodist University Press, this volume.

———, Nosal, E. A., Carlson, N. R., and Zonge, K. L., 1982, Distinguishing well casing from structural effects in electrical anomalies measured over hydrocarbons: A case history: Technical program and biographies, 52nd Annual International Society of Exploration Geophysicists Meeting and Exposition, p. 443–445. IP case history at Cowboy Field (Utah) and computer modeling suggest that both conductive alteration and conductive well-casing contamination exist in the data.

———, Ostrander, A. G., Reed, E. V., and Zonge, K. L., 1984, Structure and alteration mapping with CSAMT, CR, and TEM in Australian basins: A feasibility study: Second Australian Petroleum Geophysics Symposium, Proceedings, p. 162–178. Application of Hughes, Carlson, and Ostrander (1984) to Australian exploration.

Klein, J. D., 1983, Spectral induced polarization survey, David Field, Alberta: presented at the 36th Annual Meeting of the Midwest Society of Exploration Geophysicists, Denver, Colorado. IP case history over David Field (Alberta) shows a deep conductive anomaly and a weak, near-surface polarization anomaly; proprietary well-casing model indicates conductive anomaly is not due to well casings, although phase anomaly could be; includes general description of field and data processing techniques.

Oehler, D. Z., and Sternberg, B. K., 1982, Induced polarization for hydrocarbon exploration: Geochemical/geological interpretation: Technical program abstracts and biographies, 52nd Annual International Society of Exploration Geophysicists Meeting and Exposition, p. 445–448. Condensed version of Oehler and Sternberg (1984).

———, 1984, Seepage-induced anomalies, "false" anomalies, and implications for electrical prospecting: American Association of Petroleum Geologists Bulletin, v. 68, p. 1121–1145. Combined geochemical/electrical investigation of two sites; IP survey at Ashland Field (Oklahoma) shows near-surface resistive feature and polarizable feature over the field, which are tied by in-hole gas measurements and core analysis to carbonate cementation and pyrite, respectively; IP survey at Salt Draw (Texas) shows "false" polarization anomalies attributed to geology and not alteration, as confirmed by gas measurements and core analysis; discussion of alteration mechanisms; how to distinguish alteration anomalies from "false" anomalies.

Ostrander, A. G., Carlson, N. R., and Zonge, K. L., 1983, Further evidence of electrical anomalies over hydrocarbon accumulations using CSAMT: Expanded abstracts and biographies, 53rd Annual International Society and Exploration Geophysicists Meeting and Exposition, p. 60–63. CSAMT case history over Cowboy Field (Utah) correlates well with previous IP data; anomalies are complex in form with both resistive and conductive elements, possibly in "halo" form; results are due to alteration, not just geologic and well-casing effects.

Petrick, W. R., 1983, Interpretation of induced polarization data, *in* Electrical methods in oil and gas exploration, v. 2: Salt Lake City, Earth Science Laboratory, University of Utah Research Institute. Presents IP modeling of anomalies measured by Zonge Engineering (promotional brochure) over Lisbon Field in Utah; presents a suite of IP data from David Field, Alberta (see Klein, 1983) and illustrates some modeling of EM coupling.

Pirson, S. J., 1971, New electric technique can locate gas and oil: World Oil, Apr., p. 69–72, May, p. 72–74. Describes electrotelluric "fuel cells" due to the reducing zone in sediments over petroleum.

———, 1974, Unified magneto-electrotelluric exploration method: Oil and Gas Journal, Mar. 18, p. 138–140, Mar. 25, p. 142–148. Describes electrotelluric method and shows examples; tabulates high rates of success for the technique.

———, 1976, Prediction of hydrocarbons in place by magneto-electrotelluric exploration: Oil and Gas Journal, May 31, p. 82–86. Describes electrotelluric method and shows examples; predicts barrels of oil per acre; shows several case histories.

———, 1980, Pirson: Oil is confined in the earth by redox

potential barriers: Oil and Gas Journal, July 7, p. 153–158. Redox potentials control oil and gas accummulation, not structural or stratigraphic traps.

Powell, J. A., 1981, Electric transient surveys for hydrocarbon: Technical papers, 50th Annual International Society of Exploration Geophysicists Meeting and Exposition, v. 3, p. 1773–1794. Reverses previous explanations of Elflex data as direct hydrocarbon detection, and explains results in terms of alteration of sediments overlying hydrocarbons due to upward hydrocarbon migration; supports assertions with proof that reflections from hydrocarbons are not readily measurable, but that electrochemical effects are.

Sill, W. R., 1983, Cultural effects in induced polarization data, *in* Electrical methods in oil and gas exploration, v. 2: Salt Lake City, Earth Science Laboratory, University of Utah Research Institute. Brief discussion of cultural effects with an example of IP data over David Field in Alberta (see Klein, 1983).

Snyder, D. D., Kolvoord, R. W., Frangos, W., Bajwa, Y., Fleming, D. B., and Tasci, M. T., 1981, Exploration for petroleum using complex resistivity measurements, *in* Advances in induced polarization and complex resistivity, Tucson, University of Arizona Press, p. 209–253. IP case history over Lambert Field in Texas shows distinct low resistivity and high polarization anomalies directly over the field, suggesting an alteration plume, attributed to sulfides, due to upward hydrocarbon migration from the trap; an IP case history over a Wyoming project showed similar anomalies, and a discovery well was drilled subsequently for Millis Field; IP case history over Meridan Field in Texas showed similar anomalies.

Sternberg, B. K., Miller, D. E., and Bahjat, D. S., 1981, Electrode prospecting method providing calculable electromagnetic coupling for the indirect detection of hydrocarbon reservoirs: United States Patent 4,295,096, issued Oct. 13, 1981. Describes Schlumberger soundings optimized for hydrocarbon exploration by Conoco.

Zonge, K. L., and Hughes, L. J., 1981, The complex resistivity method, *in* Advances in induced polarization and complex resistivity: Tucson, University of Arizona Press, p. 163–208. Mention of EM coupling to detect conductive alteration.

———, Reed, E. V., and Young, G. N., 1980, The complex resistivity method: 59th Annual International Society of Exploration Geophysicists Meeting technical papers, v. 5, p. 2579. Brief discussion of EM coupling to detect conductive alteration.

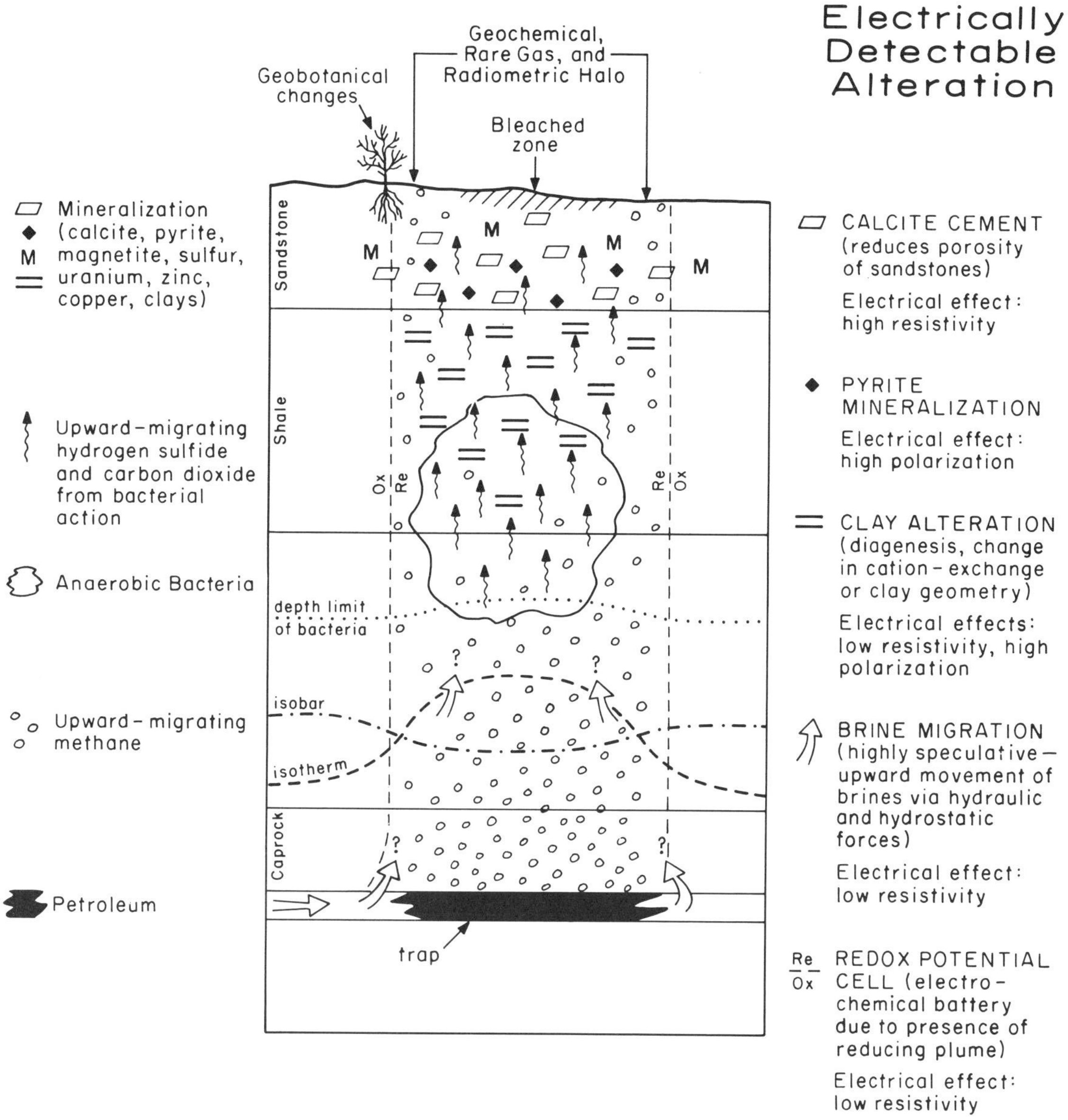

Fig. 1. Idealized illustration of the migration/alteration hypothesis. Geology is presumed to be a significant influence on the types and patterns of phenomena which are observed.

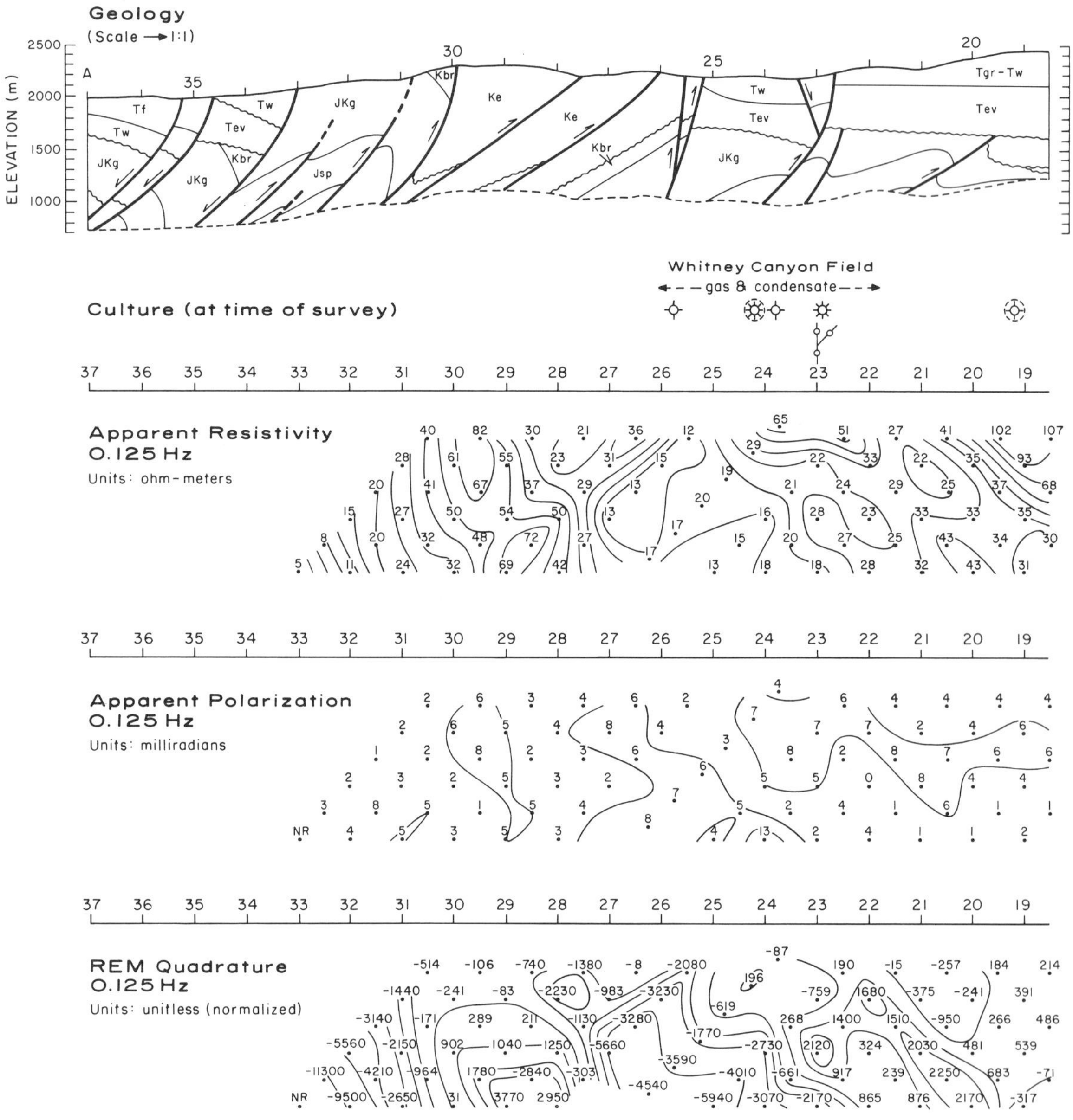

Fig. 2a. Geology, culture and field data for the IP survey across Whitney Canyon and Ryckman Creek fields. Well symbols surrounded by broken circles indicate wells in progress at the time of the survey. Pipelines are indicated by small circles connected by lines; fences are lines with superimposed x's. Formation symbols: Tf = Fowkes Formation, Tgr = Green River Formation, Tw = Wasatch Formation, Tev/Ke = Evanson Formation, Keh = Hams Fork Conglomerate, Kf = Frontier Formation, Ka = Aspen Shale, Kbr = Bear River Formation, JKg = Gannett Group, Js = Stump Sandstone, Jp = Preuss Sandstone.

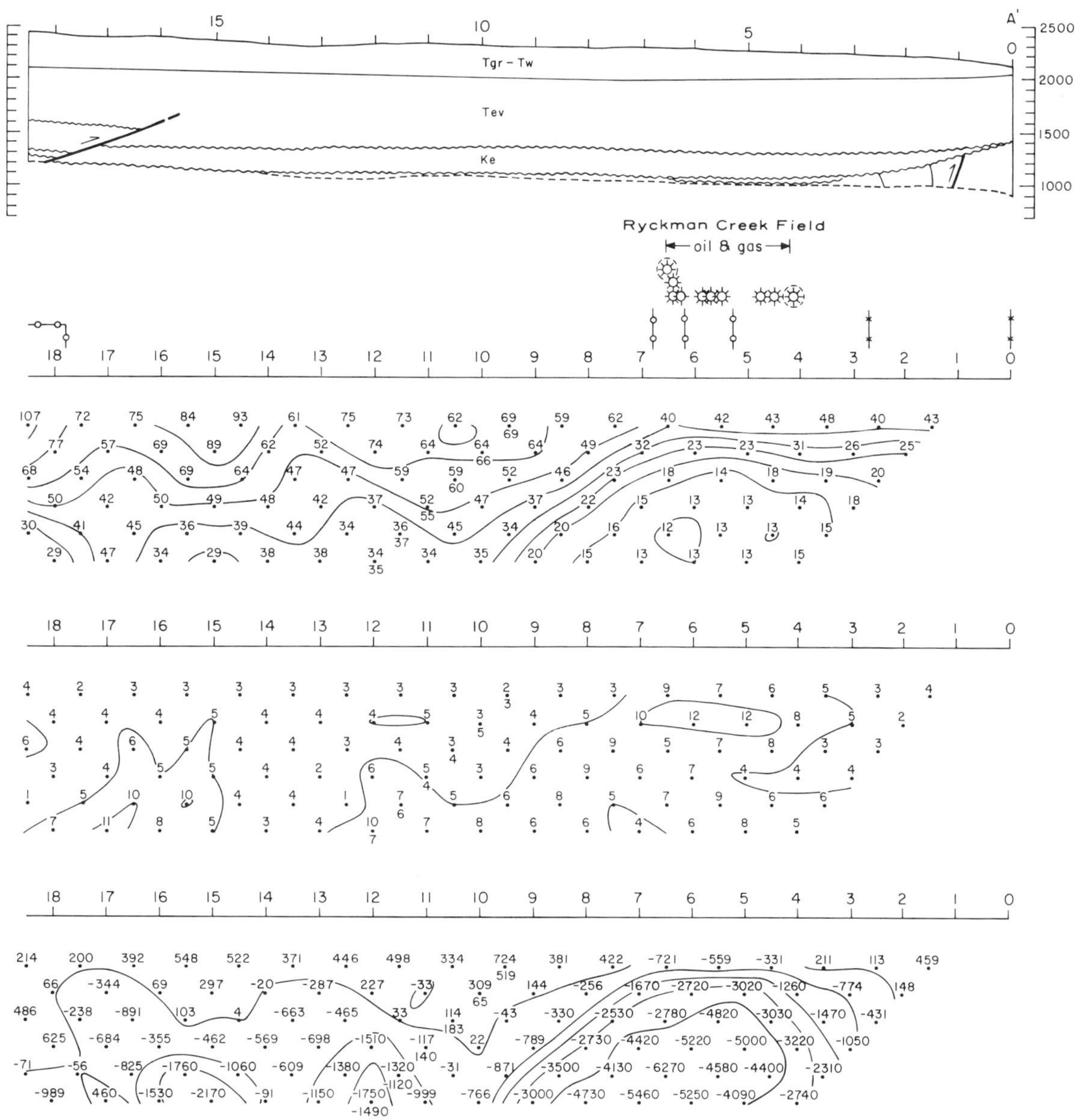
A'
2500
2000
1500
1000
Tgr – Tw
Tev
Ke
Ryckman Creek Field
oil & gas

Fig. 2b

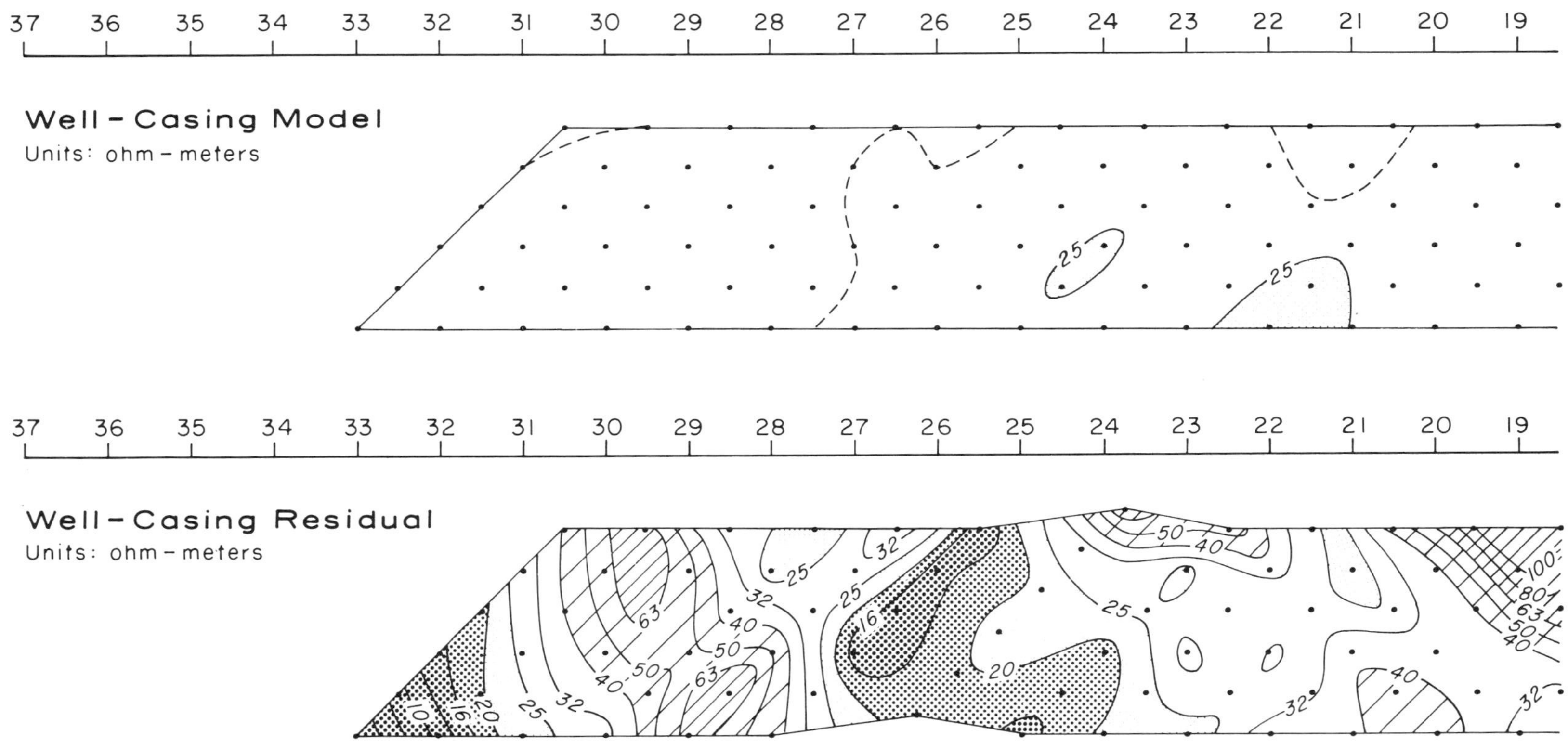

Fig. 3a. Well-casing model results for apparent resistivity data, Whitney Canyon and Ryckman Creek fields, (a) well-casing model results; (b) field data less calculated well-casing effect.

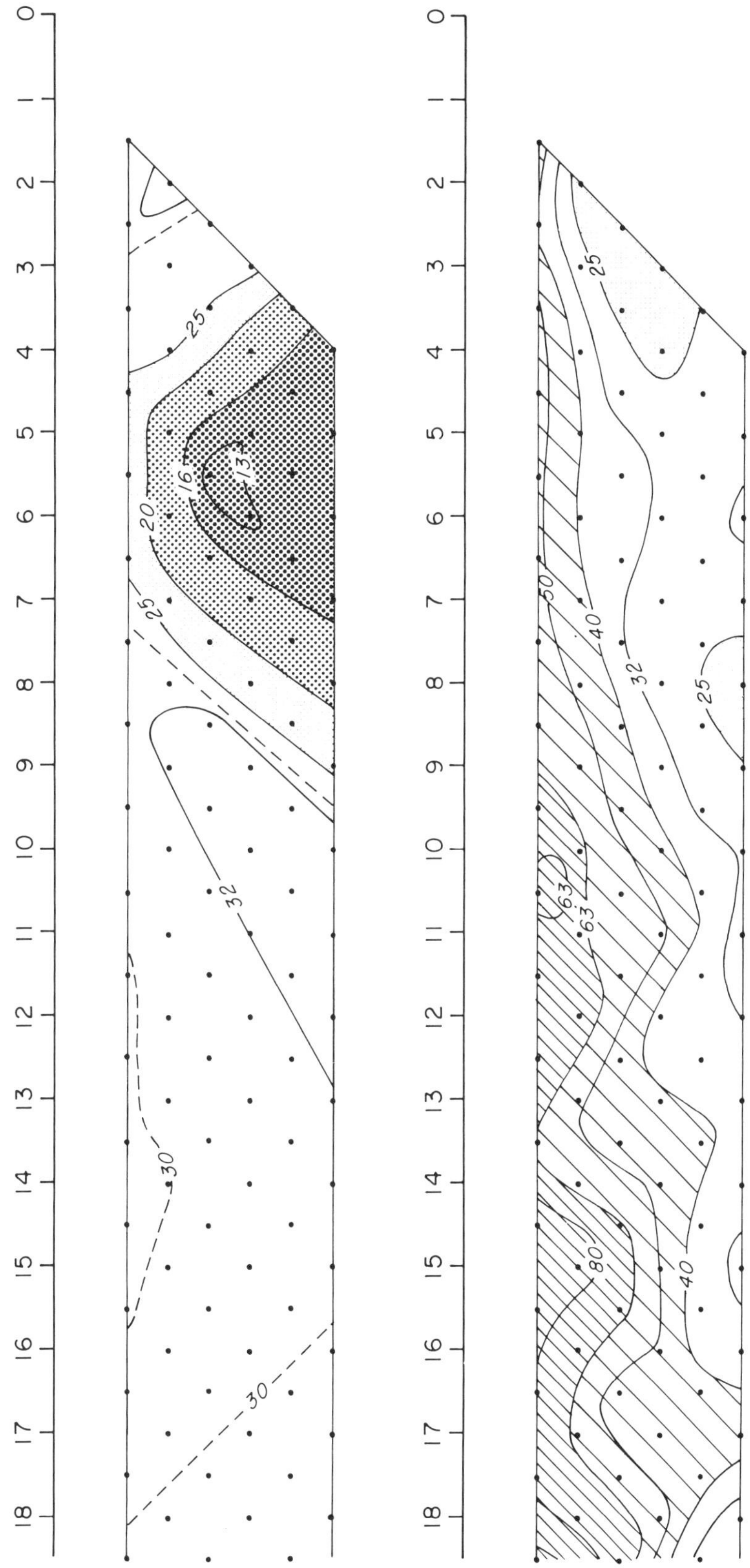
0
1
2
3
4
5
6
7
8
9
10
11
12
13
14
15
16
17
18
25
13
16
20
25
32
30
30
25
50
40
32
25
63
63
80
40

Fig. 3b

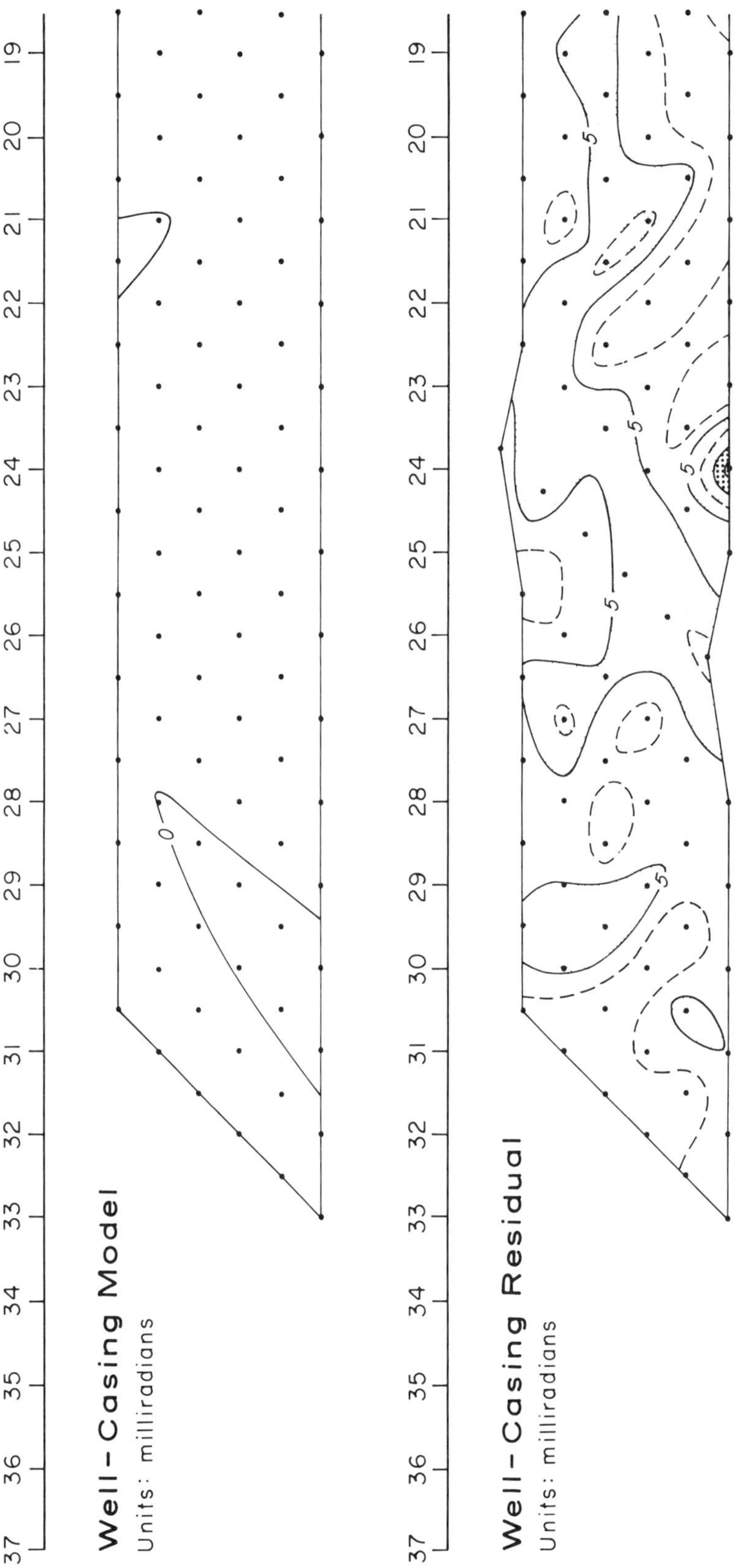

Fig. 4a. Well-casing model results for apparent polarization data, Whitney Canyon and Ryckman Creek fields, (a) well-casing model results; (b) field data less calculated well-casing effect.

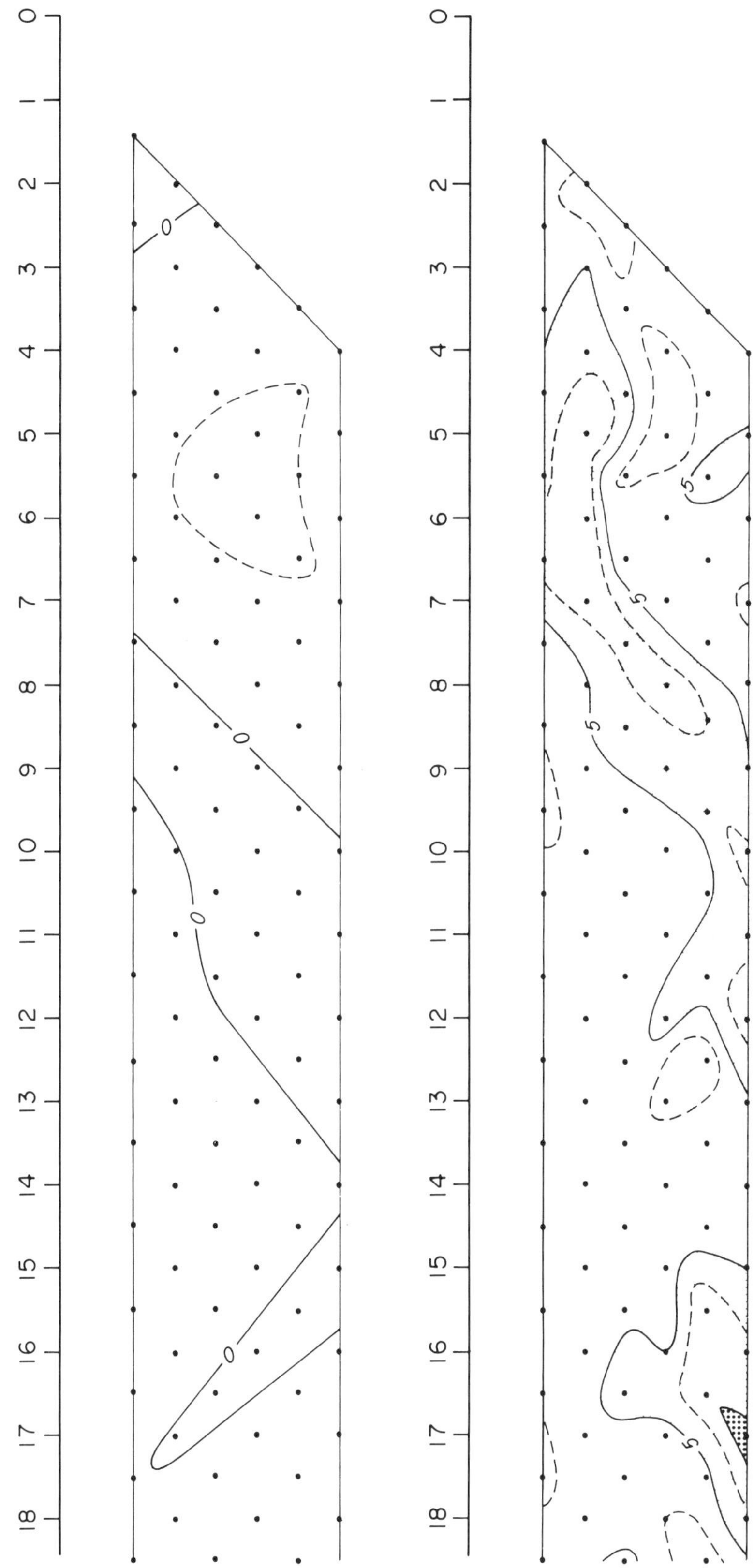

Fig. 4b

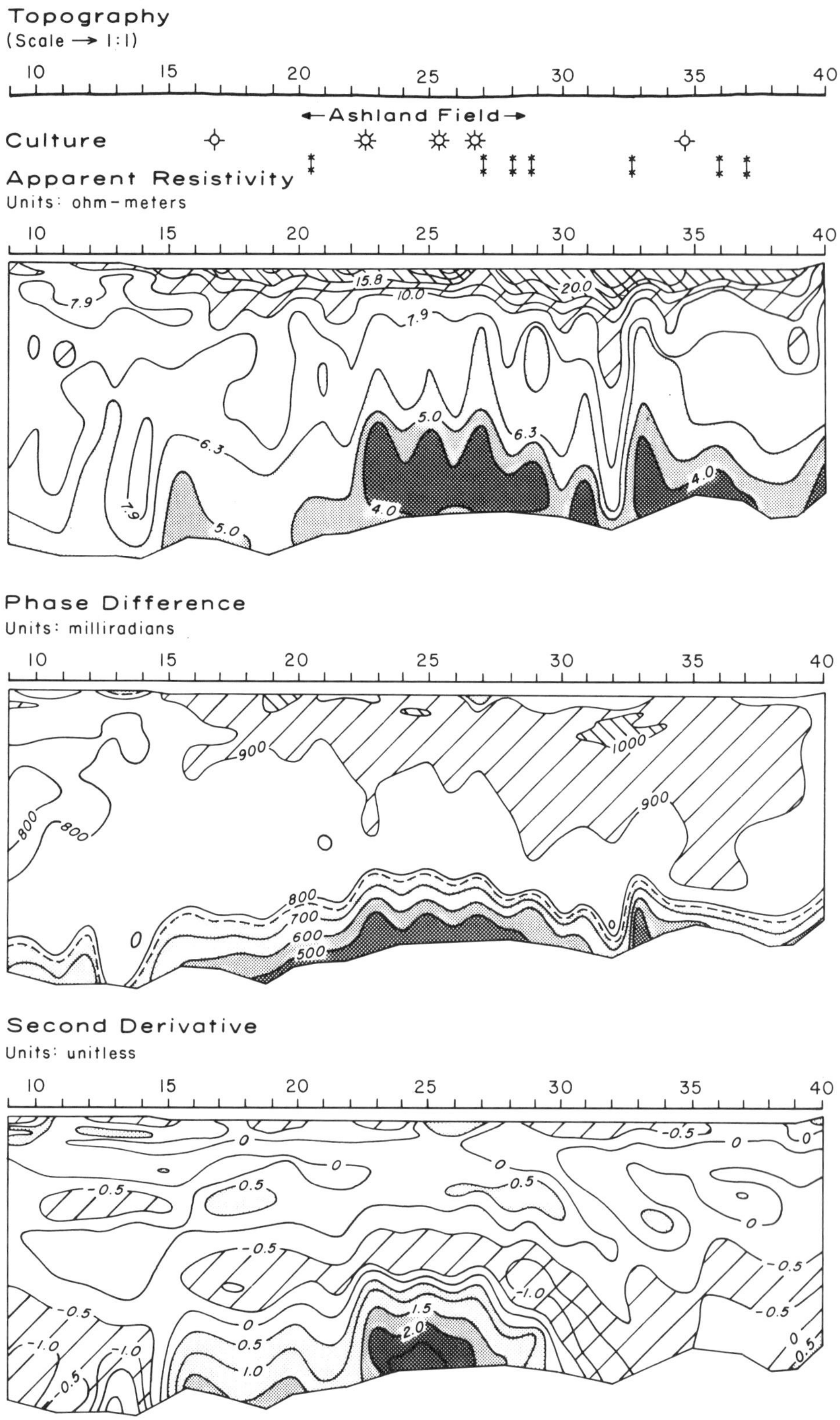

Fig. 5. Field data for the CSAMT survey across Ashland Field. Increasingly dark shading indicates increasing conductivity; cross-hatching indicates resistive areas.

EFFECTS OF HYDROCARBON LEAKAGE ON EARTH SURFACE MATERIALS

Martin D. Matthews

TEXACO USA

ABSTRACT

The detection of hydrocarbon seepage was one of the most successful of the early exploration techniques. Hydrocarbon seepage has been shown to cause a variety of late stage diagenetic effects. Early diagenetic effects, however, may also be caused by hydrocarbon maturation and migration. These early diagenetic effects may be recognizable by remote sensing and geophysical techniques.

The maturation path of kerogen, summarized by the van Krevelin diagram, indicates that oxygen-rich compounds, such as carbon dioxide and organic acids, are produced prior to peak hydrocarbon generation. These acids are responsible for the creation of secondary porosity in reservoir sands by the solution of carbonates and unstable silicates such as feldspar. The continued migration of these fluids results in the precipitation of these materials as cements farther along the migration pathway. The concentration of these early cements along the preferential hydrocarbon migration pathway can be linked to the reported relationship of high seismic velocities, concentration of iron/manganese/clay minerals and carbonates, hard drilling, and topographic relief to reservoired hydrocarbons.

INTRODUCTION

Near-surface diagenetic changes reported to be related to reservoired hydrocarbons are generally assumed to be caused by recent hydrocarbon leakage. These changes, however, may also occur early in the diagenetic history of the rocks, prior to the accumulation of significant quantities of petroleum. In fact, diagenetic changes potentially occur along the entire hydrocarbon migration path from the source rock to the reservoir and up to the surface. These changes differ along the migration pathway as the fluids interact with the changing chemical/biologic systems of the rocks, and they change with time as the composition of the products produced from maturing kerogen changes. The diagenetic changes associated with migration of early maturation products creates a prepared ground situation that is in many ways similar to the prepared ground concept in ore deposits, complete with the potential for subtle alteration patterns further along the migration path from accumulations. This paper will review a model of the interaction of early kerogen maturation and its effects on reservoir diagenesis, and suggest that many of these diagenetic processes may be recognizable over accumulations as early events rather than late. The consequences of this model will then be compared to the available information on diagenetic patterns related to petroleum accumulations.

SEEPAGE AND MINERALIZATION

Many of the early explorationists reported associations of productive areas to seeps, topographic highs, saline waters, and surface mineralization. Sawtelle (1936) reports that these features were instrumental in the discovery of 70% of the Gulf Coast salt dome fields. Although many of these observations appear to be largely unsupported by critical scrutiny, it should be remembered that many of these early explorationists were not only astute observers but also successful explorationists. There are many instances where an operational rule of thumb appears to be unfounded but upon careful and thorough study reveals new well-founded principles.

A causal relationship of near-surface hydrocarbon seepage to reservoired petroleum, while not universally accepted, is the best understood of the near-surface relationships and is a necessary building block for the others. The migration routes of macroseepage from petroleum accumulations and source rocks has been summarized by Link (1952). The extension of this relationship to microseepage is, however, less straightforward. The establishment of a spatial correlation is much more difficult because of the smaller signal-to-noise ratio, the high spatial variability caused by the presence of fractures and other permeable pathways, and sampling limitations The simplifying assumption that microseep anomalies occur vertically over the hydrocarbon accumulations limits the acceptance of microseep data. A sounder but more difficult approach is to extend the surface seepage information downward into the subsurface with geological and geophysical information, as Jones and Drozd (1983) recommend and Link (1952) did with macroseeps. The compositional correlation of microseepage with the type of underlying production (gas, condensate, oil) (Jones and Drozd, 1983) is, however, convincing evidence that a causal relationship does exist between microseepage and reservoired petroleum.

Accepting the seepage of reservoired hydrocarbons to the surface, as well as directly from source rocks, let us review the relationship of these migrating hydrocarbons with the mineral phases through which they pass. The correlation of mineralogic changes to hydrocarbon seepage has been recognized since the early days of exploration. Thompson (1933) points out that sulfur and pyrite are commonly associated with oil accumulations. He also points out that in the presence of hydrocarbon gases, gypsum and possibly anhydrite are replaced by limestone in salt dome cap rocks and in the Persian Gulf-Iraq oil belt. These limestones generally form resistant topographic highs in these areas. Indeed, one of the early rules of thumb was to look for the high ground when locating a well. McDermott (1940) reports the coincidence of secondary mineralization, such as caliche, with microseepage in the vicinity of some Texas oil fields. Rosaire (1940) expands the association of secondary mineralization with hydrocarbon seepage to include silicates.

Three factors inhibited the acceptance of a relationship between diagenesis and reservoired petroleum: the lack of a theoretical basis for the relationship between diagenetic and reservoired petroleum, nonhydrocarbon related diagenetic effects, and the newly developed seismic techniques. In Russia, however, research on surface evidence of hydrocarbon seepage continued. These results, coupled with the efforts of a few American workers, have led to a resurgence of interest in surface techniques, and in some cases to a scientific basis for a causal relationship to petroleum concentrations. Feely and Kulp (1957) demonstrated that the sulfur present in the cap rocks of Gulf Coast salt domes originated by bacterial action on the anhydrite, and that calcite replaced the anhydrite as a result of bacterial oxidation of petroleum. Kirkland and Evans (1976) used this theory to explain the topographically distinct limestone buttes in the Castile formation of the Delaware Basin. They suggested that the replacement limestones were initiated by seepage of hydrocarbons upward into the anhydrite along fractures and then extended outward parallel to bedding. The recent work of Donovan (1974), Ferguson (1975), and Donovan et al. (1979) demonstrates that the presence of hydrocarbons reduces the mineral phases present. Thus, magnetite and pyrite are produced from the reduction of ferric oxides and reddish units are bleached to a lighter color. These effects are often associated with calcite cementation produced by the oxidation of hydrocarbons.

KEROGEN MATURATION

The general changes in the composition of maturing kerogen are summarized in the van Krevelin (1961) diagram shown in Figure 1. These reaction paths require that the early maturation products be dominated by oxygen-rich compounds and that later compounds be hydrogen rich. The early products are generally characterized as mixtures of carbon dioxide, water, and methane. These are idealized end members. The actual products probably include organic acids such as acetic (Carothers and Kharaka, 1978) and oxalic (Surdam et al., 1984), which subsequently break down to these components. These end member compositions in combination with the reaction paths of the van Krevelin diagram predict a range of product yields. Reported estimates of carbon dioxide are 1%–15% (Bjorlykke, 1984), 22% (Franks and Forester, 1984), and 25% with 3% water (Laplante, 1974).

The produced organic acids and carbon dioxide (hereafter referred to as early products) are more soluble than the later produced hydrocarbons and should really migrate out of the source rock with the formation waters. The compounds are very reactive and, given sufficient flux, are capable of mobilizing and later precipitating a variety of mineral phases.

MODEL OF DIAGENESIS

Cementation in sedimentary rocks has been observed in thin sections for a considerable period of time. The recent recognition of the importance and widespread occurrence of secondary porosity in sandstone reservoirs has spurred research in this area. The importance of secondary reservoir porosity to hydrocarbon accumulation has caused conceptual models of these diagenetic processes to emphasize the creation of porosity. We will turn these observations and models around and address the question: What happens to the mineral matter that is removed from the reservoir rock?

A generalized sequence of diagenetic changes in sandstone reservoirs is diagrammatically represented in Figure 2 (summarized from Loucks et al., 1984; Schmidt and McDonald, 1979; Milliken et al., 1981; Land, 1984; Franks and Forester, 1984; and Moncure et al., 1984), using a generalization of Hunt's (1979) conception of the generation of gases from kerogen as a reference. The following explanation of the reactions and mass transport of materials represented by this sequence is based on the work of the above authors with the addition of Siebert et al. (1984), Graustein et al. (1977), and Hayes (1979). Many diagenetic events occur which are not shown in Figure 2. These will not be discussed because of their local nature.

The creation of secondary porosity requires fluid motion and a solution mechanism. The principal driving force for fluid transport is shale compaction. Clay dewatering and aquathermal expansion are generally unimportant, while hydrocarbon generation is locally important. The driving force for the diagenetic mineral reactions is the generation of early products from kerogen with a variety of undetermined mineral reactions in the shale, such as kaolinite altering to illite (Lee, 1984), being potentially important. The interaction of fluid motion and kerogen maturation causes a subsiding, compacting depositional sequence to evolve from an open system dominated regime to one dominated by closed system processes.

The open system is characterized by mass transport of kerogen maturation products and inorganic species to lower pressures (usually upward). Mass transport initiates in the region where kerogen forms early products and sufficient water flow exists to carry the products out of the shales. The solutions formed by this process continue along the migration path of the water, resulting in early cementation.

The closed system occurs under two circumstances: (1) in the region where kerogen can produce early products but is either not present in sufficient quantities and/or in a zone of reduced fluid flow as a result of bypassing; (2) below the region of production of early products from kerogen, where water drive is less and the only sources of acids are diminished production of such products from continued maturation of kerogen, unexpelled products, and those produced by mineral reactions in the shales.

The diagenetic sequence in the open system begins at a depth corresponding to the production of early products. These products react with the associated shales to dissolve as much of the carbonate as they can. Hower et al. (1976) indicates nannofossil carbonate is absent in many Gulf Coast wells below this depth. The organic acids also have the capability of complexing aluminum ions (Figure 3), thus enabling the solution of aluminosilicates (Surdam et al., 1984). Seibert et al. (1984) suggest that complexing is limited in the shale because the activity of aluminum is buffered to low values by the conversion of smectite to illite.

The waters, with their dissolved products, leave the shales and migrate toward lower pressured areas along the most permeable pathways available. As the waters enter sands, the clay mineral buffers are no longer present, and solution of the most soluble aluminosilicates (dominantly plagioclase) is enabled by the complexing of the organic acids. Carbonates may be either dissolved or precipitated in these sands, depending on the amount previously dissolved. Petroleum emplacement occurs shortly after the development of secondary porosity. Lee (1984) reports that the emplacement of gas arrested the precipitation of illite in the Rotliegendes sands of the North Sea. Continued migration of these fluids results in quartz precipitation and eventually in the precipitation of feldspar overgrowths, as well as other aluminosilicates in the lower temperature section, as the organic acids break down to carbon dioxide, water, and methane. This conversion to carbon dioxide and changes in pressure and temperature cause irregular precipitation and solution of carbonate along the migration path of the water. Complete pore blockage by carbonate cement in selected sands occurs over a zone of secondary porosity development in the Mobile-Texaco Citnalta 1-59 well on the Scotian Shelf (Figure 4, Schmidt and McDonald, 1979). These cements are early diagenetic events which predate early kerogen maturation in that section. As burial continues, the carbonate and feldspar previously precipitated may be redissolved as they pass downward into the region of active solution and are then recycled upward. If insufficient early products are available at that time, or if cementation is so complete that the flow path no longer goes through a particular sand, these cements will pass downward below the zone of production of organic acids and into the lower closed system regime.

This model suggests that most diagenetic calcite has been precipitated by the action of early products derived from the maturation of kerogen. If this is true, they should be isotopically depleted in carbon 13. While isotopic carbonate analysis data in Milliken et al. (1981) indicate that depleted compositions occur only in the zone of carbon dioxide production, relatively few measurements are reported above that zone. The few measurements above the zone of carbon dioxide production reported by Land (1984) show depletion. The recycling of carbonate by carbon dioxide from kerogen would be expected to deplete the carbonate successively each time it was recycled, resulting in very negative carbon 13 values. Insufficient data is available to test this hypothesis. The isotopic effects of the organic acids on the carbonate precipitated are presently unknown.

The zone of closed circulation located within and above the zone of early products generation is typified by closed system reactions. Small amounts of acids locally dissolve and precipitate carbonate species and some silicates. The general lack of organic acids will limit the feldspar and other aluminosilicate reactions. The carbonate reactions will proceed without significant interaction with organically derived carbon and thus will not be isotopically depleted.

The diagenetic sequence in the closed system below the zone of early product generation is characterized by in-place albitization of feldspars and precipitation of late calcite cements. These late calcite cements appear to be isotopically similar to marine carbonates, suggesting that they have not interacted with carbon dioxide from kerogen. This appears to argue against the hypothesis of transport of isotopically depleted carbonates down into this zone. The common explanation is that at the bottom of the zone of carbon dioxide production from kerogen, the carbon released is no longer depleted and thus the isotopic composition is reset and maintained in the closed system.

SPATIAL VARIABILITY OF DIAGENESIS RELATED TO MIGRATION

Seibert et al. (1984) indicate that secondary porosity in sands is essentially ubiquitous above the zone of early product generation from kerogen. He and Surdam et al. (1984) also point out that the zones of highest primary porosity generally have the greatest development of secondary porosity. This suggests a spatial mix of open and closed systems above the zone of kerogen early production generation. The detection of this spatial mix may be a clue to the subsurface pathway of hydrocarbon migration. The diagenetic model, developed to explain the creation of secondary porosity in sandstone reservoirs, suggests that there are two potential causes for spatial variability in the amount of both secondary porosity at the reservoir level and cementation farther along the migration pathway (open system versus upper closed system and degree of transport in the open system).

The upper closed system contains an average variability in porosity, compaction, and cementation. Depositional porosity is continually decreased by compaction. In situ recrystallization further decreases porosity in the sands aided by local mass transport from the shales to the sands.

In the open system (region of significant diagenetic mass transport) the reservoirs have well-developed secondary porosity. The materials removed from them are distributed farther along the migration path as local areas of increased cementation. Compared with the upper closed system, this section should have greater porosity variation with more porosity at the lower reservoir levels and less in the upper levels. An end member of the open system is the preferential migration pathway, a network of interconnecting high permeability channels between a high pressured region and a low pressured region (the path of least resistance and maximum flux). This pathway is composed of a network of high depositional and fracture permeability aided by the structuring of the depositional units which focuses flow (Roberts, 1980). Along this pathway the periodic upward recycling of carbonate and feldspar should be maximized. This section should exhibit the best secondary porosity development at the reservoir level. The shallower region of porosity reduction should be extensive and complete due to the extensive upward recycling.

The region of porosity reduction in the open system should differ from that in the upper closed system by:

1. Increased carbonate cementation.
2. Depletion of carbon 13 due to repeated interaction of the calcite with kerogen-derived carbon.
3. Increased occurrence of iron and manganese due to the complexing action of the organic acids.
4. Increased occurrence of diagenetic clay minerals (the composition of these clay minerals being homogenous in an area but variable between areas, composition being determined by the activity of the cations present).
5. Increased bulk density, caused by the precipitation of transported dissolved and complexed material.
6. Increased grain-to-grain contact area due to cementation.
7. High cohesion due to cementation.
8. High seismic velocity due to the increased bulk density and grain-to-grain contact areas.
9. Topographic relief due to increased mechanical and chemical stability.

The recognition of the areas of high cementation upward along the water migration path is based on the events in sandstones and hindered by our lack of knowledge of diagenesis in shales. It is possible that the majority of the cementation may occur in or just below the shales. If the cementation is distributed over a thick vertical sequence, it will be more subtle but perhaps more likely to be detected than if it is concentrated in one or more thin zones. In either case, different techniques of detection would be appropriate.

The recognition of these regions of high cementation is also complicated by the overlap in the degree of porosity reduction/cementation in the open system with that in the closed system. Although both systems may have regions of intense porosity reduction, the open system diagenetic pattern is expected to be more complete and spatially systematic than that of the closed system. Further complications occur during uplift. At this time, late diagenetic processes may overprint these early patterns, in some cases reinforcing them and in other cases obliterating the older patterns and developing new ones.

COMPARISON OF THE MODEL WITH PUBLISHED DATA

The existence of large regions displaying lateral variability in secondary porosity caused by the migration of early maturation products is reported by Loucks et al. (1984) and Franks and Forester (1984) in the central Texas Gulf Coast Wilcox and Frio sections. The sands in this approximately 150-mi-wide strip have higher permeabilities (Figure 5) and secondary porosity than those to the northeast and southwest. This area also contains higher concentrations of carbon dioxide in the reservoir gases and is undersaturated with respect to calcite (Figure 6). More than 50% of the secondary porosity is due to dissolution of calcite and feldspar. This dissolved material must have gone somewhere and should be further along the migration path.

Carbonate diagenesis has been related to reservoired hydrocarbons and their leakage by many workers (Thompson, 1933; McDermott, 1940; Rosaire, 1940; Donovan, 1974; and others). The early workers noted the relationship while the later workers suggested it was related to surface bacterial action (Feely and Kulp, 1957). The recent work in reservoir sands suggests that although surface processes may be involved, they may represent a remobilization of an earlier relationship, or in some cases the earlier relationship itself. Isotopically depleted carbon in carbonates is dominantly reported in caliche, but Land (1984) reports its subsurface occurrence in Frio sands. Increased iron and manganese have been reported by many workers (Donovan, 1974; Ferguson, 1975; Donovan and Dalziel, 1977). These concentrations may be due either to late near-surface effects or to earlier migration events that have only recently been exposed. Few studies have attempted to relate clay mineral occurrence at shallow depths to deeper reservoired hydrocarbons, as suggested by Rosaire (1940). Collins (1980) interpreted an anomalously high clay and iron spectroscopic signature over the Wyoming Patrick Draw and Table Rock fields, using an airborn spectroradiometer. Paylor (1983) studied the surface formations at Table Rock and found an increase in clay content over the field (Figure 7) as well as increased iron mineralization (hematite and pyrite). The increased clay concentration was accompanied by an alteration of illite to smectite (Figure 8). This zone is not currently seeping significant quantities of light hydrocarbons like that at nearby Patrick Draw. There are, however, light hydrocarbons in the surface material which are released by mechanical and thermal disaggregation (Matthews et al., 1984). This suggests that alteration at Table Rock may have been a deeper, earlier diagenetic event, and alteration at Patrick Draw may be recent.

High cohesion over subsurface productive structure indicated by a decrease in the rate of drilling (Figure 9) was recognized by many early drillers (Rosaire, 1938). This was often related to silification in the units above production. This phenomenon is usually not recognized today because of modern drilling techniques. This high cohesion and chemical stability is also responsible for the topographic highs related to fields in some areas.

Locating prospects from seismic reflection profiles commonly utilizes the simplifying assumption that lateral changes in velocity either do not exist or are unimportant. The fact that this assumption is not always correct can be affirmed by many explorationists who have drilled time highs only to find themselves dry and lower on the structure than they predicted. Rosaire (1938) reports several instances where high interval velocities exist over structural accumulations but not off them. High seismic velocity in the Mesozoic-Tertiary sediments of the Spitsbergen basin have been related to porosity reduction due to cementation (Elverhoi and Gronlie, 1981).

CONCLUSIONS

Published conceptual models of reservoir diagenesis state that the early generation products of kerogen breakdown (carbon dioxide and organic acids) are responsible for the creation of secondary porosity in sandstones. These same models suggest that the leached material is precipitated as early cements in overlying sands. It is suggested that this zone of early cementation is concentrated along the preferential migration pathway from the reservoir to the surface (Figure 10). This theoretical zone of cementation explains the high velocity zones, concentration of iron/manganese/clay minerals and carbonates, hard drilling, and topographic relief (Figure 11) that have been reported to occur over fields. This zone of cementation should also complicate the distribution of surface hydrocarbon seepage. The detection of these early diagenetic zones offers the potential to recognize a part of the preferential migration pathway, formed by early kerogen maturation products, that was later utilized by the migrating hydrocarbons.

REFERENCES

Bjorlykke, K., 1984, Formation of secondary porosity: How important is it?: American Association of Petroleum Geologists Memoir 37, p. 277–286.

Bornhauser, Max, and Bates, S. W., 1938, Geology of Tepetate Oil Field, Acadia Parish, Louisiana: American Association of Petroleum Geologists Bulletin, v. 22, p. 285–305.

Carothers, W. W., and Kharaka, Y. K., 1978, Aliphatic acid ions in oil-field waters: Implications for origin of natural gas: American Association of Petroleum Geologists Bulletin, v. 62, p. 2441–2453.

Collins, W., 1980, Airborn infrared spectroradiometer survey of Patrick Draw, Wyoming: New York, Geophysical Environmental Research, Inc. (GPO Box 340, New York, N. Y. 10001), 49 p.

Donovan, T. J., 1974, Petroleum microseepage at Cement, Oklahoma: Evidence and migration: American Association of Petroleum Geologists Bulletin, v. 58, p. 429–446.

———, Dalziel, M. C., 1977, Late diagenetic indicators of buried oil and gas: United States Department of the Interior, Geological Survey, Open-File Report 77-817, 44 p.

———, Forgey, R. L., and Roberts, A. A., 1979, Aeromagnetic detection of diagenetic magnetite over oil fields, American Association of Petroleum Geologists Bulletin, v. 63, p. 245–248.

Elverhoi, A., and Gronlie, G., 1981, Diagenetic and sedimentological explanation for high seismic velocity and low porosity in Mesozoic-Tertiary sediments, Svalbard region: American Association of Petroleum Geologists Bulletin, v. 65, p. 145–153.

Feely, R. W., and Kulp, J. L., 1957, Origin of Gulf Coast Salt-Dome sulfur deposits, American Association of Petroleum Geologists Bulletin, v. 41, p. 1802–1853.

Ferguson, J. D., 1975, Surface alteration and mineralization of Permian Red Beds overlaying several oil fields in Southern Oklahoma: Masters thesis, Oklahoma State University, 95 p.

Franks, S. G., and Forester, R. W., 1984, Relationships among secondary porosity, pore-fluid chemistry and carbon dioxide, Texas Gulf Coast: American Association of Petroleum Geologists Memoir 37, p. 63–79.

Graustein, W. C., Cromack, K., Jr., and Sollins, P., 1977, Calcium Oxalate: Occurrence in soils and effect on nutrient and geochemical cycles: Science, v. 198, p. 1252–1254.

Hays, J. B., 1979, Sandstone diagenesis: The hole truth: Society of Economic Paleontologists and Mineralogists Special Publication 26, p. 127–139.

Hower, J. E., Eslinger, M. E., Hower, M. E., and Prerry, E. A., 1975, Mechanism of burial metamorphism of argillaceous sediments, 1: Mineralogic and chemical evidence: Geological Society of America Bulletin, v. 87, p. 725–737.

Hunt, J. M., 1979, Petroleum geochemistry and geology: San Francisco, W. H. Freeman, 617 p.

Jones, V. T., and Drozd, R. J., 1983, Predictions of oil or gas potential by near-surface geochemistry: American Association of Petroleum Geologists Bulletin, v. 67, p. 932–952.

Kirkland, D. W., and Evans, R., 1976, Origin of limestone buttes, Gypsum Plain, Culberson County, Texas: American Association of Petroleum Geologists Bulletin, v. 60, p. 2005–2018.

Land, L. S., 1984, Frio sandstone diagenesis, Texas Gulf Coast: A regional isotopic study: American Association of Petroleum Geologists Memoir 37, p. 47–62.

LaPlante, R. E., 1974, Hydrocarbon generation in Gulf Coast tertiary sediments: American Association of Petroleum Geologists Bulletin, v. 58, p. 1281–1289.

Lee, M., 1984, Diagenesis of the Permian Rotliegendes sandstone, North Sea: K/Ar, O^{18}/O^{16}, and petrologic evidence: Ph.D. dissertation, Case Western Reserve University, 346 p.

Link, W. K., 1952, Significance of oil and gas seeps in world oil exploration: American Association of Petroleum Geologists Bulletin, v. 36, p. 1505–1541.

Loucks, R. G., Dodge, M. M., and Galloway, W. E., 1984, Regional controls on diagenesis and reservoir quality in the lower tertiary sandstones along the Texas Gulf Coast: American Association of Petroleum Geologists Memoir 37, p. 7–45.

Matthews, M. D., Jones, V. T., Richers, D. M., 1984, Remote sensing and surface hydrocarbon leakage: International Symposium on Remote Sensing of the Environment, Proceedings, p. 663–670.

McDermott, E., 1940, Geochemical exploration (soil analysis): American Association of Petroleum Geologists Bulletin, v. 24, p. 859–881.

Milliken, K. L., Land, L. S., and Loucks, R. G., 1981, History of burial diagenesis determined from isotopic geochemistry, Frio Formation, Brazoria County, Texas: American Association of Petroleum Geologists Bulletin, v. 65, p. 1397–1413.

Moncure, G. K., Lahann, R. W., and Siebert, R. M., 1984, Origin of secondary porosity and cement distribution in a sandstone/shale sequence from the Frio Formation (Oligocene): American Association of Petroleum Geologists Memoir 37, p. 151–161.

Paylor, E. D., II, 1983, Investigation of surface features interpreted from remote sensing data of the table rock gas field, Sweetwater County, Wyoming: Ph.D. dissertation, University of Wyoming, 81 p.

Roberts, W. H., III, 1980, Design and function of oil and gas traps, *in* Roberts, W. H., III, and Cordell, R. J., eds., Problems in Petroleum Migration: American Association of Petroleum Geologists Studies in Geology, no. 10, p. 217–240.

Rosaire, E. E., 1938, Shallow stratigraphic variations over Gulf Coast Structures: Geophysics, v. 3, p. 96–121.

———, 1949, Geochemical prospecting for petroleum: American Association of Petroleum Geologists Bulletin, v. 24, p. 1400–1433.

Sawtelle, G., 1936, Salt Dome statistics: American Association of Petroleum Geologists Bulletin, v. 20, p. 726–735.

Schmidt, V., and McDonald, D. A., 1979, The role of secondary porosity in the course of sandstone diagenesis: Society of Economic Paleontologists and Mineralogists Special Publication 26, p. 175–207.

Seibert, R. M., Moncure, G. K., and Lahann, R. W., 1984, A theory of framework grain dissolution in sandstones: American Association of Petroleum Geologists Memoir 37, p. 163–175.

Surdam, R. C., Boese, S. W., and Crossey, L. J., 1984, The chemistry of secondary porosity: American Association of Petroleum Geologists Memoir 37, p. 127–149.

Thompson, A. B., 1933, The economic value of surface petroleum manifestations: London, First World Petroleum Congress, Proceedings, p. 241–250.

Van Krevelen, D. W., 1961, Coal: New York, Elsevier, 514 p.

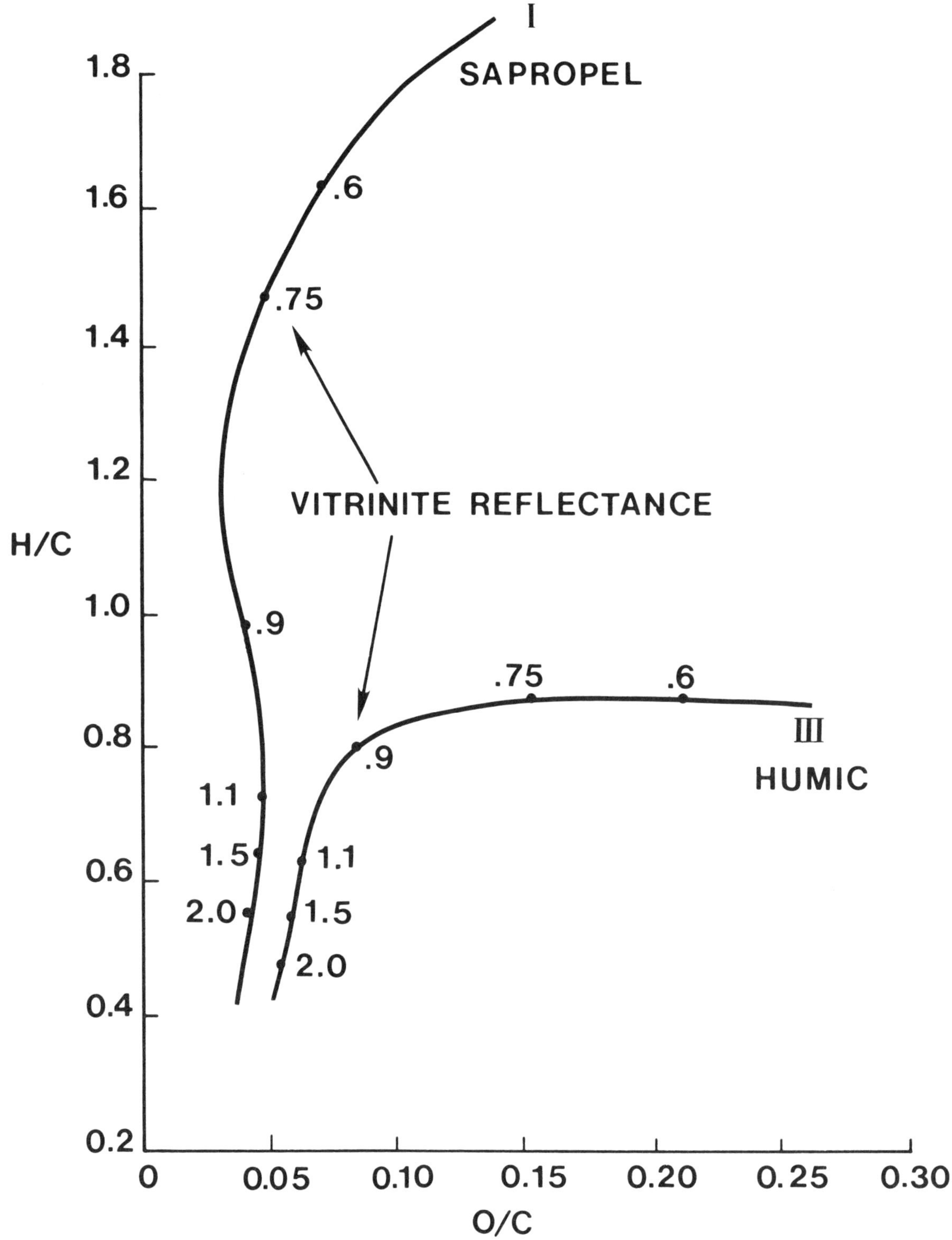

Fig. 1. Compositional changes of maturing kerogen.

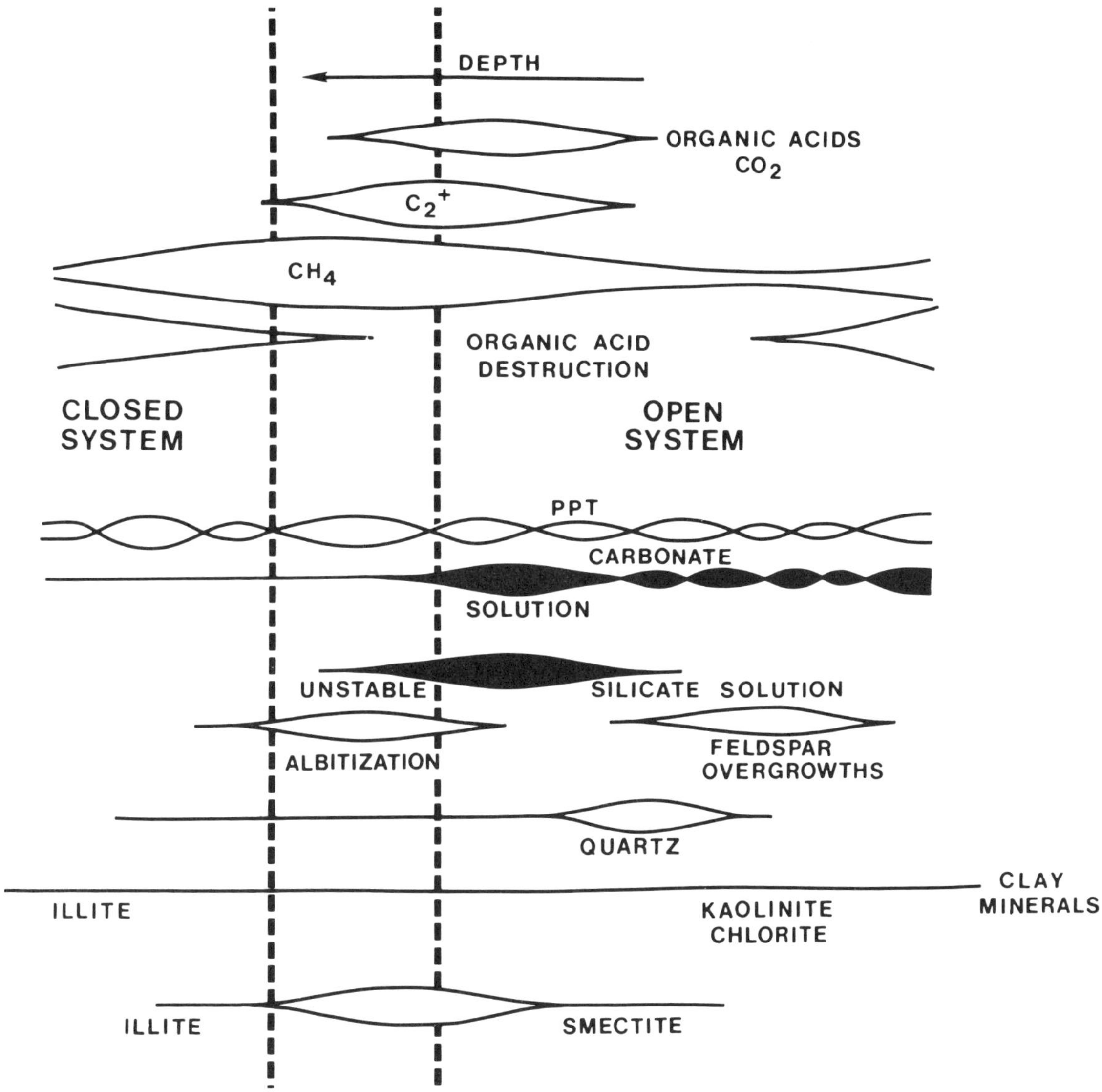

Fig. 2. Generalized diagenetic sequence in sandstone related to kerogen maturation products.

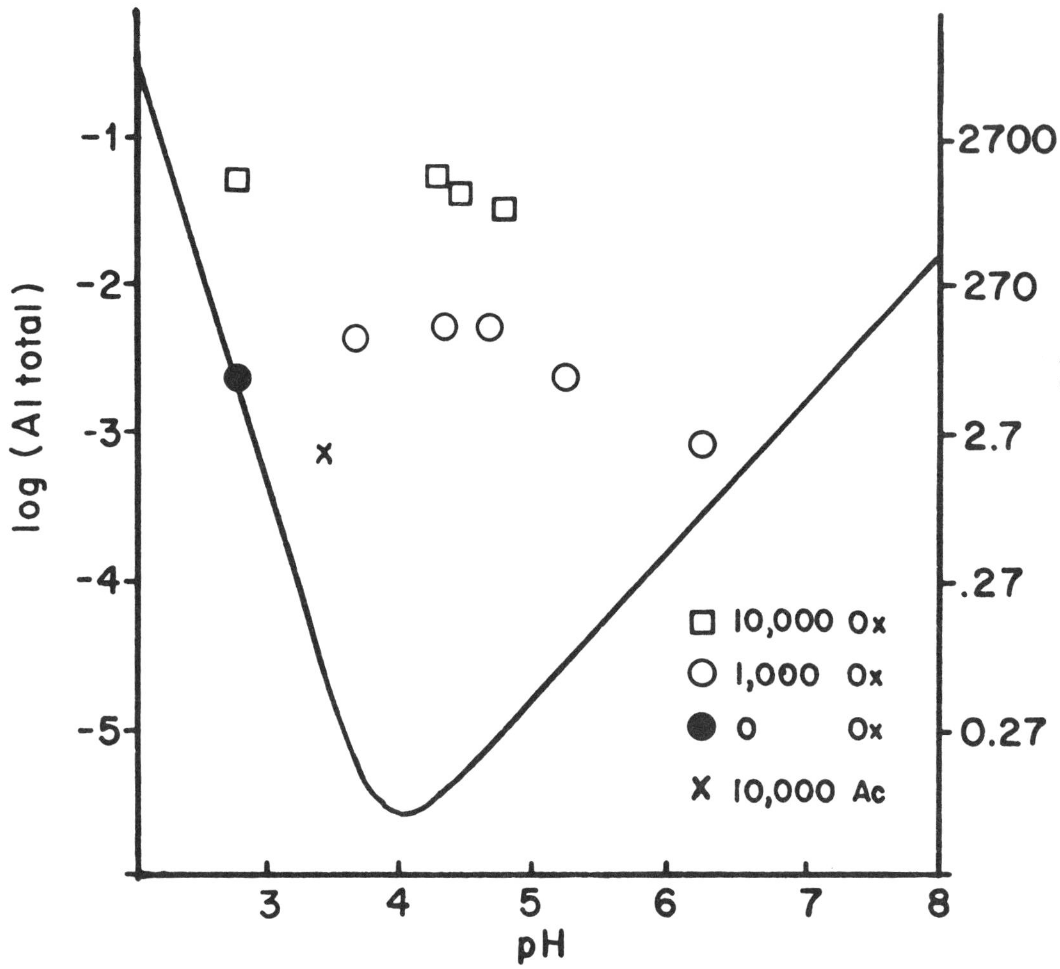

SURDAM, BOESE, AND CROSSEY (1984)

ALUMINUM SOLUBILITY 100 °C

Fig. 3. Effect of oxalate and acetate ions on aluminum solubility in water.

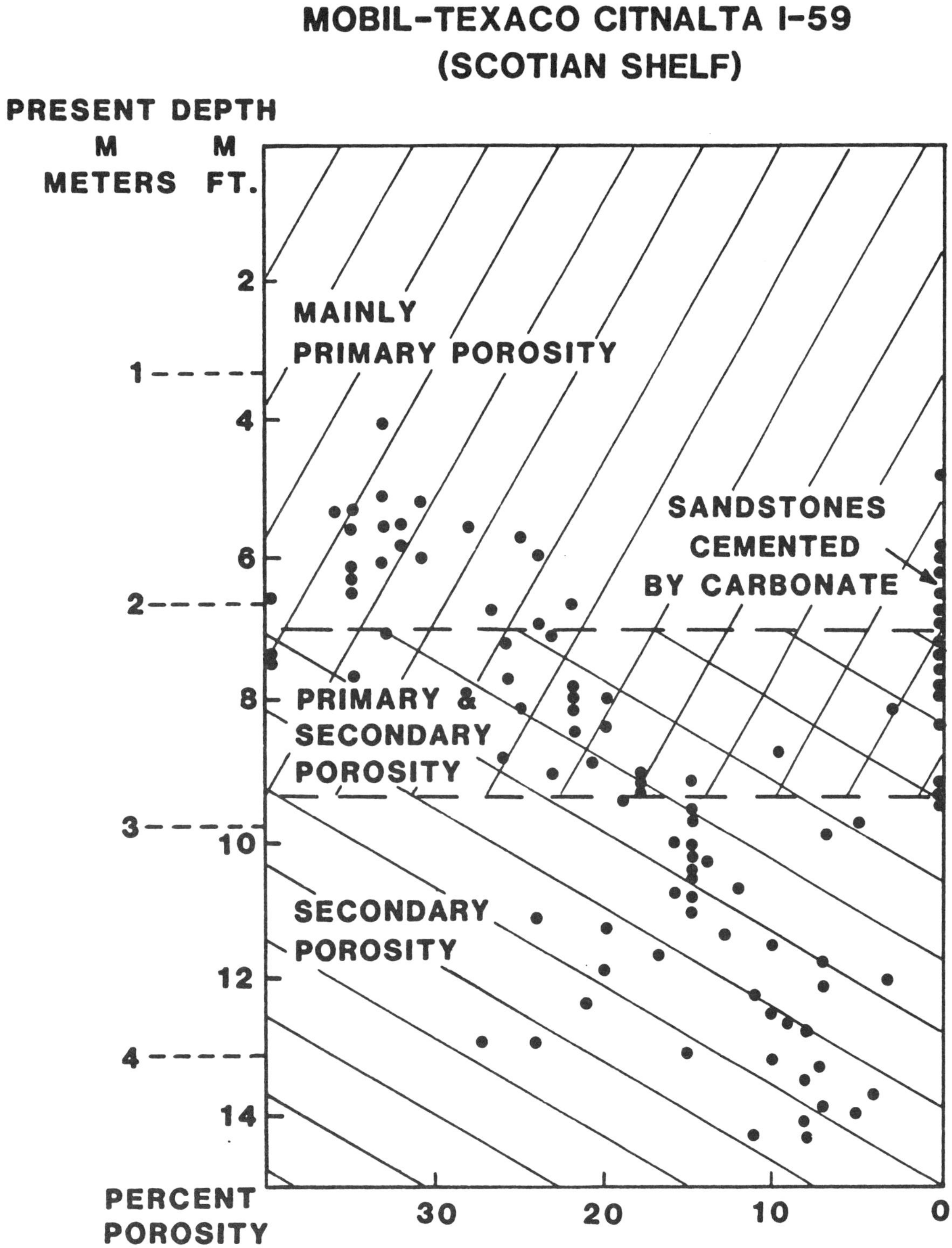

SCHMIDT AND MCDONALD (1979)

Fig. 4. Porosity modification with depth in the Mobil-Texaco Citnalta 1-59 well on the Scotian Shelf.

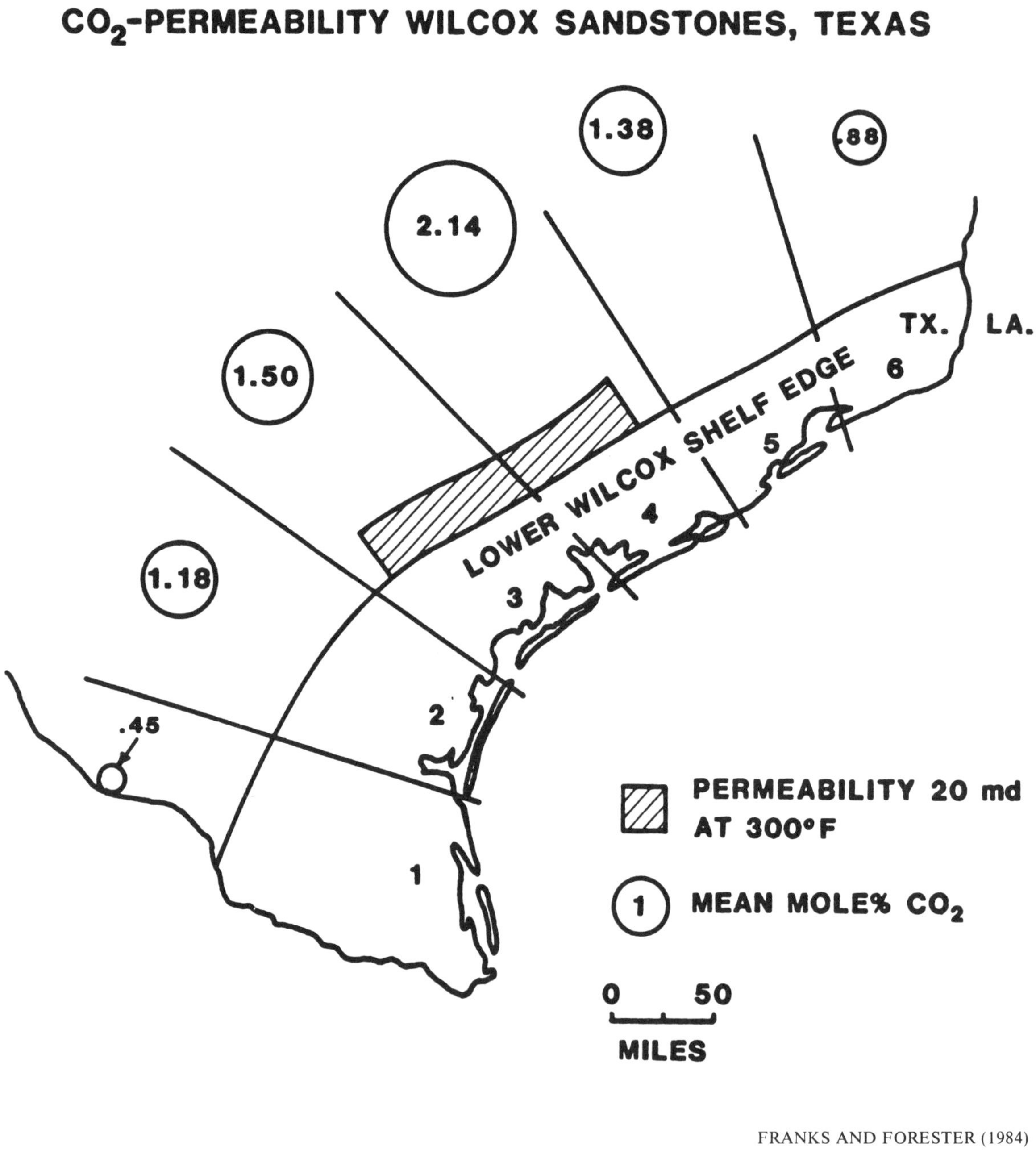

Fig. 5. Permeability and CO_2 concentration in the Wilcox formation of Texas.

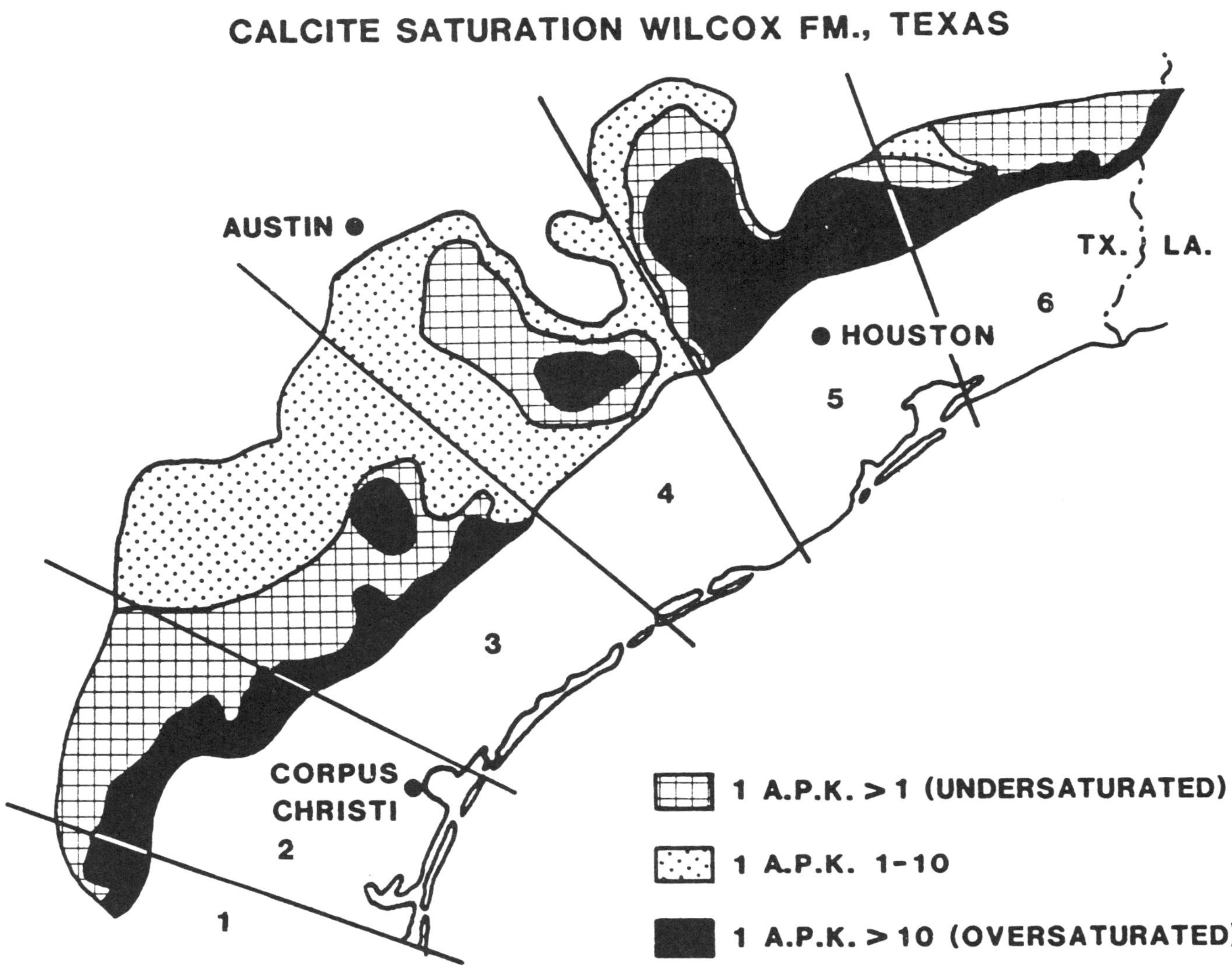

FRANKS AND FORESTER (1984)

Fig. 6. Calcite saturation in the Wilcox formation of Texas.

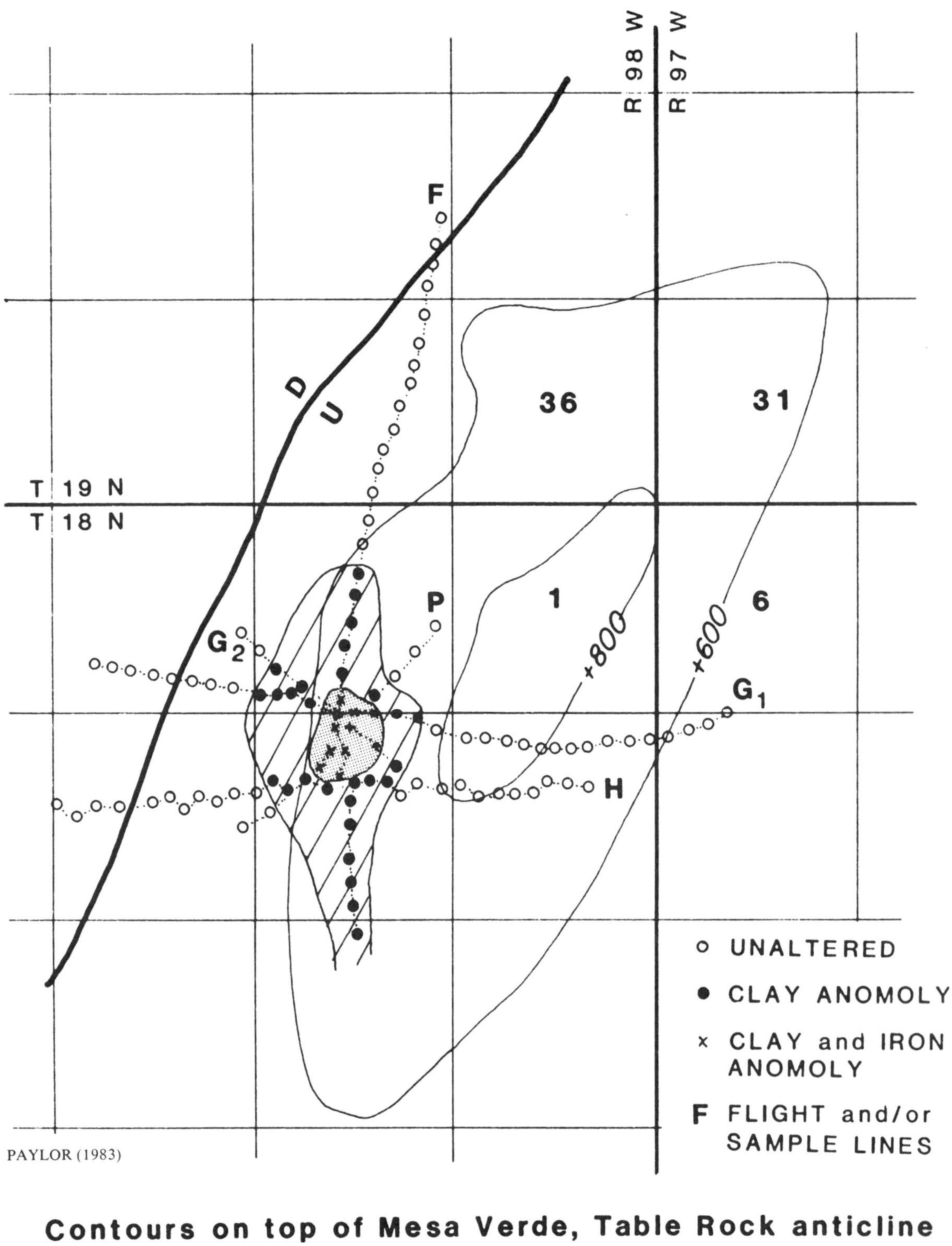

Fig. 7. Surface anomalous clay and iron concentrations over the Table Rock anticline, Wyoming, as detected by airborn spectrometry.

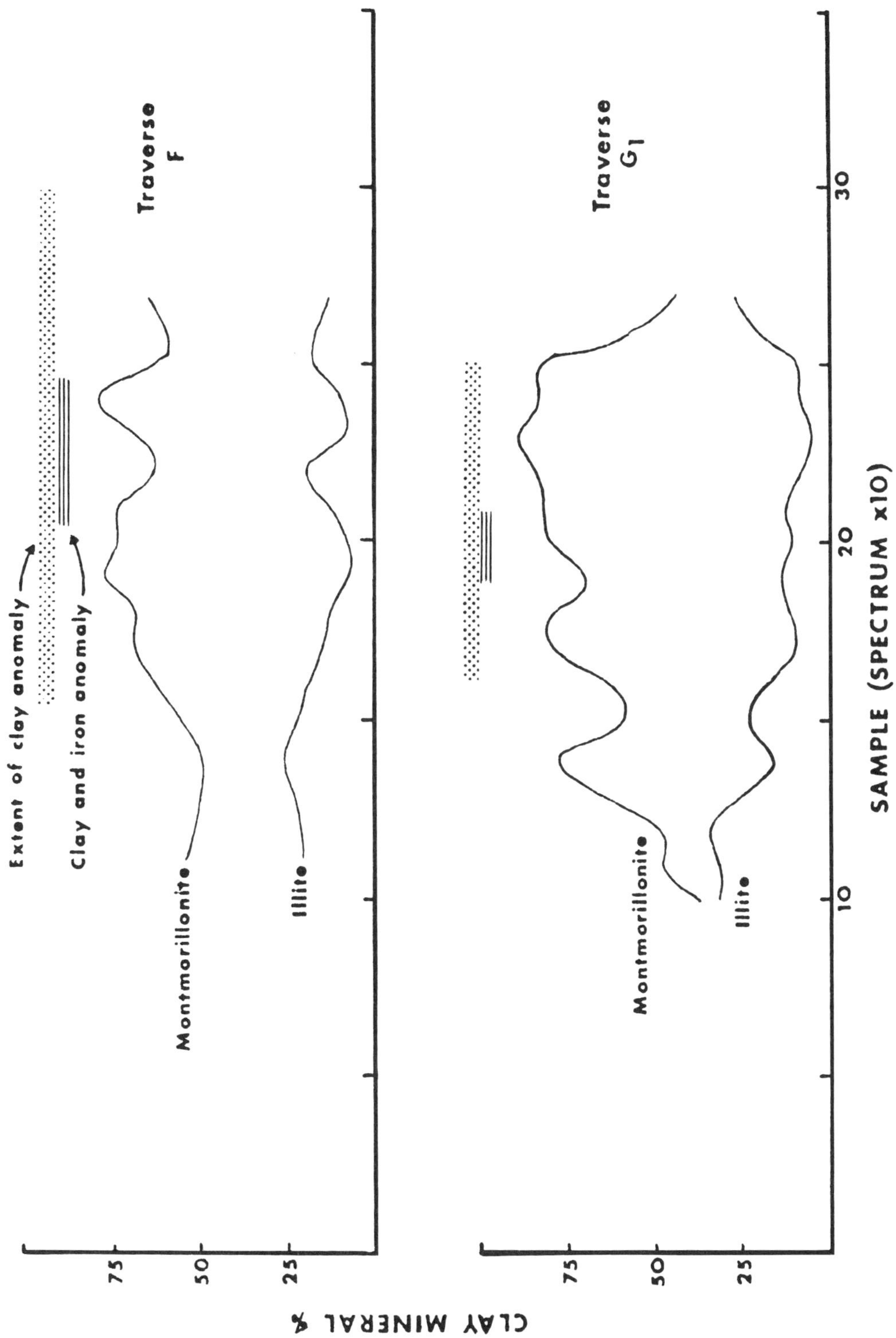

PAYLOR (1983)

Fig. 8. Illite and montmorrillonite concentrations along lines F and BI of Figure 7.

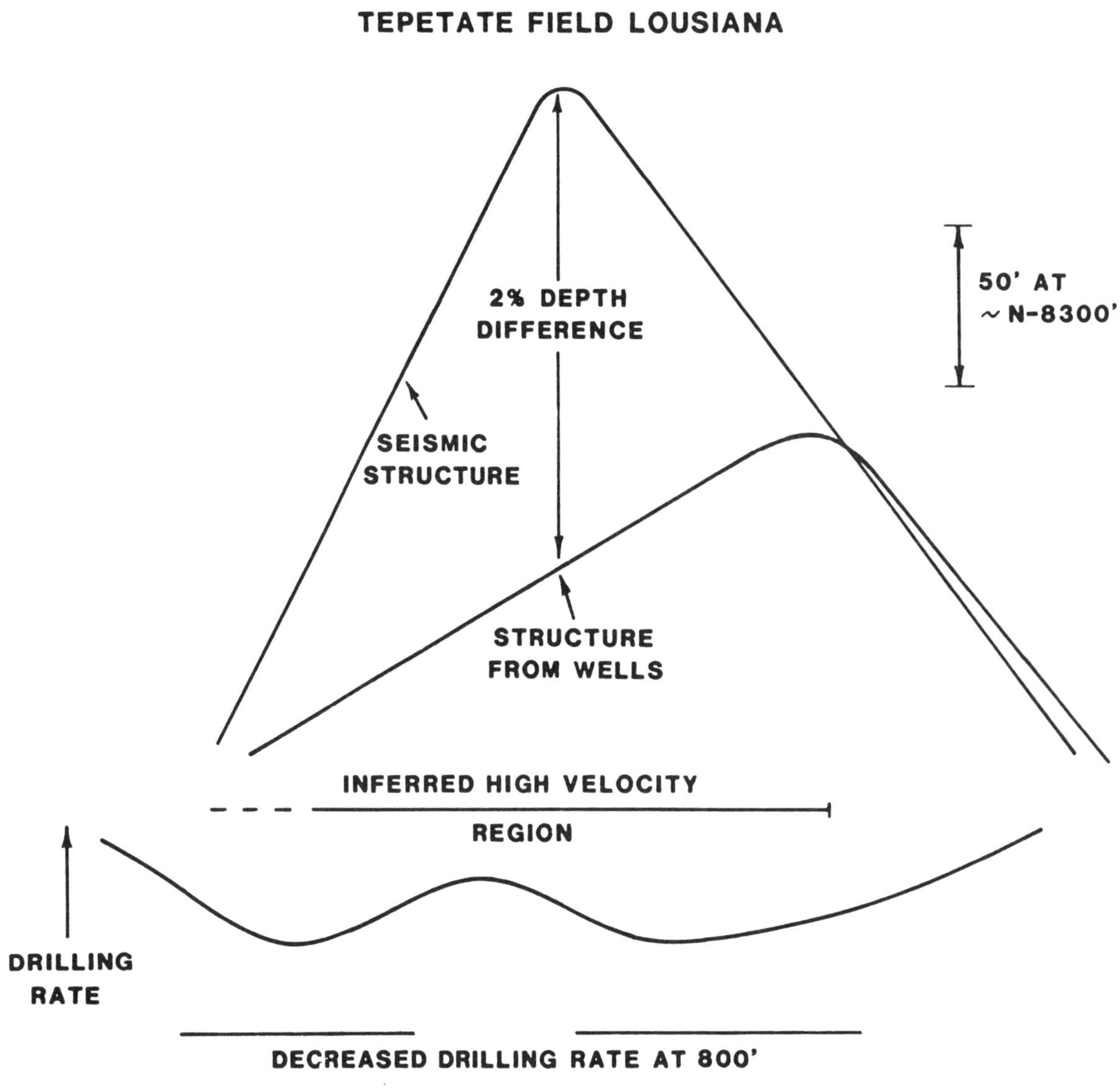

MODIFIED FROM BORNHAUSER AND BATES (1938)

Fig. 9. Relationship at Tepetate Field, Louisiana, between decreased drilling rate and a high velocity, near-surface, zone. The existence of this zone is inferred from the difference between true depth and a depth estimated from seismic.

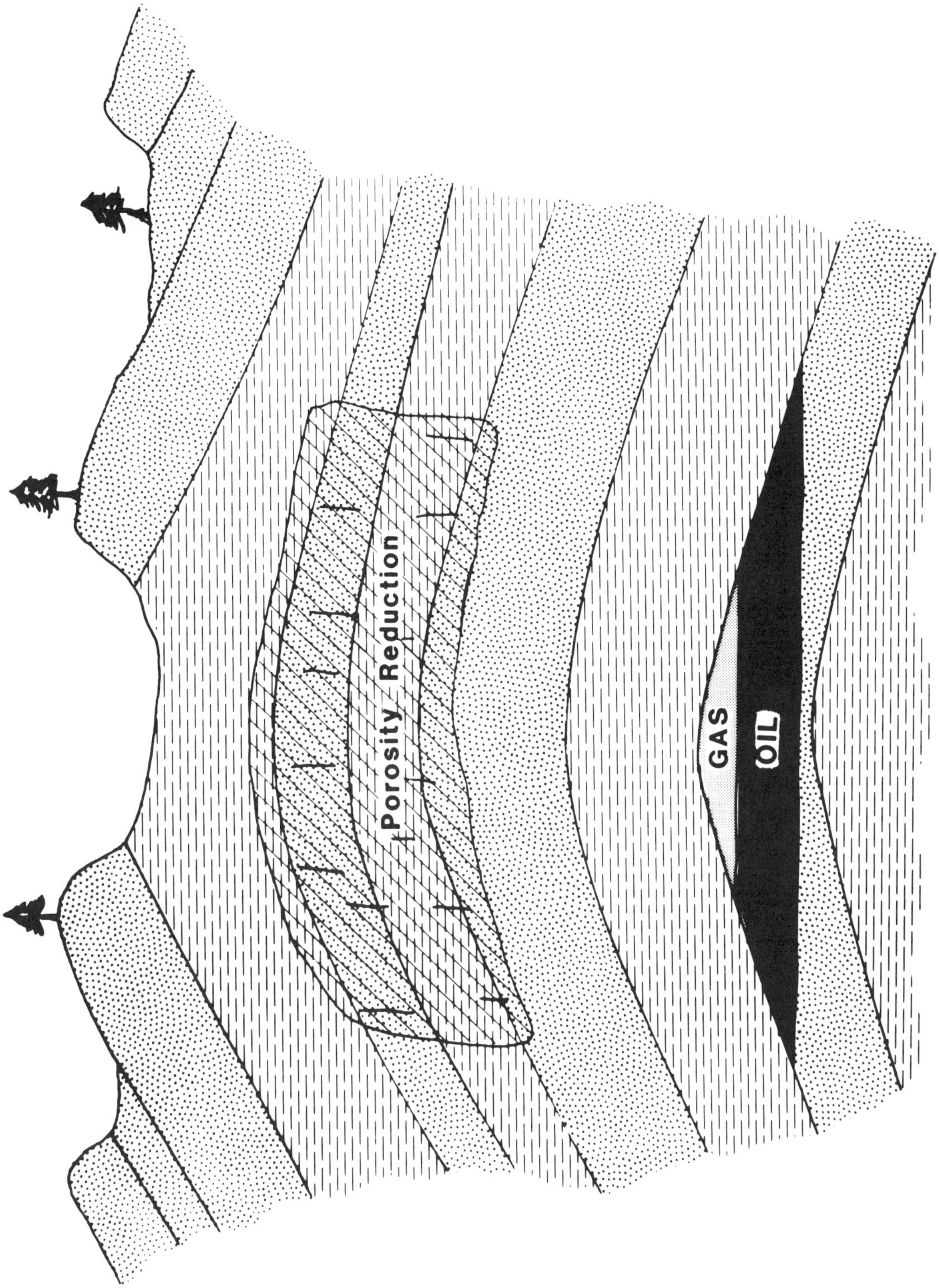

Fig. 10. Conceptual zone of porosity reduction due to cementation along a preferential fluid migration path.

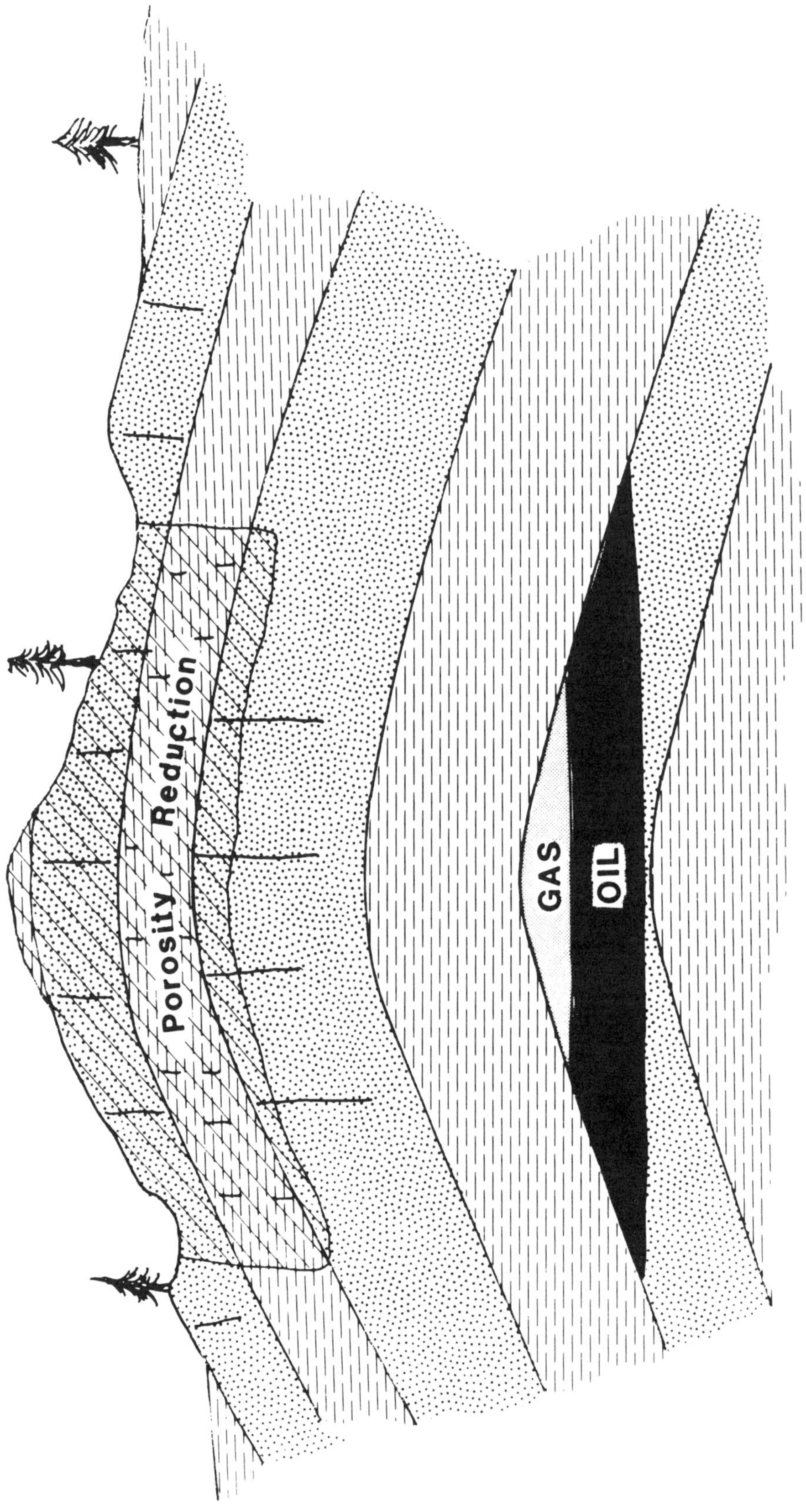

Fig. 11. Exposure of an earlier cementation event (as formed in Figure 10), due to erosion.

INNOVATIVE DIRECTIONS OF MAGNETOELECTRIC EXPLORATION

Ronald M. Herzfeld

INDEPENDEX, INC.

ABSTRACT

Magnetoelectric (ME) exploration is a geophysical science combining the principles of magnetic and electric measurements. The ME process is based on the discovery that producing and potentially productive oil and/or gas fields usually have associated with them vortices of electrotelluric (ET) currents which find their origin in the structural or stratigraphic reservoirs of oil and gas, and in the overlying lithologic environment which has been physicochemically modified by escaping reservoir waters and hydrocarbons.

The vertical ET currents generated in the earth by the "fuel cell" effect of oil and gas accumulations give rise to anomalies in the static earth magnetic field. Detection and measurement of the intensities, distribution, and direction of the ET currents are accomplished by measuring the magnetic intensities, directions, and signs of the perturbations created in the earth's magnetic field.

The magnetic perturbations may be measured directly and intentionally by performing total intensity magnetic surveys, either by means of sampling in the air, on the ground, or at sea.

In the data processing procedure, line-integral calculations of the total magnetic field are performed on one or more closed profiles. Through this procedure, it is possible to map the vertical flux of the ET currents in intensity and in direction, and to provide a surface indication of potential areas or trends favorable for the accumulation of hydrocarbons.

Recent experimental work in depth projection has established the possibilities of computer-depth determination, utilizing the principles of ME line-integral calculations, so that possible depth ranges of hydrocarbon accumulations might be derived.

INTRODUCTION

This paper is written in memory of Dr. Sylvain J. Pirson, a man who dedicated the greater portion of his life to the development of a process whereby oil and gas accumulations could be delineated by surface measurement. In his quest for a direct oil and gas finding method, Dr. Pirson investigated many avenues of thought, each contributing to the understanding of how oil and gas might be found in location, extent, depth, quantity, and productivity. Geochemical, radioactive, and electrical techniques were all evaluated, and each established without question the existence of environmental modifications above an oil field. Through the study of electric logs, SP shale-base-line drifts were observed (Pirson, 1981), which led to the theory that these drifts were a measure of the electrotelluric (ET) current flux that flows vertically within the sediments traversed by the borehole.

It was also observed that magnetic field perturbations were generated by the ET current flow (Pirson, 1977), and it was then realized that from the study of the earth's magnetic field, localized disturbances, when properly filtered, could relate to oil and gas accumulations.

The concept of magnetoelectric exploration establishes that the vertical ET currents generated in the earth by the "fuel cell" effects of oil and gas accumulations give rise to anomalies in the static earth magnetic field. This "fuel cell" effect is manifested by an oxidation-reduction potential formed at the interface of: (1) an oil-saturated environment, being reduced by the excess of electrons, and (2) a water-saturated environment, being oxidized and deficient of electrons.

In the data-processing procedure used in ME exploration, the effects of basement rock structure and of rock magnetic polarization are theoretically removed, and only the residual magnetic part related to vertical electric currents remains.

The process of ME exploration provides for the line integration of the earth's magnetic field around a closed circuit, which can be used for the determination of the intensity of the subsurface ET currents, in direction and in magnitude, so that their distribution below the surface of the earth can be measured.

The mapping and contouring of the polarity and of the intensity of such currents therefore depict the subsurface "fuel cell" anomalies, and thereby delineate the favorable areal extent of potential oil and gas accumulations (Herzfeld, 1985). They likewise denote areas that are unfavorable.

ME SURVEY: A PREDEVELOPMENT STUDY

An area in SE Illinois, situated in Edwards County, was surveyed to determine the potential for hydrocarbons in a leaseblock located in section 28 T1N R10E. Figure 1 shows the drilling activity prior to an ME ground magnetic survey. A survey was subsequently conducted over section 28 and a 1-mi area around the section. A station density pattern of ¼ mi was maintained. An anomalous condition was found to exist in a N-S axis through section 28, as seen in Figure 2. Immediately following the ME survey, development of the section commenced and resulted in seven Rosiclaire (Mississippian) wells, two O'Hara wells, one McClosky well, and three dry holes, as exhibited in Figure 3. It was apparent that the Rosiclaire trend followed the N-S axis of the ME survey and was confined within the limits of the 15 ma/ac (milliamp per acre) intensity value contour. The discovery well was completed in the SE NE SW 28-1N-10E flowing 180 bo/d from the Rosiclaire pay through perforations at 3,284–94 ft.

ME SURVEY OF AN ABANDONED FIELD

An ME survey was conducted over the Griffiths Field located in Clay County, Kansas, and situated in section 3-T9S-R4E. The field was abandoned in 1981, after producing 95,053 bbls of oil. A field statistical breakdown is shown in Table A. The Griffiths Field produced from the Mississippian chert, from fractured porosity along the axis of a plunging nose. The updip wells had no porosity development and the downdip wells dipped into water. The structural point of the top of the Mississippian chert horizon is shown in Figure 4. A ground magnetic survey was conducted in and around the environs of the Griffith Field, the results of which are seen in Figure 5. The stations were spaced at an approximate ½ mi density pattern, except in the center of section 3, where existing pipe was intentionally avoided in an attempt to avoid the influence of possible spurious readings. The adjusted total intensity values of the magnetic survey were viewed as a magnetic high in the north half of section 3. The magnetic data were processed using the ME procedure, and a surface integration map was created, as presented in Figure 6. An N-S axis anomaly was defined, with the productive wells situated within the boundary of the 10 ma/ac contour. A depth investigation study was conducted on this project, and a point at location X was used to investigate the change of milliamp current flow at various depth levels. Graph I was prepared from a computer printout of the Griffiths Field and surrounding acreage. At location X there was a sharp increase in intensity value directly overlying the horizon of the top of the productive zone. It appears that the intensity values are greatest directly over the productive horizon penetrated.

TABLE A

GRIFFITHS FIELD
Section 3, T9S R4E
Clay County, Kansas

Discovery date	May 30, 1957
Operator	Pure Oil Company
Well	#1 F.L. Griffiths
Location	C NE NW Sec. 3-9S-4E
Producing formation	Mississippian (Burlington-Keokuk) Chert
Initial production (Pot.)	Pump 86.5 bbls oil; 26 bbls water/day
Initial reservoir pressure	685 psi (DST)
Method of discovery	Subsurface geology
Nature of trap	Stratigraphic entrapment
Reservoir thickness	22–29 ft
Average depth	1,920 ft
Productive zone thickness	4 ft plus fractures
Porosity (log)	18% to 35% where porous and permeable
Water saturation	Variable
Water datum	−644 ft (related to permeability more than structure)
Productive area	Approximately 200 acres
Drive mechanism	Inadequate water drive
Oil Character	27° API @ 60°F
Gas	Too small to measure
Completion technique	Perforations
Treatment	Acidfrac; 6,000 gals frac acid, 6 gals Mar-flo, 6000# sand
Total producers drilled	4 wells; 1 abandoned, 8/1/59 (4,091 BO)
Cumulative production	95,053 bbls oil
Year field abandoned	1981

ME SURVEY OF A PRODUCING FIELD

An ME survey was conducted over the Nimrod Field located in Eastland County, Texas, situated in sections 94 and 95, block 3, H&TC Survey. The field produces from the shallow Crosscut sand, at a depth of approximately 1,300–1,350 ft. Figure 7 is a structure map at the top of the Crosscut-sand porosity. The wells in the NE portion of section 94 produced from the Caddo lime formation at approximately 2,900 ft. An ME surface integration map, Figure 8, depicts the "fuel cell" effect over the Nimrod Field area. The productive limits of the field are situated within the boundary of the surface integration results. The definition of this anomalous area is probably rather regional, since the ground survey had a wide station density pattern of ½ mi to 1 mi. It nevertheless provided an adequate delineation of the "fuel cell" area and a distribution of the ET current flux.

A geologic cross section was created, through the wells shown in Figure 8. It was noted that there was an apparent shale-base-line drift on the SP curve of the electric logs of the field area. This drift is presented in Figure 9. The log on the left exhibited a strong water-gradient to the left, from the base of the log upward. This log is of a dry hole, with a wet horizon at the Crosscut, near the edge of the field. The other log exhibited an oil-gradient drift to the right, which is from a producing well. The vertical drifts present an example of the flow of electrotelluric currents (Pirson, 1981).

A computer depth study was conducted over the field area, which provided greater understanding for the ME depth projection interpretations (Pirson, 1982). Location A, on Figure 10, was chosen to demonstrate the vertical intensity differences at

TABLE B

MAGNETOELECTRIC EXPLORATION DEPTH PROJECTION LEVELS
(Estimated Depth in Feet from Surface)

	Grid Unit Dimensions[1]						
Level	330′	660′	1320′	2000′	2640′	3960′	5280′
Z 1	80	165	330	500	660	990	1300
2	120	231	460	700	920	1386	1700
3	165	330	660	1000	1320	1980	2700
4	230	462	924	1400	1850	2772	3700
5	330	660	1320	2000	2640	3960	5300
6	410	825	1650	2500	3300	4950	6700
7	460	924	1848	2800	3700	5544	7500
8	530	1056	2110	3200	4220	6336	8500
9	580	1155	2310	3500	4620	6930	9300
10	610	1221	2440	3700	4880	7326	9900
11	660	1320	2640	4000	5280	7920	10700
12	690	1386	2770	4200	5540	8316	11200
13	740	1485	2970	4500	5940	8910	12000
14	780	1551	3100	4700	6200	9306	12500
15	810	1617	3230	4900	6470	9702	13000
16	840	1683	3370	5100	6730	10098	13600
17	875	1749	3500	5300	7000	10494	14100
18	940	1881	3760	5700	7520	11286	15200
19	960	1914	3830	5800	7660	11484	15500
20	990	1980	3960	6000	7920	11880	16000
Surface Requirements							
Acres	490	1960	7840	18000	31360	70560	125440
Sq Mi	.77	3.06	12.3	28.0	49.0	110.3	196.0
Mi Sq	.88	1.75	3.5	5.3	7.0	10.5	14.0

[1]ESTIMATED DEPTHS: We currently feel as though our depth projection work is still in the experimental phase of development, but offer this work as a potential evaluation process.

TABLE C

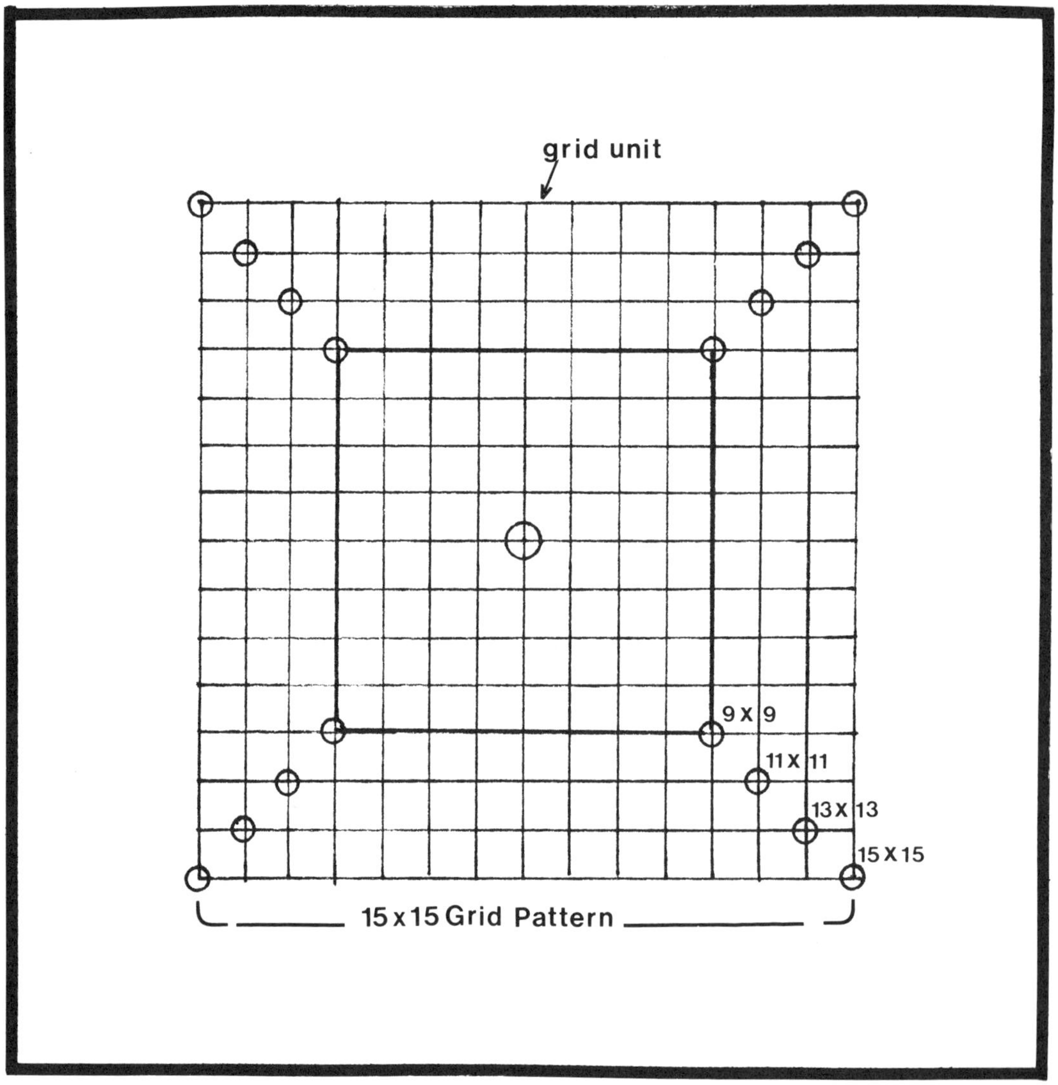

Our computer program is designed for a grid coordinate system of 15×15 grid points, as shown above. Each square, or grid unit dimension, can represent any scaled dimension. If each grid unit is 1,000 ft, then the distance on each side of the 15×15 grid pattern is 14,000 ft; and, if each grid unit is 1 mi, then the distance is 14 mi.

The 15×15 grid is superimposed over an adjusted, contoured total intensity map generated from magnetic measurements, and grid coordinates are designated at each grid corner by interpolating the contours of the total intensity map to the grid corners. The coordinates are then fed into the computer system for computation. During the process of integration, the deeper projections lose actual coverage area, as shown on the following table:

Depth Level	Actual Coverage
Z1–Z5	13×13 grid pattern
Z6–Z11	11×11 grid pattern
Z12–Z20	9×9 grid pattern

In essence, what may have been an initial 14 mi×14 mi coverage is reduced to 8 mi×8 mi at the Z12–Z20 levels, so it is desirable to center the project area for maximum results.

various depth horizons. The area was gridded for computer input. Tables B and C provide descriptive information concerning ME depth projections, based on a 15×15 grid pattern. A 9×9 grid pattern has also been developed for smaller areas. A 2,000 ft grid pattern was first studied at Nimrod; and on Graph II it was observed a shallow indication was present. The area directly above the Crosscut zone and another above the Caddo has a strong buildup in intensity, compared with other horizons. The same data were used, but interpolated to achieve a 500 ft grid pattern, which is smaller in areal extent and which fit within the 2,000 ft grid pattern, as outlined in Figure 10. From Graph III it was observed that the area directly over the Crosscut zone had a sharp increase in intensity. This discovery demonstrated that in the 2,000 ft grid pattern, the intensity buildup was at the third level of projection (Z3); whereas, in the 500 ft grid pattern, the buildup was at the thirteenth level (Z13), each level corresponding approximately to the same depth.

The depth projection capability of ME exploration may be valid, but the work is still considered to be very experimental and will require continued theoretical and empirical testing and evaluation.

CONCLUSIONS

Thorough knowledge of ME is still elusive; nevertheless, there have been substantial correlations between ME anomalies and actual productive results (Herzfeld, 1984). ME exploration has only begun to scratch the surface in fitting itself into the complex dynamics of the vertical migration process. The expansion of today's scientific understanding compels innovative investigation into magnetoelectric exploration.

REFERENCES

Herzfeld, R.M., 1984, Magneto-electric exploration:Development of a statistical track record: Oil and Gas Journal, February 13, 1984, p. 125–131.

——, 1985, Oil and Gas Exploration: A good place to start: World Oil, April 1985, p. 79–80.

Pirson, S.J., 1977, Geologic Well Log Analysis (2nd ed.): Houston, Gulf Publishing Company, chap. 1.

——, 1981, Significant advances in magnetoelectric exploration, *in* Gottlieb, B. M., ed., Unconventional methods in exploration for petroleum and natural gas II: Dallas, Southern Methodist University Press, p. 169–196.

——, 1982, Progress in magneto-electric exploration: Oil and Gas Journal, October 11, 1982.

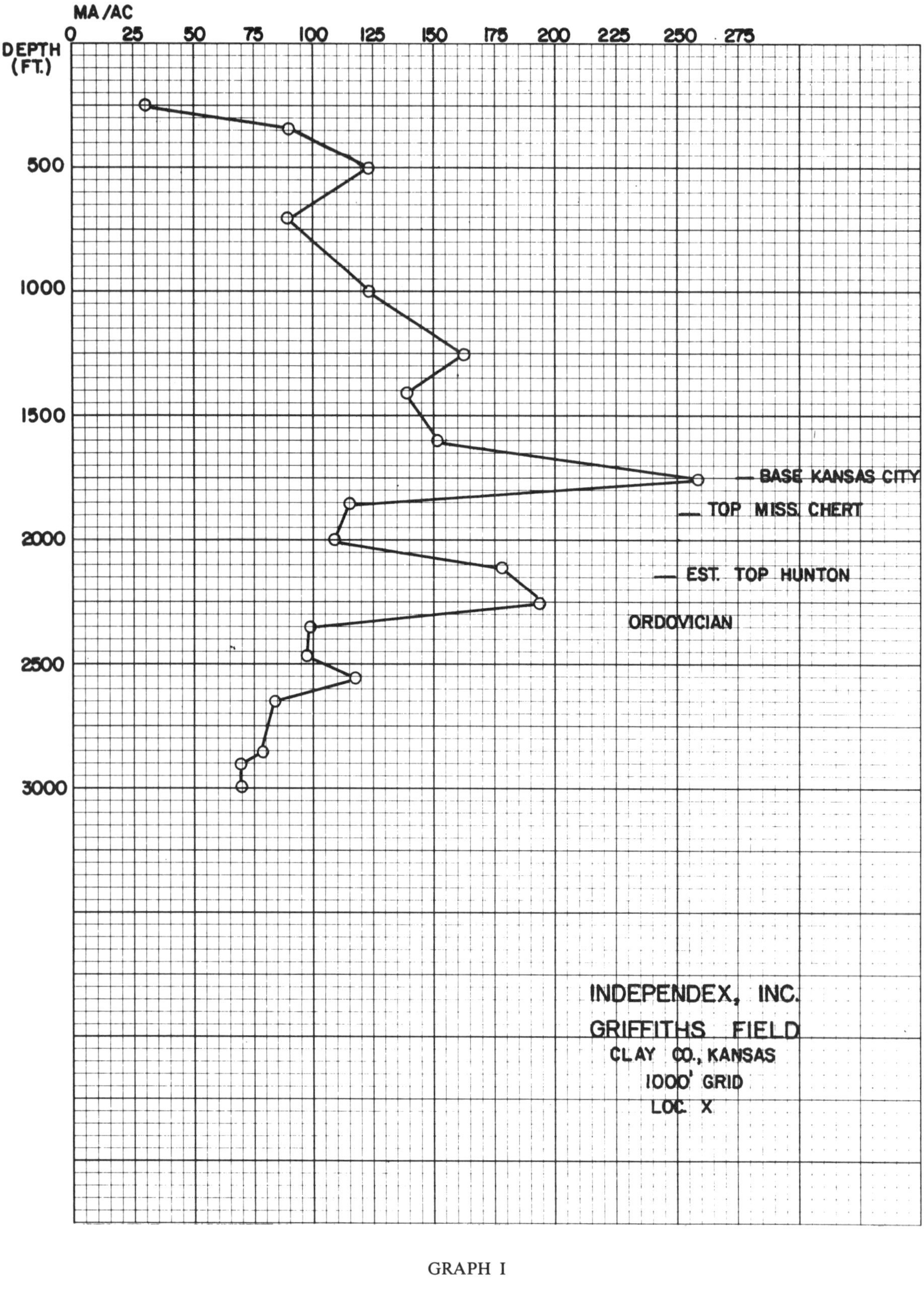
MA /AC
0
25
50
75
100
125
150
175
200
225
250
275
DEPTH (FT.)
500
1000
1500
2000
2500
3000
BASE KANSAS CITY
TOP MISS. CHERT
EST. TOP HUNTON
ORDOVICIAN
INDEPENDEX, INC.
GRIFFITHS FIELD
CLAY CO., KANSAS
1000' GRID
LOC. X

GRAPH I

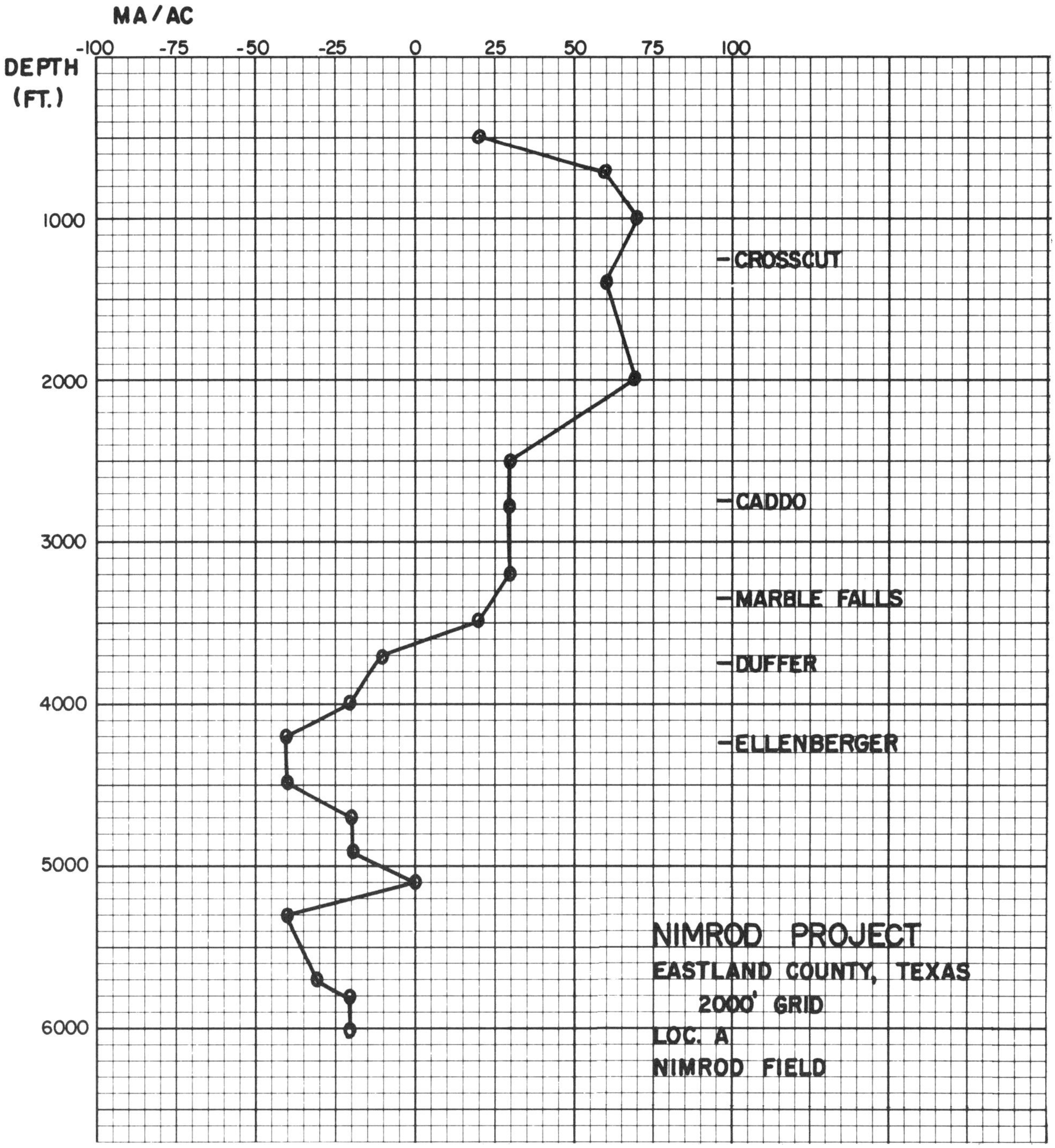
MA/AC
-100
-75
-50
-25
0
25
50
75
100
DEPTH (FT.)
1000
2000
3000
4000
5000
6000
CROSSCUT
CADDO
MARBLE FALLS
DUFFER
ELLENBERGER
NIMROD PROJECT
EASTLAND COUNTY, TEXAS
2000' GRID
LOC. A
NIMROD FIELD

GRAPH II

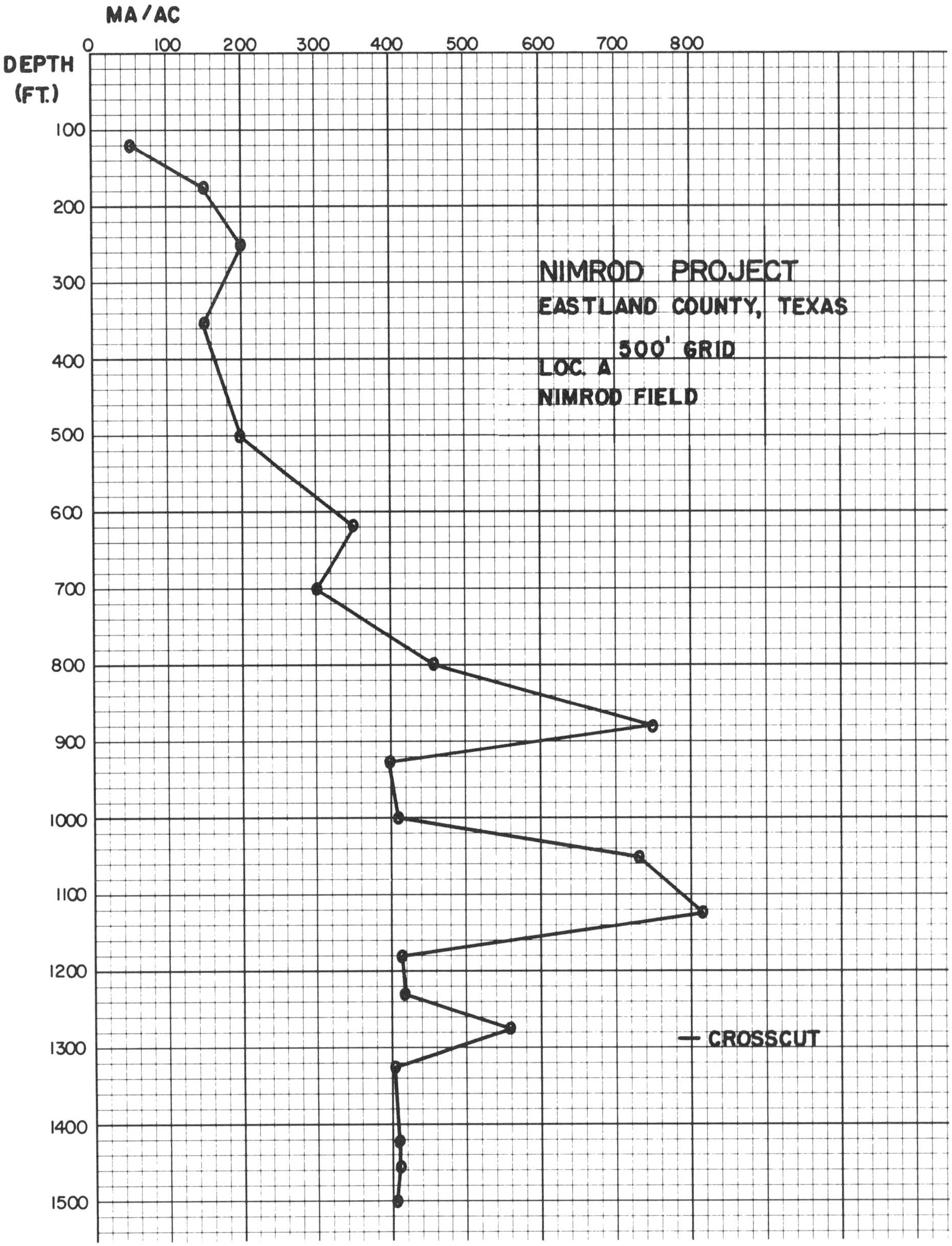
MA/AC
0
100
200
300
400
500
600
700
800
DEPTH
(FT.)
100
200
300
400
500
600
700
800
900
1000
1100
1200
1300
1400
1500
NIMROD PROJECT
EASTLAND COUNTY, TEXAS
500' GRID
LOC. A
NIMROD FIELD
CROSSCUT

GRAPH III

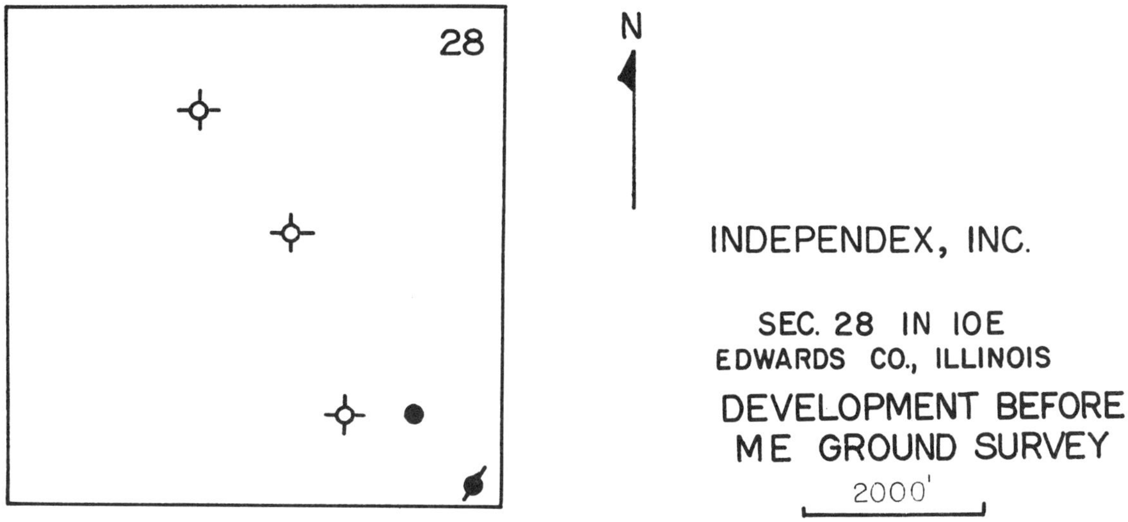

Fig. 1. Drilling activity prior to ME survey, section 28.

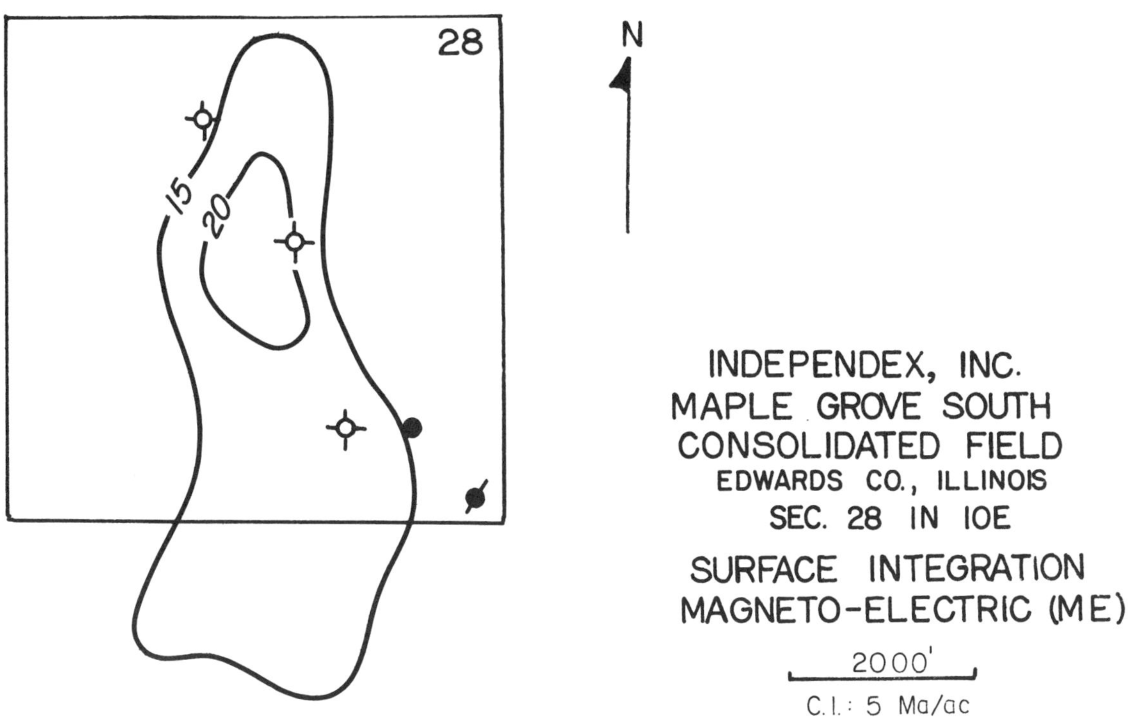

Fig. 2. ME anomaly, section 28.

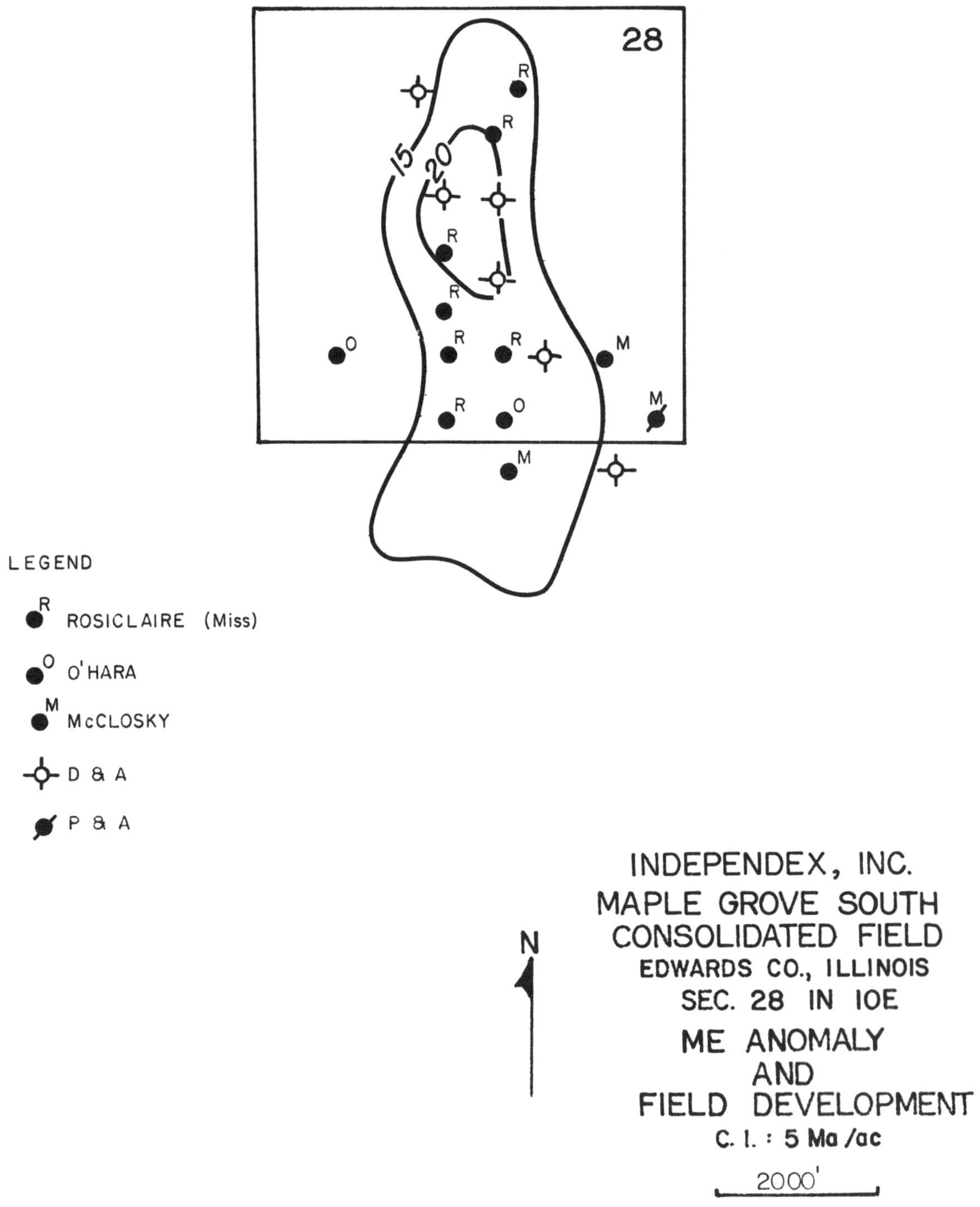

Fig. 3. ME anomaly and field development, section 28.

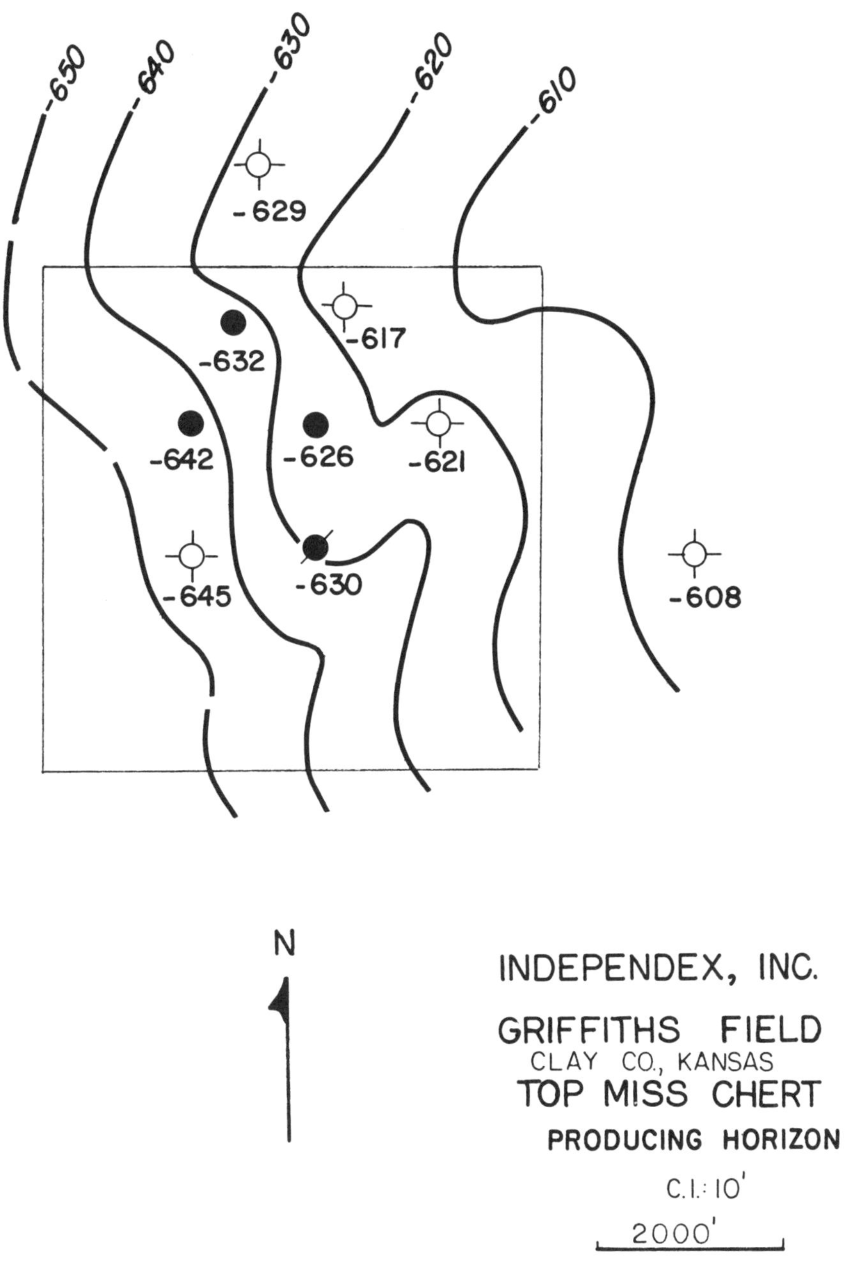

Fig. 4. Top Mississippian chert, Griffiths Field.

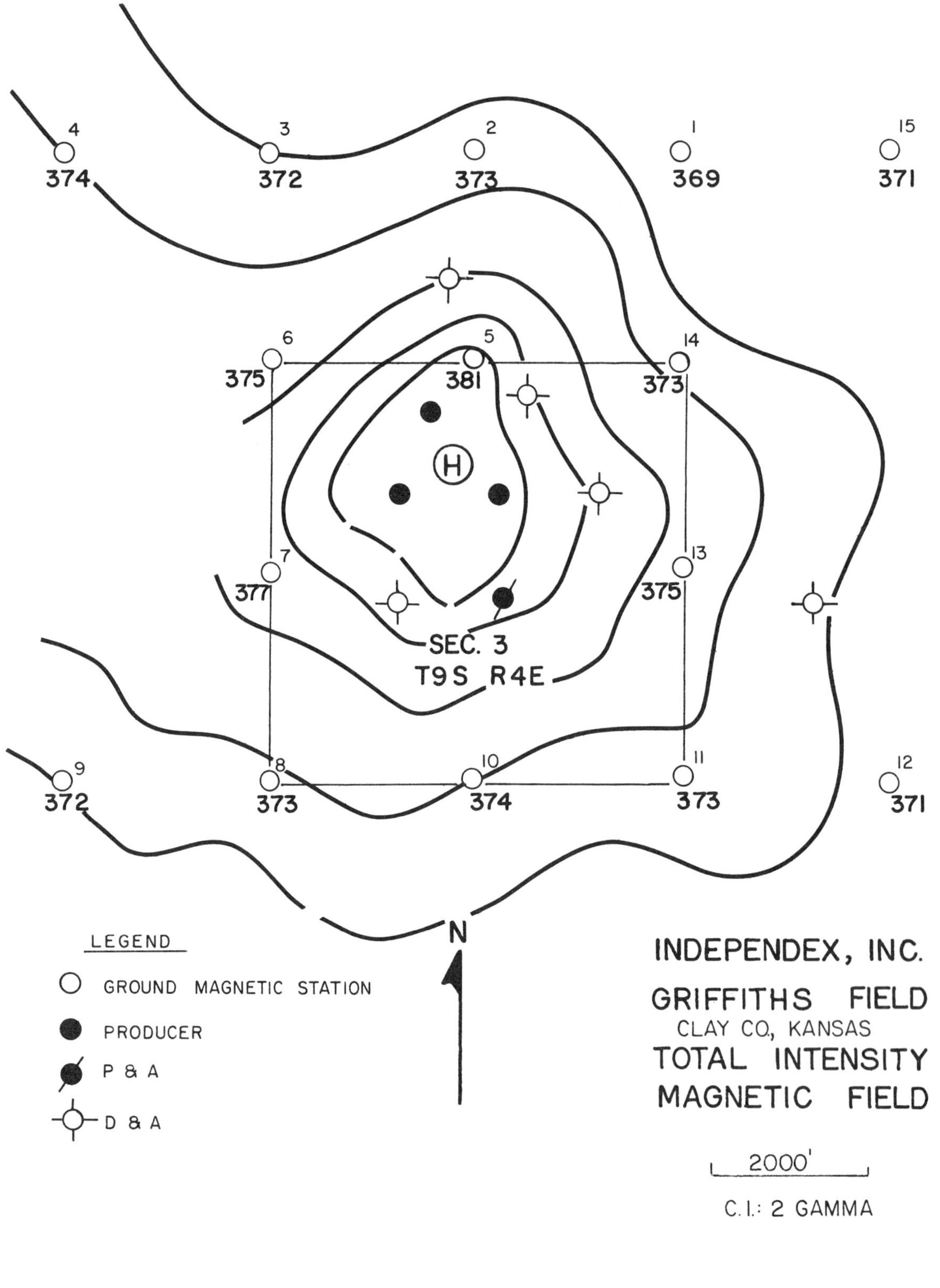

Fig. 5. Total intensity magnetic map, Griffiths Field.

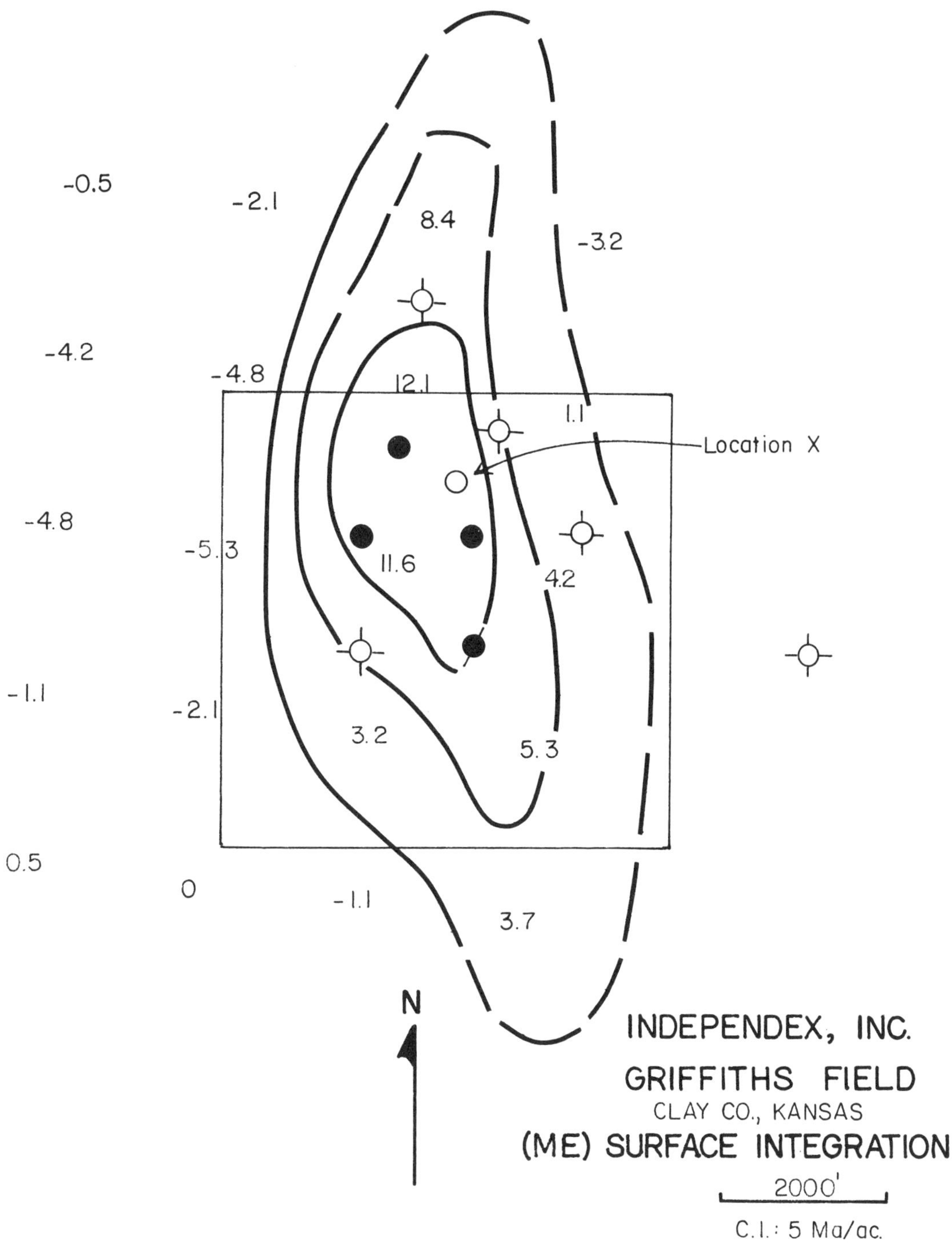

Fig. 6. ME surface integration, Griffiths Field.

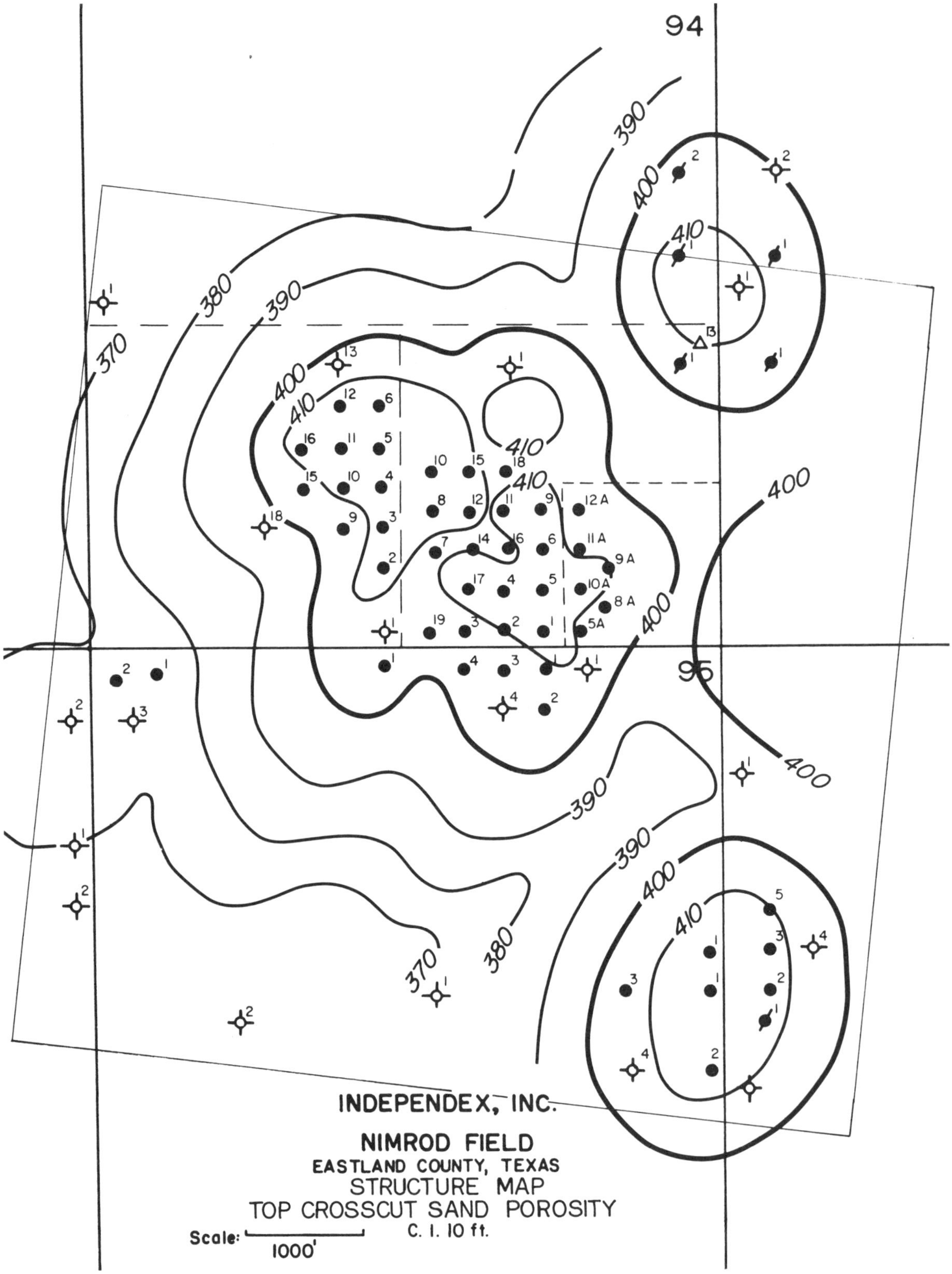

Fig. 7. Top Crosscut sand porosity, Nimrod Field.

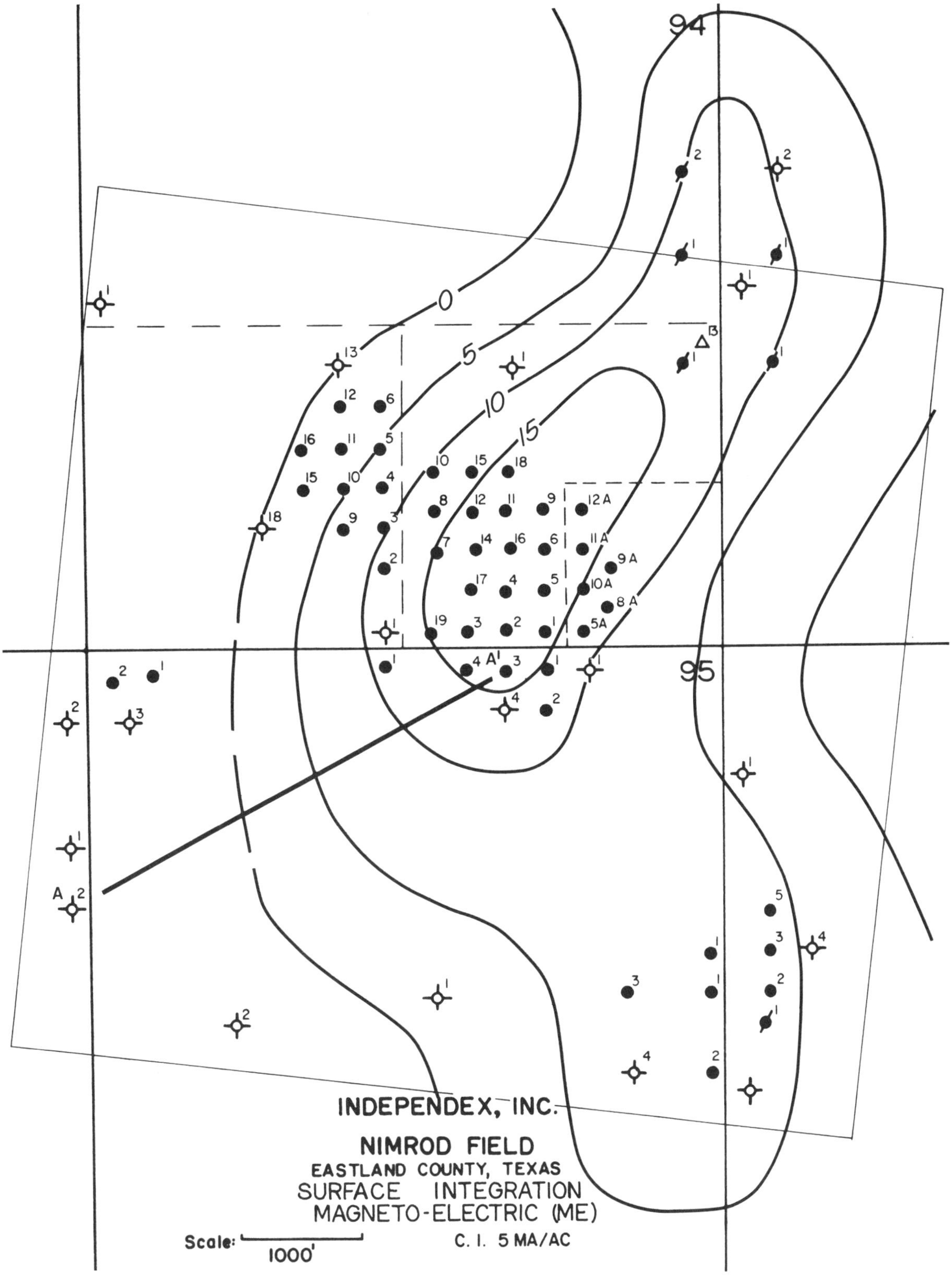

Fig. 8. ME surface integration, Nimrod Field.

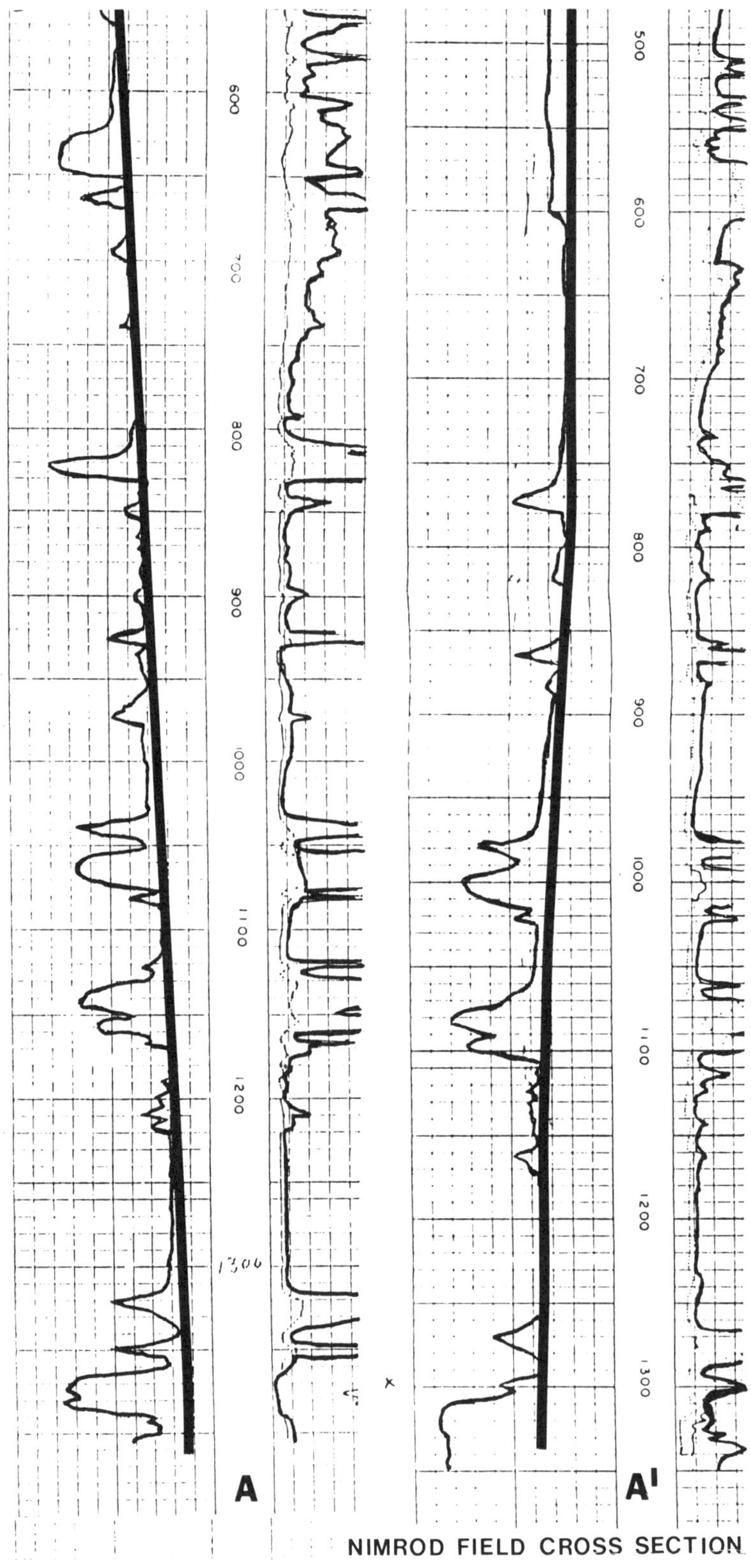

Fig. 9. SP shale-base-line drift, section A–A′, Nimrod Field.

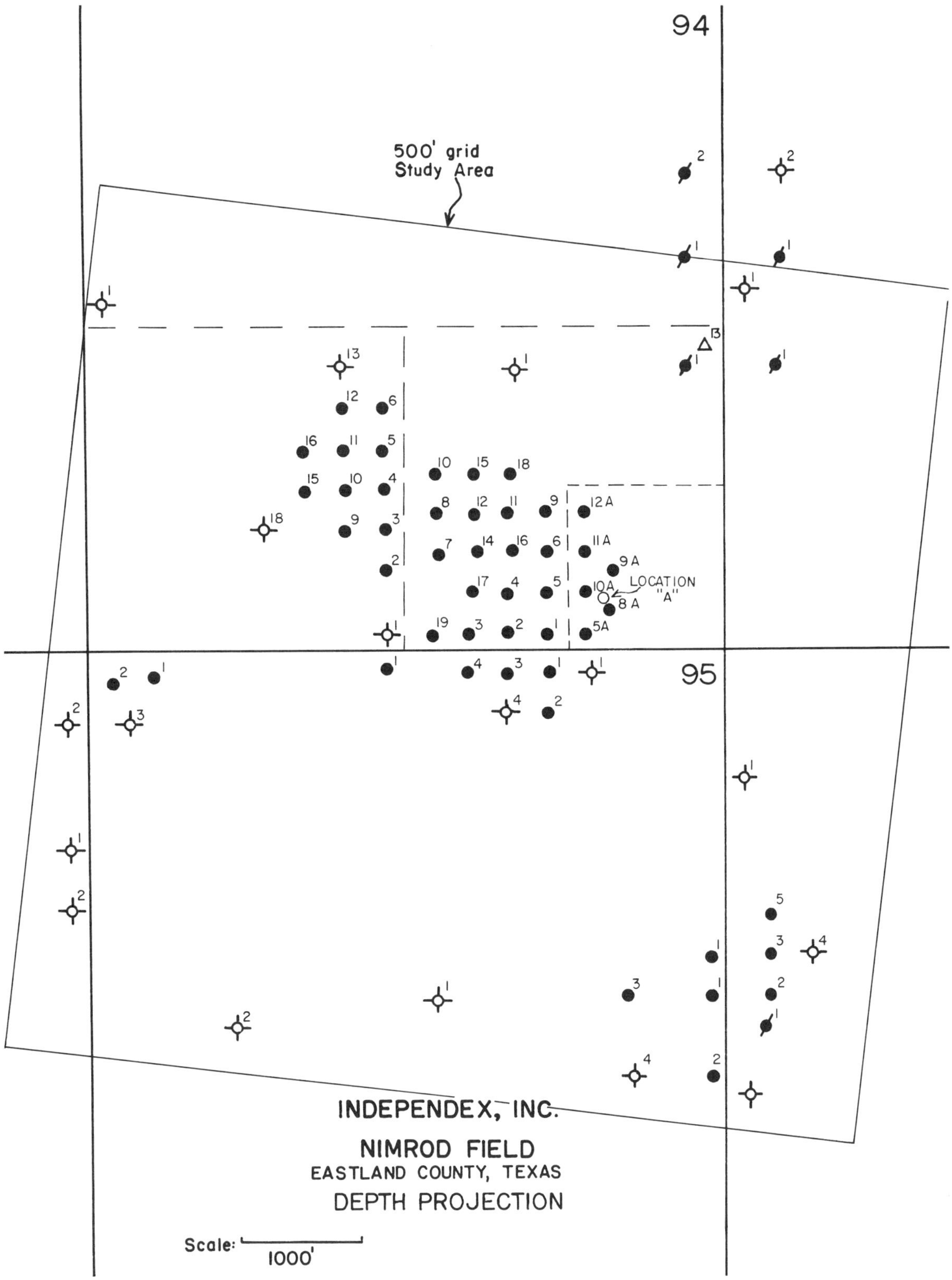

Fig. 10. Location "A", depth projection, Nimrod Field.

THE MAGNETOELECTRIC PHENOMENON REVIEWED AND EVALUATED AS AN EXPLORATION TOOL TO DETECT HYDROCARBON ACCUMULATIONS

Carlos Aiken and T. C. Yang

UNIVERSITY OF TEXAS AT DALLAS

Caroline Lambert

EG&G GEOMETRICS, INC.

ABSTRACT

It has been suggested that hydrocarbons migrating toward the surface of the earth create magnetic anomalies. Two separate hypotheses relate to the source of such magnetic anomalies. The magnetoelectric method (ME) of Sylvain J. Pirson suggests the existence of active vertical telluric currents thought to be caused by the movement of charged particles in the presence of vertically migrating hydrocarbon microseepage. The moving, charged particles induce a measurable magnetic field. The Donovan effect reflects the magnetic consequence of diagenetic magnetite that is produced within a sedimentary section as a result of the chemical reaction of the vertically migrating hydrocarbons with host rocks.

A synthetic model of a magnetic anomaly induced by a vertical current source has been generated to determine the shape as well as the magnitude of an ME related anomaly. The magnetic effect of a symmetrical current model produces a broad, low amplitude anomaly as a high-low pair pattern, if measured by a total intensity survey. Alternatively, diagenetic magnetic effects usually have shorter wavelenghts than ME effects and do not produce any unique form. The amplitude of both effects can be as high as 10 gammas.

Two procedures are developed and tested for the ME analysis of 2-D magnetic data assuming the existence of the vertical currents. The first procedure is Pirson's magnetoelectric integration technique designed to generate vertical telluric currents from magnetic data by means of line integrals. The second procedure is a magnetoelectric inversion technique designed to solve the geometry of the vertical current by applying a generalized linear inversion principle.

The integration technique is theoretically valid only when applied to the proper direction and magnitude of the current-induced horizontal magnetic field. The magnetoelectric inversion technique allows calculation of the vertical current from any measured magnetic anomaly and appears to reproduce the current density pattern more accurately.

Both procedures are tested upon synthetic models as well as upon ground magnetic data obtained at a test site near Tucumcari, New Mexico. Airborne magnetic data from a National Uranium Resource Evaluation (NURE) survey and from ground magnetic data observed in the same area are analyzed in an attempt to provide more information about the source of the magnetic anomalies. Intermediate wavelength anomalies with sources in the sedimentary section are found which have a form similar to ME-type anomalies. However, very high frequency anomalies indicating diagenetic sources are also found. Whether the ME-type anomalies are actually diagenetically related, or whether both ME and diagenetic effects exist independently, is not known. These possibilities must be considered.

INTRODUCTION

In recent years the search for more subtle structures associated with hydrocarbon deposits has stimulated the development of unconventional prospecting techniques. Many unconventional exploration techniques are designed to identify hydrocarbons directly or to identify the effects of hydrocarbons indirectly as alterations in the sediments. Among the indirect detection tools in hydrocarbon exploration are the magnetoelectric method (Pirson, 1976) and the diagenetic magnetite method (Donovan et al., 1979). Both methods utilize magnetic anomalies; however, the theoretical and rational source behind each method is distinctively different.

Pirson (1976) examined well data over several oil fields and found a common phenomenon in most of the wells. A positive self-potential (SP) gradient existed between the deposit at depth and the surface of the earth. Pirson called this shale-base-line drift "the oil gradient" in contrast to the "water gradient," and concluded that there should be an electrical potential drop over the hydrocarbon deposits measured at the surface above the reservoir. Different types of electrical phenomena do appear to be associated with hydrocarbon accumulations and have been measured by a number of researchers (Carlson et al., 1982; Hughes et al., 1983; Haren, 1984). Induced polarization (IP) effects (Sternberg and Oehler, 1984) and conductive resistivity anomalies (Carlson et al., 1982) have been reported over some oil fields. Haren (1984) reported significant negative SP anomalies measured over hydrocarbon deposits in Australia. Proprietary data available to the authors from industrial sources would seem to confirm this. Each of these several methods appears to detect a column above a hydrocarbon deposit reflecting either active vertical hydrocarbon migration or else an alternation zone above the deposits.

If, as Pirson suggested, an electrical potential is related to active hydrocarbon migration, and such phenomena are seen over oil fields, then perhaps the vertical telluric current flux above the deposit is capable of producing a magnetic field significant enough to observe. Maxwell's equations of electromagnetism show that there exists a relationship between such telluric currents and perturbations in the earth's magnetic field. Thus, it should be possible to deduce both the direction and the magnitude of the vertical telluric currents from magnetic anomalies, as shown by Pirson (1981, 1982) and Herzfeld (1984).

Another magnetic related phenomenon was recognized by Donovan et al. (1979) and has become known as the diagenetic magnetite, or Donovan, effect. It is on occasion referred to, with less precision, as the micromagnetic effect. In a study of the Cement Field, Oklahoma, Donovan suggested that diagenetic effects of the seeping hydrocarbons caused the reduction of hydrated ferric oxides and hematite to form magnetite in the near-surface zone. These near-surface magnetites generate high-frequency magnetic anomalies over hydrocarbon deposits. Such high-frequency magnetic anomalies could act as specific indications of of the presence of hydrocarbons. This is a possible cause of magnetic anomalies over hydrocarbon deposits, which may serve as an alternative to the Pirson-Herzfeld magnetoelectric (ME) phenomenon.

This study considers the presence of the magnetic anomalies over, and directly related to, hydrocarbon deposits. More specifically, it considers their possible relationship to the magnetoelectric (ME) phenomenon. This phenomenon is reviewed and evaluated as an exploration tool to detect hydrocarbon accumulations. It will be examined theoretically and applied to both synthetic and real data.

PREVIOUS WORK

Magnetics are usually thought to be one of the methods used indirectly to detect oil and gas deposits (Steenland, 1965). Most sedimentary rocks are expected to have low magnetic content and, therefore, low magnetic susceptibilities. Igneous or metamorphic rocks usually have more magnetic material and, therefore, high magnetic susceptibilities. Magnetic anomalies are usually considered to be caused by inhomogeneities in the compostion of these basement rocks or by structural or topographic relief of the basement surface. An important goal of traditional magnetic interpretation is to determine the depth to the basement rocks and, therefore, the thickness of the sediments. Features of interest are the local relief in the basement surface related to structural relief in the overlying sediments which could be favorable for the accumulation of oil and gas in structural traps.

The conventional magnetic method is generally used in a reconnaissance mode in large and relatively unknown areas. Areas of interest based upon traditional magnetics would then be selectively tested by other geophysical methods, usually seismic. Recently, magnetic anomalies in sedimentary rocks themselves have been used to study intrasedimentary structure (Friedberg, 1976). The newer unconventional magnetic methods used in hydrocarbon exploration, as discussed here, are designed to detect parameters other than structure. They are either directly related to hydrocarbon seepage or to alterations within sediments brought about by oil and gas migration or accumulation.

Unconventional Magnetic Methods

Vertical migration of hydrocarbons was first introduced by Laubmeyer (1933) in Germany and had been studied by a number of different researchers (Davidson, 1982). Dalziel and Donovan (1984) suggested that by effusion, solution, and gas diffusion hydrocarbons can escape from their reservoir. Other studies by Roberts (1980) and Davidson (1982) suggested that the most common or fundamental mechanism for hydrocarbon to migrate vertically upward is by solution or colloidal suspension. This occurs when hydrocarbons are dissolved or suspended in water that is expelled from shales (or cap rock) under the compacting effect of the weight of the rock overburden. As the rising water escapes, it gradually releases the hydrocarbons still in solution or colloidal suspension as a consequence of both temperature and pressure decline at shallower depths. Compared to the shallower ambient water, the deeper ascending water is reduced, thus forming a chimney or funnel of reduced rocks which extends to the surface of the earth.

As a result of hydrocarbon microseepage, it is observed (Donovan et al., 1979; Duchscherer, this volume) that a variety of chemical reactions occur above the oil and gas accumulation. The two main alterations are the oxidation of hydrocarbons and the reduction of iron. As early as the 1920s, researchers had noticed alteration of the surface rocks over an oil field in southwestern Oklahoma (Reeves, 1922). This alteration resulted in a color change with accompanying carbonate cement on most of the topographically expressed structure. Similar alterations were described at another structurally controlled oil field (Harton, 1960). A comprehensive study of the surface alteration of the Cement Anticline was conducted by Donovan (1974), who explained these diagenetic features as evidence of buried hydrocarbons at depth and suggested that color changes were due to the reduction of iron by migrating hydrocarbons. Ferguson (1977), in his study of four oil fields in southwestern Oklahoma, showed that alteration did extend into the subsurface. He stated that pyrite mineralization occurred over areas of hydrocarbon accumulations and that the pyrite was formed by the action of hydrocarbon sulfide upon the abundant iron oxide. Donovan et al. (1979) further suggested that the presence of hydrocarbons provides adequate reducing conditions to mobilize iron from minerals following its reduction to a ferrous state. The ferrous ions can be precipitated as magnetite or pyrite, depending on the environment (Donovan et al., 1979).

$$Fe^{++} \text{ (ferrous iron)} — Fe_3O_4 \text{ (magnetite)} \qquad (1)$$
$$— FeS_2 \text{ (pyrite)}$$

Donovan et al. (1979) identified high-frequency (short wavelength) magnetic anomalies from an aeromagnetic survey over the Cement Field, Oklahoma. These anomalies were interpreted as reflecting the presence of abundant near-surface magnetite formed by the reduction of hydrated iron oxides and/or hematite as a direct result of petroleum microseepage. The magnetic contrast between typical sedimentary rock of low magnetic susceptibility and those rocks locally enriched in magnetite results in distinctive high-frequency, low-amplitude magnetic anomalies. The pattern is generally seen as zones or clusterings of small anomalies having wave lengths measured in hundreds of feet (Wold, 1984). Saunders and Terry (1985) studied more than 130,000 sq mi, including more than 100 known oil and gas fields, and have found diagenetic magnetite-type anomalies over more than 90% of them. Traditionally, near-surface responses have usually been considered as noise and have been disregarded. Sternberg and Oehler (1984) point out that in order for the magnetite to form, certain near-surface geologic conditions must be met: (1) hydrocarbons from the reservoir must be able to seep to the near-surface in sufficient quantity and (2) the near-surface section must contain porous, iron-bearing host rock. This would agree with electrical prospecting results which indicate that different types of electrical anomalies could occur over different petroleum areas, depending on local geological phenomena.

Pirson and Pirson (1976, U.S. Patent: 3,943,436), by repeated laboratory experiments, simulated the condition of a reservoir seepage by injection of hydrocarbons in a shaly rock (sandstone or limestone) and created reduced conditions in such rocks; i.e., an excess of electrons. Essentially, the hydrocarbon accumulations were viewed as negative poles, while the surrounding massive sediments which had not been subjected to environmental modifications, were considered as oxidized (having an excess of positive charges) with respect to the reduced hydrocarbons. Pirson then concluded that this formed a redox fuel-cell, or electrochemical coupling, between these two environments. In a conductive medium such as the earth, the electrotelluric current (ET) will flow in the external circuit from the oxidized region to the reduced region. This effect will last as long as the vertical migration continues. In order to solve the problem, he used a line integral method to calculate the vertical current density from magnetic data. Herzfeld (1984) evaluated the process of magnetoelectric exploration over twenty oil fields and concluded that he could predict the occurrence of hydrocarbons. Any magnetic approach applied in a producing area would be subject to possible cultural contamination due to oil rigs, pumps, etc.

Haren (1984) examined well data over a number of oil and gas fields in Australia and reported that significant negative SP anomalies occurred above hydrocarbon deposits. He claimed that these anomalies could be caused by large-scale redox (oxidation-reduction) cells formed in a plume or chimney above trapped hydrocarbons. These active cells have slightly altered the lithology above hydrocarbon traps. He concluded that the mechanism is electrochemical in nature, using energy from the hydrocarbons that escape past the cap rock which, by nature, is an imperfect seal. This resembles Pirson's ME effect.

Other researchers (Carlson et al., 1982; Sternberg and Oehler, 1984) also claimed that the geochemical alterations above hydrocarbon accumulations are frequently detectable by electrical techniques. These researchers found that the electrical anomalies associated with hydrocarbon accumulations could be measured, but that the actual forming mechanism or mechanisms remained in question. Sternberg and Oehler (1984) reported that induced polarization (IP) effects were present over some oil fields but not over others, while conductive resistivity and residual electromagnetic anomalies (Carlson et al., 1982) are more consistently found. These anomalies may occur over certain accumulations, but not necessarily together. All these methods reflect the conductivity of a chimney, yet more than a single mechanism may be at work and each mechanism may vary with the local geology. However these electrical anomalies may differ in their specific results, all of them represent measurements at shallow depths. Perhaps because of this they tend, in general, to agree with both Donovan's diagenetic effect and the Pirson-Herzfeld ME effect.

MAGNETOELECTRIC AND DIAGENETIC MAGNETITE METHODS

Magnetoelectric Method

Theoretically, the electrotelluric current (ET) measured at the surface of the earth can be derived from the perturbations in the earth's magnetic field. The objective of magnetoelectric exploration is to recover such vertical telluric current from magnetic anomalies that may be associated with it.

ME Theory

By virtue of Maxwell's electromagnetic equations and only considering the static condition, we have:

$$\int \vec{B} \cdot d\vec{l} = \mu_o i \qquad (2)$$

where
$\vec{B}$: induced magnetic field
i : current flow in the wire
$d\vec{l}$: length element
μ_o : relative permeability $= 4\pi \times 10^{-7}$ Tm/A

which is Ampere's law. It states that the line integral of the magnetic field around a closed loop results in the net current flux enclosed by the path. The equivalent, or rather the differential form, is defined by the Biot-Savart law (Fig. 1):

$$d\vec{B} = \frac{\mu_o}{4\pi} \frac{i\, d\vec{l} \times \vec{r}}{r^2} \qquad (3)$$

where
$d\vec{B}$: induced magnetic field
r : distance between current to observation point P
$\vec{r}$: vector pointing from dl to P

This is the differential contribution set up by current element idl at point P. The direction of $d\vec{B}$ is that of the vector $d\vec{l} \times \vec{r}$ (where $\times$ defines a cross product), into the plane of the paper at right angles. We find the resultant field B at point P by integrating elements idl that make up the arbitrary current distribution. For B near a long straight wire, $d\vec{B}$ for every current element has the same direction, so the resultant B is simply the summation of all $d\vec{B}$ and we get

$$B(r) = \frac{\mu_o i}{2\pi r} \qquad (4)$$

where r now represents the horizontal distance.

This is the magnetic induction expressed in gammas at a distance of r m from a long straight wire carrying a current of i in amperes. The formula, in principle, is valid only for an infinitely long wire. In practice, it is valid within 1% but only if r is less than 7% of the length of the straight portion of the wire (Larrain and Corson, 1975).

For the giant fuel cell the horizontal current components cancel each other, assuming an isotropic medium, so that only vertical components are considered. An image point is at negative infinity, and assuming r is relatively small with respect to z (the depth to the top of the hydrocarbon deposit), then the formula becomes one half of the original value for an infinitely long wire (eq. 4). So by definition, the induced magnetic anomaly is a function of current and horizontal distance.

Vertical Current Model and the Corresponding Induced Magnetic Anomalies

In order to generate a model, the magnitude of the vertical current must be known. Ohm's law states

$$V = IR = I\frac{KL}{A} \qquad (5)$$

where
R : resistance
V : potential difference
K : resistivity
L : length
A : area
I : current

The following values were adopted assuming a possible ME effect observed by Pirson (1981). Data has been taken from wells in the Denver-Julesberg Basin and provide an average magnitude of the ME phenomena:

voltage gradient	:	15 mv/500 ft
resistivity	:	5 ohm-m
area	:	1 acre

In this case then, the corresponding current density is approximately 80 ma/acre, using Pirson's units.

A current density model is constructed by using these values. Figure 2 shows the current density distribution, with the darker areas representing higher values of a current moving downward with a cylindrical pattern, and an upward current farther away from the center. A profile through the center is also shown. A synthetic induced magnetic field model is constructed based on this pattern of current (Fig. 3). Figure 4 shows the induced magnetic anomalies that can be recorded by a total magnetic intensity survey in the northern magnetic latitudes. Note that the shape of such a high-low pair is quite similar to that of the Y components (Fig. 5) of Figure 3, only the magnitudes are smaller. So the ME anomaly measured by a total intensity survey is a biaxially symmetric high-low pair instead of being symmetrical, as might normally be expected for this particular mode. An east-west profile across the center of the map is also presented (Fig. 4). Note that from west to east, the magnetic field increases to a maximum, then drops to zero then goes to a minimum and finally increases again, which really represents the change in direction of the magnetic flux which is tangential to both the earth and the current column.

For the above set of parameters, the maximum change in amplitude is approximately 4 gammas. The magnitude of the magnetic anomaly depends on the size of the reservoir (i.e., the horizontal dimension of the active current). For a reservoir of radius of 1 mi (= 1609.4 m), the maximum magnetic anomaly will be 10 gammas, but magnitude does not relate to the reservoir characteristics, such as its thickness, etc. In general, the magnetic anomaly induced by vertical current that can be detected would be of the order of 10 gammas if all components of the magnetic field are measured separately. But since most modern surveys only record the total field intensity, then the inclination also needs to be taken into consideration. At latitude 35°N, 104°W, the average total field is approximately 52,500 gammas and the inclination angle is 62.5° (Nettleton, 1978). If there is an induced magnetic anomaly of 10 gammas, then the resultant earth's total field is 52,505 gammas measured in a total intensity survey (Fig. 6). The residual anomaly is therefore 5 gammas instead of 10 gammas. With most modern magnetometer sensitivities such amplitudes are easily detectable. The magnitude can also be affected by other conditions. The actual measured anomaly is the vector summation of the earth's magnetic field and the induced magnetic anomalies. Since the induced magnetic anomaly has a direction tangent to a circle around the current, we can only recover a portion of the induction, and the resultant depends on the geometry. Taking the above synthetic model as an example, this model has a maximum value of approximately 4 gammas, while under the best conditions the measured value in a total intensity survey is less than 2 gammas. The current may not be uniform along the vertical path, as assumed by Pirson, but may be distorted if a high conductive zone is encountered which will reduce the magnetic induction. As shown in Figure 7, the magnetic induction will decrease dramatically if the distance increases between the active current and the observer.

From Pirson's model (Fig. 8) the size of the current distribution is approximately the dimension of the reservoir and is directly above it. Then the induced magnetic anomalies will have wavelengths of at least the dimension of the reservoir. Generally the magnetic ME effects can be seen as a relatively long wavelength and small amplitude (compared with diagenetic magnetite effects) and is affected significantly by the local geology but not necessarily by the actual quantity of hydrocarbons.

METHODS USED IN CALCULATING ME

Two ME approaches, the integration method presently in use and an inversion method developed herewith, are investigated as methods which may be applied to magnetic anomalies in order to determine possible current density distributions, assuming the phenomenon exists.

ME Integration Method

An integration method based on Ampere's Law had been applied previously by Pirson (1976) and is examined here. The method is best explained by Figure 9. In performing a line integral around the path through the grid, the net current enclosed inside the cell is equal to the result of the line integration.

If the grid cell is oriented (angle between grid and the magnetic north is zero) and only the east-west paths are considered (Pirson, 1976) the integral is

$$\int \vec{B} \cdot d\vec{l} = \frac{(D + A)}{2} - \frac{(B + C)}{2} \qquad (6)$$

Figure 10 shows the ME current density map generated from the synthetic magnetic model (Fig. 4) by the integration method. The high amplitude has been recovered, but the geometry of the pattern is poorly defined. This is no surprise, because the north and south parts have been neglected. Also, the magnitude recovered is only 10% of the original model because, aside from the above problems, a trapezoidal approximation is not sufficient. Another way to look at the integral is to note that Equation (6) can also be expressed as

$$\int \vec{B} \cdot d\vec{l} = \frac{(A - B)}{2} + \frac{(D - C)}{2} \qquad (7)$$

This is similar to a first order derivative along the X-direction (east-west direction). This operator has a tendency to look for gradients in the X-direction. For any anomalies, even those caused by sources other than vertical currents, gradients along the X-direction will be amplified on the current density map. The results of the integration method and a first order derivative was compared and similar trends are seen (Fig. 11 and 12) in Clark County, Kansas. The general shape and trends are very close.

There are some other problems with this method. Since the current paths near the surface are horizontal instead of vertical, in Figure 8, can this particular equation still be applicable? Since the horizontal paths canceled each other out in an isotropic medium, the contribution came only from vertically downward current. By definition of Biot-Savart's law, this current will generate magnetic induction that can be measured at the surface of the earth, and thus the integral will not vanish. Pirson (1981) claimed that by downward continuation of the magnetic data he was able to obtain the current density at different depths. The induced magnetic anomaly is not a function of depth, however, if the current is uniform along its vertical path, but rather a function of horizontal distance. A reasonable explanation is that the magnetic anomalies he dealt with may not be due to vertical current only, but rather contained some other information because other effects have not been removed properly. Pirson also claimed that by performing the line integral along larger and larger paths, he was able to recover the current density at deeper and deeper depths; he therefore claimed he could determine the depth of the source, i.e., the depth to the top of the hydrocarbon deposit.

Let us consider two simple cases, as shown in Figure 13: (a) with the integral along path B representing the average current density in the area; (b) if Ib = 0, then integrals along both paths A and B will give the same result. The conclusion is that the integral along larger and larger paths will not give information about the depth of the source.

ME Inversion Method

Geophysics uses physical theory to obtain a mathematical connection between observations and parameters. A forward problem is the calculation of theoretical values of would-be observations corresponding to the specific values of a model's parameters. Inversion theory is the quantitative study of the solution for the physical properties of the inaccessible regions of the earth's interior and can only be deduced from geophysical observations.

Generalized linear inversion may be applied if a geophysical system which can be expressed as

$$Y = AX \tag{8}$$

where Y : a column vector composed of N observed data (magnetic anomaly)
X : a column vector composed of M unknown parameters (current density)
A : an N×M matrix

To solve this equation we can perform singular value decomposition (Lanczos, 1961) of A in the form

$$A = U \Lambda V^T \tag{9}$$

where U and V are both orthogonal matrices, Λ is the diagonal matrix, and T denotes transposition.

Once A is obtained, the solution of AX = Y is

$$X = HY \tag{10}$$

where H is the inverse matrix of A

$$H = V \Lambda^{-1} U^T$$

In the present approach the main goal is to find out the current density distribution as well as its magnitude. Figure 14 is the current density, a symmetrical model having been chosen. An arbitrary shape is used for an intitial guess (Fig. 15) required in the inversion process. Positive is vertically downward. Each grid cell represents the current density J = (I/A) inside the cell of area A. Darker shades represent more positive current density (J) values. Figure 16 shows the calculated result after seven iterations. The mean square error is almost zero. In a real case, since some noise is involved in the data, and because the forward formula (eq. 4) is only an approximation, the inversion routine may not give a definite result, but rather one of the possible solutions.

TRINCHERA CREEK QUADRANGLE, NEW MEXICO : AN APPLICATION

General Geology

Northeastern New Mexico lies in the Great Plains, ranging in altitude from 1,200 m to 1,400 m. The general geology of this area is recorded in a geologic column involving Precambrian, upper Paleozoic, Mesozoic, and Cenozoic rocks. The Precambrian is overlain by varying thicknesses of limestones with interbedded sandstone and shale of Pennsylvanian age. A nearly complete section of sedimentary rocks is present from Pennsylvanian time. Deep drill holes in northeastern New Mexico penetrate sandstone, shale, and limestone of Mesozoic and Paleozoic ages that are several thousand feet thick. Beginning with the Paleozoic, these rocks are relatively flat-lying (Bates, 1946; Judson, 1953). A study of drill-hole, magnetic, and gravity data had been conducted by Andreasen et al. (1962) to determine the topography and lithology of the buried Precambrian rocks. A Precambrian surface was determined by magnetic studies and the control of the drill-hole data (Fig. 17). The Precambrian basement is a complex of metamorphic rocks that are predominantly schists at a depth approximately 500 m (1,500 ft) below sea level or 2 km below the surface. A recent study by Steinmetz (1978), who integrated deep well data (5,500 m), also suggests similar depths to the basement. The Trinchera Creek 7-1/2 min topographic sheet represents a portion of the northern shelf of the Tucumcari Basin, with Precambrian basement 300-400 m below sea level.

Magnetic Models

Two aeromagnetic profiles taken from National Uranium Resource Evalauation (NURE) data in this study site were examined to investigate the topography of the basement rock and to gain information about the regional (i.e. nonhydrocarbon related) anomalies. Figure 18 shows the location of these two lines inside the test site. A 2-1/2 D magnetic inversion technique (Lai, 1984) was used to model a body with a magnetic susceptibility contrast of 0.0025 cgs. Figures 19 and 20 show east-west and north-south lines respectively. The calculated models from inversion show consistent results at their intersection. The tops of the anomalous bodies lie aproximately 2 km below the datum for both models, which is 0.6 km below sea level, and which agrees with the results by Andreasen et al. (1962) and Steinmetz (1978). The dominant magnetic source within this area is body at a depth of 2 km below the surface of the earth, in the crystalline basement. The corresponding magnetic anomaly then may suggest the regional anomaly, which in turn can be utilized as a reference for regional-residual separation.

Magnetoelectric Processing

An initial ground total intensity magnetic survey was conducted in 1984. Over 300 stations covered an area of approximately 36 sq km with a data spacing of 60 m along north-south lines separated at 150 m. The diurnal variations were recorded at a base station and subtracted from each observation. Figure 21 shows the data location and Figure 22 the magnetic contour map. The effects caused by basement or deeper sources must be removed so that what remains is of interest and can be considered as caused by shallow sources such as the ME or the Donovan effects. In addition to the knowledge received from the models discussed above, Fourier trend surface analysis (James, 1966) was used to determine the regional anomaly. The wavelengths of the trend surface are determined by map size (30×30 km) and data distribution. The first three harmonics of the double Fourier series fits frequencies from 0 to 0.1 cycle/km, which corresponds to wavelengths between infinity and 10 km (Fig. 23). By subtracting the regional trend from the total intensity map, the residual values can be calculated.

Figure 24a shows the residual magnetic map with a grid interval of 0.1 km. This map consists of only relatively low amplitude and short wavelength anomalies. A trend surface fit is performed on the residual data (Fig. 24b) which results in a smooth version of Figure 24a. It displays anomalies of amplitudes less than 10 gammas, and maximum wavelengths of 5 km. It does not necessarily mean that those anomalies are related to the ME effect. Compared, however, with the synthetic model (Fig. 4) there is a partial high-low pattern similar in shape and wavelength in the west central part of the figure. The gridded residual data are then used for further processing. Figure 25 shows the result as derived from the integration method, and as indicated previously it is sensitive to changes in the east-west direction, the effect of which is to show north-south trends. The darker area indicates higher values, and generally, there is a positive anomaly trend from west to southeast on the map. Figure 26 shows the result of the data using the ME inversion method. The number of unknowns is 64 and the number of data is 302, so it is an overconstrained probem. The mean square error is less than 1.45 gammas, compared with data values ranging from −10 to 10 gammas, or less than 8%.

There are qualitative similarities between these two results, especially with regard to the large positive anomalies. A northwest-southeast trend in the western portion of the area together with some isolated anomalies in the east-central part of the site are observed in both figures. Certifying these as ME anomalies or determining their sources would require further geological, geophysical, and geochemical studies.

Diagenetic Magnetite Analysis

An additional total intensity ground magnetic survey was conducted in the spring of 1985 to provide closer station spacing (15 m or 50 ft) in order to examine even shorter wavelength anomalies. Two lines of data were collected with one north-south line and one east-west line (also shown on Fig. 20). They intersect at the center of the site. A total of approximately 1,200 stations were recorded. Cultural effects in this undeveloped area are at a minimum; and pipelines, fences, etc., were avoided by a careful planning of the station locations both prior to and during the survey.

Correlation between Average Depth and Power Spectrum

By assuming that the earth consists of a number of independent ensembles of rectangular, vertical-sided parallelepipeds, Spector (1968) derived a relationship between the frequency band and the average depth of those ensembles. The average depth can be derived by the slope of the power spectrum.

Figure 27a shows a synthetic forward magnetic model generated by a 2-1/2 D magnetic forward program (Lai, 1984). Figure 27b shows the corresponding power spectrum. Note that the dominating factor is the shallow body, but the power spectrum shows responses from both depths. The average depths calculated from the slope of power spectrum are 0.4 km and 6.5 km, which are very close to the model. Using this concept, an analysis may be performed to indicate the sources of the anomalies. High-pass and band-pass filters are also used to enhance certain anomalies.

Figures 28 and 29 show the original data and power spectra for the second ground survey lines. Despite the apparent noise, both power spectra suggest two groups of sources. The dominant one is of lower frequency while the higher frequency group seems to be more complicated. The average depths calculated from the power spectra for both groups is approximately 6 km and 0.6 km respectively. The 6 km depth is probably incorrect because Spector (1968) suggested that at least a 10:1 ratio of the magnetic profile with respect to the depth is required in order to investigate the source depth. Figures 30 and 31 show band-pass filtered results. After removal of the low frequency anomaly, the remaining high frequency anomalies have major contributions from anomalies with wavelengths of two groups: (a) between 1 km and 2 km and (b) less than 500 m.

Are these anomalies related to diagenetic magnetic effects? There is similarity, but perhaps the 1–2 km wavelength anomalies appear longer than those seen over other diagenetic magnetite related to hydrocarbon deposits. If the high frequency anomalies are examined (Fig. 30, 31), there is a predominance of such anomalies to the south and west. The ME-type anomalies are also concentrated to the south and west.

The conclusion is that ME-type magnetic anomalies appear to exist in the residual magnetic map (Fig. 24a) and are inferred in its smoothed version (Fig. 24b). There are also apparent shallow diagenetic-type sources seen along the profiles (Fig. 28–31). Ultimately the ME processing does not discriminate between the types of sources of the magnetic anomalies. It will define a current density regardless of the actual source of the anomaly. Therefore, diagenetic-related anomalies, as well as any other magnetic anomaly, will result in a "current density." The separation of these various types of anomalies (as to their sources) must be done carefully.

CONCLUSIONS

Under ideal conditions (homogeneous, isotropic, etc.), i.e., assuming the existence of vertical current with uniform intensity along its path, the magnitude of the magnetoelectric (ME) effect on magnetic anomalies is on the order of 10 gammas with a wavelength of about twice the width of the current pattern. However, any inhomogeneity in the earth, such as a highly conductive zone, will significantly alter the shape and magnitude of the current and, therefore, will alter the magnetic anomaly.

The earth's total magnetic field is a vector that is the summation of all its components. If we measure only the total intensity of the earth field, the magnitude as well as the shape of the ME effect will also be altered just by the change in the inclination itself.

The ME integration method is valid only when the characteristics of the horizontal component of the magnetic anomaly are recorded and all contributions along the integral path are considered. As applied presently, the integration method is not very accurate. The ME inversion method may be applied to total intensity data and may recover the current density much more accurately.

There are other mechanisms that may be operative, jointly or separately, with the ME oil field fuel-cell hypothesis or with the diagenetic magnetite hypothesis, and various effects of each may be superimposed on the other.

Both the ME effect and the diagenetic effect could exist over hydrocabon deposits simultaneously, depending on specific geological conditions. Theoretically it is possible to separate these two effects, since (a) the ME effect usually has a longer wavelength and a characteristic form and (b) the ME related magnetic anomalies induced by vertical telluric currents effect the horizontal component. In practice basement, diagenetic and ME magnetic anomalies could have overlapping wavelengths and amplitudes. Other geophysical and geological information would be necessary in eliminating the regional effects that are thought to be caused by sources other than the hydrocarbon deposit. Geochemical anomaly correlations would be especially helpful in determining the existence of the reducing condition.

Self potential anomalies would appear to have to exist if the ME phenomenon does exist. Although SP anomalies have been reported, they have not been consistently found. This could be due to (1) the nonexistence of ME phenomenon, (2) the existence of SP and ME effects only over some hydrocarbon deposits, and (3) the greater difficulty in observing and detecting SP anomalies while the ME effects are more easily observed.

To investigate ME phenomena further it is suggested that the horizontal component of a magnetic reading be recorded, rather than the total intensity. Recorded in this manner, the line integration method may also be applicable, but Pirson's original integration formula would have to be modified and might well provide greater accuracy.

This study assumes that the vertical current related to hydrocarbon microseepage does exist. Electrochemical studies are needed for further substantiation of Pirson's observations and experiments.

REFERENCES

Andreasen, G. E., Kane, M. F., and Isidore, Z., 1962, Aeromagnetic and gravity studies of the Precambrian in northeastern New Mexico: Geophysics, v. 27, p. 343–358.

Bates, R., 1946, Northern New Mexico: Developments in 1945: American Association of Petroleum Geologists Bulletin, v. 30, p. 958–962.

Bracewell, R. N., 1978, The Fourier Transform and Its Applications: New York, McGraw-Hill.

Breiner, S., 1973, Applications manual for portable magnetometers: Sunnyvale, California, EG&G Geometrics, 46 p.

Carlson, N. R., Hughes, C. T., and Zonge, K. L., 1982, Hydrocarbon exploration using induced polarization, apparent resistivity and electromagnetic scattering: Geophysics, v. 47, p. 1118.

Dalziel, M. C., and Donovan, T. J., 1984, Correlations of suspected petroleum-generated biogeochemical and aeromagnetic anomalies, *in* Davidson, M. J., and Gottlieb, B. M., eds., Unconventional methods in exploration for petroleum and natural gas III: Dallas, Southern Methodist University Press, p. 59–68.

Dahlquist, G., et al., Numerical methods: Englewood Cliffs, New Jersey.

Davidson, M. J., 1982, Toward a general theory of vertical migration: Oil and Gas Journal, June 21, p. 288–300.

Donovan, T. J., 1974, Petroleum microseepage at Cement, Oklahoma: Evidence and mechanism: American Association of Petroleum Geologists Bulletin, v. 58, p. 429–446.

———, Forgey, R. L., and Roberts, A. A., 1979, Aeromagnetic detection of diagenetic magnetite over oil fields: American Association of Petroleum Geologists Bulletin, v. 63, no. 2, p. 245–248.

Duchscherer, Wm., Jr., 1986, Geochemical hydrocarbon prospecting, this volume.

Ferguson, J. D., 1977, The surface alteration and mineralization of Permian red beds overlying several oil fields in southern Oklahoma: Master's thesis, Oklahoma State University, p. 95.

Friedberg, J. L., 1976, The two-gamma fault: For the Society of Exploration Geophysicists 46th Annual International Meeting, October, 1976.

Grant, F. S., and West, G. F., 1965, Interpretation theory in applied geophysics: McGraw-Hill International Series in the Earth Sciences.

Haren, R. J., 1984, An evaluation of electrical methods applied to petroleum exploration in Australia: Exploration Geophysics, Australia, v. 15, p. 197–212.

Harton, B. H., 1960, Stratigraphy of Cement pool and adjacent area, Caddo and Grady Counties, Oklahoma: American Association of Petroleum Geologists Bulletin, v. 44, p. 210–226.

Herzfeld, R. M., 1984, Magneto-electric exploration: Development of a statistical track record: Oil and Gas Journal, February 13, 1984, p. 125–131.

Hughes, L. J., Carlson, N. R., and Zonge, K. L., 1983, Distinguishing well casing from structural effects in electrical anomalies measured over hydrocarbons: Geophysics, v. 48, p. 472.

James, W. R., 1966, Fortran IV program using double Fourier series for surface fitting of irregularly spaced data: Kansas Geological Computer Contribution 5.

Judson, S., 1953, Depressions of northern portion of Southern High Plains of eastern New Mexico: Geological Society of America Bulletin, v. 61, p. 253.

Lai, S. F., 1984, Generalized linear inversion of 2.5 D gravity and magnetic anomalies: Ph.D. diss., University of Texas, Dallas.

Lanczos, C., 1961, Linear Differential Operator: Princeton, New Jersey, D. Van Nostrand.

Larrain, P., and Corson, D. R., 1975, Electromagnetic Fields and Waves, 2d ed.: San Francisco, W.H. Freeman.

Laubmeryer, G., 1933, A new geophysical prospecting method, especially for deposits of hydrocarbons: Petroleum, v. 29, no. 18, p. 1–4.

Nettleton, L. L., 1978, Gravity and magnetics in oil prospecting, McGraw-Hill International Series in the Earth and Planetary Sciences.

Parker, R. L., 1977, Understanding inverse theory, Annual Review of Earth and Planetary Science, v. 5, p. 33–64.

Pedersen, L. B., 1977, Interpretation of potential field data: A generalized inversion approach, Geophysical Prospecting, v. 25, p. 199–230.

Penrose, R., 1955, A generalized inverse for matrices, Cambridge Philosophical Society, Proceedings, v. 51, p. 406–413.

Pirson, S. J. and Pirson, J. E., 1976, Line integral method of magneto-electric exploration: U.S. patent 3943436, Mar. 6, p. 17.

Pirson, S. J., 1981, Significant advances in magneto-electric exploration, *in* Gottlieb, B. M., ed., Unconventional methods in exploration for petroleum and natural gas II: Dallas, Southern Methodist University Press, p. 169.

———, 1982, Progress in magneto-electric exploration: Oil and Gas Journal, October 11, 1982.

Reeves, F., 1922, Geology of the Cement Oil Field, Caddo County, Oklahoma: U.S. Geological Society Bulletin, v. 726, p. 41–85.

Roberts, W. H., 1980, Design and function of oil and gas traps, *in* Roberts, W. H., III, and Cordell, R. J., eds., Problems of petroleum migration: American Association of Petroleum Geologists Studies in Geology, no. 10, p. 217–240.

Saunders, D. F., and Terry, S. A., 1985, Onshore exploration using the new geochemical and geomorphology: Oil and Gas Journal, September 16, 1985, p. 126–130.

Spector, A., 1968, Spectral analysis of aeromagnetic data: Ph.D. diss., University of Toronto.

Steenland, N. C., 1965, Oil fields and aeromagnetic anomalies: Geophysics, v. 30, no. 5, p. 706–739.

Steinmetz, R., 1978, Statistical summary of wells drilled below 18,000 ft in West Texas and east New Mexico: American Association of Petroleum Geologists Bulletin, v. 62, p. 853–863.

Sternberg, B. K., and Oehler, D. Z., 1984, Electrical methods for hydrocarbon exploration, *in* Davidson, M. J., and Gottlieb, B. M., eds., Unconventional methods in exploration for petroleum and natural gas III: Dallas, Southern Methodist University Press, p. 188–201.

Wiggins, R. A., 1972, The general linear inverse problem: Implications of surface waves and free oscillations for earth structure: Reviews of Geophysics and Space Physics, v. 10, p. 251–285.

Wold, J. R., 1984, Aeromagnetic detection of anomalies produced by chemical alteration over oil-gas deposits: Sunnyvale, California, EG&G Geometrics.

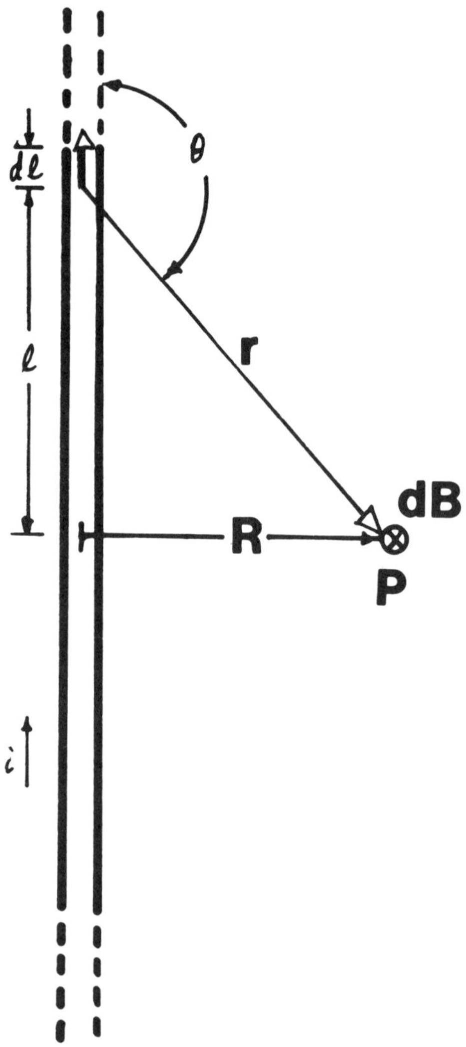

Fig. 1. Biot-Savart Law. Vertical cross section of current line (i) and magnetic induction (B).

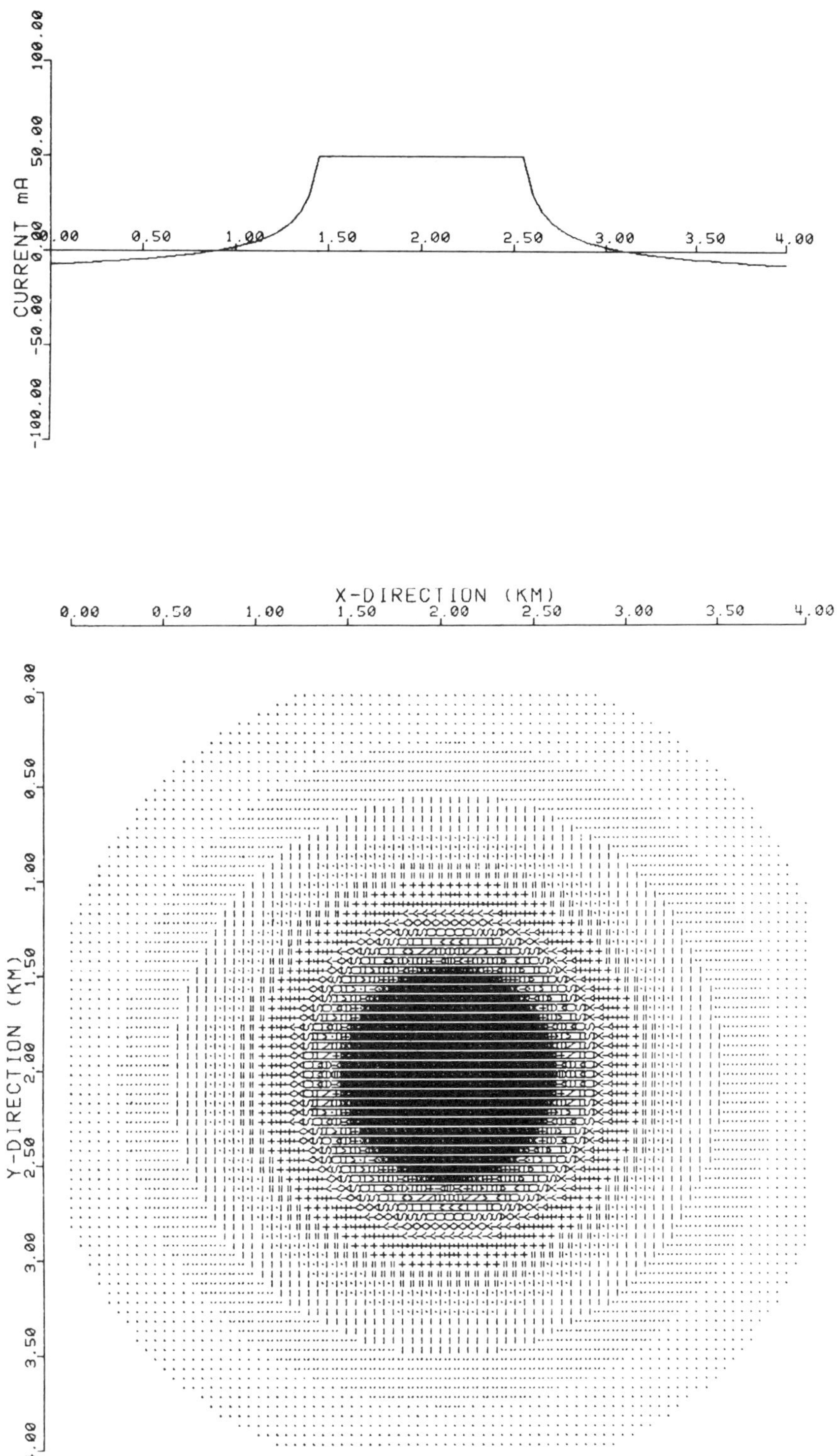

Fig. 2. Vertical current density model. Based on Pirson's (1981) parameters. Upper figure is a profile through the center.

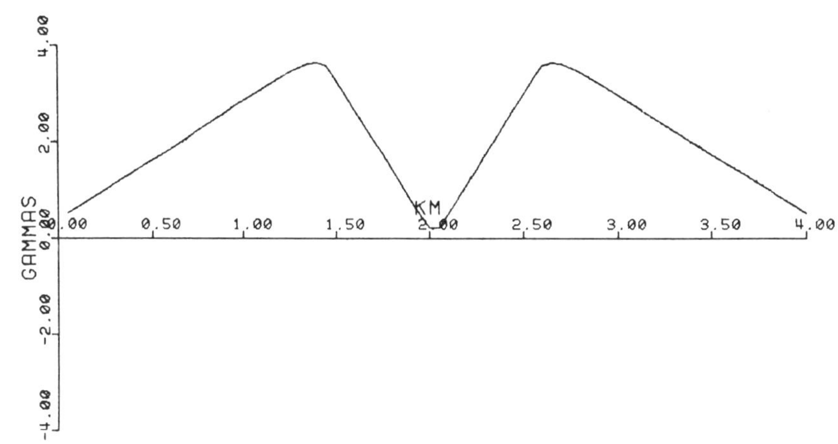

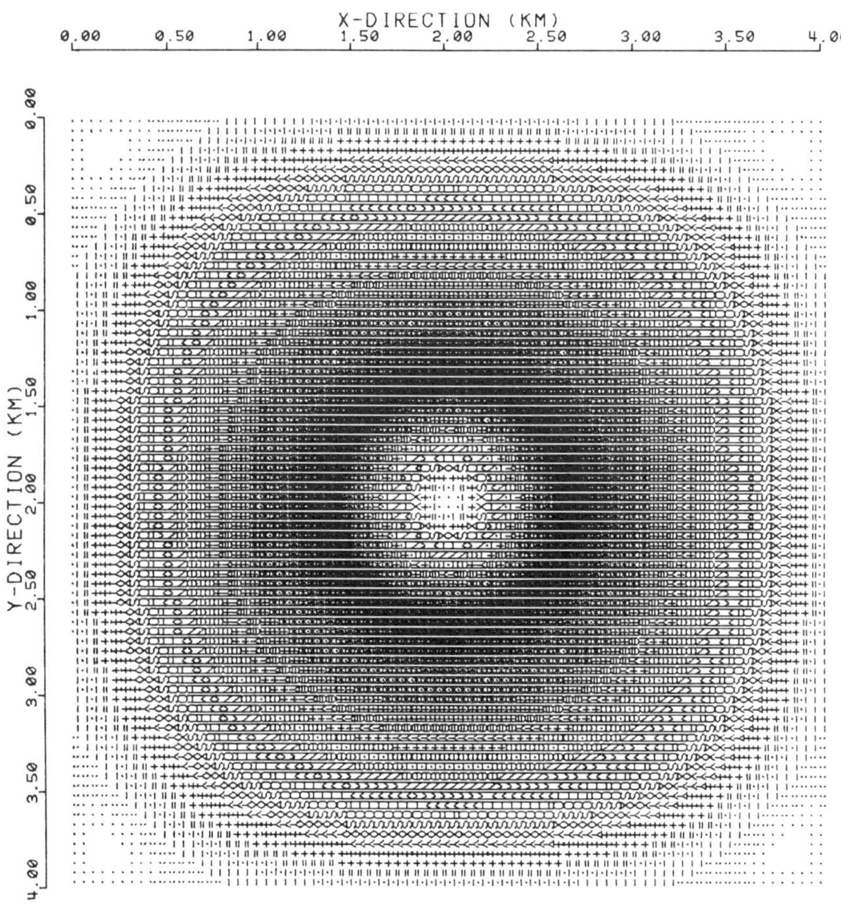

Fig. 3. Intensity of the horizontal magnetic field regardless of direction induced by current. Darker = greater values.

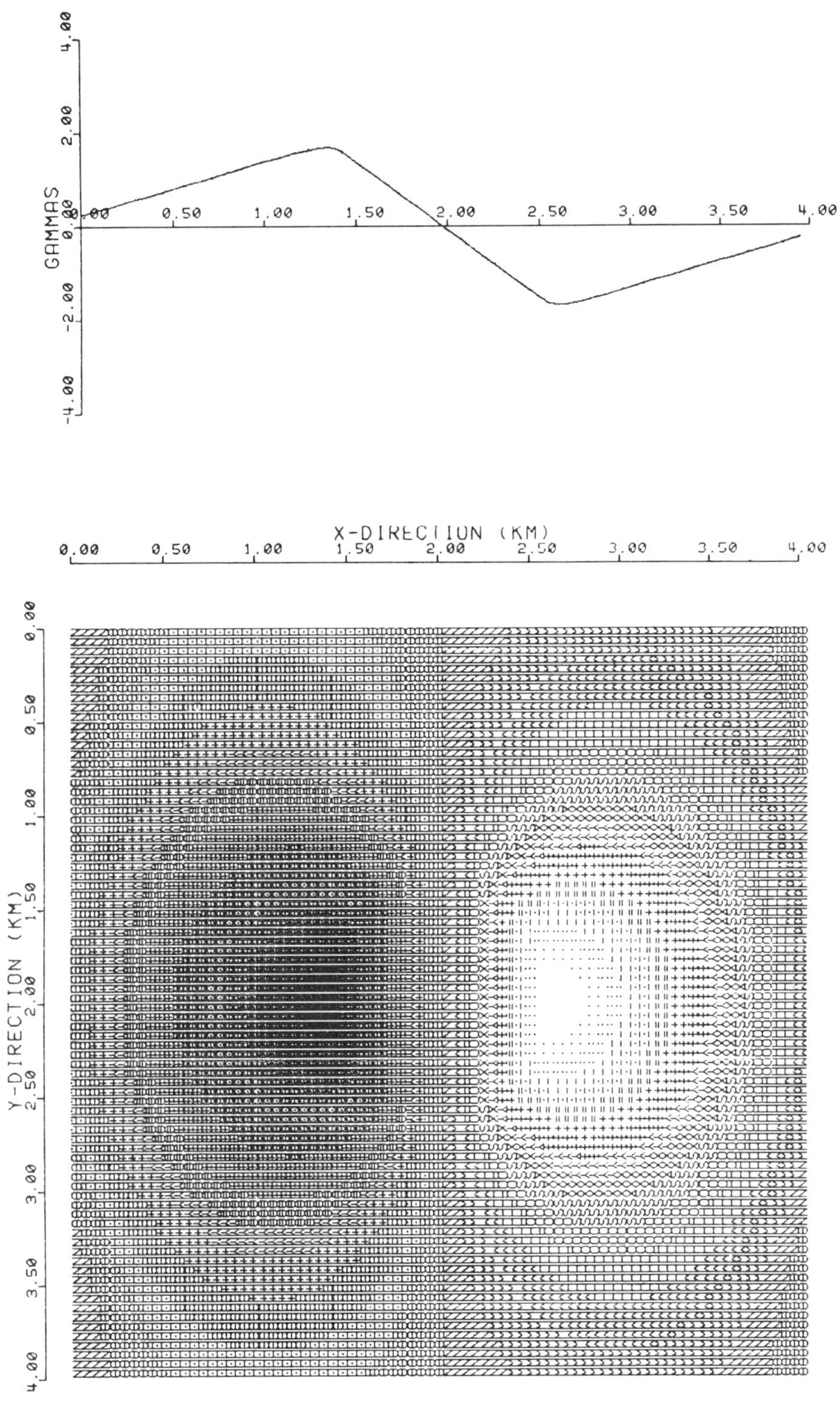

Fig. 4. Magnitude of total intensity of current-induced magnetic field of Fig. 2. Observed by total intensity instrument in northern magnetic latitudes, with an inclination 62.5°. Positive Y is north.

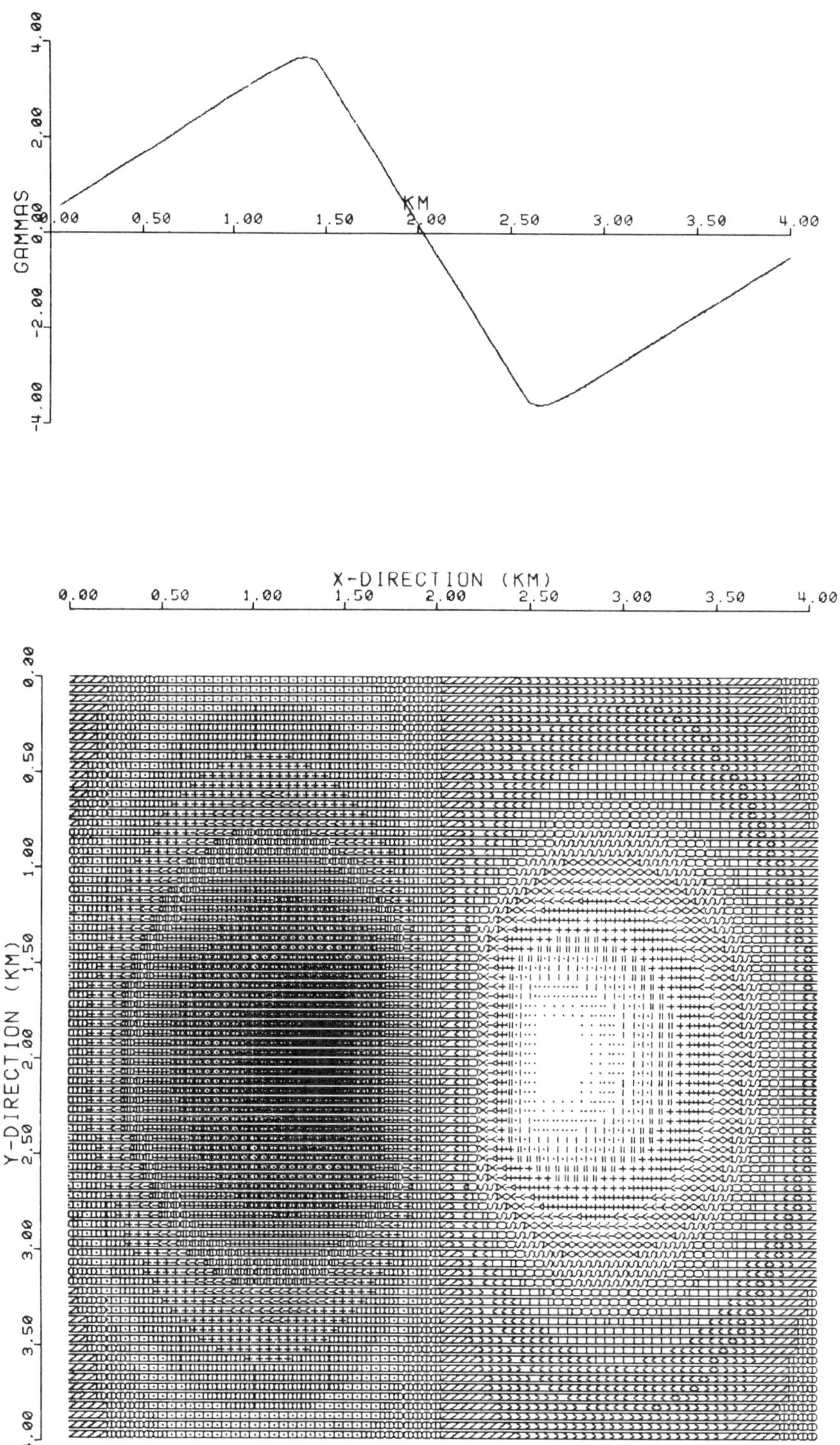

Fig. 5. Magnitude of north component of current-induced magnetic field of Fig. 3 as observed at magnetic equator. This is equivalent to total intensity.

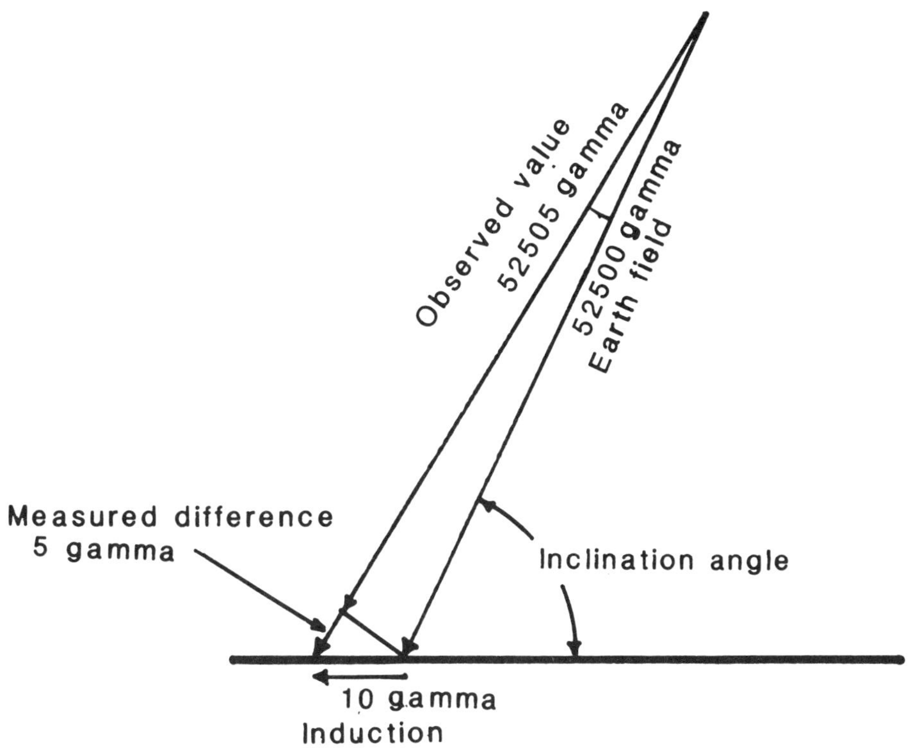

Fig. 6. Vector summation of magnetic field (total intensity).

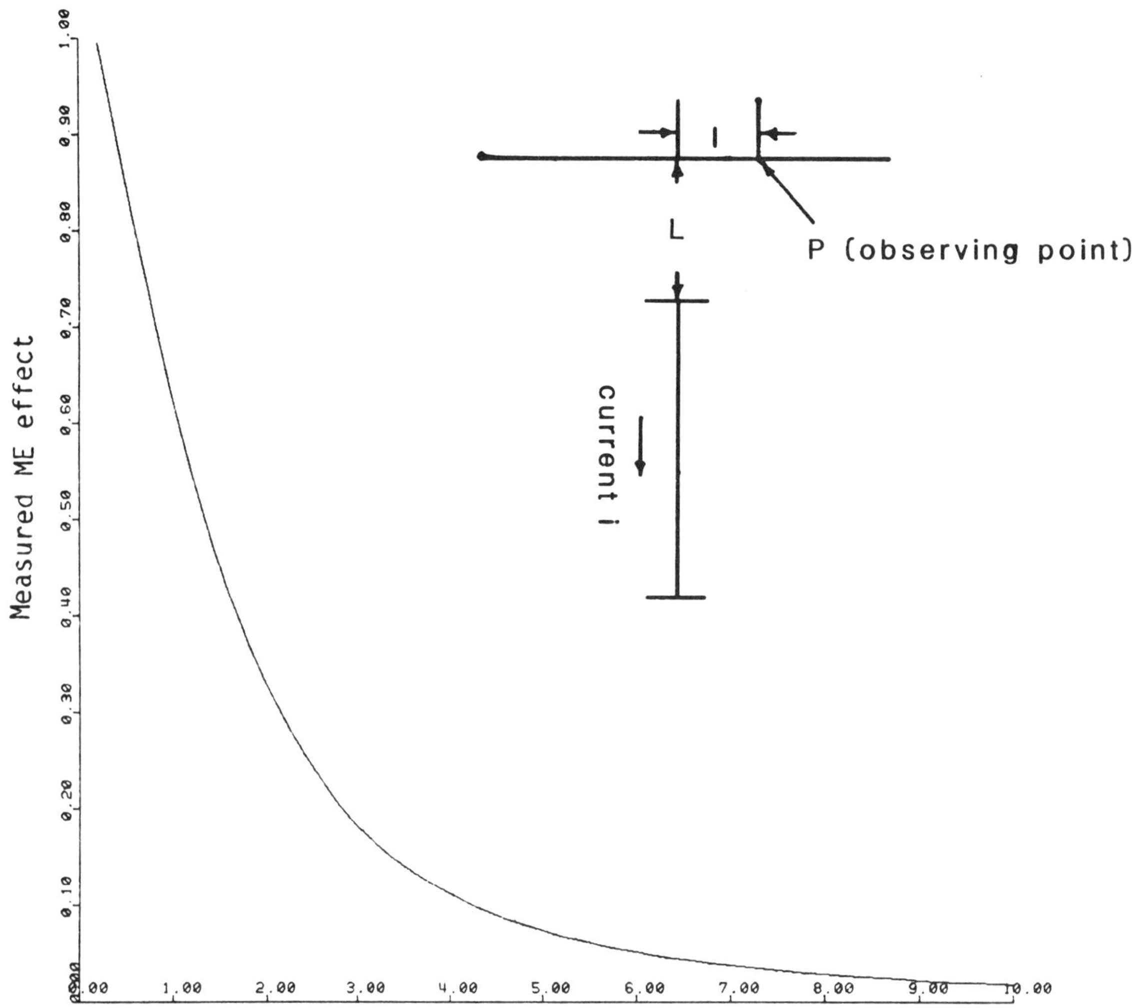

Fig. 7. ME magnetic field versus vertical distance. Vertical axis is normalized ME effect. Horizontal axis is the ratio of the depth to the current source with the horizontal distance to the center of the current source.

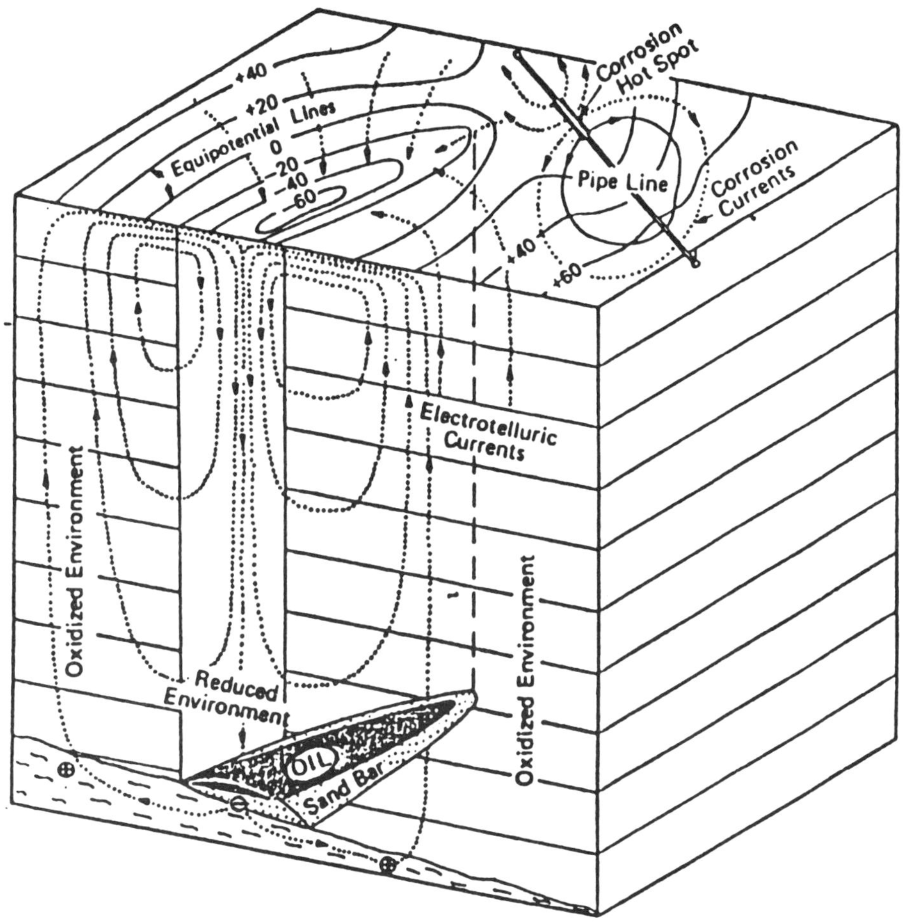

Fig. 8. Pirson's magnetoelectric model (Pirson 1976).

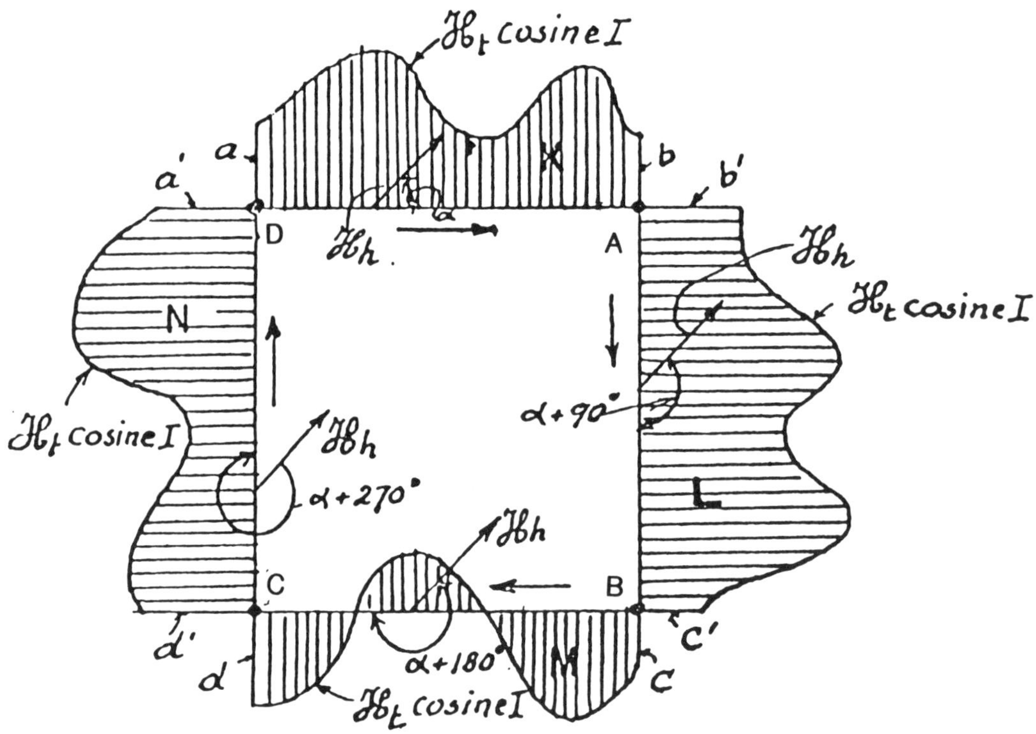

Fig. 9. Integration method (Pirson, 1976). α = declination and I = angle between path of integration and magnetic north.

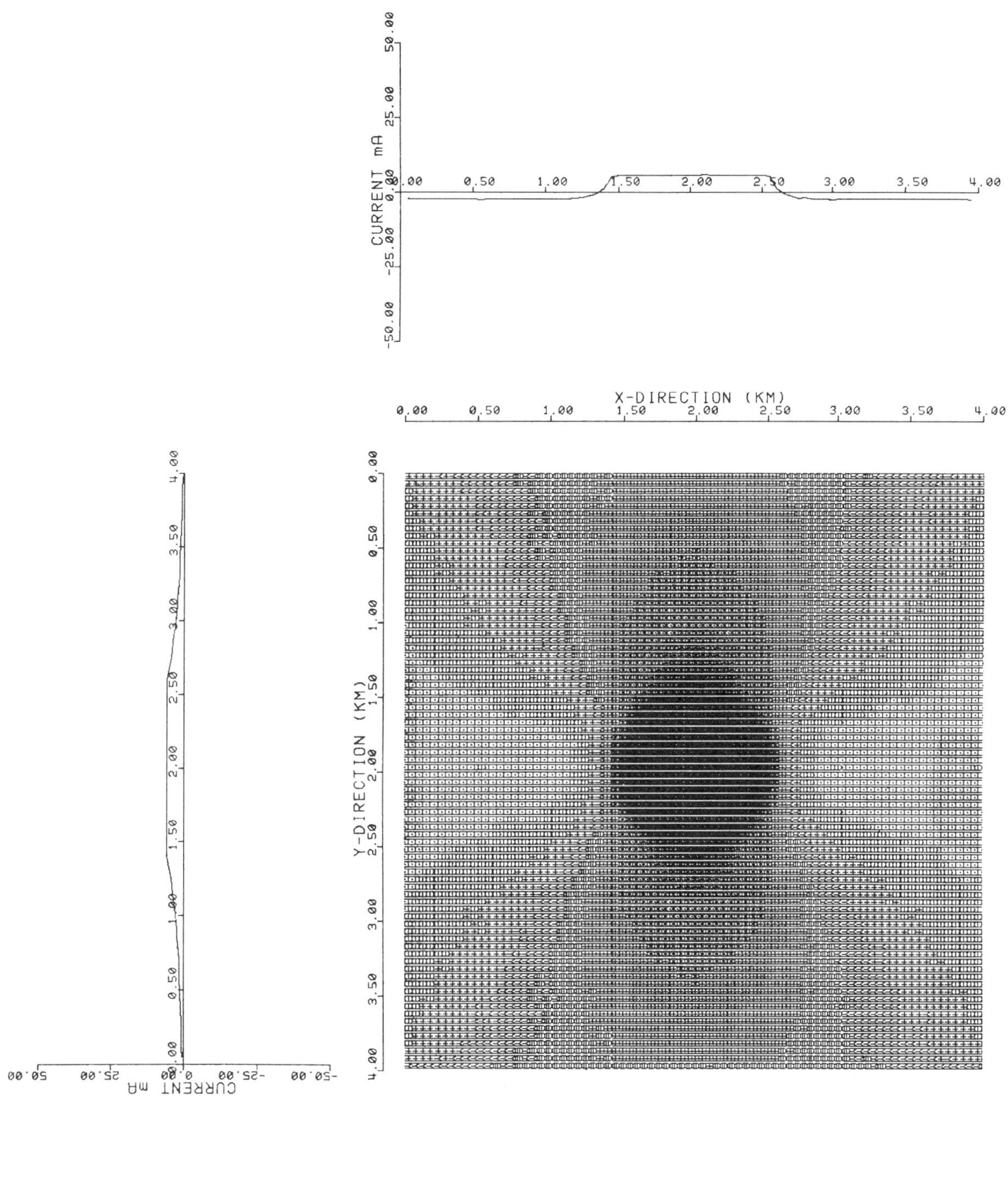

Fig. 10. ME current anomaly calculated by integration method of Pirson (1976). Applied to anomaly of Fig. 4. Compare with original current density shown in Fig. 2. East-west (upper diagram) and north-south (left diagram) are profiles through center of figure.

Fig. 11. Vertical current density generated by integration method on magnetic anomalies in Clark County, Kansas.

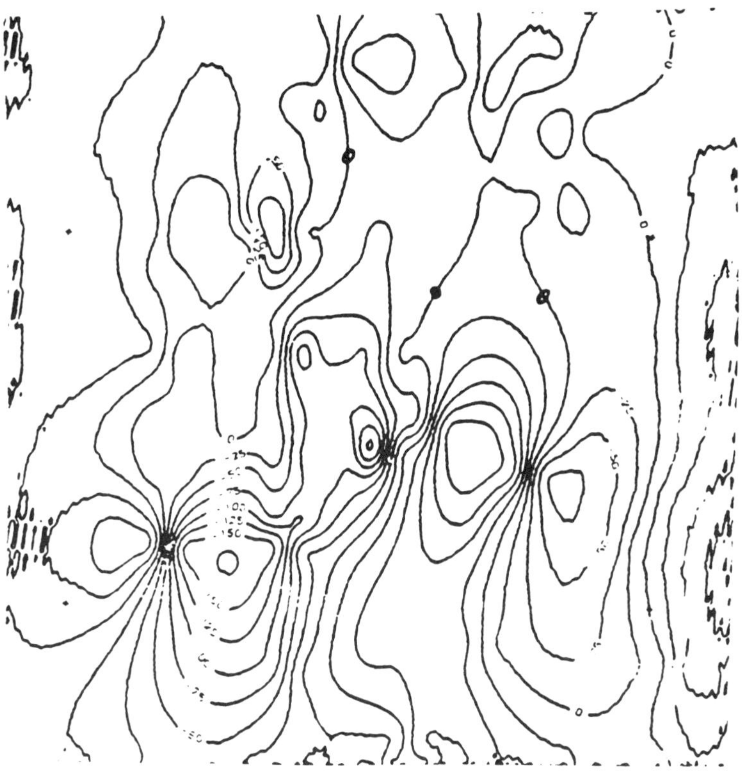

Fig. 12. First order derivative along east-west direction of magnetic anomalies of Clark County, Kansas. Note similarity to Fig. 11.

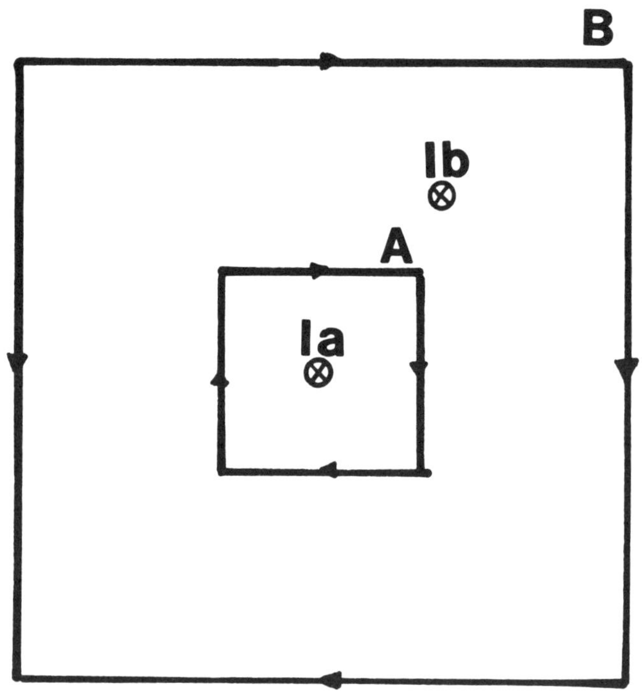

Fig. 13. Integration paths A and B applied to magnetic induction from current 1a and 1b, respectively.

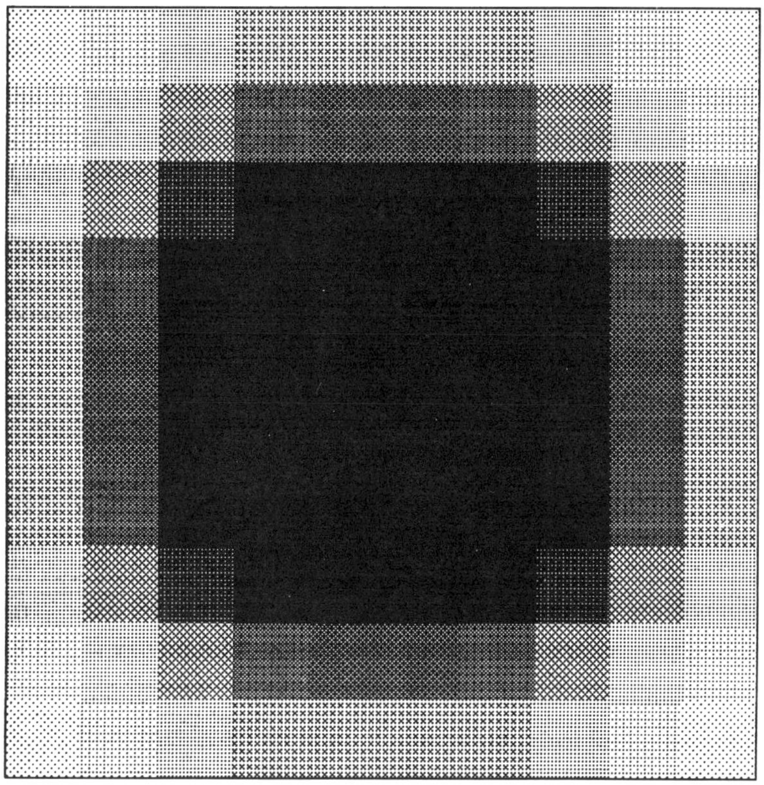

Fig. 14a. Vertical current density model to be used for inversion. White = maximum.

Fig. 14b. Magnetic anomaly from the vertical current density model in Fig. 14a.

Fig. 15. Initial guess of the possible current density related to the magnetic anomaly in Fig. 14b.

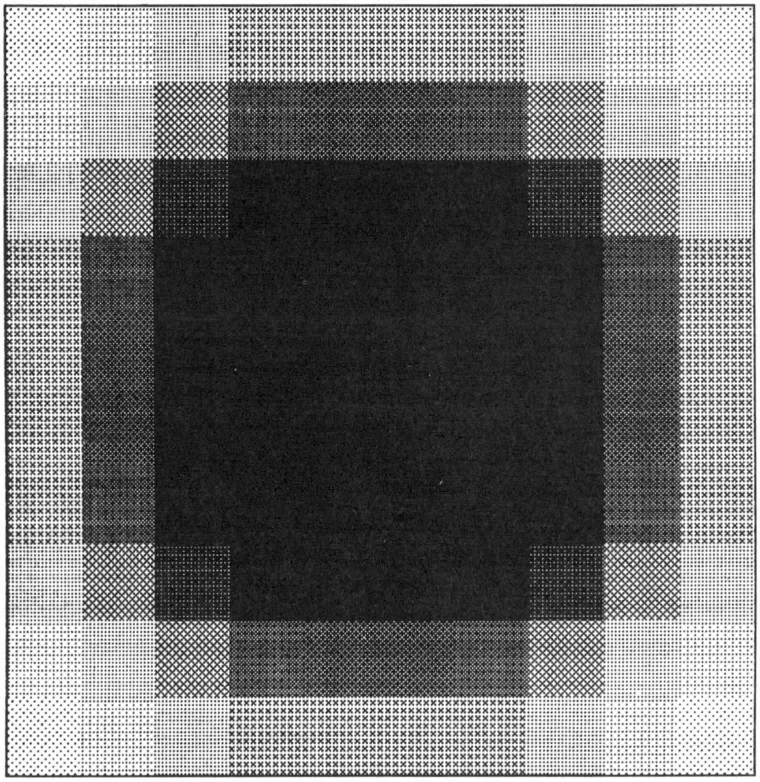

Fig. 16. Inversion solution for vertical current density calculated from magnetic anomaly shown in Fig. 14b. (Note the similarity with original current density in Fig. 14a.)

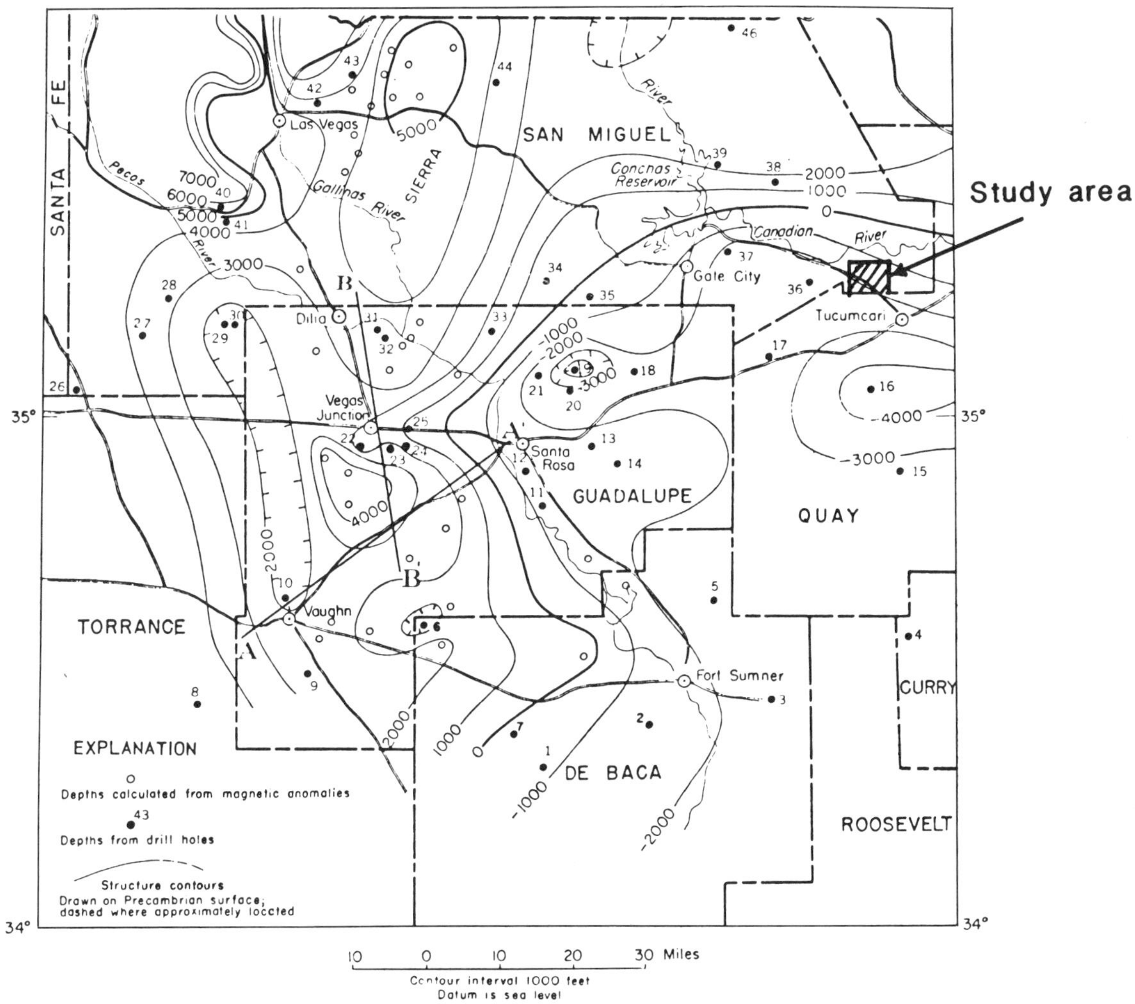

Fig. 17. Pre-Cambrian surface map of northeast New Mexico (Andreasen et al., 1962).

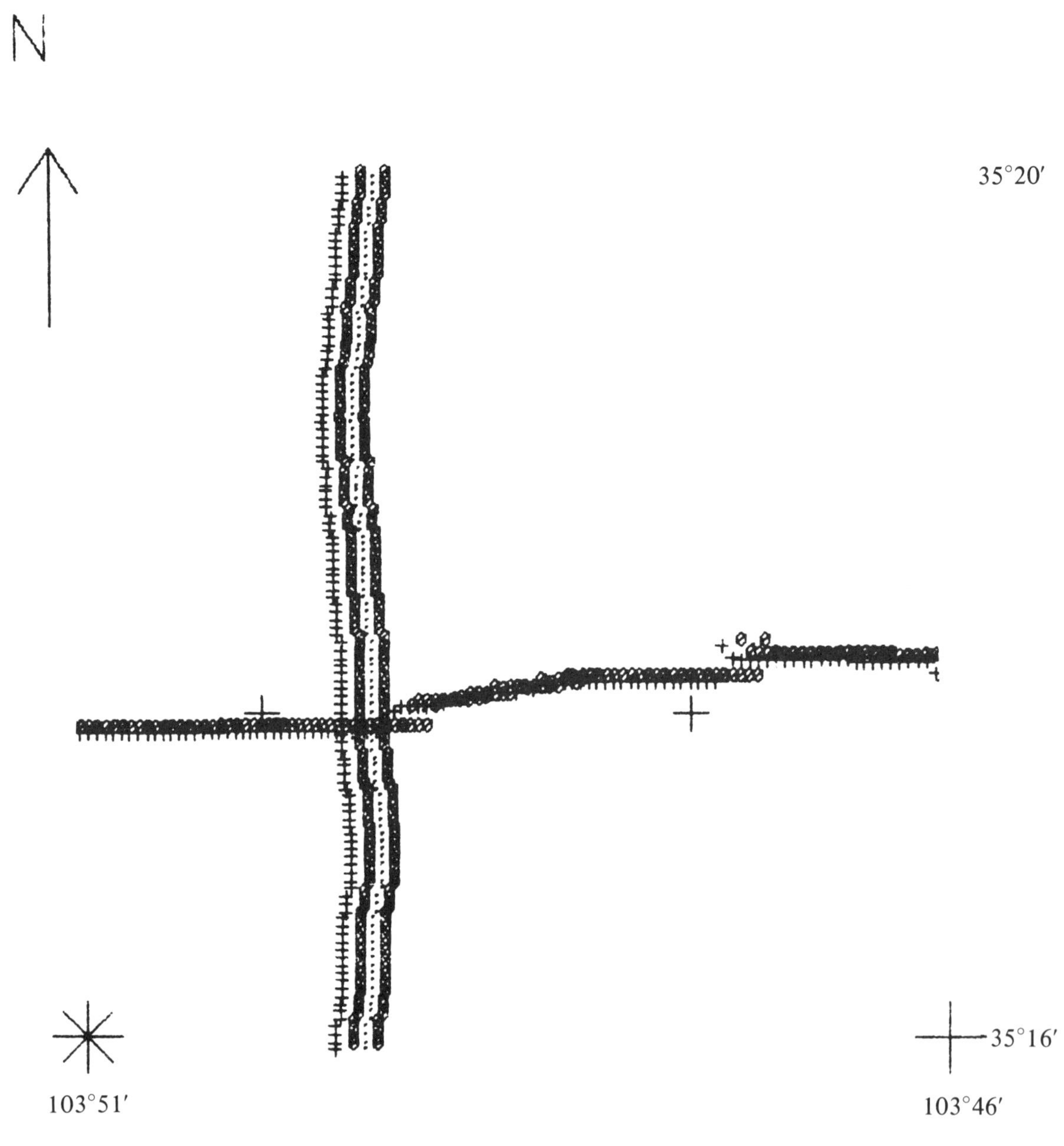

Fig. 18. Location of NURE data.

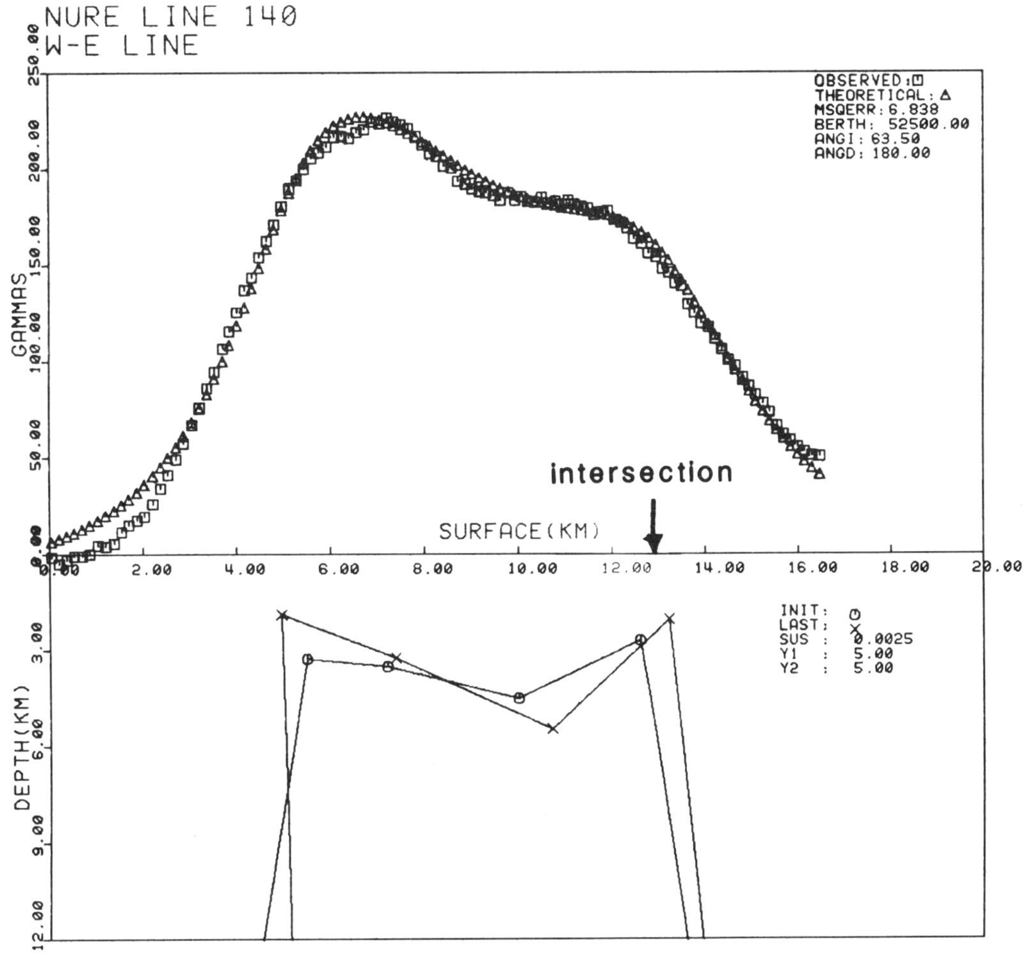

Fig. 19. Inverse 2-1/2 D magnetic model of west-east NURE line.

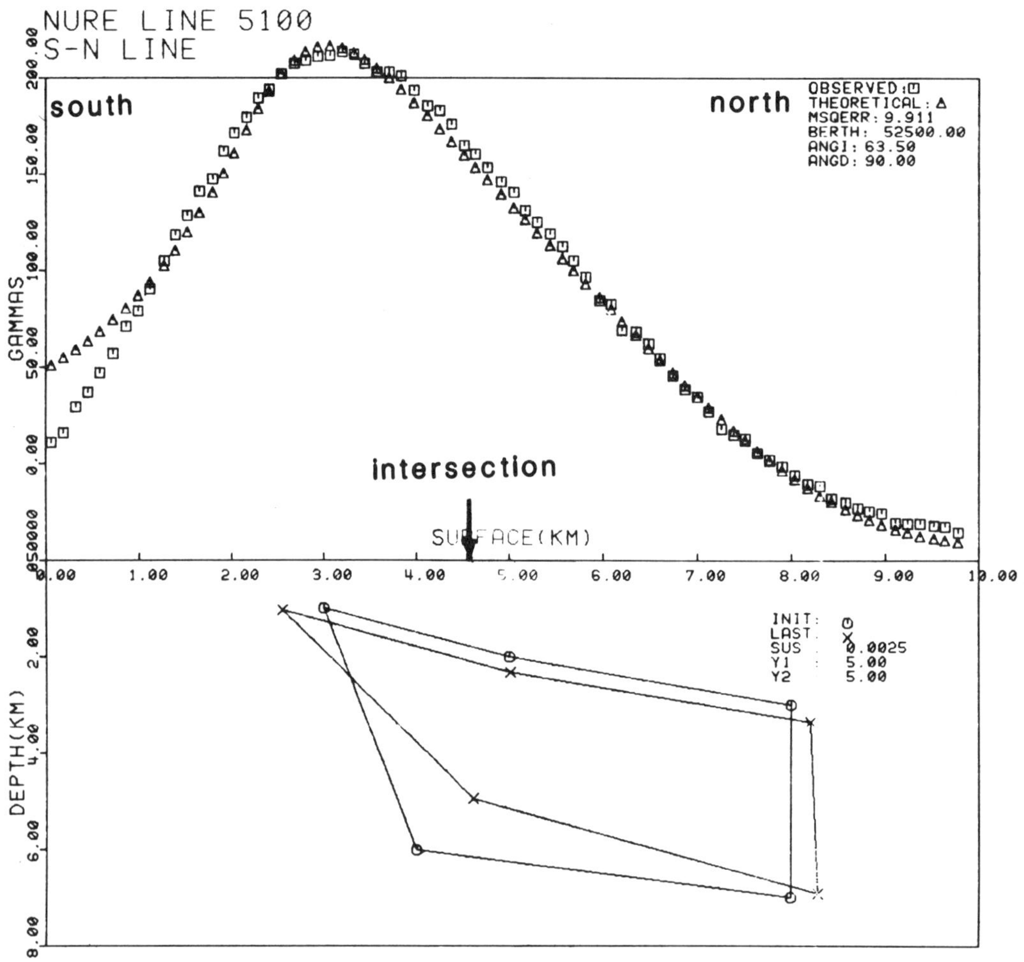

Fig. 20. Inverse 2-1/2 D magnetic model of south-north NURE line.

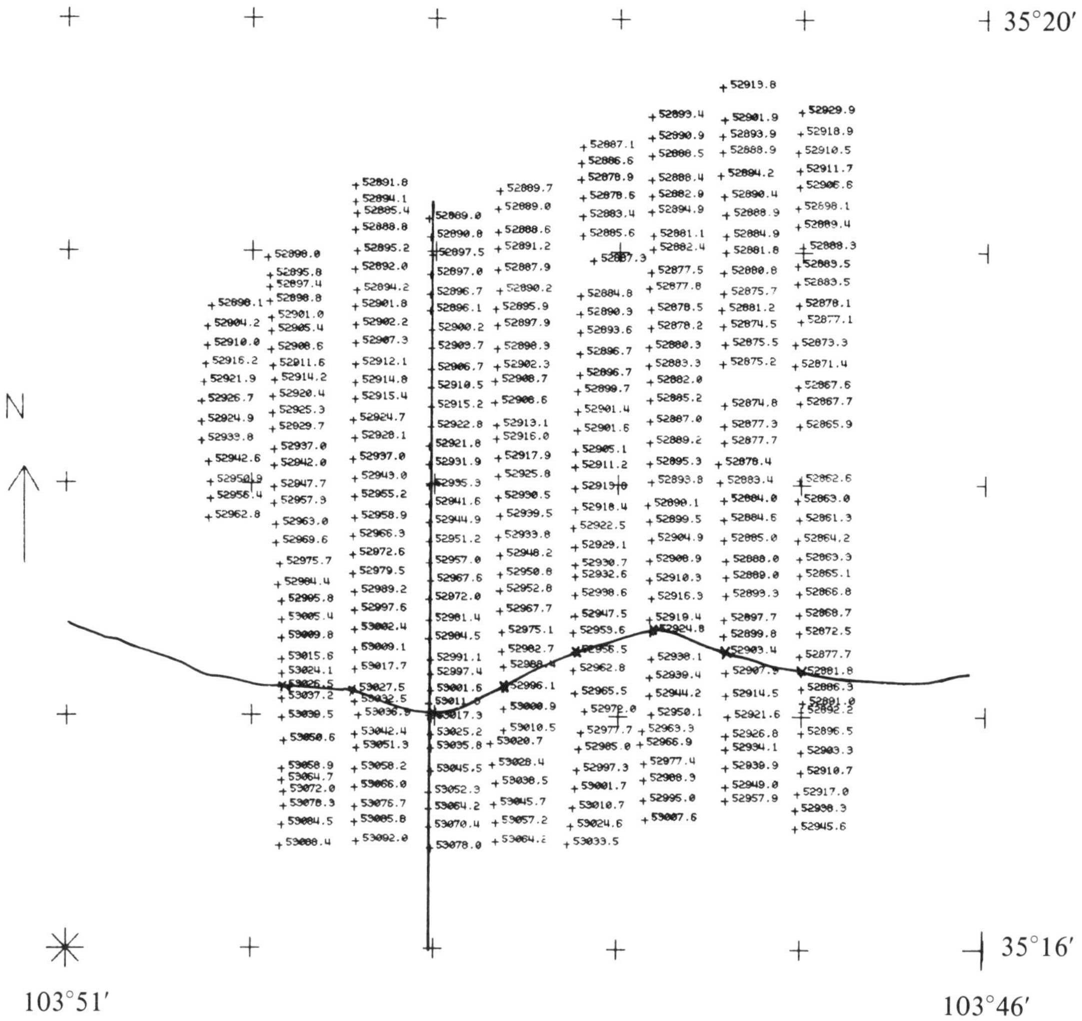

Fig. 21. Ground magnetic data observed in 1984 and 1985, Trinchera Creek Quadrangle, New Mexico. Solid lines indicate more detailed data observed in 1985. Observations by Olaf Aiken and C. T. Yang. Values are total intensity.

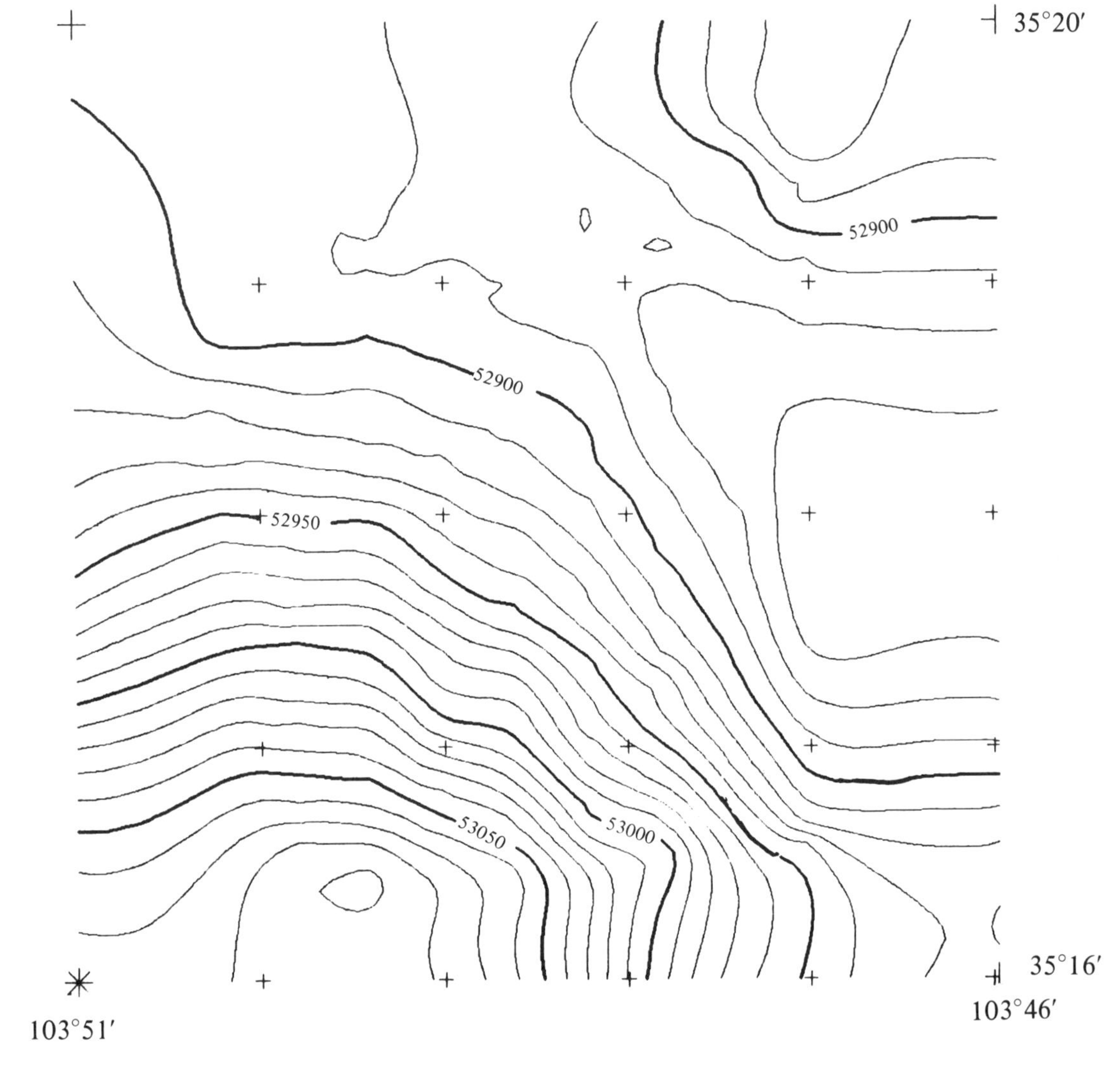

Fig. 22. Total intensity magnetic map, Trinchera Creek Quadrangle, New Mexico. Contour interval = 10 γ.

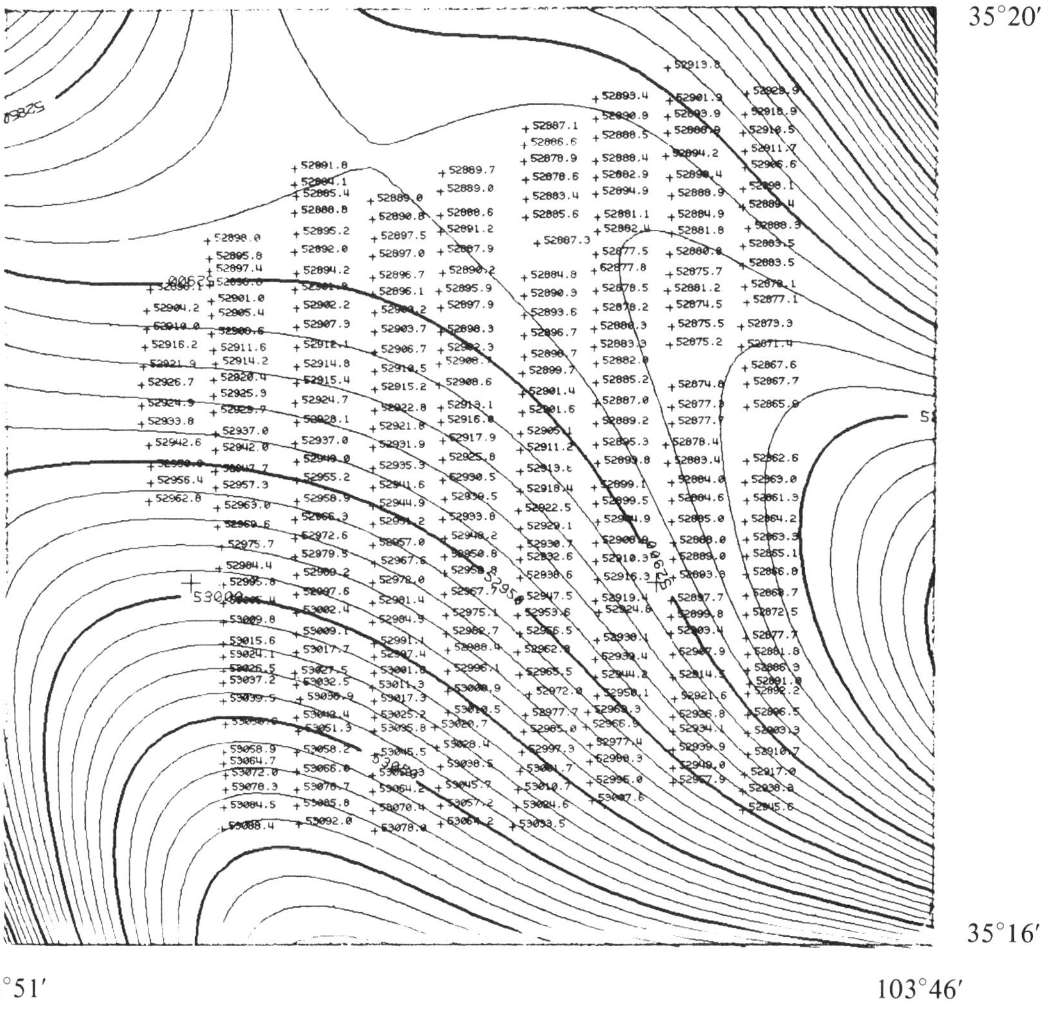

Fig. 23. Trend surface of total intensity magnetic data, Trinchera Creek Quadrangle, New Mexico. Contour interval = 10 γ.

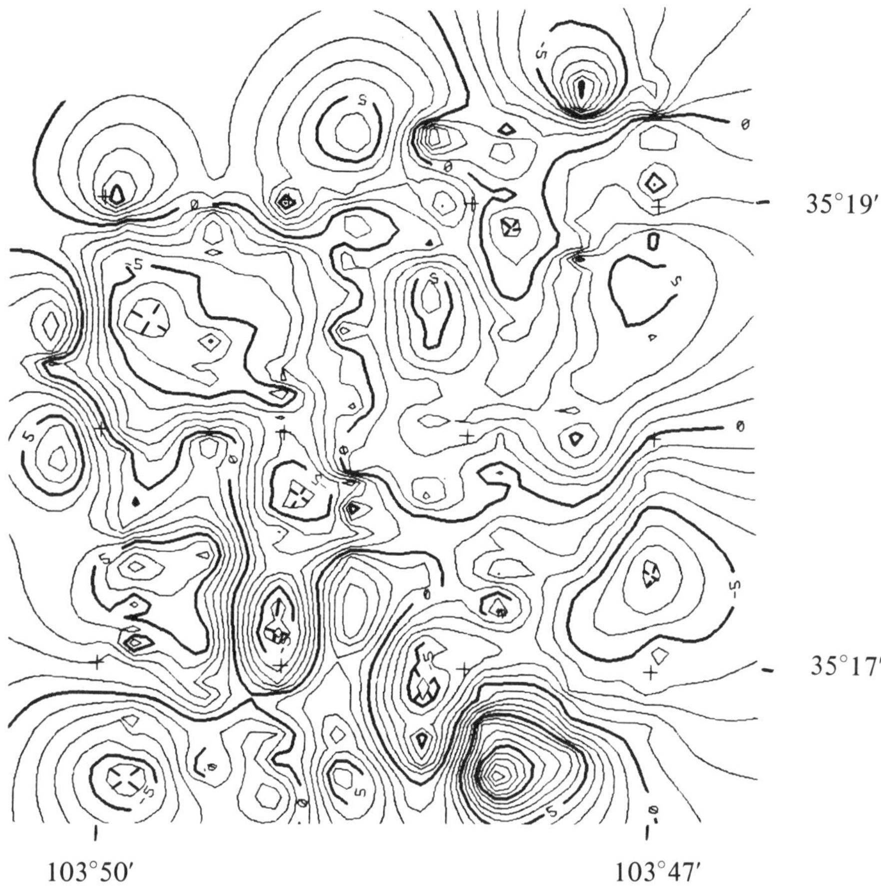

Fig. 24a. Residual total intensity magnetic map, Trinchera Creek Quadrangle, New Mexico. Trend surface of Fig. 23 removed.

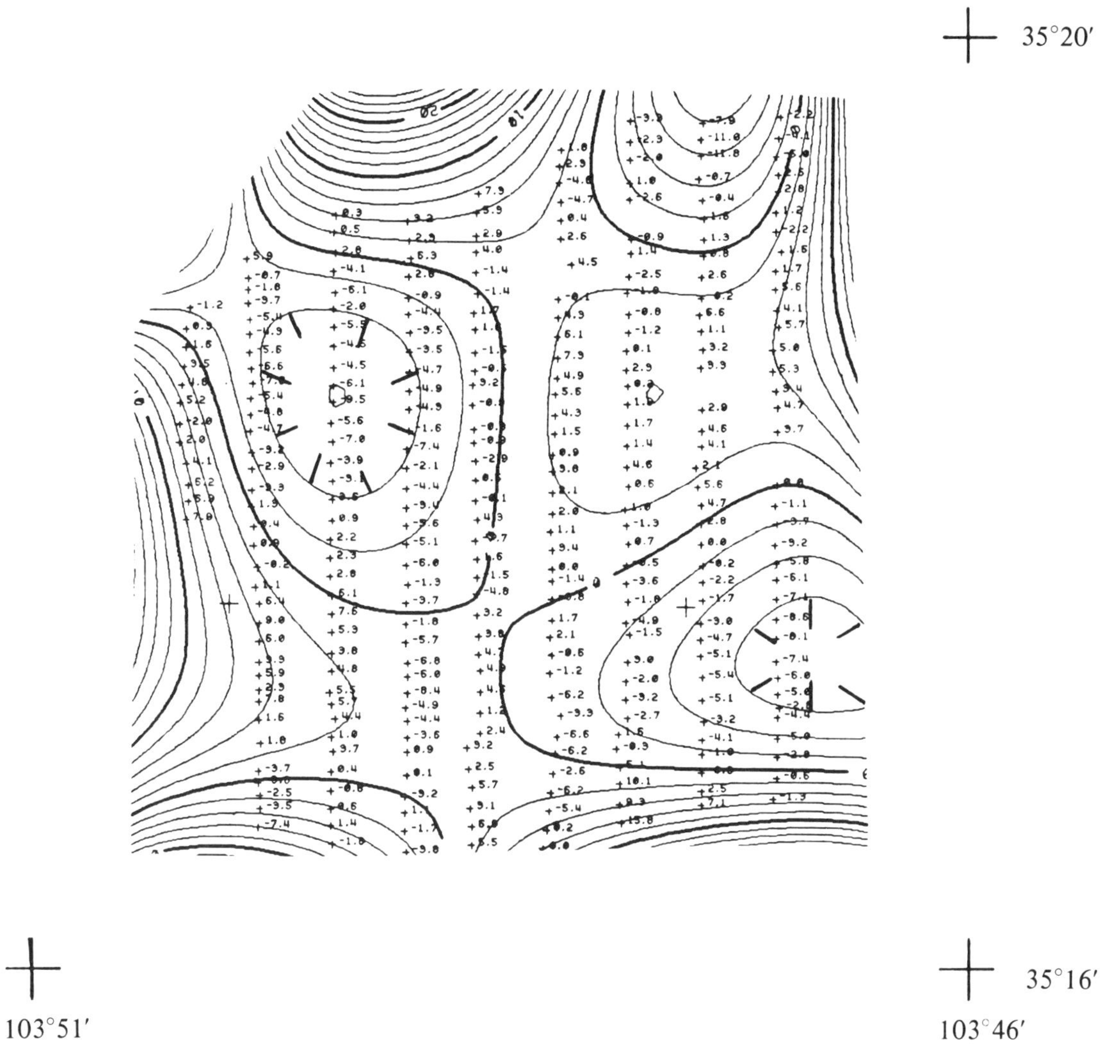

Fig. 24b. Trend surface of residual magnetic data, Trinchera Creek Quadrangle, New Mexico. Contour interval = 2 γ.

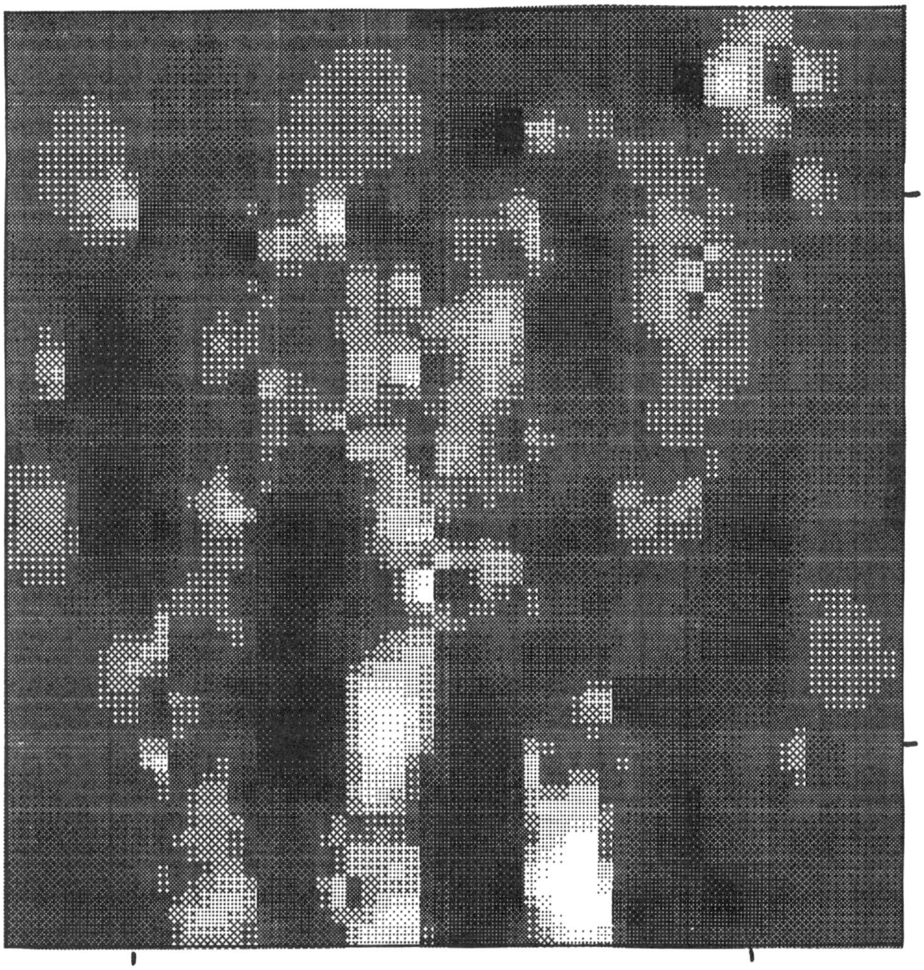

Fig. 25. Vertical current density calculated by integration method applied to residual magnetic data of Fig. 24a, Trinchera Creek Quadrangle, New Mexico. Grid interval 0.1 km. Black = maximum.

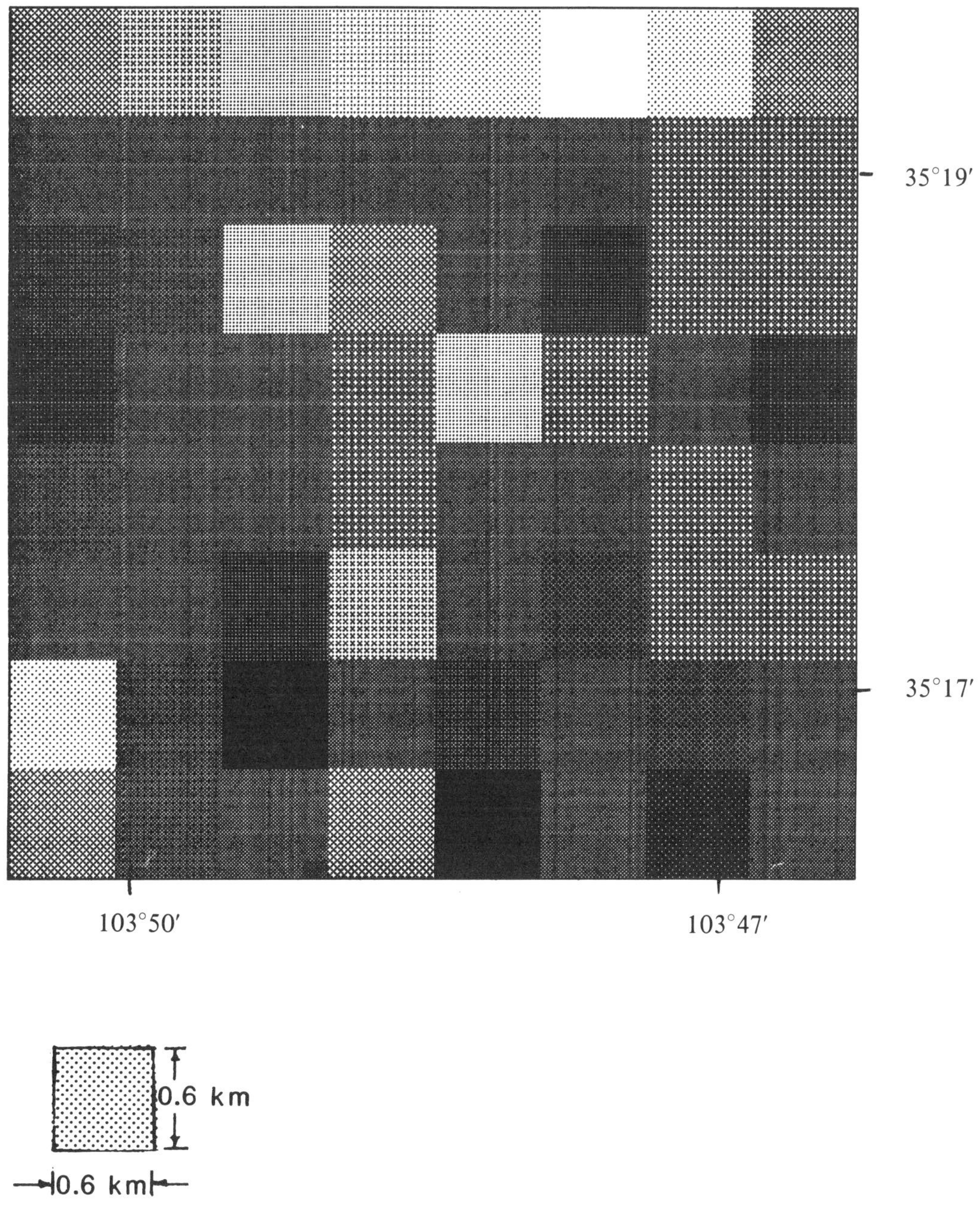

Fig. 26. Vertical current density calculated by inversion method applied to residual magnetic data of Fig. 24a, Trinchera Creek Quadrangle, New Mexico. Black = maximum.

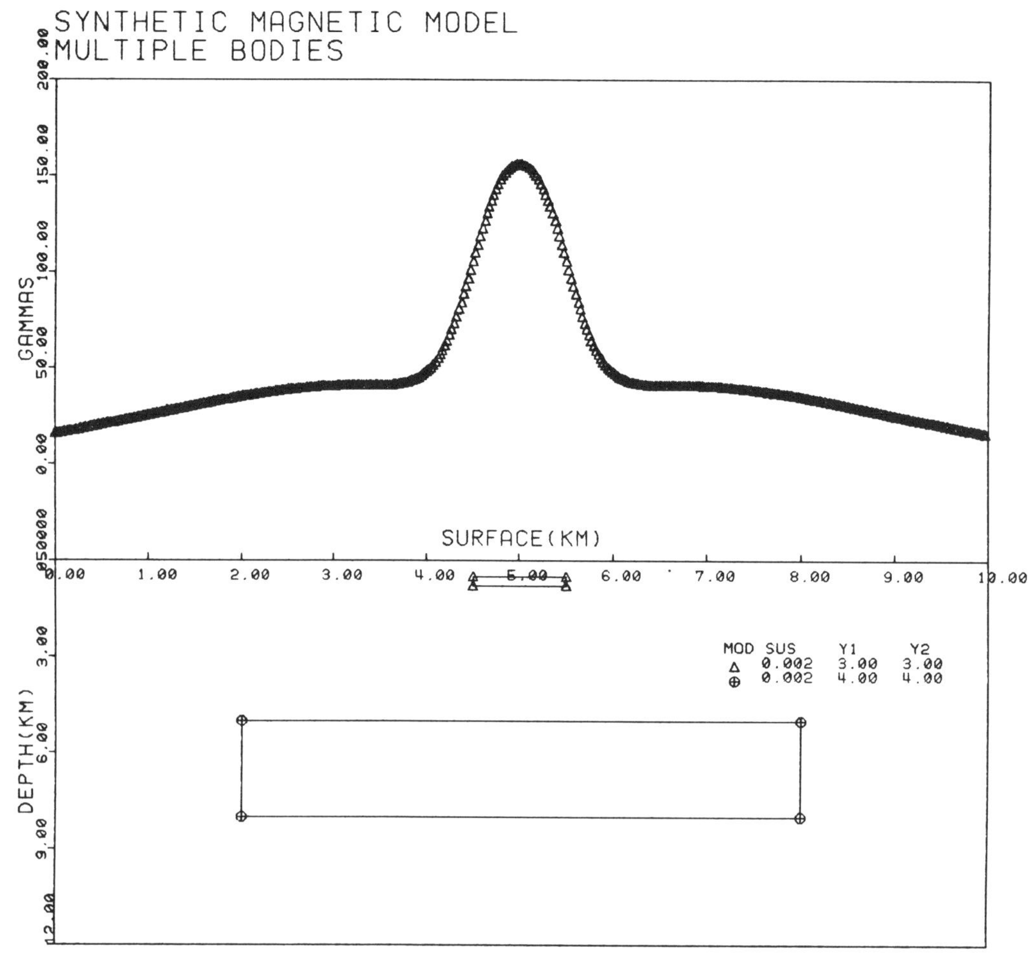

Fig. 27a. Synthetic magnetic model.

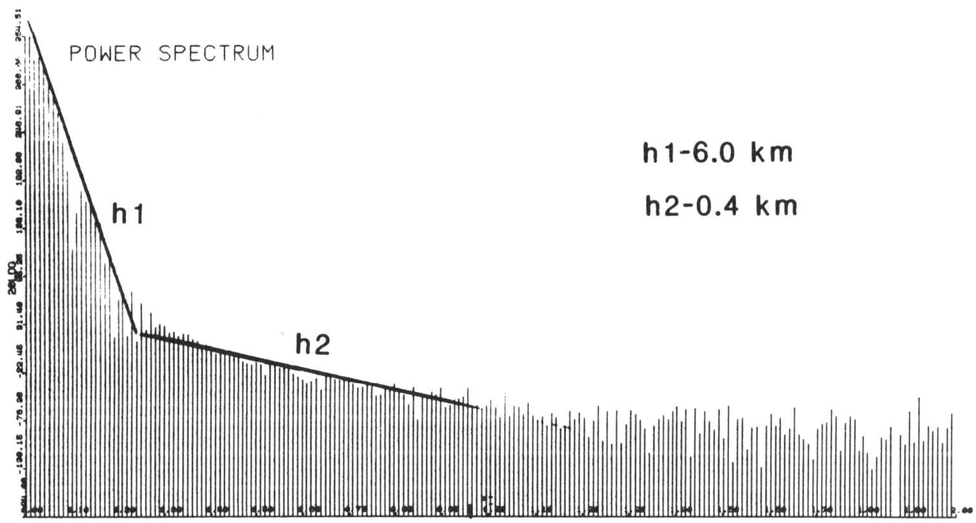

Fig. 27b. Power spectrum of magnetic anomaly of synthetic magnetic model shown in Fig. 27a. Note separation of depths to sources indicated by changes in slopes of spectrum.

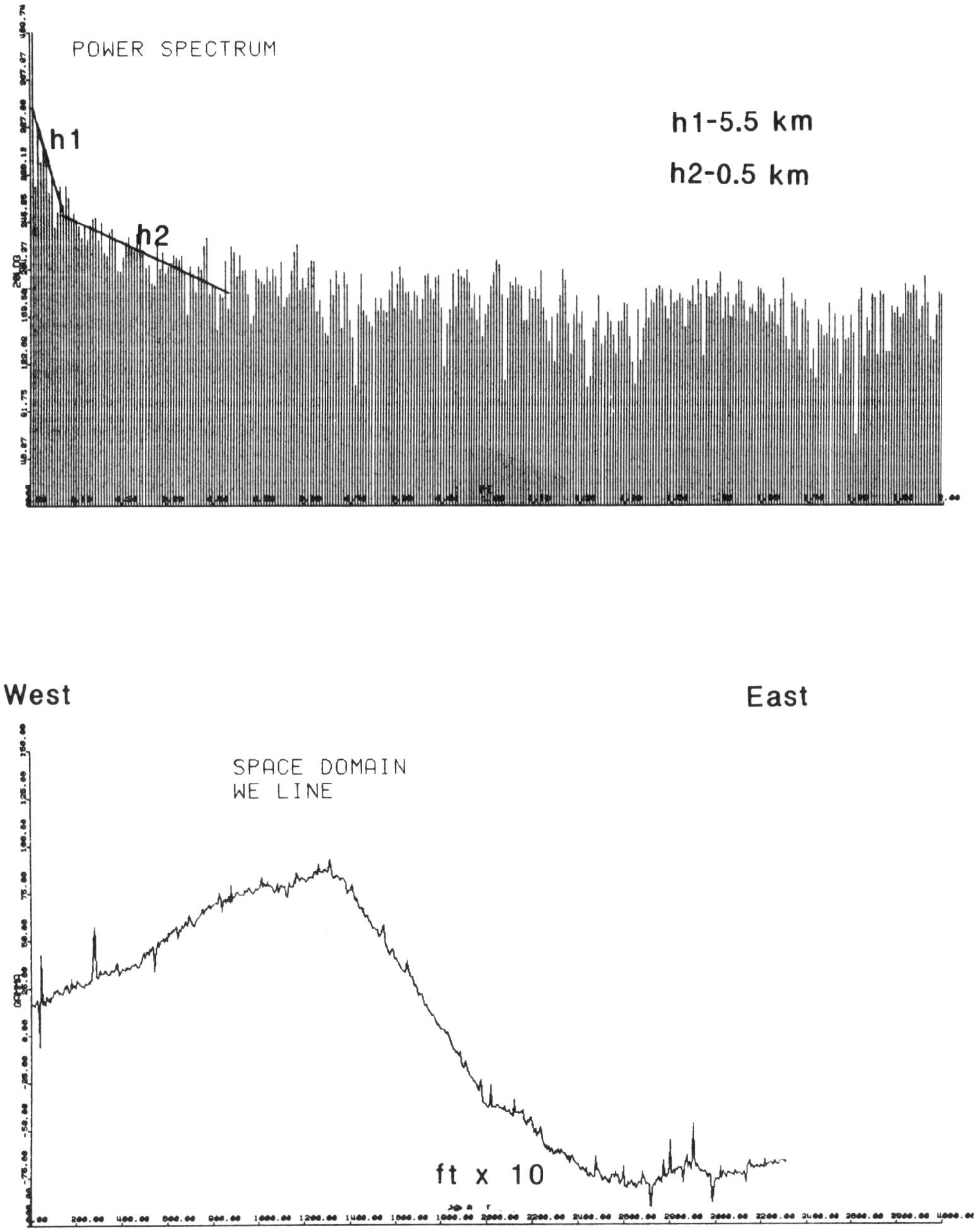

Fig. 28. Magnetic data and its corresponding power spectrum of west-east ground magnetic line, Trinchera Creek Quadrangle, New Mexico.

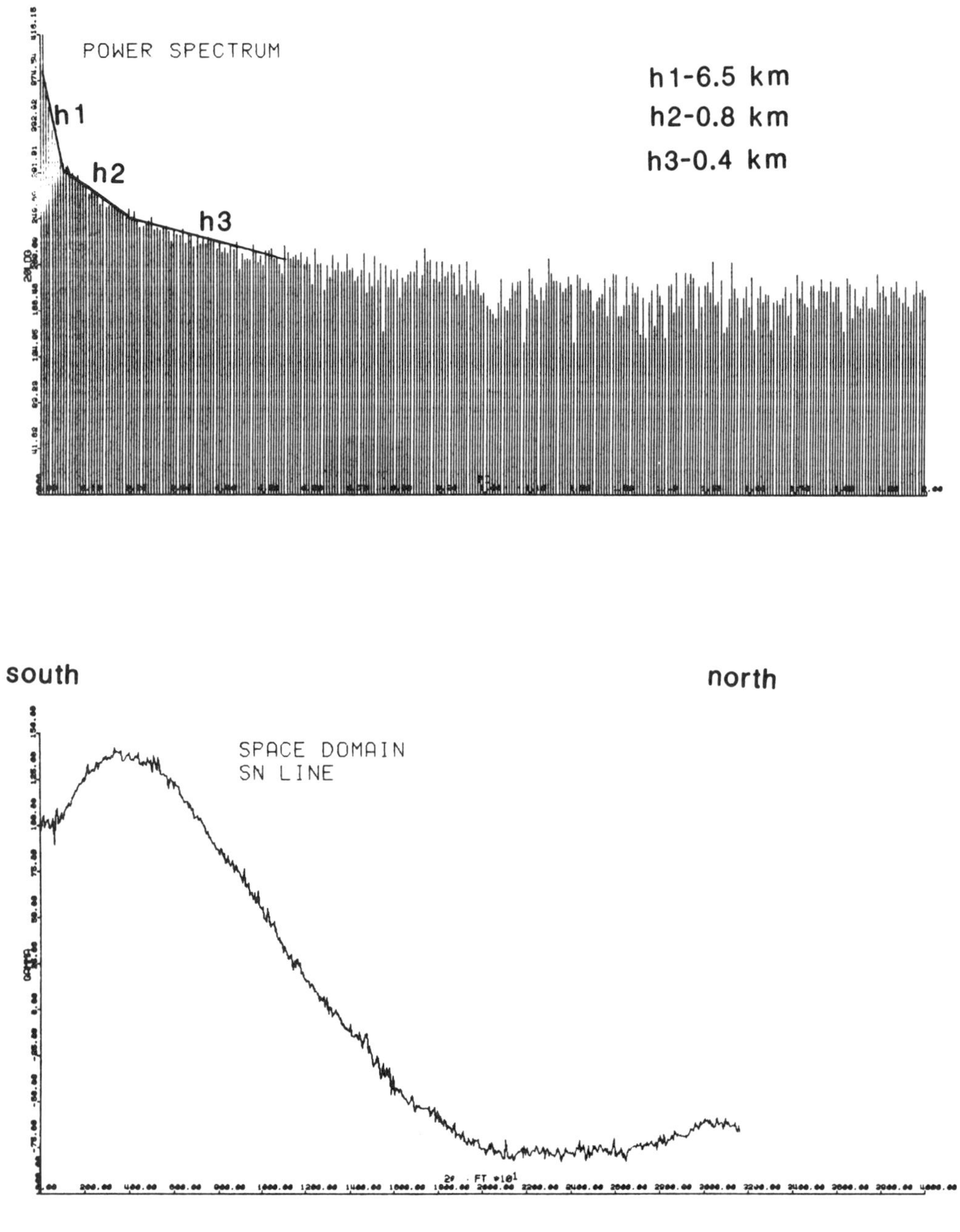

Fig. 29. Magnetic data and its corresponding power spectrum of south-north ground magnetic line, Trinchera Creek Quadrangle, New Mexico.

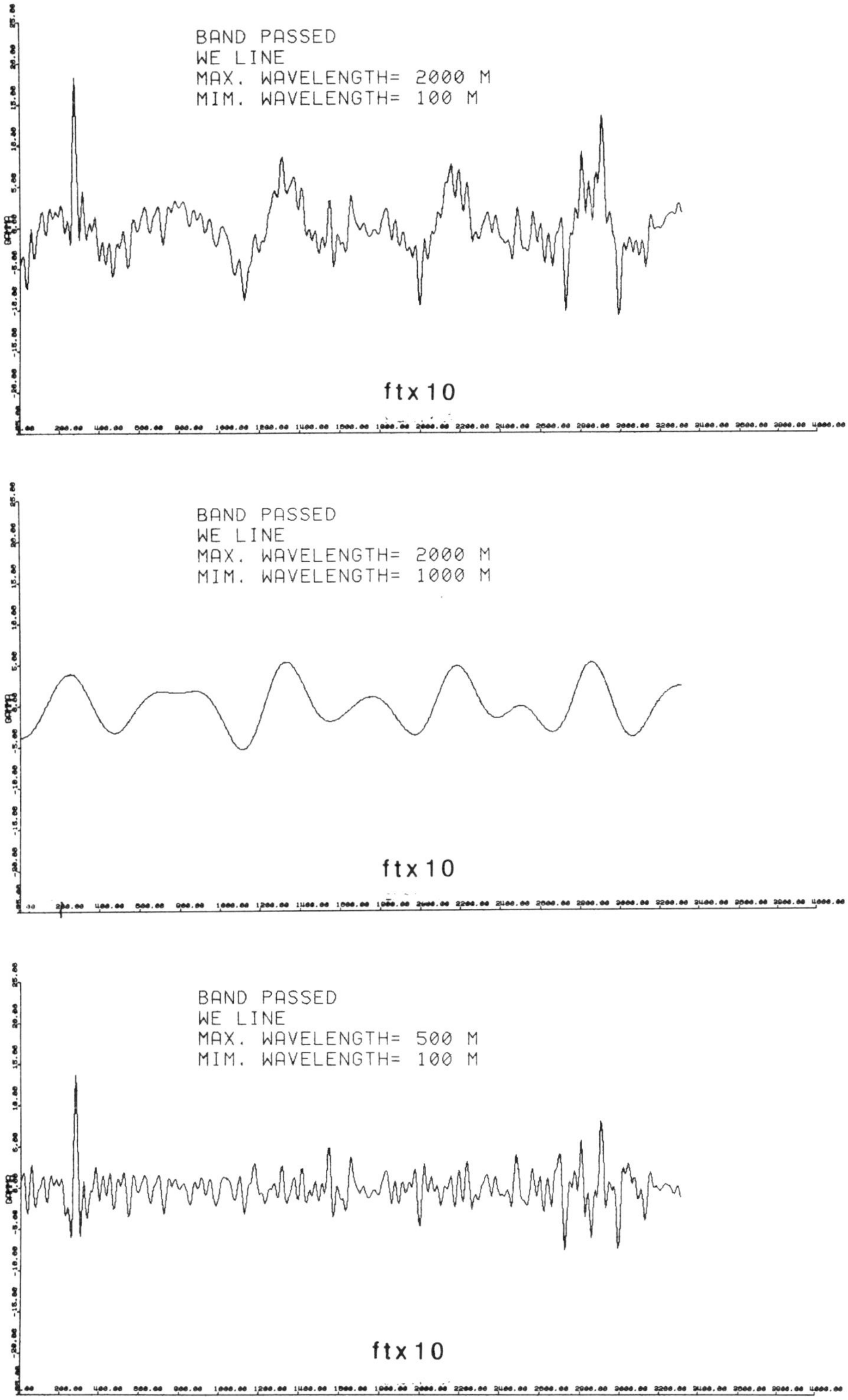

Fig. 30. Band-pass filtered results of west-east ground magnetic line, Trinchera Creek Quadrangle, New Mexico.

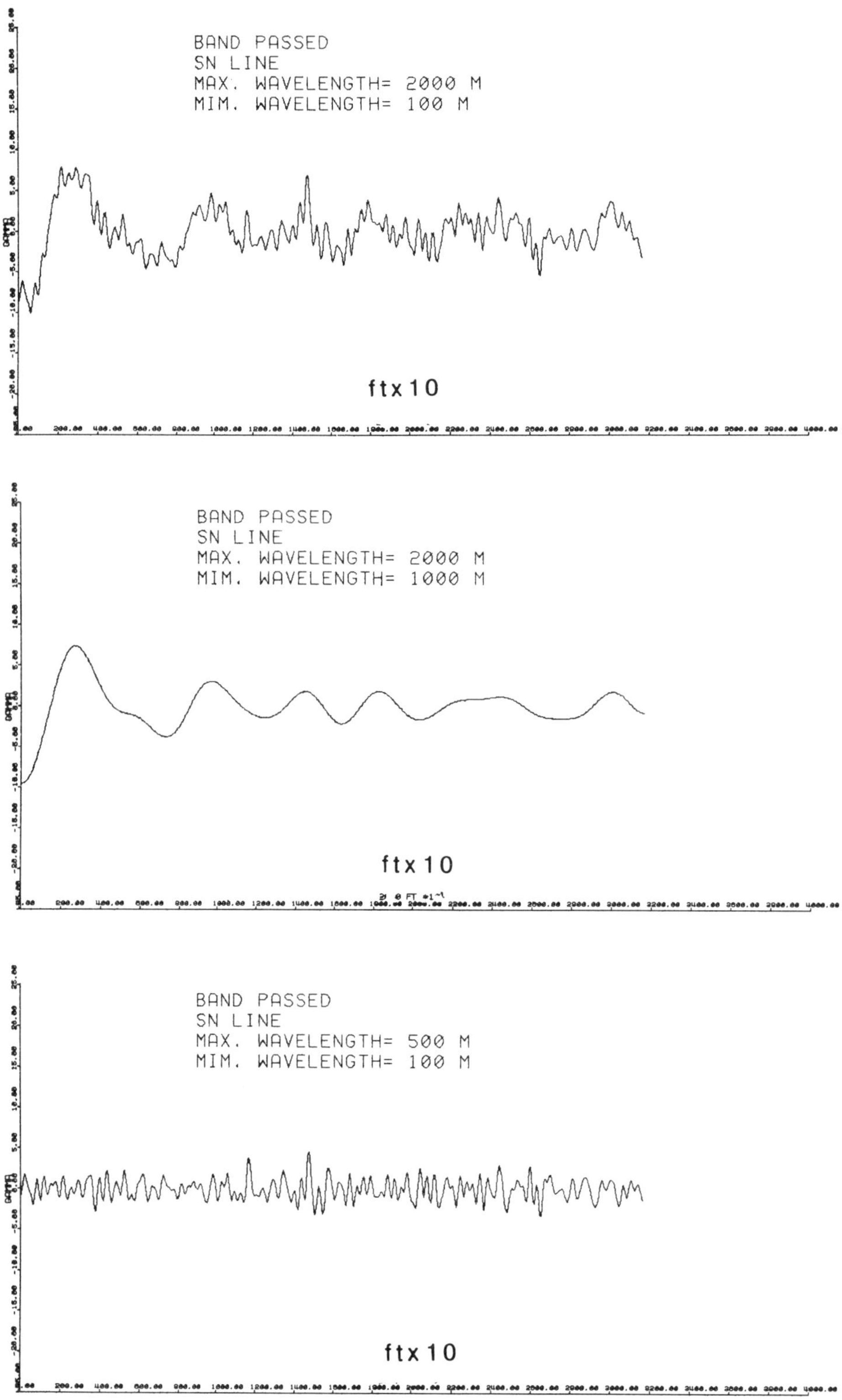

Fig. 31. Band-pass filtered results of south-north ground magnetic line, Trinchera Creek Quadrangle, New Mexico.

HORIZONTAL GRADIENT INTERPRETATION OF LOW-ALTITUDE CESIUM VAPOR MAGNETOMETER DATA COMPARED WITH MEASUREMENT OF ROCK MAGNETIC SUSCEPTIBILITY OF DRILL CUTTINGS

Robert S. Foote

GEOSCIENCE & TECHNOLOGY, INC.

ABSTRACT

Iron compounds are abundant in soil and rock materials throughout the earth's crust. Some of these iron compounds are in solution in subterranean waters, and those that exist in the rocks above petroliferous deposits have been observed to be enriched with concentrations of the ferrous ion (Horvitz, 1943). Insertion of oxygen into these ferrous ion waters from the penetration of oxygen-rich meteoric waters gives a plausible explanation for the anomalous creation of diagenetic magnetic material in the broad interface region between the fluids of the lower reducing zone and the introduced oxygen-rich surface waters. Identification of geographic areas having magnetic material in the near-surface is possible by analysis of low altitude airborne high resolution aeromagnetic data, and by measurement of the rock magnetic susceptibility of drill-hole cutting samples. Many examples comparing results of horizontal gradient intensity (HGI) analysis of high resolution aeromagnetic data with drill hole rock magnetic susceptibility measurements are shown for those drill tests where appropriate library samples are available.

PURPOSE OF THE STUDY

Local regions having anomalous magnetic intensities at shorter wavelengths have been observed (Donovan et al., 1979; Foote, 1982). Airborne magnetometer sensors used to acquire most of the more recent airborne magnetic field data have been the proton precession type which uses kerosine as the proton source. Data from this system are modulated by yaw/pitch/roll motion accelerations of the sensor, and by the system electronic noise, the results of which can amount to many gammas of erroneous data. The purpose of this study is to demonstrate the correlation of data from a low-noise cesium vapor airborne magnetometer system (Airmag Surveys, Inc.), where the total airborne noise is greatly reduced. Removal of this overall airborne "noise" together with a high sampling rate allows identification of low-level near-surface magnetic anomalies which may result from diagenetic alteration. Confirmation of the HGI anomaly can then be made by magnetic mass susceptibility, χ_ρ, measurements of drill hole rock cuttings over and outside of oil and gas areas. All oil and gas fields found to be anomalous from analysis of airborne data have been shown to be anomalous in χ_ρ where shallow rock cuttings samples are available.

MEASUREMENT SYSTEM AND METHODS

All airborne magnetic field data have been acquired using the cesium vapor stinger-mounted sensor having a sampling rate of four samples/sec. Magnetic field data are processed to remove longer wavelength modulations caused by the effect of the distant magnetic basement rocks, and to remove the magnetic effect of cultural objects such as pipelines, drill pipe and oil field equipment, highways, railways, etc. Horizontal gradient intensity analysis of the data then gives intensity quantification to those anomalies estimated to be generated by diagenetic magnetic material, presumed to consist largely of magnetite. The magnetic susceptibility measuring equipment has the sensitivity in cgs units of approximately 5×10^{-7}, with the ability to measure masses greater than about 1 g of the average sedimentary sample, depending upon the measuring environment conditions.

RESULTS OF DATA ANALYSIS

Comparison of the rock magnetic susceptibility with the results of HGI analysis of high quality airborne magnetometer data has been made in various geologic regions. These comparisons have been limited only because of the lack of availability of desired shallow rock cuttings samples. The Midland Sample Library supplied most of the samples for these measurements. Other libraries supplying samples for these measurements are the University of Texas Sample Library, Oklahoma University Sample Library, and the Amstrat Sample Library.

PECAN STATION OIL FIELD AREA, TOM GREEN COUNTY, TEXAS

The Pecan Station Oil Field area is presented in Figure 1 and indicates the HGI analysis anomaly limits around the field. The area was surveyed using a ¼-mi flight-line spacing according to the methods desribed. The results of the horizontal gradient intensity analysis are shown in Figure 2. The magnetic susceptibility as a function of sample depth for the two available oil well cuttings sample sets are shown in Figure 3. Neither of these sets has samples available more shallow than 100 ft. Oil well 1, Figure 3, indicates a high value extending to below 500 ft, while oil well 2 shows only a very minimum shallow anomaly. Dry holes, samples 3 and 4, indicate no anomalous χ_ρ at shallow depths. Flight Line 7060 (Figure 2) overflies oil well 2 and indicates no anomaly from the airborne data in the region of the oil well. This is confirmed by the lack of magnetic material in the rocks from well 2, Figure 3.

LIPAN CREEK OIL FIELD AREA, TOM GREEN COUNTY, TEXAS

The results of airborne magnetic field data analysis show the Lipan Creek Oil Field area to be anomalous (Figure 1). Rock magnetic susceptibility data for the two available oil wells, well 5 and well 6, show the near-surface anomaly to extend downward to about 200 ft (Figure 4). No dry holes provided samples above 225 ft in the vicinity of the oil field.

KWB OIL FIELD AREA, TOM GREEN COUNTY, TEXAS

The results of HGI analysis of the airborne survey of the KWB Oil Field area of western Tom Green County, Texas, give the contiguous anomaly of Figure 5. The oil drilling results for the KWB area are given in Figure 6, indicating the comparison of the HGI anomaly with the drilling activity. The χ_ρ for the only available oil well with sampling to the surface, oil well 1, located in the southeastern section of the field, indicates in Figure 7 a strong anomaly above 150 ft. Dry hole 2, two miles south of the KWB Oil Field, shows no shallow χ_ρ anomaly, nor does another more distant dry hole, number 3 (shown only on Figure 7).

MT. SUSAN OIL FIELD AREA, TOM GREEN COUNTY, TEXAS

The χ_ρ of the shallow rock samples for the discovery well in the Mt. Susan Oil Field near Cristoval, Texas, is compared with the χ_ρ measurement of rock samples from a nearby dry hole (Figure 8) and shows as anomalous only the χ_ρ of the shallow samples less than a depth of 100 ft over the oil well.

PATTERSON CANYON OIL AND GAS FIELD AREA, SAN JUAN COUNTY, UTAH

The χ_ρ measurements were made only in the Patterson Canyon field area, San Juan County, Utah. Even though shallow rock cuttings samples above 600 ft in well 2 are not available,measurement of χ_ρ of the deeper rock cuttings samples from the producing well shows alteration to a depth of 900 ft, compared to the much lower or nonexistent alteration in the nonproducing test (well 2) 1 mi southeast (Figure 9).

SOUTH PEEK FIELD AREA, ROGER MILLS COUNTY, OKLAHOMA

Rock cuttings samples were made available from the drilling of shot holes along two seismic lines over the South Peek Field area, Roger Mills County, Oklahoma (Figure 10). Samples were gathered at 10-ft intervals to a depth of 160 ft. Figure 11 gives the results of χ_ρ measurement of the rock samples from line 1, and indicates maximum χ_ρ to be no deeper than 10–20 ft below the surface, and gives maximum integrated alteration in the region of the gas well having initial production of 7.1 mmcf (Figure 10, SP225). Similar results for line 2 are given in Figure 12. The integrated anomalous χ_ρ intensities are also shown in Figure 10, which includes the outline of the cesium vapor magnetometer airborne data horizontal gradient anomaly. This HGI anomaly includes the known production area with initial productions given. The detail analysis results of the ¼-mi spaced flight line data (Figure 13) correlate with the approximate boundary of the known field.

CEMENT OIL FIELD AREA, CADDO COUNTY, OKLAHOMA

Of the many oil and gas wells drilled within the Cement Oil Field, only a very few have shallow rock cuttings samples available from sample libraries. Results of χ_ρ measurement of three oil wells found having shallow samples available, which are located along the fault axis of the field, show that the deepest alteration (Ohio #9 Wade) extends to a depth of about 300 ft (Figure 14). This oil well is located in the high HGI results of the flight data (Figure 15) and is within the eastern anticlinal dome along the Cement fault line. Oil well 3 (Mobil #1 Royce), located 1 mi east of well 1, indicates a lesser diagenetic alteration depth and a lesser HGI value (Figure 15). Oil well 2, the only well having rock cuttings available to the near-surface, indicates a maximum alteration at about 70 ft, and a lesser HGI value (Figure 14). The only dry hole cuttings found having shallow samples are at the western edge of the field (Figure 15) and indicate no shallow anomaly (Amerada #1 Clift and Amerada #1 Rogers, Figure 14). Evaluation of χ_ρ for the two core sample sets provided by R. Forgey, U.S. Geological Survey, and reported in their open file report (Forgey et al., 1982), are included in the data of Figure 14.

SUMMARY

These results of horizontal gradient intensity analysis of low altitude cesium vapor airborne magnetometer data show a positive correlation with anomalous near-surface magnetic susceptibility measurement of shallow drill hole rock cuttings from above oil and gas fields, and also indicate a positive correlation with the lack of anomalous χ_ρ results in shallow samples from nonproductive areas. Proper horizontal gradient analysis of high resolution airborne measurement of the earth's magnetic field indicates that anomalous near-surface distributed magnetic field intensities exist over known petroliferous areas which have been shown to have correspondingly anomalous shallow rock magnetic susceptibilities. All anomalous regions discovered as a result of airborne surveys can be further evaluated for the presence of the higher-than-normal shallow rock magnetic susceptibility as a criterion for enhancement of prospective drill site locations by measurement of χ_ρ for rock cuttings to a depth of at least several hundred feet. Magnetic susceptibility anomalies are believed to be controlled by the penetration of oxygen-rich ground water and its effect upon vertically migrating ferrous ions.

REFERENCES

Donovan, T. J., Forgey, R. L., and Roberts, A. A., 1979, Aeromagnetic detection of diagenetic magnetite: American Association of Petroleum Geologists Bulletin, vol. 63, no. 2, p. 245–248.

Foote, R. S., 1984, Significance of near-surface magnetic anomalies, *in* Davidson, M. J., and Gottlieb, B. M., eds., Unconventional methods in exploration for petroleum and natural gas III: Dallas, Southern Methodist University Press, p. 12–24.

Forgey, R. L., et al., 1982, Remanent magnetization and magnetic susceptibility of two oriented cores, Cement Oil Field, Oklahoma: A reference data set: U.S. Geological Survey Open-File Report 82-848.

Horvitz, L., 1943, Geochemical prospecting: U.S. Patent No. 2,310,291.

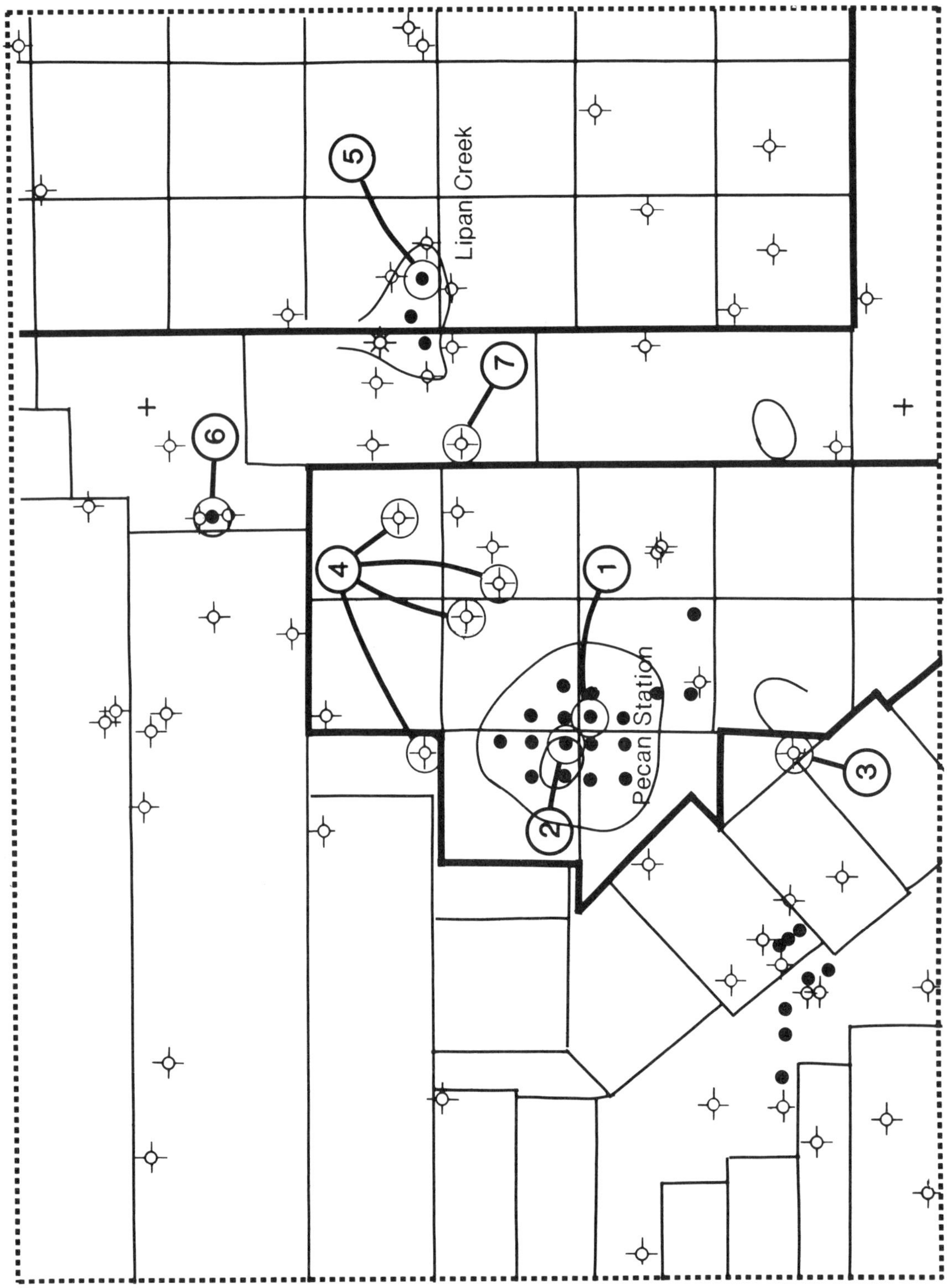

Fig. 1. Pecan Station/Lipan Creek Oil Field area, Tom Green County, Texas.

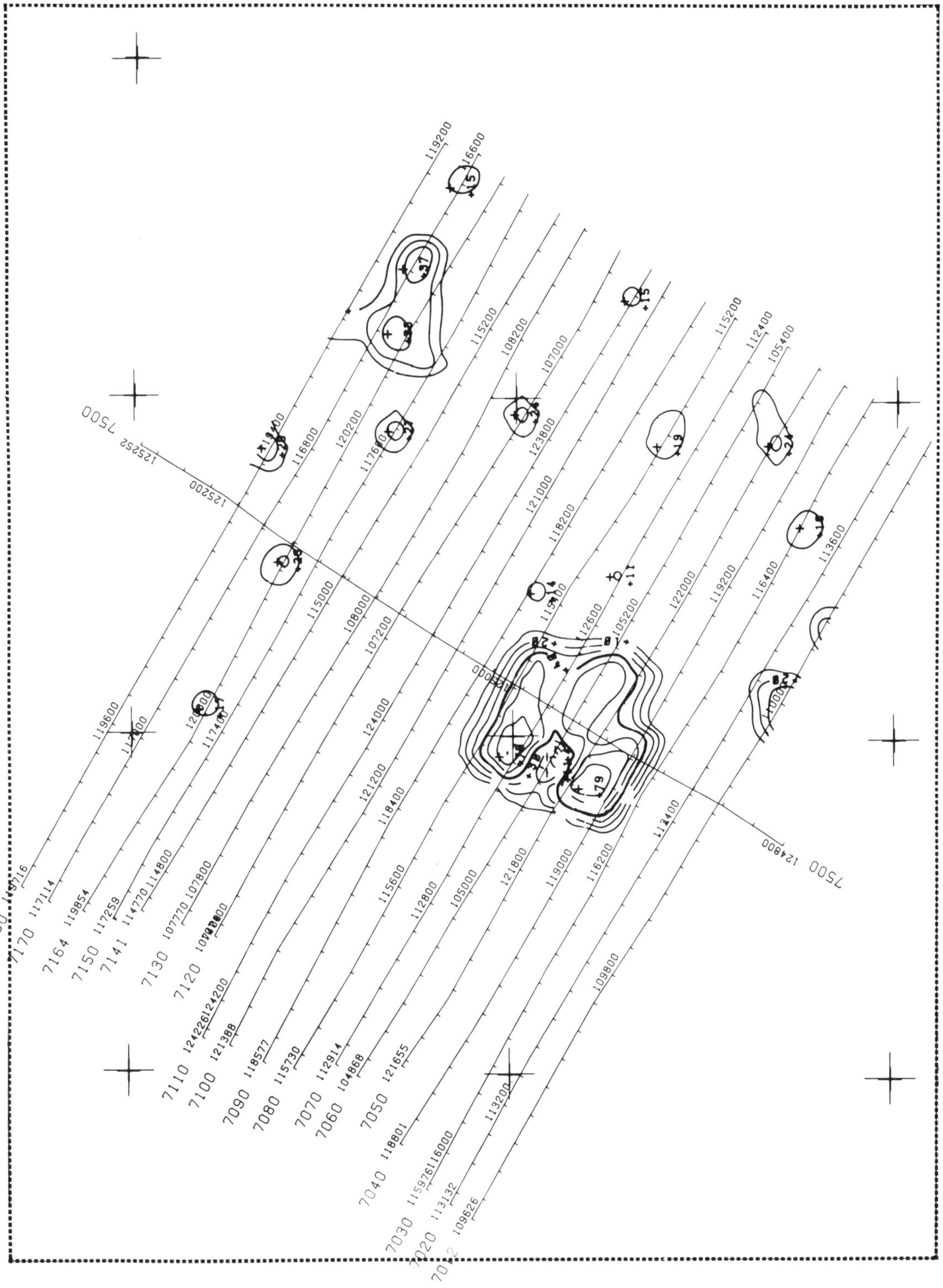

Fig. 2. Pecan Station/Lipan Creek Oil Field area, Tom Green County, Texas. Ce vapor magnetometer horizontal gradient analysis.

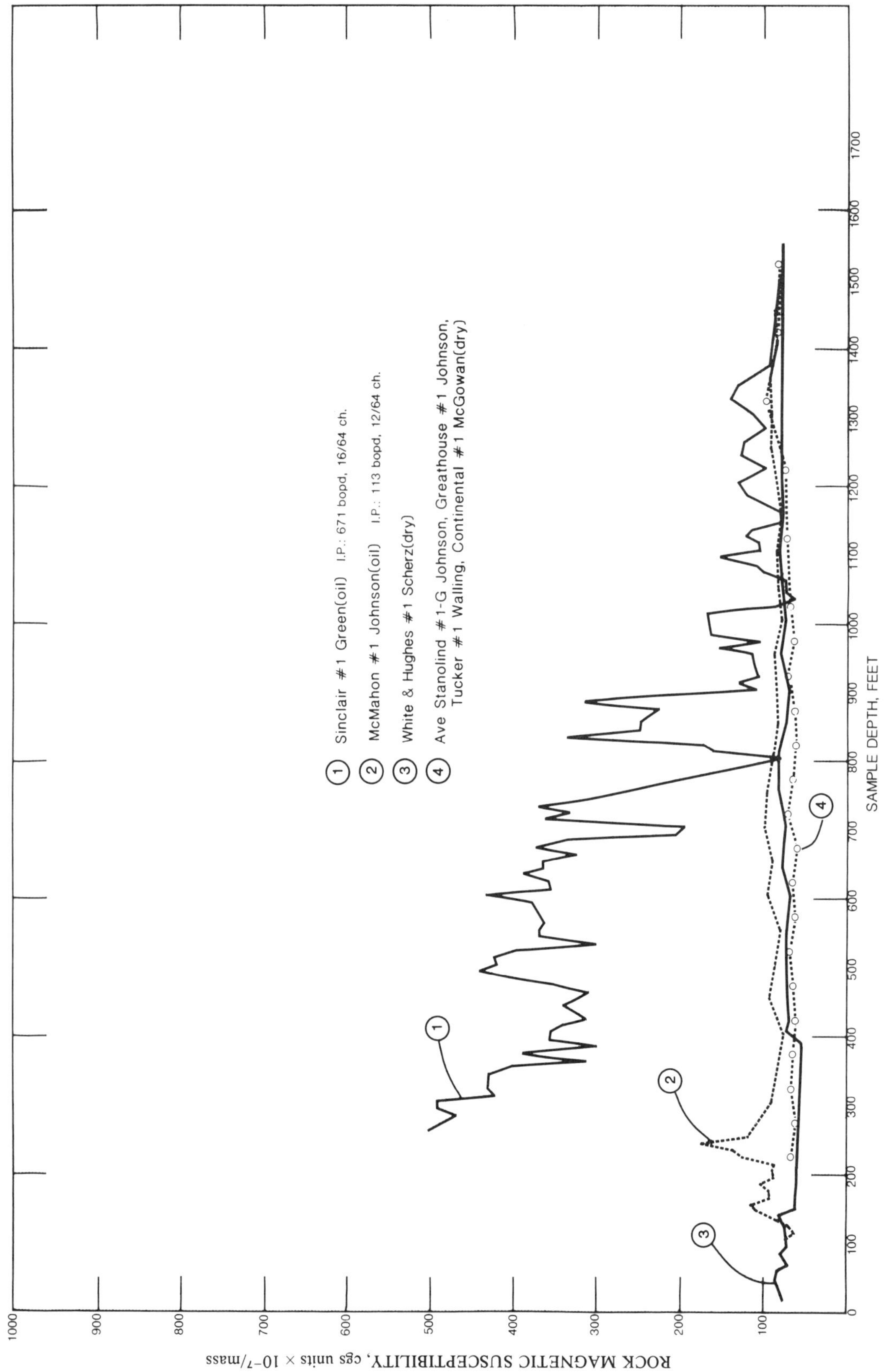

Fig. 3. Pecan Station Oil Field area, Tom Green County, Texas. Drill cuttings rock magnetic susceptibility as function of depth.

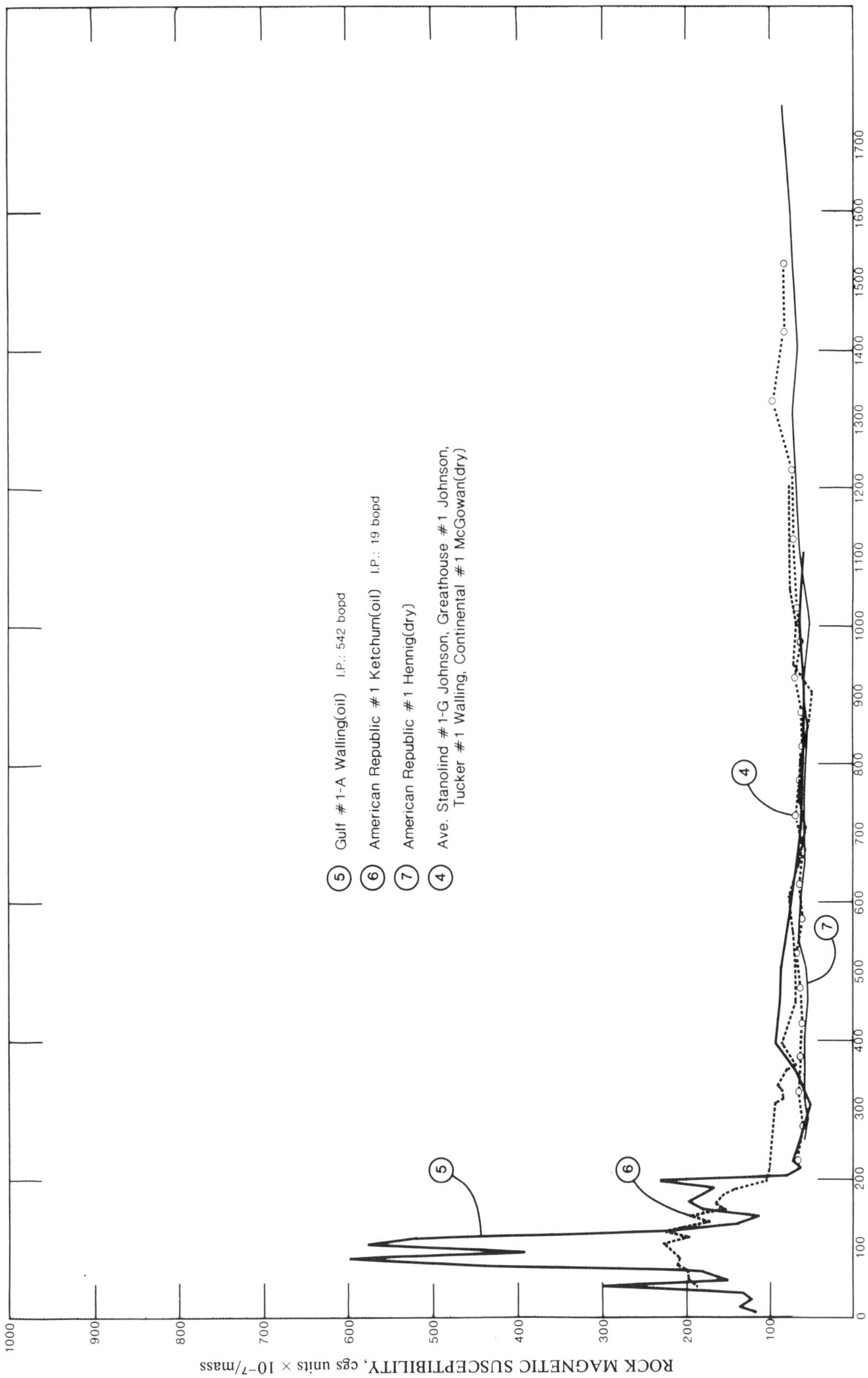

Fig. 4. Lipan Creek Oil Field area, Tom Green County, Texas. Drill cuttings rock magnetic susceptibility as function of depth.

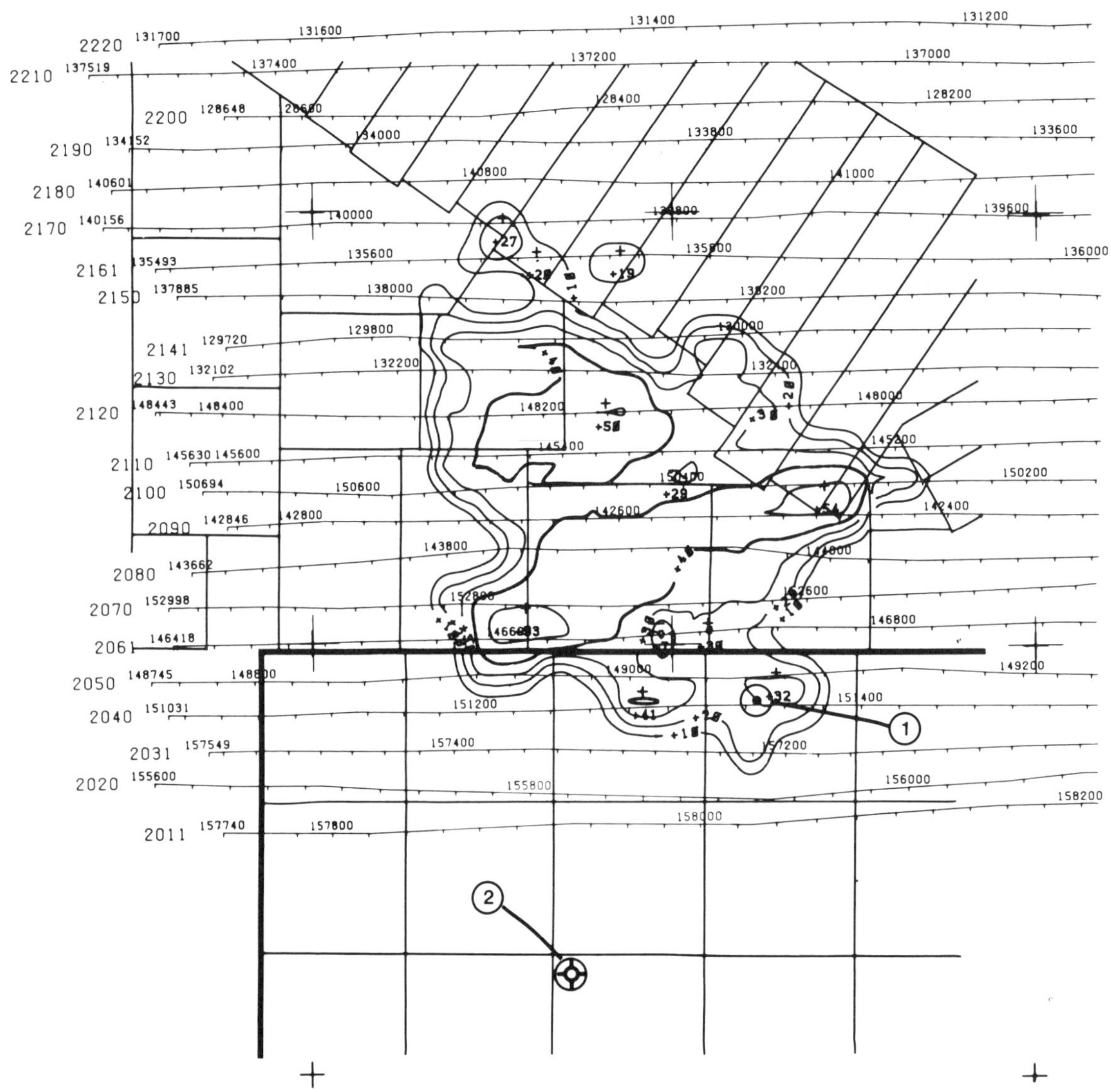

Fig. 5. KWB Oil Field area, Tom Green County, Texas. Ce vapor magnetometer horizontal gradient analysis.

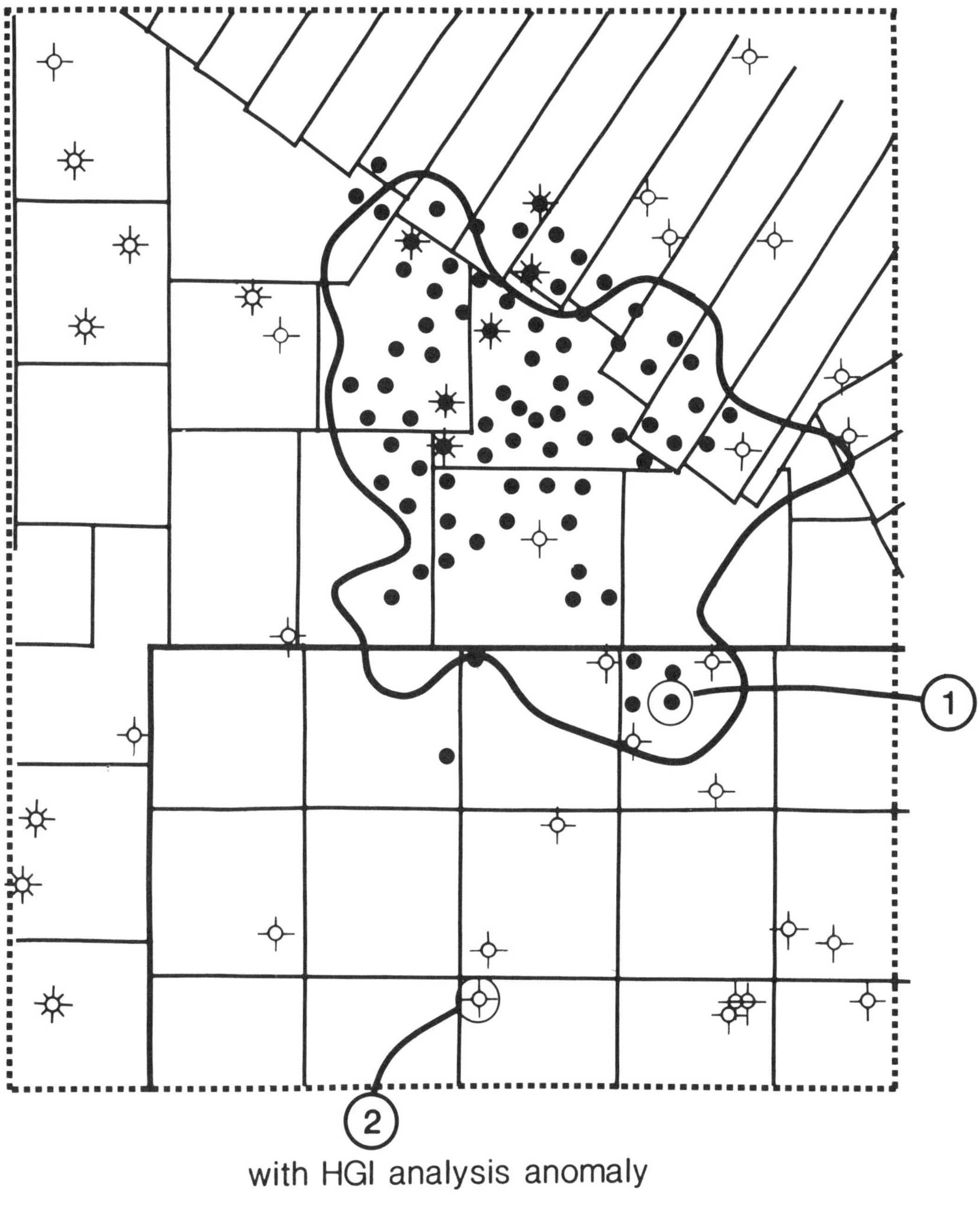

Fig. 6. KWB Oil Field area, Tom Green County, Texas.

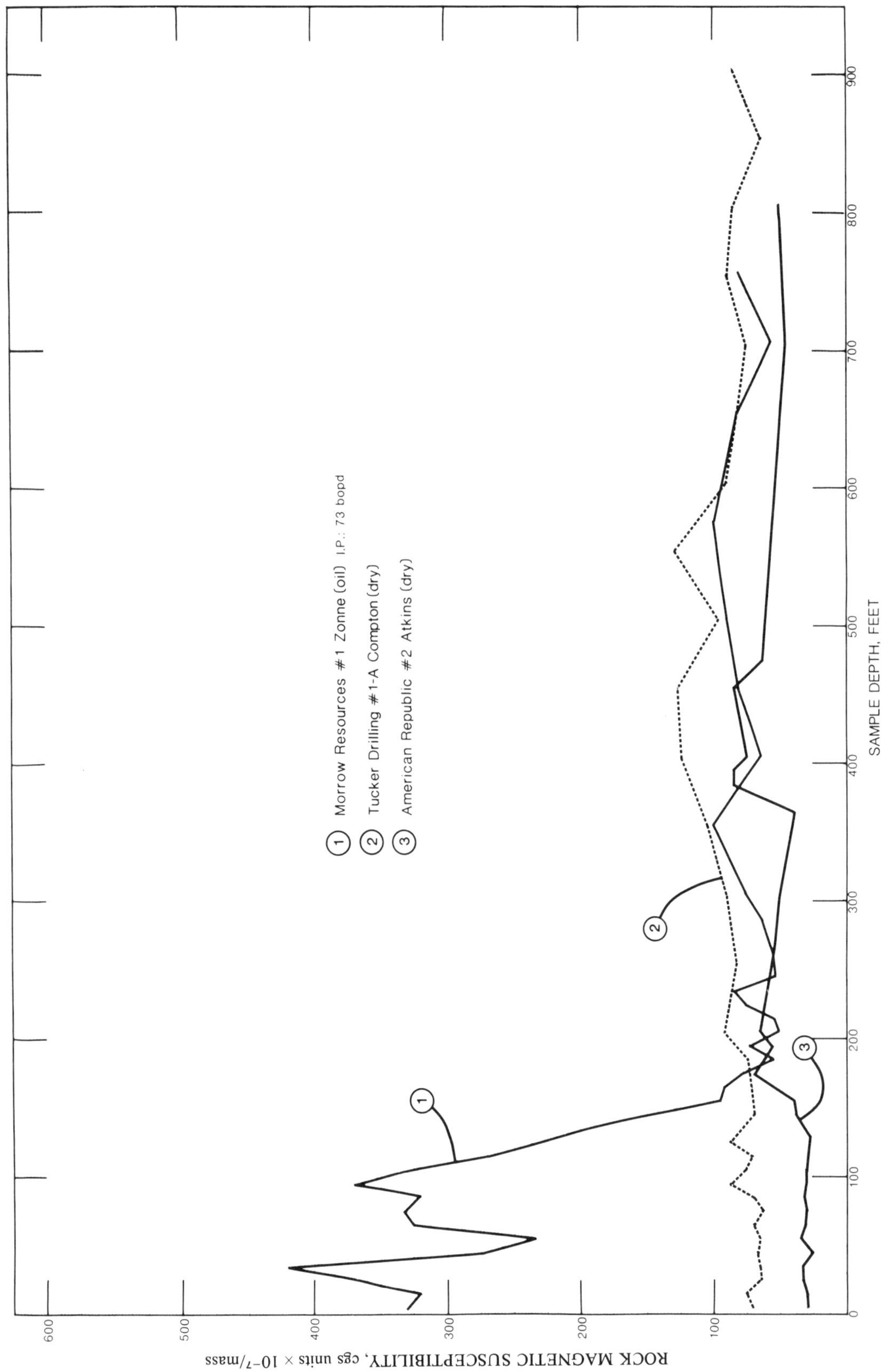

Fig. 7. KWB Oil Field area, Tom Green County, Texas. Drill cuttings rock magnetic susceptibility as function of depth.

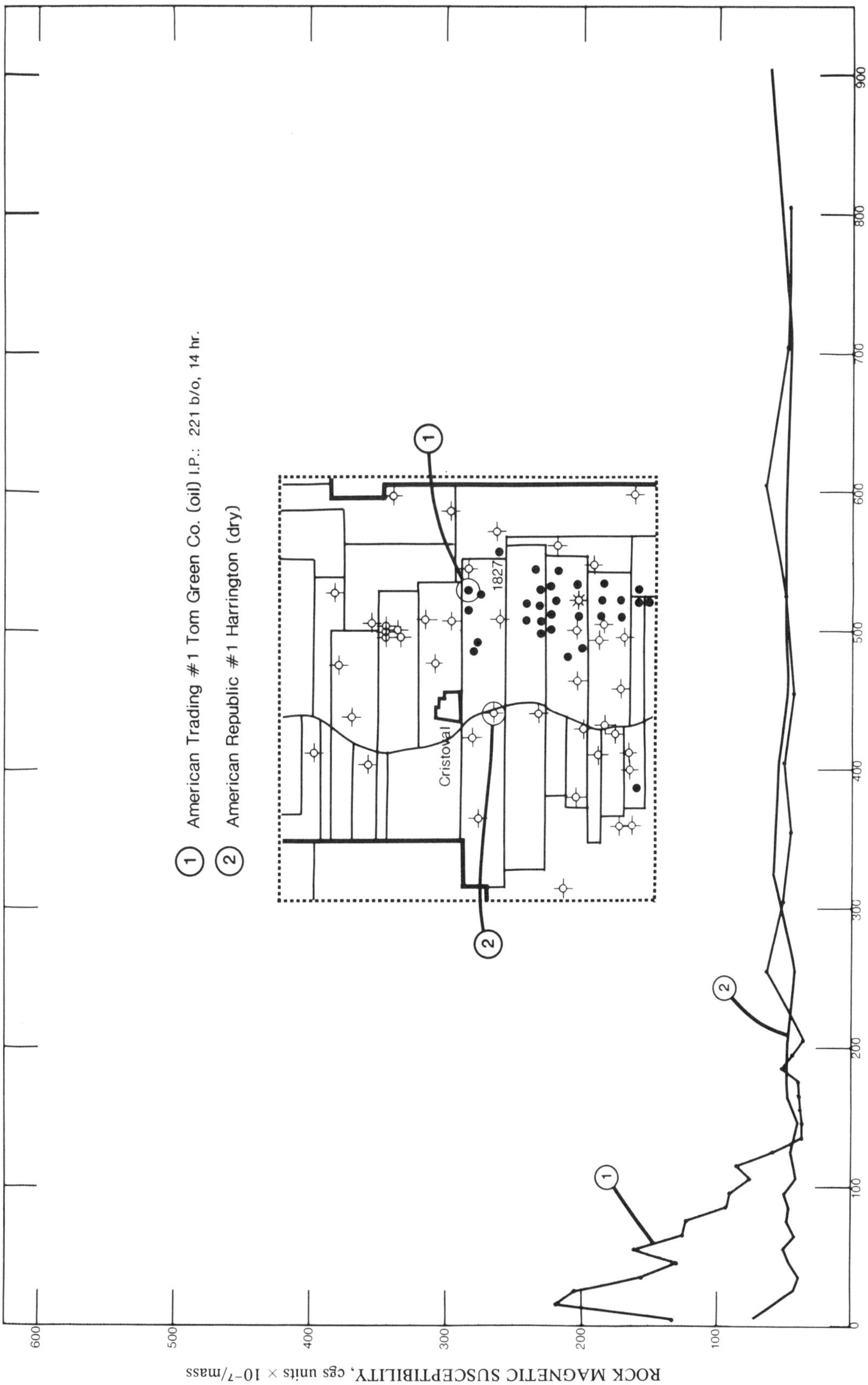

Fig. 8. Mt. Susan Oil Field area, Tom Green County, Texas. Drill cuttings rock magnetic susceptibility as function of depth.

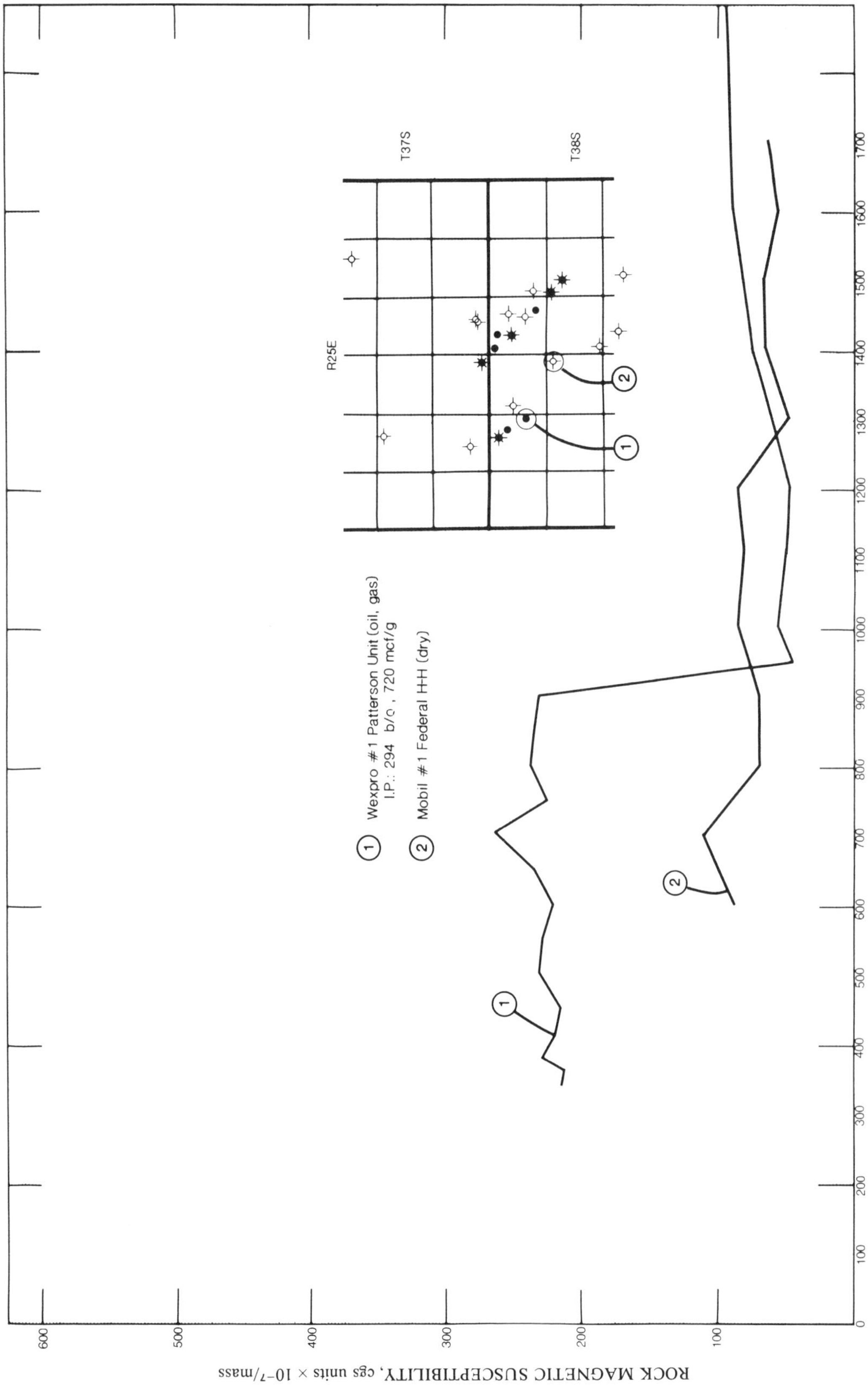

Fig. 9. Patterson Canyon Oil Field area, San Juan County, Utah. Drill cuttings rock magnetic susceptibility as function of depth.

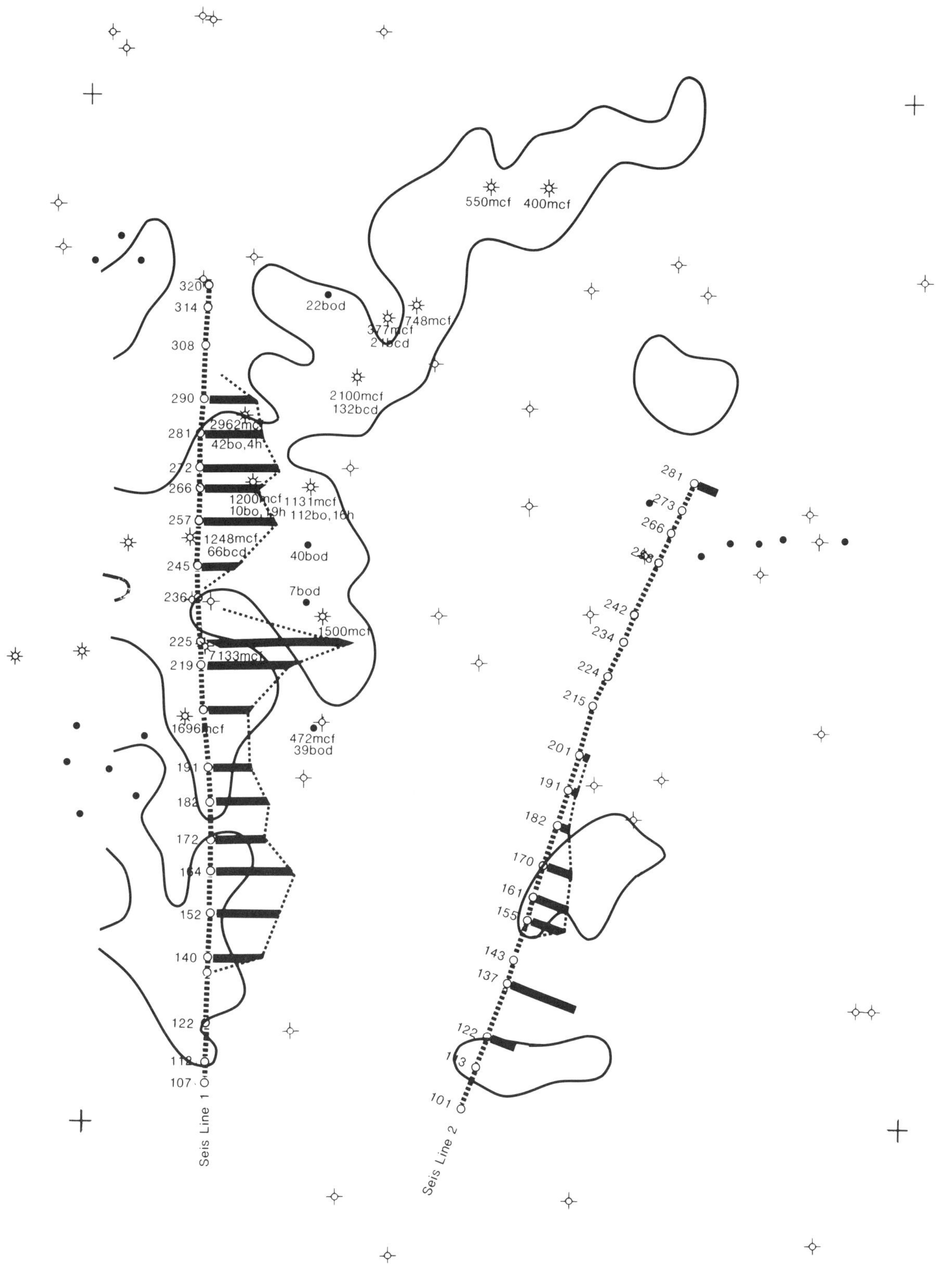

Fig. 10. South Peek Field area, Roger Mills, County, Oklahoma. Seismic shot hole lines with results of HGI analysis.

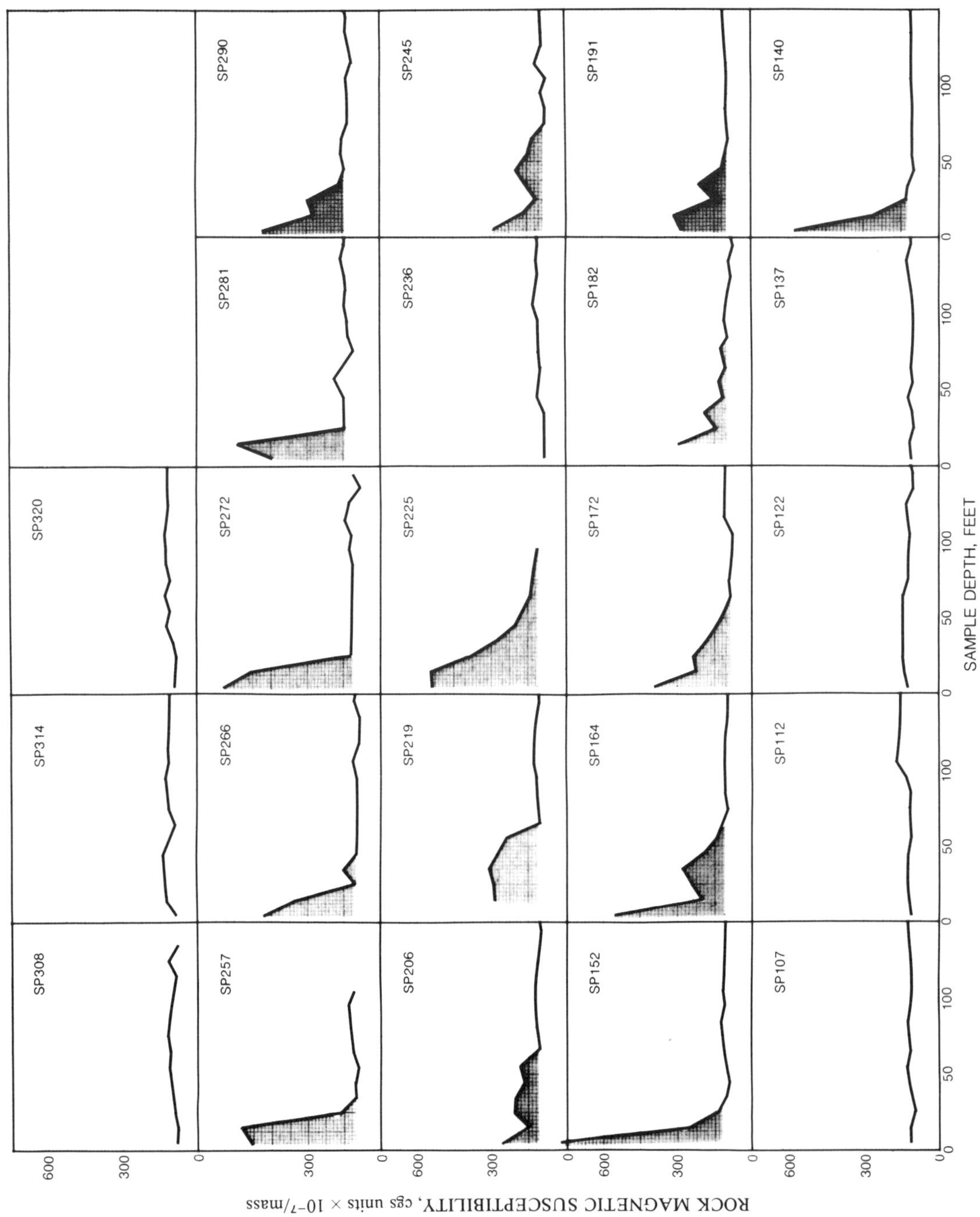

Fig. 11. Rock magnetic susceptibility of 10-ft interval seismic shot hole cuttings samples as function of depth: Line 1.

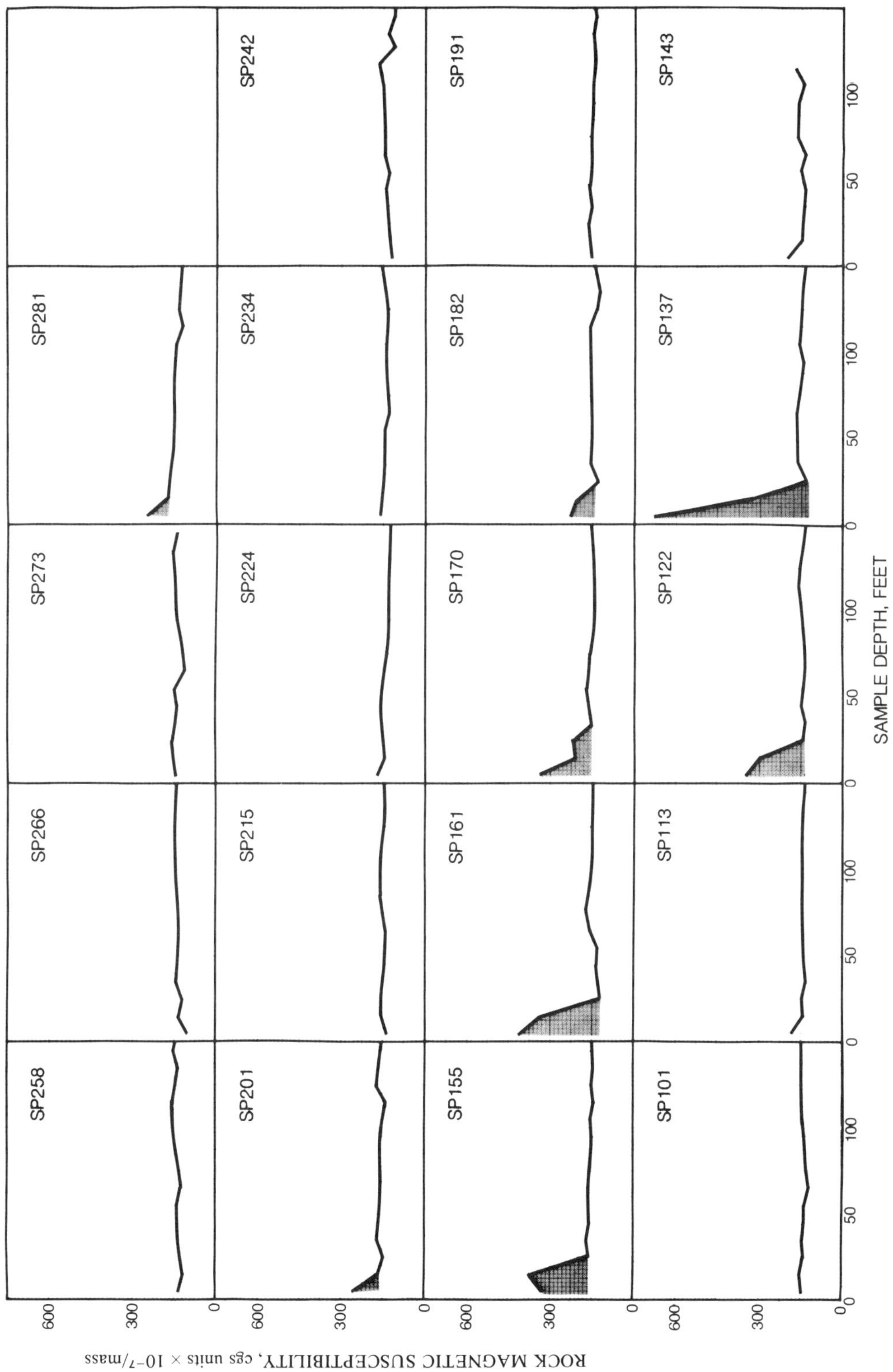

Fig. 12. Rock magnetic susceptibility of 10-ft interval seismic shot hole cuttings samples as function of depth: Line 2.

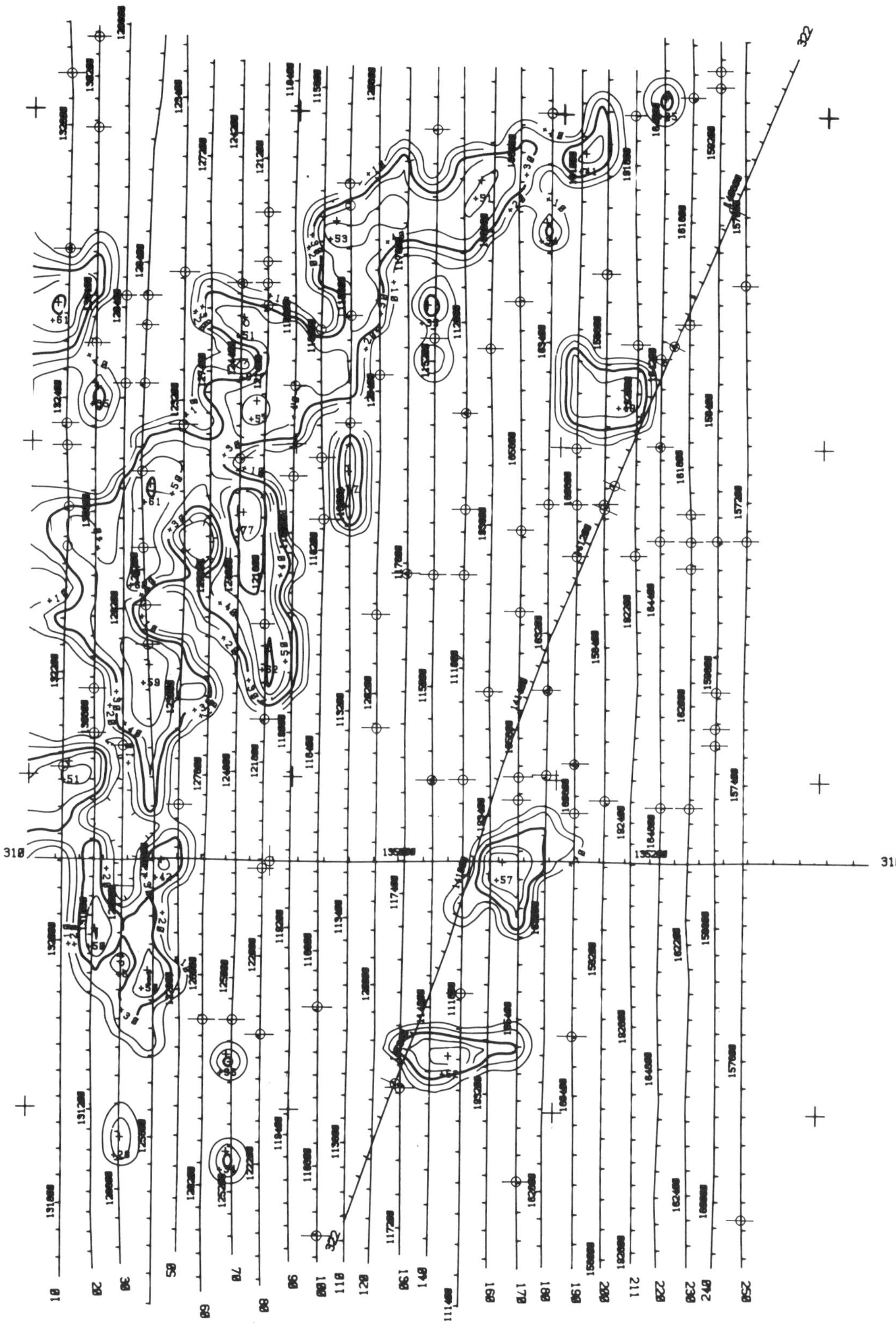

Fig. 13. South Peek Field area, Roger Mills County, Oklahoma. Ce vapor magnetometer horizontal gradient analysis.

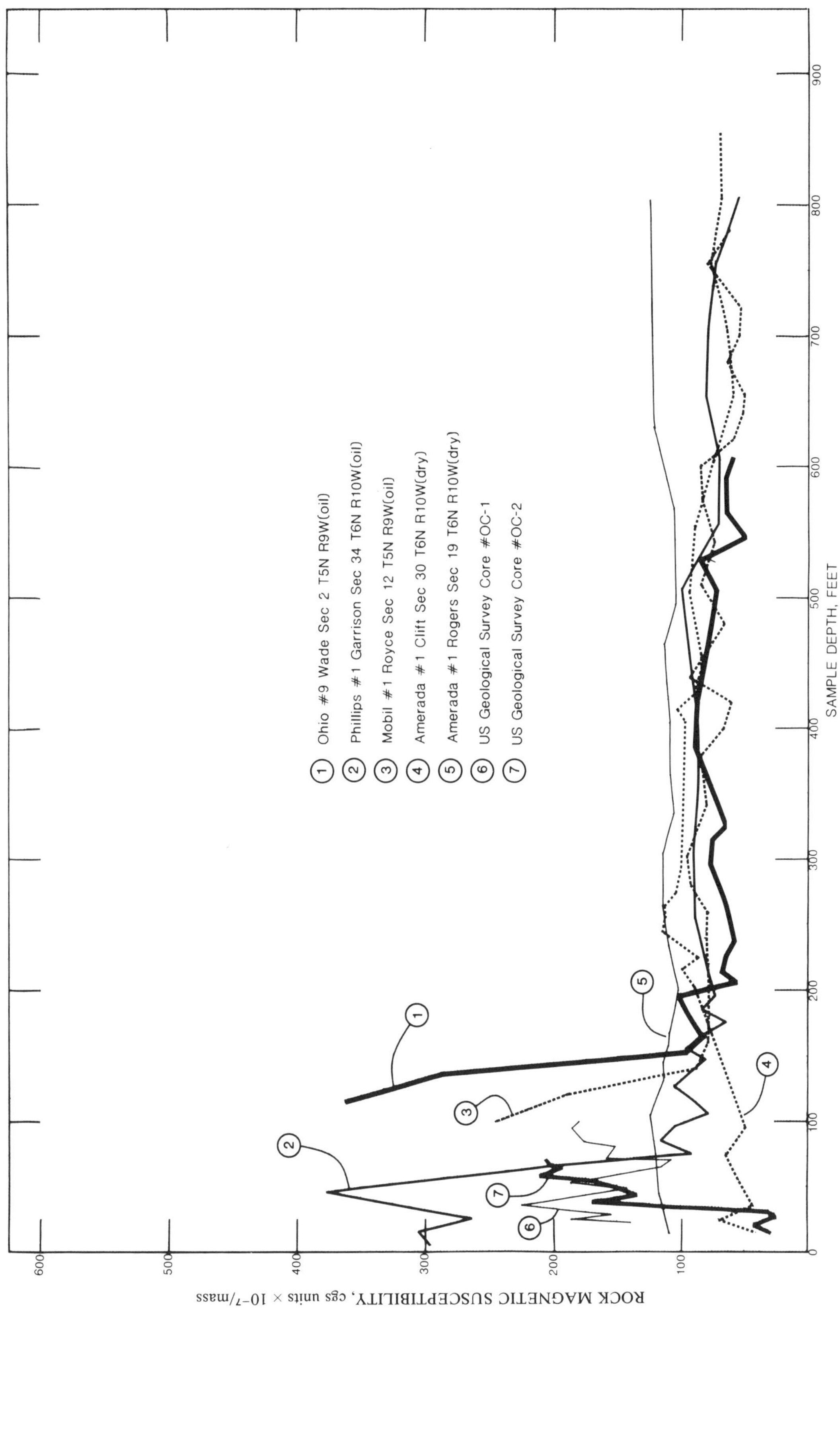

Fig. 14. Cement Oil Field area, Caddo County, Oklahoma. Drill cuttings rock magnetic susceptibility as function of depth.

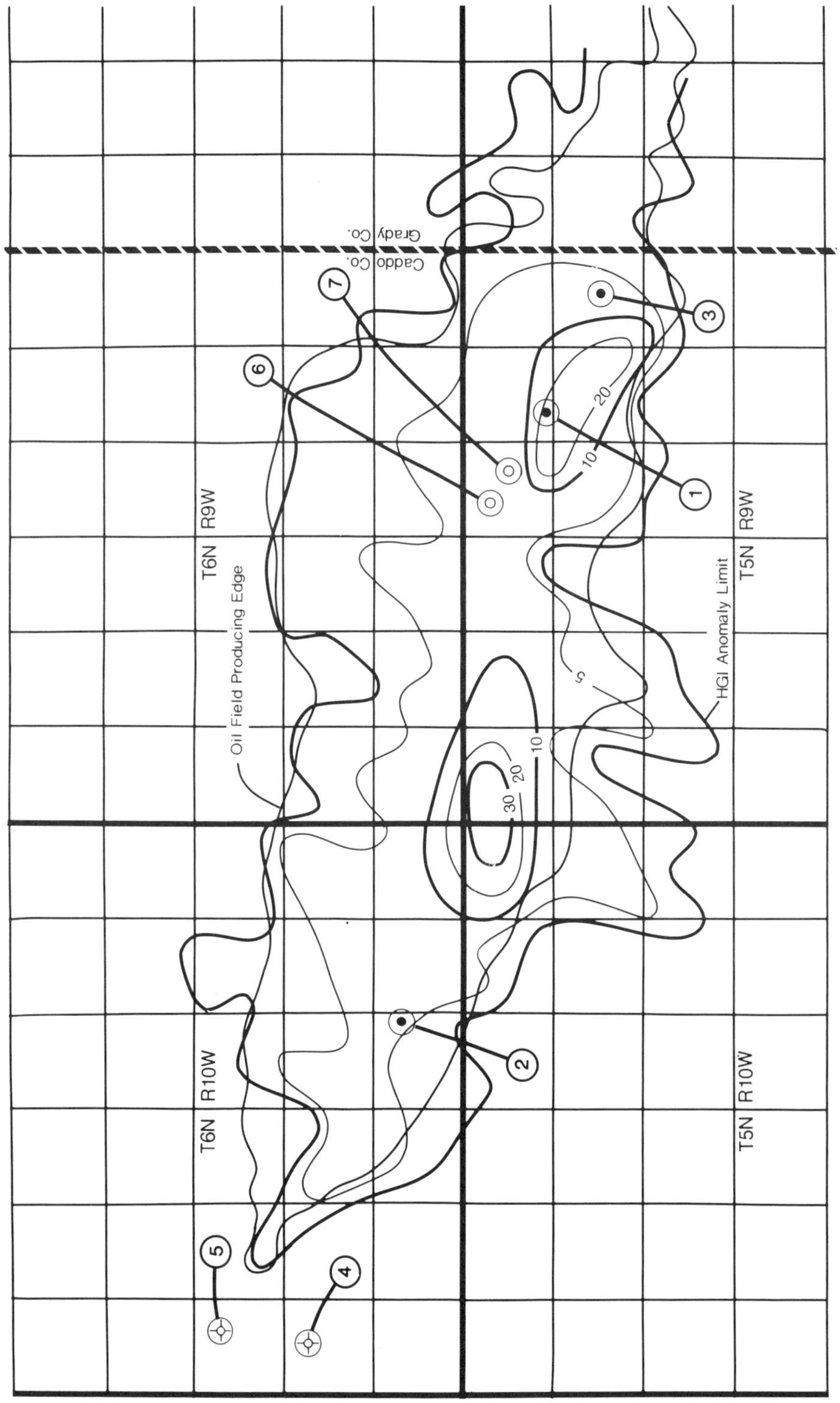

Fig. 15. Cement Oil Field area, Caddo County, Oklahoma. Ce vapor magnetometer horizontal gradient analysis.

A SURFACE VITRINITE REFLECTANCE ANOMALY RELATED TO THE BELL CREEK OIL FIELD, MONTANA, U.S.A.

Charles E. Barker and Mark J. Pawlewicz

U.S. GEOLOGICAL SURVEY

ABSTRACT

Vitrinite reflectance measurements from surface samples of Upper Cretaceous to Paleocene coal and coaly particles in mudrock show anomalously high values over the Bell Creek Oil Field. The mean random vitrinite-reflectance (R_m) increases to a peak of 0.8–0.9% over the field relative to background values of about 0.25%. There is no evidence that the thermal or burial history has varied across the field, and R_m should be constant. The limited topographic relief (less than 300 m; 1,000 ft) over a shallow-dipping homoclinal structure indicates that the anomaly is not due to erosion exposing more mature rocks. Coal clinkers are rare in this area, and altering R_m by widespread heating from burning the thin coal lenses is unlikely. Temperature surveys in wells indicate a thermal anomaly of less than 10° C (18° F) across the field. Paleotemperatures of about 100° C (200° F) would be necessary to produce the observed R_m difference over the field. Thermal maturation evidence, however, indicates no significant temperature anomaly exists above the reservoir because: (1) R_m increases slowly from 0.3% in the shallow subsurface to 0.5% in the oil production zone at 1,400 m (4,500 ft) depth; (2) the Muddy Sandstone which forms the oil reservoirs in the area has a consistent regional R_m of 0.4–0.6%; and (3) the peak burial temperature of the Muddy Sandstone is a relatively low 35–41°C.

Coal oxidation enhanced by petroleum-oxidizing microbes is a plausible explanation for this R_m anomaly. Soil cultures from above the field show significant petroleum-oxidizing microbial activity. Microbial activity affecting the coal is indicated by enrichment of oxygen and carbon content with minor depletion in hydrogen content. Laboratory experiments on lignite innoculated with fungi caused changes in the aromatic structures like that produced by heating. This altered lignite may also have a slightly higher R_m. The limited time involved in producing these changes in the laboratory suggests that over the extended time available in the weathering zone, significant changes in R_m are possible from microbial activity.

Microseepage of petroleum from reservoirs supports colonies of petroleum-oxidizing microbes. These bacteria require certain hydrocarbons supplied by microseepage to establish a population and to reproduce. Once established, petroleum-oxidizing bacteria can utilize other hydrocarbon materials through a cooxidation process. The microbes are concentrated above the reservoir because of vertical microseepage to the surface. The metabolic activities of petroleum-oxidizing microbes apparently cause an increased aromaticity of the vitrinite, and consequently increased reflectance over the field.

A consequence of the large variation of R_m across the Bell Creek area is that rank data from surface samples are not as reliable as previously thought, and they should be interpreted with caution. Our data suggest that source rock evaluation and erosion estimates based on R_m from surface samples may have significant error if taken near areas of hydrocarbon microseepage. The R_m anomaly above this oil field also indicates that surveys of coal rank can be used to explore for petroleum.

INTRODUCTION

Coal and coaly organic matter in rock from surface samples above the Bell Creek Oil Field, Montana, show significant differences in apparent thermal maturity. The sedimentary rocks cropping out over this field apparently have been affected by hydrocarbon microseepage (Dalziel and Donovan, 1984; Roeming and Donovan, 1985), which is also recorded in the coal. Similar lateral variations of rank have been noted in subsurface but not surface studies of thermal maturity. Urban and Harding (1979), in a subsurface study of vitrinite reflectance (mean random value, R_m) over the Lenora Gas Field, found R_m was lower over the field compared to R_m off the field. Grayson and Groth (1978) found anomalous palynomorph maturation in 150 m (500 ft) deep boreholes above the Peoria Oil Field, which was attributed to thermal effects. Thermal maturation indicators may have a variable response to late diagenesis in the subsurface. Our study investigates the effect of hydrocarbon microseepage and weathering on vitrinite reflectance in coal exposed at the surface.

Previous studies indicate that the effect of weathering on exposed coal seams is poorly understood. Chandra (1962) concluded that there is no significant difference in the vitrinite reflectance (mean random value, R_m) of weathered and unweathered coal samples from the same seam. Marchioni (1983) reported that in a bituminous coal seam (R_m = 0.9–1.5%) vitrinite reflectance decreased at the surface by about 0.1–0.2%, but that in a subbituminous coal seam (R_m = 0.5%) vitrinite reflectance increased by 0.05–0.1%. In a similar study, Ingram and Ristidt (1984) found no significant difference in R_m if measurements are taken away from weathered features, and concluded that R_m is a reliable indicator of rank.

Near-surface samples of coal seams from the Bell Creek area show both macro- and microscopic changes due to weathering. The lignite has developed extensive cleat (microfractures) and variation in polishing hardness in vitrinite common in weathered coal (Stach et al., 1982; Marchioni, 1983). Weathering of Bell Creek coal is commonly so intense that samples often disintegrate into a fine powder during handling. Very thin coal seams may be so altered that they take on the appearance of black soils. Although extensive and severe, this pervasive near-surface alteration and weathering generally does not affect the reflectance of vitrinite in areas where there is no apparent hydrocarbon microseepage.

Various surface geochemical exploration methods for detecting the effects of hydrocarbon microseepage have been tested in the Bell Creek area: iron concentration in sage leaves and pine needles (Dalziel and Donovan, 1984), carbon isotopes in carbonate cement from sandstone exposures (M.C. Dalziel, written communication, 1983), ethane-oxidizing microbe activity (Miller, 1976), aeromagnetic detection of diagenetic magnetite (Donovan et al., 1979; Dalziel and Donovan, 1984), and soil chemistry (Roeming and Donovan, 1985). These methods have had some success in defining the production limits of the Bell Creek Oil Field (Philp and Crisp, 1982). As discussed in the following section, the numerous thin coal seams which outcrop over the oil field, the simple structure, low topographic relief, and the shallow stratigraphic trap make the Bell Creek area a low-noise geological environment to test the response of R_m to uniform burial diagenesis but laterally variable late diagenetic conditions.

GEOLOGY OF BELL CREEK OIL FIELD

The Bell Creek Oil Field is a shallow (1,400 m; 4,500 ft) oil and gas reservoir in a northwest-dipping, low-angle homocline on the northeast flank of the Powder River Basin, Montana (Fig. 1). The reservoir is the Cretaceous Muddy Sandstone which is interpreted to be a barrier bar that interfingers updip with lagoonal mudrocks (Berg and Davies, 1968; McGregor and Biggs, 1970). There is no local structural control on the oil accumulation, and the trap is formed by a lateral facies change. The reservoir sand consists of fine-grained arenite that averages 6 m (20 ft) thickness with a porosity up to 33% and permeability up to 13,500 md (Berg and Davies, 1968).

The Upper Cretaceous sediments cropping out in the Bell Creek area consist of gray mudrock, brown carbonaceous shale, coal, and buff friable sandstone of the Hell Creek Formation (Fig. 2). The Tertiary Fort Union Formation conformably overlies Cretaceous rocks and is similar in lithology. The Hell Creek and Fort Union Formations are thought to have been deposited in a fluvial environment (Robinson et al., 1964; Bryson and Bass, 1973).

VITRINITE REFLECTANCE SURVEY OVER THE BELL CREEK OIL FIELD

Surface samples of mudrock and coal were taken over an area 64 km (40 mi) long and 40 km (25 mi) wide surrounding the Bell Creek Oil Field. A major portion of Carter County east of the field was not sampled because a landowner would not allow us access. Over 1,100 samples were taken over five years. Only about 200 of these samples had coal or humic organic matter in rock, exposed at a reasonable depth 0.5 m (1.5 ft), that were amenable for R_m analysis. Coal samples were air dried, ground to minus 48 mesh, and mounted on glass slides for reflectance analysis (Baskin, 1979). Dispersed sedimentary organic matter in mudrock and sandstone samples was concentrated for reflectance analysis (Barker, 1982). Some coal samples rapidly expanded and disintegrated when prepared with water-base polishing compounds (alcohol solved this problem). All reflectance data were measured by M.J. Pawlewicz to minimize the differences of vitrinite selection between operators.

Computer contouring of the R_m data (Fig. 3) indicates an anomaly roughly outlining the production limits of the Bell Creek Oil Field. The anomaly peaks at about 0.8–0.9% over the oil production field. This is three-to-four times the background R_m of about 0.25%. This change is significant because the typical standard deviation for 75 reflectance readings is 0.1% Rm, and the precision of measurement is about +/− 0.05% in the range of 0.2–1.0% R_m. A few isolated, spuriously high R_m values occur across the study area. We do not believe that R_m anomalies outlined by a relatively isolated, single value are significant, and four data points were not used.

PHYSICAL AND CHEMICAL EFFECTS ON VITRINITE REFLECTANCE

Thermal

R_m responds primarily to increasing burial temperature, and thus can be correlated to its maximum burial temperature (Tissot and Welte, 1984). Because of these factors, R_m as a function of depth is proportional to the maximum geothermal gradient attained during burial. R_m measured on disseminated coaly organic-matter from cores of the Muddy Sandstone from the Bell Creek reservoir and non-petroleum-bearing areas (Fig. 4) increases from the surface background reflectance of about 0.25% to about 0.5 ± 0.1% at 1,500 m (5,000 ft) in the reservoir (Figs. 5 and 6). The similar R_m in the Muddy Sandstone across the area suggests it has experienced a uniform thermal condition over its burial history.

Thermal anomalies in the Powder River Basin with the potential to produce local increases of R_m are not found close to the Bell Creek area. Daniel and Cole (1983) report a R_m high in subsurface samples related to an igneous intrusion into the Mississippian Heath Formation of central Montana. Knirsch (1980) found that the geothermal gradient increased from a background of about 25° C/km to 45° C/km in a few wells penetrating the Madison Limestone in the Edgemont area, Black Hills, South Dakota. The heat source in this case was concluded to be the local effect of a thermally conductive Precambrian granite. These types of thermal anomalies are not considered to be important at Bell Creek because the R_m data here do not vary significantly in the subsurface and there is no evidence for local igneous intrusions.

There is a close relationship between increased temperature and petroleum reservoirs (Roberts, 1979). Warm ascending fluid can heat the surrounding sediment, causing a temperature anomaly of a few degrees Celsius in the vicinity of the thermal plume. Thermal plumes have been shown to increase R_m along with quartz dissolution, potassic feldspar alteration (to sericite), and calcite cementation in the sedimentary rocks (Cassan et al., 1981). These extensive diagenetic changes have not been observed at Bell Creek Oil Field (Almon and Davies, 1979). Further, the observed change of R_m from a background of about 0.25% to 0.8–0.9% over the field would require a temperature anomaly on the order of 100° C (200° F) (Barker and Pawlewicz, in press). Subsurface thermal anomalies over Bell Creek and other oil fields in the Rocky Mountain area (Meyer and McGee, 1985) are less than 10° C (18° F). Both higher temperature and reflectance would be expected at depth if the Bell Creek reservoir was related to ascending warm fluids. The surface R_m anomaly (up to a peak of 0.9%) above the field, and 0.5% at the reservoir level, indicates a reverse maturation gradient (and geothermal gradient) if temperature alone determined R_m above Bell Creek. The Muddy Sandstone in this area is now at maximum burial and presumably at peak temperature (Burrows et al., 1984). Temperature in the Bell Creek reservoir is now at 35–41° C (95–105° F) (Winters and Williams, 1969). This present day temperature regime has produced only slightly increased temperatures over the Bell Creek Oil Field and is inadequate to explain the R_m anomaly. The shallow subsurface rocks below the field have a much lower R_m (about 0.3%; Fig. 6) than the surface coals, indicating that the R_m anomaly at Bell Creek is related to the current erosional surface and not to differences in thermal history.

Burial Diagenesis

Field and laboratory studies have shown that in petroleum-rich rocks, R_m is lower than expected when compared to R_m in adjacent petroleum-poor rocks (McFarland, 1982; Thompson, 1982). This suppression effect would cause relatively low R_m over a production field like that reported by Urban and Harding (1979). Hydrocarbon microseepage in the surface environment above the Bell Creek reservoir is apparently related to an increase in R_m, and there appears to be no supression effect.

Hutton and Cook (1980), Newman and Newman (1982), Price and Barker (1985), and others have shown that variation of maceral composition in coal and sedimentary organic matter can influence R_m. These studies indicate that certain depositional and(or) diagenetic environments can produce a hydrogen-rich coal or sedimentary organic matter that has lower R_m values than those expected from the thermal history of the sediment. However, the depositional system was apparently uniform across the Bell Creek area (Bryson and Bass, 1973;

Robinson et al., 1964), and we expect little variation in the source and composition of the coal or sedimentary organic matter.

Late Diagenesis

The most intense near-surface effects that can increase vitrinite reflectance result from heating by brush and coal fires. Brush fires heat or burn the coal, increasing its reflectance. The short duration of brush fires and the poor thermal conductivity of soil suggest heating from this source would rapidly decline with depth. Most samples were taken at a depth of about 0.5 m (1.5 ft), and we assume brush fires had little influence on our R_m data. The burning of coal seams could cause more widespread heating, but clinker is scarce in this area and apparently coal fires were not common. Bauer (1972) studied coal clinker to the south of Bell Creek where the affected rocks are distinguished by a reddish, often pseudo-igneous appearance, and form obvious erosion-resistant bluffs. Rocks with these characteristics are rare in the Bell Creek area. The coal seams are also high-ash, thin-lenticular beds that would be difficult to ignite (Bauer, 1972), and their burning would not affect more than a small area. The effect of intense heating is also detectable under the microscope during reflectance analysis. Burning or heating with oxygen present forms highly reflectant rims on the coal grains [if temperature exceeds 150° C (300° F)] that may attain an R_m of 10% or more in some cases (Stach et al., 1982). Coal alteration by intense heating or burning in the near-surface zone is detectable during both sampling and analysis. We feel that these processes have not influenced our R_m data because any affected coal was simply not measured.

Structure

Deformation of a sedimentary sequence after a reflectance gradient has been established in the sediment could give surface vitrinite reflectance anomalies after erosion. Erosion of a simple homoclinal structure like that at Bell Creek would not produce the observed R_m anomaly (Fig. 7). Erosion and dissection of flat or shallowly dipping beds could also produce surface vitrinite reflectance anomalies by the exposure of more mature rocks in valleys. This type of surface vitrinite reflectance anomaly would be centered over topographic lows and in general controlled by surface form and sample locality elevation. Surface R_m values do not appear to be controlled by sample locality elevation as shown by the near-random correlation ($r = -0.2$) between these two variables (Fig. 8).

In summary, the thermal, diagenetic, and physical conditions are not responsible for the surface R_m anomaly related to the Bell Creek Oil Field. We suggest here that petroleum-oxidizing microbes are capable of degrading coal and increasing R_m in the near-surface zone where hydrocarbon microseepage is occurring.

MICROBIAL ALTERATION OF PETROLEUM

The decomposition of hydrocarbon by microbes is the most important process in oil degradation (Zobell, 1964). Saturated aliphatic hydrocarbons undergo three types of reactions in soil and in cultures of microorganisms: oxidation of the terminal carbon atom, oxidation of the C_2 carbon atom, and dehydrogenation (Ross, 1977). Biodegradation also involves the removal of the light hydrocarbons. Microorganisms such as *Pseudomonas methanica* can oxidize ethane, propane, butane, and other short-chain alkanes during concomitant growth on methane (McKenna and Kallio, 1965). There also appears to be a specificity characteristic of bacteria, where a one-atom change in some *n*-alkanes can be the difference between total or *no* bacterial degradation. Of the common classes of hydrocarbons, straight-chain *n*-alkanes are most easily degraded by microbes, though the rate of decomposition is usually decreased drastically by branching on the *n*-alkane chain. Connan (1984) concluded that the biodegradation of alkanes covers a range of molecular weights from C_1 to $C_{27\text{-}35}$ and that the process starts with both alkane and aromatic fractions. In laboratory tests, van der Linden and Thijsse (1965) have shown that a microorganism grown on a particular hydrocarbon will attack related hydrocarbons of a more complicated structure but cannot oxidize them completely.

These studies demonstrate the ability of bacteria to oxidize a wide range of hydrocarbons and, in some cases, an extreme selectivity in the hydrocarbon species. The most important features of bacterial hydrocarbon degradation relative to coal alteration are: (1) there is an adaptibility to different hydrocarbon food sources; (2) there is a preference for light (less than C_{18}) straight-chain *n*-alkanes; (3) the lightest hydrocarbons may be completely oxidized; (4) *n*-alkanes tend to be oxidized before aromatics; (5) straight chain alkanes are strongly preferred over branched alkanes; and (6) side chains can be removed from condensed aromatic molecules. These studies clearly indicate the potential for petroleum-oxidizing bacteria to alter coal chemically to a compound composed of aromatic nuclei with bridging aliphatic side chains (Tissot and Welte, 1984). The bulk-chemical signature of petroleum degradation by microbes is oxygen and carbon fixation with a minor change in hydrogen content (Zobell, 1964).

The surface control of hydrocarbon-oxidizing microbes, related to the Bell Creek production field, (Fig. 9), apparently arises from the rapid decrease in hydrocarbon microseepage away from the reservoir (Philp and Crisp, 1982). Hydrocarbon-oxidizing bacteria are abundant in the near-surface zone above petroleum reservoirs (Soli, 1957; Sealey, 1974). These organisms are capable of oxidizing a wide range of organic material (Zobell, 1946). According to Schwendinger (1968), many microbes which normally utilize other kinds of organic compounds can quite rapidly adapt their metabolism to hydrocarbons. In an environment of hydrocarbon microseepage, certain types of bacteria are apparently able to adapt toward petroleum utilization (Soli, 1957). After microbes with these characteristics are localized above an oil field by the hydrocarbon microseepage, they could degrade coal by cooxidation (L.R. Brown, Mississippi State University, personal communication; McKenna and Kallio, 1965; Raymond et al., 1967). Cooxidation involves a population of microbes that need certain hydrocarbon(s) to establish a population and reproduce. Once established, they are able to oxidize other compounds.

MICROBIAL ALTERATION OF COAL

Petroleum-oxidizing microbes possess a unique class of enzymes known as oxygenases, which carry out the initial step in coal transformation by incorporating atmospheric oxygen into the molecule (McKenna and Heath, 1976). Fungal breakdown of the molecules involves extracellular enzymes, whereas bacteria utilize only products leached from coal, or they excrete free extracellular proteins which adsorb onto the coal particles and may have some catalytic activity (Fakoussa, 1984). A possible mechanism could involve enzymes cleaving straight-chain and branched hydrocarbons attached to the aromatic clusters that compose the vitrinite molecule. The loss of these side-chains from the coal would increase the aromaticity of the vitrinite, and consequently its reflectance.

The microbial degradation of coal has been demonstrated in the laboratory. Cohen and Gabriele (1982) were able to reduce lignite to a black liquid by innoculating it with fungi. The methane-oxidizing bacteria *Methylomonas methanica*, when injected into a coal bed, showed an intensification of activity (Abramov et al., 1977). Similarly, Bazarova et al. (1980) found that humic acids from lignite increased the growth rate of *Pseudomonas putida* bacteria. Kuznetsov et al. (1963) state that investigations of the formation and occurrence of lignite deposits indicate no obstacles to the activity of microbes in decomposing and humifying plant remains. Microbes appear to be active in lignite and bituminous coals but have only a minor role in higher rank coal (Zobell, 1964).

The effect of microbial coal degradation on R_m was experimentally tested using the technology of Cohen and Gabriele (1982). Lignite was innoculated by fungi and held under incubation for about one week (samples prepared by M.S. Cohen). Reflected light microscopy indicated weathering in both the innoculated and control sample, with the coal particles having rounded edges and polishing relief due to variable hardness. R_m of the innoculated sample was 0.36%. The uninnoculated control sample had a R_m of 0.35%. The small change in R_m may be significant considering that it occurred in less than one week. Infrared spectroscopy indicated significant modification of conjugated aromatic clusters similar to that found in lignite after heating (Cohen and Gabriele, 1982). These experiments establish that microbial alteration of lignite can produce effects in the sample similar to thermal maturation.

GEOCHEMICAL EVIDENCE FOR MICROBIAL ALTERATION OF BELL CREEK COAL

Oxygen and carbon fixation in the Bell Creek coal, like that observed in biodegraded oils, is indicated from Rock-Eval pyrolysis. Increased oxygen content of the coal (Fig. 10) roughly corresponds with the R_m anomaly (Fig. 3), and microbial activity (Fig. 9) over the Bell Creek Field. The coals in this same area are also enriched in carbon (Fig. 11), and there is minor decrease in hydrogen (Fig. 12). A greater carbon content in a coal normally would be interpreted as a rank increase over the field. It is expected, however, that oxygen and hydrogen would both decrease with maturation if the lignite was following a normal evolution path with increasing rank (Durand and Monin, 1980). Further, gas chromatography analysis of Bell Creek lignite and shale indicates they are thermally immature (Fig. 13), as shown by their strong odd/even preference and the peak *n*-alkane signal near C_{30} even with an increase in R_m from 0.3% to 0.8% (Fig. 14). The gas chromatography patterns for the Bell Creek coal are anomalous in that they show little or no extractable *n*-alkanes in the range of C_{10} to C_{20} with increasing R_m (Fig. 13). Hollerbach and Hagemann (1981) show that with increasing rank there is a decrease in the alkane odd/even preference, a shift in the peak *n*-alkane to shorter chains, and a general flattening of the chromatogram (Fig. 14). Thus, the increased carbon content in coal over the field is not due to thermal maturation, as suggested by R_m increase.

The change of C-H-O content in coal above the Bell Creek Oil Field cannot be attributed to simple weathering. The chemical composition of weathered coal when compared to unaltered equivalents shows a moderate-to-strong decrease in carbon, a moderate loss of hydrogen, and a strong increase in oxygen (Stach et al., 1982; Durand and Monin, 1980; Marchioni, 1983). Simple weathering of coal in the Fort Union Formation in Wyoming follows this pattern. Hatch and Affolter (1981) report a strong increase in oxygen content with moderate depletion of carbon and moderate hydrogen loss. Further, the results of Marchioni (1983) suggest that weathered coal shows 10%–20% relative change in R_m, not the 300%–400% increase observed above Bell Creek.

DISCUSSION

We assert that microbial activity above Bell Creek Oil Field alters the surface coal and that the consequent change in R_m is significant. The generally accepted causes of rank increase do not explain the increase in R_m above the field to three-to-four times greater than background R_m. These coals do not show the chemical changes due to weathering observed in areas apparently not affected by hydrocarbon microseepage. Rather they show oxygen and carbon enrichment with minor hydrogen depletion consistent with microbial degradation of the coal.

Regardless of the cause, the R_m anomaly related to an oil field indicates that surveys of coal rank can be used to explore for petroleum. A surface coal rank survey over the Powder River Basin is now being conducted to test this hyopothesis. However, this type of survey may only be useful in areas where low rank coal is exposed at the surface. High rank coals show decreased or no microbial activity compared to that in lignite (Zobell, 1964).

Another important consequence of the large variation of R_m across the Bell Creek area is that rank data from surface samples are not as reliable as previously thought, and they should be interpreted with caution. R_m is often used to estimate source rock maturity and depth of erosion. R_m is considered to be the most stable indicator of thermal maturity in surface samples. This is not true of samples collected over the Bell Creek Oil Field. The surface R_m values observed in the Bell Creek area would normally be interpreted as thermally immature (less than 0.6% R_m) off the field and at near-peak oil generation (0.9% R_m) over the field—across a surface that has experienced similar maximum temperature and burial history. Dow (1977) indicated that the intersection of a R_m profile with the surface in uneroded sections occurs at about 0.2% R_m. R_m varies from about 0.25% to 0.8–0.9% across essentially flat-lying beds that must have undergone nearly the same burial and subsequent erosion. The rank profile from borehole Gary Oil Newell 27–21 (Fig. 6) suggests essentially no erosion at the observed background of 0.22% R_m surface intercept to several kilometers of erosion indicated by the peak surface intercept of 0.9% R_m.

CONCLUSIONS

The data presented in this paper support the following conclusions:

1. Vitrinite reflectance shows significant variability across the Bell Creek area. The R_m anomaly related to the Bell Creek Field is three-to-four times higher than the background value of about 0.25%. This variation is apparently related to differences in near-surface, late-diagenetic conditions producing enhanced oxidation of coal in an environment of hydrocarbon microseepage.
2. Coal oxidation enhanced by petroleum-oxidizing microbes is the most plausible explanation for this R_m anomaly. Soil cultures taken from above the field show significant petroleum-oxidizing microbe activity. Microbial activity is indicated by enrichment of oxygen and carbon in the coal. Vitrinite reflectance of fungi-innoculated lignite may show an increase in R_m after less than one week of incubation.
3. The large range of R_m across the Bell Creek area indicates that maturity data from surface samples should be interpreted with caution. The surface R_m values observed at Bell Creek would normally be interpreted as thermally immature (less than 0.6% R_m) off the field and at near-peak oil generation (about 1.0% R_m) over the field—across a surface that has experienced similar maximum temperature and burial history. Depth of erosion estimated from R_m may be grossly in error.
4. Surface surveys of R_m can be useful in exploring for subtle hydrocarbon reservoirs. The surface R_m anomaly over the Bell Creek reservoir is related to the producing field.

ACKNOWLEDGMENTS

M. C. Dalziel collected the initial samples for this study and without her help the study would not have continued. We would like to thank these geologists assisting in the sampling program: W. Page, S. Roeming, B. Crysdale, C. Sanderson, and J. Murphy. James Hartshorne, Gary Oil, Inc., arranged for us to stay in Belle Creek. Sample new wells being drilled in the field with Jon Schwarz, Gary Oil, Inc., and L. R. Brown, Mississippi State University, provided background information on bacterial degradation, and soil samples innoculated with microbes. M. S. Cohen, University of Hartford, provided a sample of lignite altered by microbes.

REFERENCES

Abramov, F. O., Malashenko, Y. R., Myakenkyi, V. I., Kurdysh, I. K., Shevelyev, H. A., Lytvynov, P. S., and Voloshyn, M. V., 1977, Vital activity of methane oxidizing bacteria after their filtration through coal pore structures: Mikrobiol Zh, v. 39, p. 290–293.

Almon, W. R., and Davies, D. K., 1979, Regional diagenetic trends in the Lower Cretaceous Muddy sandstone, Powder River Basin: Society of Economic Paleontologists and Mineralogists, Special Publication, v. 26, p. 379–400.

Barker, C. E., 1982, A rapid method for concentrating sedimentary organic matter for vitrinite reflectance analysis: Journal of Sedimentary Petrology, v. 52, p. 663–664.

———, and Pawlewicz, M. J., in press, The correlation of vitrinite reflectance with maximum temperature in humic organic matter: *in* Buntebarth, G., and Stegena, L., eds., Paleogeothermics, Berlin, Springer-Verlag.

Baskin, D. K., 1979, A method of preparing phytoclasts for vitrinite reflectance analysis: Journal of Sedimentary Petrology: v. 49, p. 633–635.

Bauer, L. P., 1972, The alteration of sedimentary rocks due to burning coal in the Powder River Basin, Campbell County, Wyoming: Unpublished masters thesis, Texas Technical University, 65 p.

Bazarova, O. V., Kucher, R. V., Dzumedzei, N. V., and Petrik, G. K., 1980, Effect of humic acids from brown coal of the Aleksandriiskoe deposit Ukrainian-SSR USSR on the growth of Pseudomonas-putida: Mikrobiol Zh (Kiev), v. 42, p. 155–157.

Berg, R. R., and Davies, D. K., 1968, Origin of Lower Cretaceous Muddy Sandstone at Bell Creek Field, Montana: American Association of Petroleum Geologists Bulletin, v. 52, p. 1888–1898.

Bryson, R. P., and Bass, N. W., 1973, Geology of Moorehead coal field, Powder River, Big Horn, and Rosebud counties, Montana: U.S. Geological Survey Bulletin 1338, 116 p.

Burrows, S. M., Savin, S. M., and Aronson, J., 1984, Oxygen isotope evidence for the conditions of diagenesis of the Muddy Sandstone, east flank of the Powder River Basin, Montana and Wyoming: Geological Society of America, abstracts and programs, v. 16, p. 460.

Cassan, J. P., Garcia, M. C., Fritz, B., and Tardy, Y., 1981, Diagenesis of sandstone reservoir as shown by petrographical and geochemical analysis of oil bearing formations in the Gabon basin: Bulletin Centres Recherces Exploration-Production Elf Aquitaine, v. 5, p. 113–135.

Chandra, D., 1962, Reflectance and microstructure of weathered coals: Fuel, v. 41, p. 185–193.

Cohen, M. S., and Gabriele, P. D., 1982, Degradation of coal by the Fungi Polyporus versicolor and Poria monticola: Applied and Environmental Microbiology, v. 44, p. 23–27.

Connan, J., 1984, Biodegradation of crude oils in reservoirs: *in* Brooks, J., and Welte, D., eds., Advances in Petroleum Geochemistry, v. 1, p. 299–335.

Dalziel, M. C., and Donovan, T. J., 1984, Correlations of suspected petroleum-generated biogeochemical and aeromagnetic anomalies, Bell Creek Oil Field, Montana: *in* Davidson, M. J., and Gottlieb, B. M., eds., Unconventional Methods in Exploration for Petroleum and Natural Gas III: Dallas, Southern Methodist University Press, p. 59–69.

Daniel, J. A., and Cole, G. A., 1983, Vitrinite reflectance of coals from the Heath Formation, Central Montana (abstract): American Association of Petroleum Geologists Bulletin, v. 67, p. 1333.

Donovan, T. J., Forgey, R. L., and Roberts, A. A., 1979, Aeromagnetic detection of diagenetic magnetite over oil fields: American Association of Petroleum Geologists Bulletin, v. 63, p. 245–248.

Dow, W. G., 1977, Kerogen studies and geological interpretations: Journal of Geochemical Exploration, v. 7, p. 79–99.

Durand, B., and Monin, J. C., 1980, Elemental analysis of kerogens: *in* Durand, B., ed., Kerogen: Paris, Editions Technip, p. 113–142.

Fakoussa, R. M., 1984, Partial degradation of coal by microorganisms (abstract): Systematic and Applied Microbiology, v. 5, p. 256.

Grayson, J. F., and Groth, P. K. H., 1978, Mineral prospecting by organic diagenesis: United States Patent no. 4,093,420. 8 p.

Hatch, J. R., and Affolter, R. H., 1981, Weathering-induced chemical changes in surface coals from Paleocene Fort Union Formation, Red Rim area, Sweetwater and Carbon Counties, Wyoming (abstract): American Association of Petroleum Geologists Bulletin, v. 65, p. 936.

Hollerbach, A., and Hagemann, H. W., 1981, Organic geochemical and petrological investigation into a series of coals with increasing rank: International conference on coal science, Proceedings, Dusseldorf, 1981. Essen: Gluckauf GmbH, 1981, p. 80–85.

Hutton, A. C., and Cook, A. C., 1980, influence of alginite on the reflectance of vitrinite from Joadja, NSW, and some other coals and oil shales containing alginite: Fuel, v. 59, p. 711–714.

Ingram, G. R., and Rimstidt, J. D., 1984, Natural weathering of coal: Fuel, v. 63, p. 292–296.

Knirsch, K. F., 1980, Investigations of possible sources of heat in geothermal waters at Edgemont, South Dakota: Geological Society of America, abstracts and programs, v. 12, no. 6, p. 277.

Kuznetsov, S. I., Ivanov, M. V., Lyalikova, N. N., 1963, Introduction to Geological Microbiology: New York, McGraw-Hill, 252 p.

Marchioni, D. L., 1983, The detection of weathering in coal by petrographic, rheologic and chemical methods: International Journal of Coal Geology, v. 2, p. 231–259.

McFarland, D. R., 1982, Organic-organic and organic-mineral interactions in coal and kerogen: Unpublished masters thesis, Colorado School of Mines, 300 p.

McGregor, A. A., and Biggs, C. A., 1970, Bell Creek field, Montana: A rich stratigraphic trap: American Association of Petroleum Geologists Memoir 14, p. 128–146.

McKenna, E. J., and Kallio, R. E., 1965, The biology of hydrocarbons: Annual Reviews of Microbiology: v. 19, p. 183–208.

———, and Heath, R. D., 1976, Biodegradation of polynuclear aromatic hydrocarbons by soil and water microorganisms: University of Illinois, Urbana, Water Resources Center, research report 113, p. 1–25.

Meyer, H. J., and McGee, H. W., 1985, Oil and gas fields accompanied by geothermal anomalies in the Rocky Mountain region: American Association of Petroleum Geologists Bulletin, v. 69, p. 933–945.

Miller, G. H., 1976, Microbial surveys help evaluate geological, geophysical prospects: Oil and Gas Journal, v. 74, p. 192–202.

Newman, J., and Newman, N. A., 1982, Reflectance anomalies in Pike River coals: Evidence of variability in vitrinite type, with implications for maturation studies and "Suggate Rank": New Zealand Journal of Geology and Geophysics, v. 25, p. 233–243.

Philp, R. P., and Crisp, P. T., 1982, Surface geochemical methods used for oil and gas prospecting—a review: Journal of Geochemical Exploration, v. 17, p. 1–34.

Price, L. C., and Barker, C. E., 1985, Suppression of vitrinite reflectance in amorphous rich kerogen—A major unrecognized problem: Journal of Petroleum Geology, v. 8, p. 59–84.

Raymond, R. L., Jamison, V. W., and Hudson, J. O., 1967, Microbial hydrocarbon co-oxidation: Applied Microbiology, v. 15, p. 857–865.

Roberts, W. H., III, 1981, Some uses of temperature data in petroleum exploration: *in* Gottlieb, B. M., ed., Unconventional Methods in Exploration for Petroleum and Natural Gas II: Dallas, Southern Methodist University Press, p. 8–49.

Robinson, C. S., Mapel, M. J., and Bergendahl, M. H., 1964, Stratigraphy and structure of the northern and western flanks of the Black Hills uplift, Wyoming, Montana, and South Dakota: U.S. Geological Survey Professional Paper 404, 134 p.

Roeming, S. S., and Donovan, T. J., 1985, Correlations among hydrocarbon microseepage, soil chemistry, and uptake of micronutrients by plants, Bell Creek Oil Field, Montana: Journal of Geochemical Exploration, v. 23, p. 139–162.

Ross, C. P., Andrews, D. A., and Witkind, I. R., compilers, 1958, Geologic map of Montana: Montana Bureau of Mines and Geology, scale 1:500,000, 2 sheets.

Ross, R. H., 1977, Environmental interactions: *in* Braunstein, H. M., Copenhaver, E. D., and Pfuderer, H. A., eds., Environmental, Health, and Control Aspects of Coal Conversion: An Information Overview: Oak Ridge National Laboratory Publication ORNL/EIS-95, v. 2, p. 6-1 to 6-31.

Schwendinger, R. B., 1968, Reclamation of soil contaminated by oil: Journal of the Institute of Petroleum, London, v. 54, p. 182–197.

Sealey, J. Q., 1974, A microbiological method of prospecting for petroleum, parts I and II: Oil and Gas Journal, v. 72, p. 98–102, and p. 142–146.

Soli, G. G., 1957, Micro-organisms and geochemical methods of oil prospecting: American Association of Petroleum Geologists Bulletin, v. 41, p. 134–145.

Stach, E., Mackowsky, M.-Th., Teichmuller, M., Taylor, G. H., Chandra, D., and Teichmuller, R., 1982, Coal petrology, third ed.: Berlin, Borntraeger, 535 p.

Thompson, K. F. M., 1982, Classification and thermal history of petroleum based on light hydrocarbons: Geochimica et Cosmochimica Acta, v. 47, p. 303–316.

Tissot, B. P., and Welte, D. H., 1984, Petroleum formation and occurrence, second ed.: Berlin, Springer-Verlag, 699 p.

Urban, J. B., and Harding, R., 1979, Hydrocarbon occurrence as a function of thermal alteration of organic material (abstract): American Association of Petroleum Geologists Bulletin, v. 63, p. 544.

Van der Linden, A. G., and Thijsse, G. J. E., 1965, The mechanism of microbial oxidations of petroleum hydrocarbons: *in* Nord F. F., ed., Advances in Enzymology and Related Subjects of Biochemistry, v. 27, p. 469–546.

Winters, J. C., and Williams, J. A., 1969, Microbiological alteration of crude oil in the reservoir: American Chemical Society, Division of Petroleum Chemistry, Symposium on Petroleum Transformation in Geologic Environments, September 7–12, New York City, Paper PETR 86, p. E22–E31.

Yen, T. F., ed., 1976, The genesis of petroleum and microbiological means for its recovery: Microbiology Group Symposium, Institute of Petroleum, Birmingham, England, 82 p.

Zobell, C. E., 1946, Action of microorganisms on hydrocarbons: Bacteriological Review, v. 10, p. 1–49.

Zobell, C. E., 1964, Geochemical aspects of the microbial modification of carbon compounds: *in* Columbo, U., and Hobson, G.dD., eds., Advances in Organic Geochemistry, 1962: New York, Macmillan, p. 339–356.

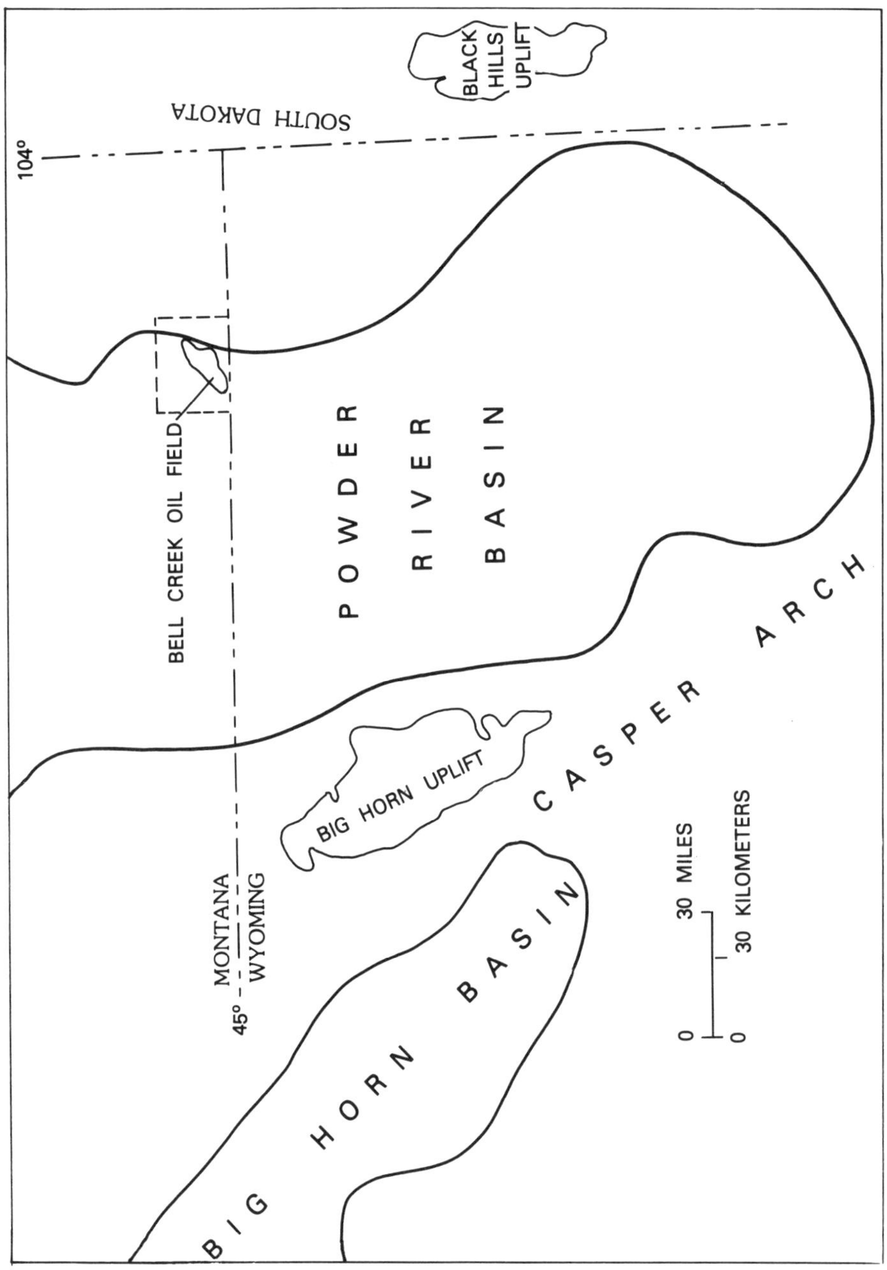

Fig. 1. Bell Creek Oil Field, Powder River Basin, Montana.

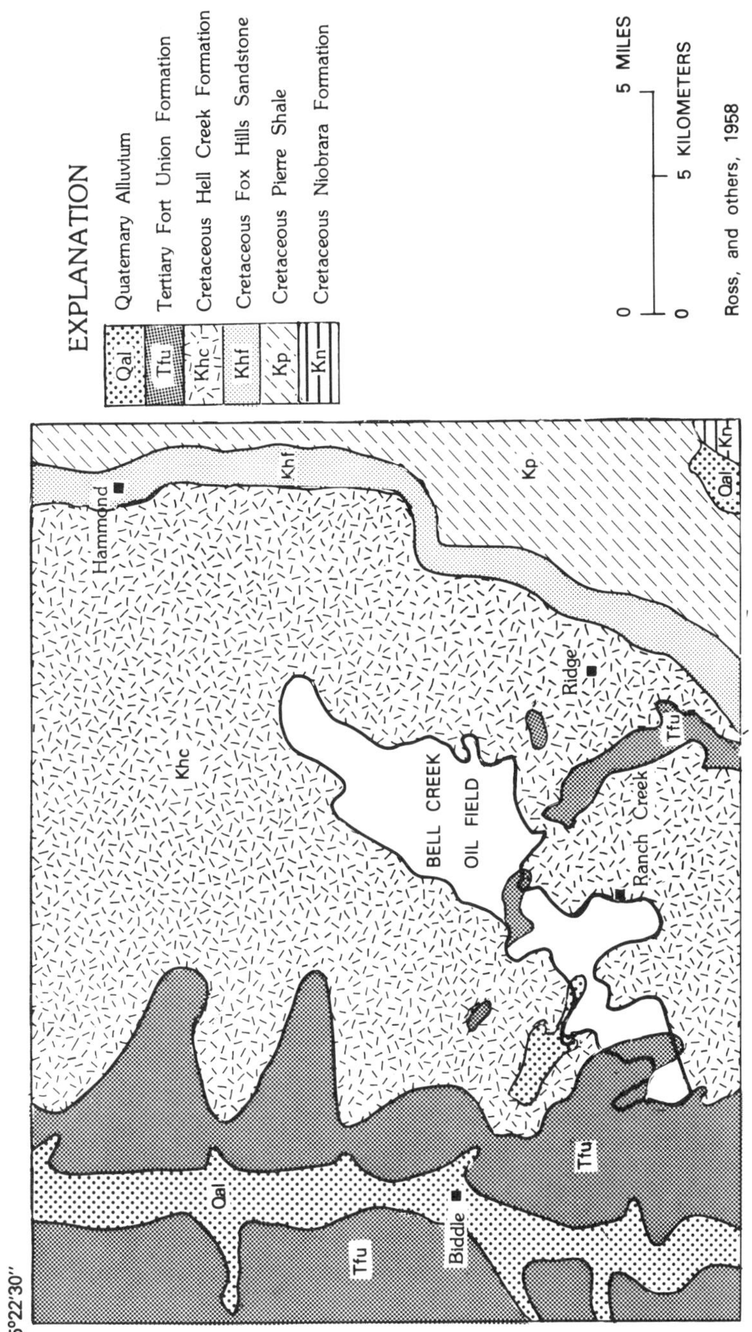

Fig. 2. Geology of the Bell Creek oil area, east-central Powder River Basin, Montana and Wyoming.

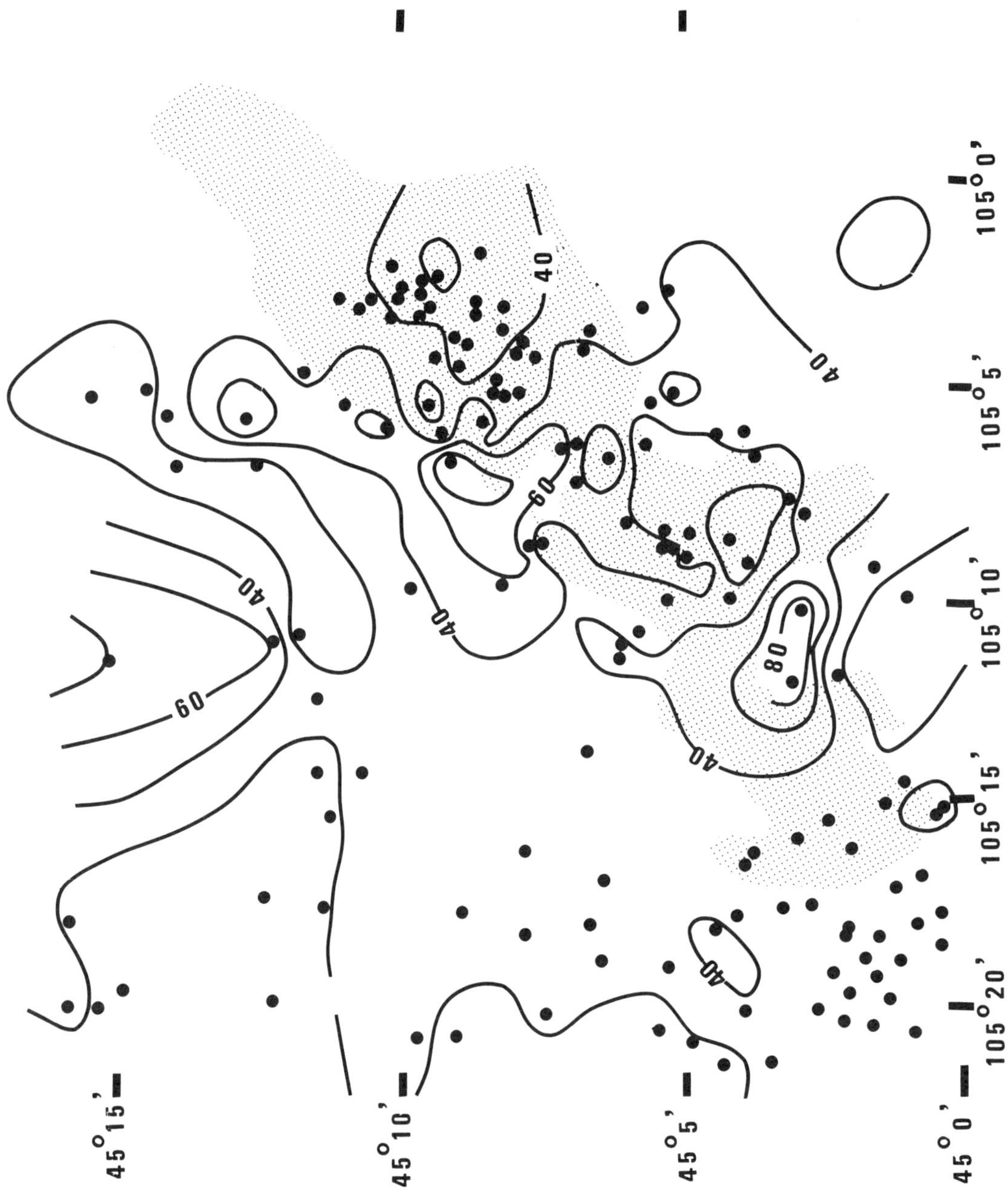

Fig. 3. Surface vitrinite reflectance survey over the Bell Creek Oil Field, Montana. Contour interval is R_m (in percent) multiplied by ten. Solid dots represent sample localities. The stippled area is the Bell Creek production field. R_m is the average of seventy-five measurements (given enough vitrinite) under oil immersion. In some samples much less vitrinite was available and in a few cases (about fifteen samples) less than twenty reflectance measurements were made. Reflectance was measured in random orientation without rotation to maximum reflectivity.

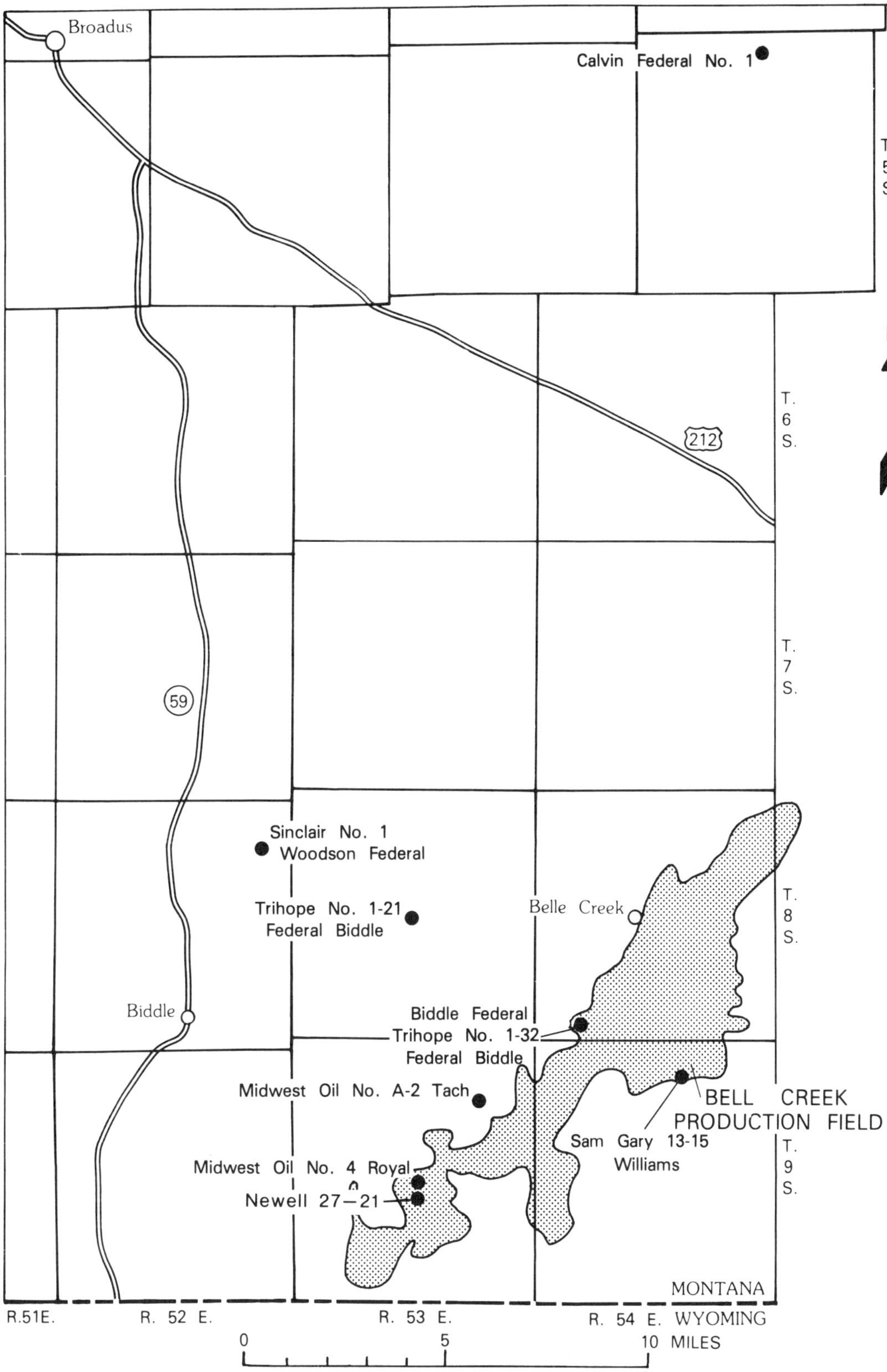

Fig. 4. Location of boreholes selected for vitrinite reflectance study in the vicinity of Bell Creek Oil Field, Montana. Data shown in Figures 5 and 6.

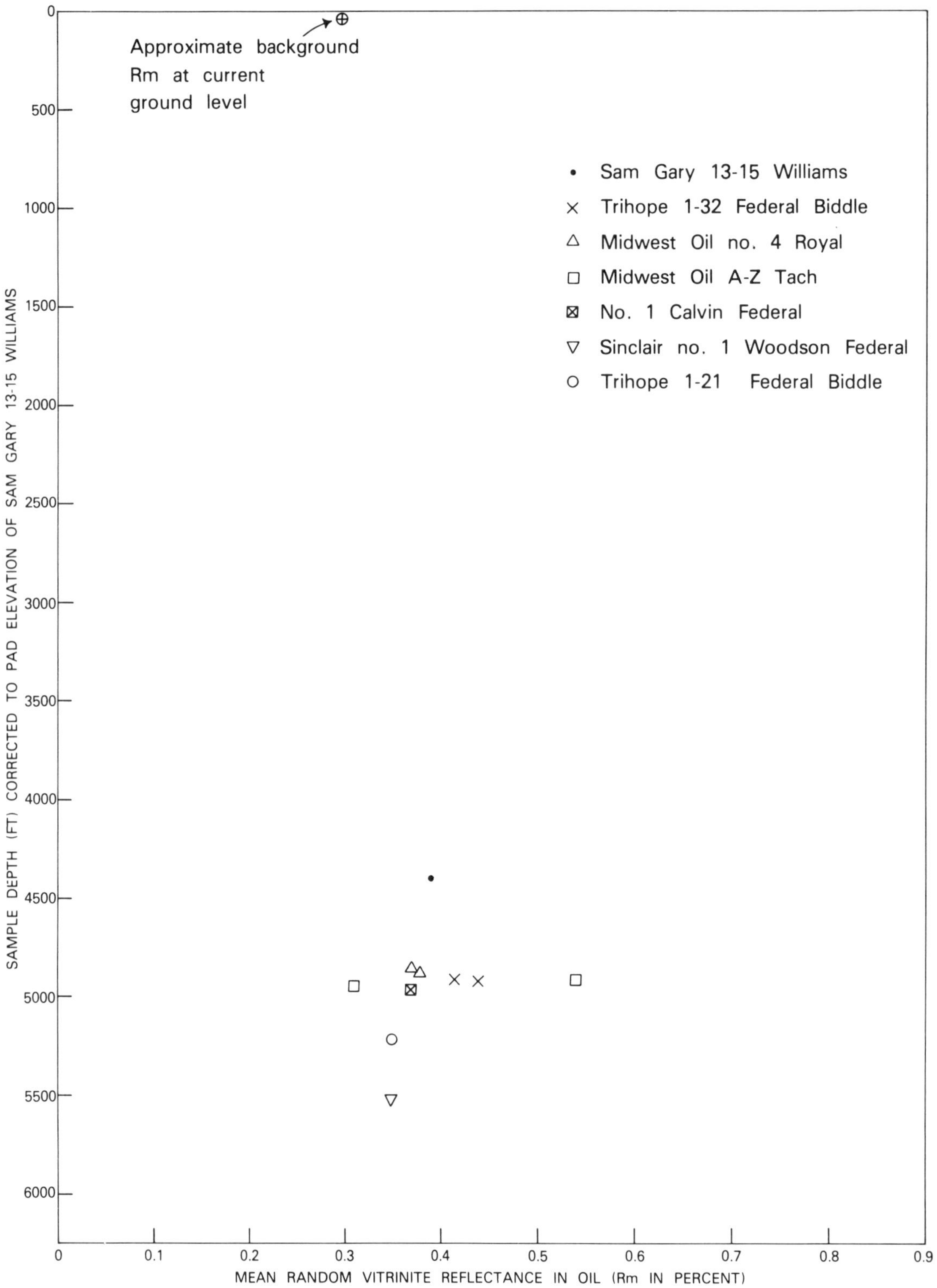

Fig. 5. Vitrinite reflectance from the Muddy Sandstone, Bell Creek area, Montana. The Muddy Sandstone forms the Bell Creek reservoir-rock. R_m data from coaly organic matter within core samples of Muddy Sandstone. The sample depth in each well was corrected to the pad elevation of borehole Gary 13-15 Williams. This correction allows comparison of R_m data between wells. No rock samples were available to us outside of the reservoir levels in these boreholes.

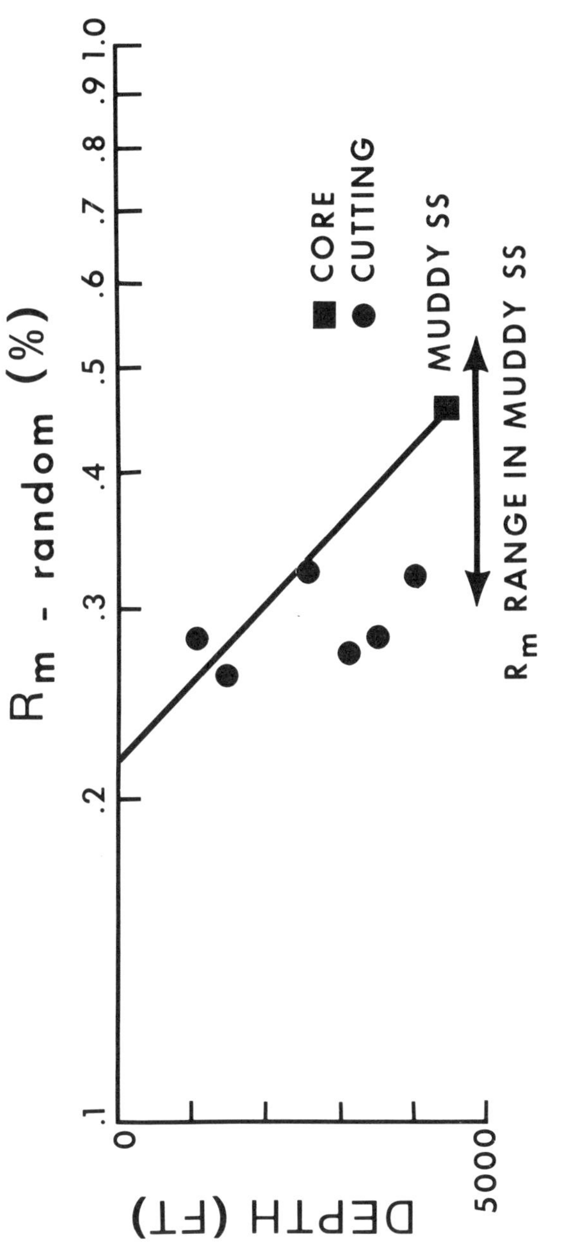

Fig. 6. Vitrinite reflectance versus depth in borehole Gary Oil Newell 27-21, Bell Creek Oil Field, Montana.

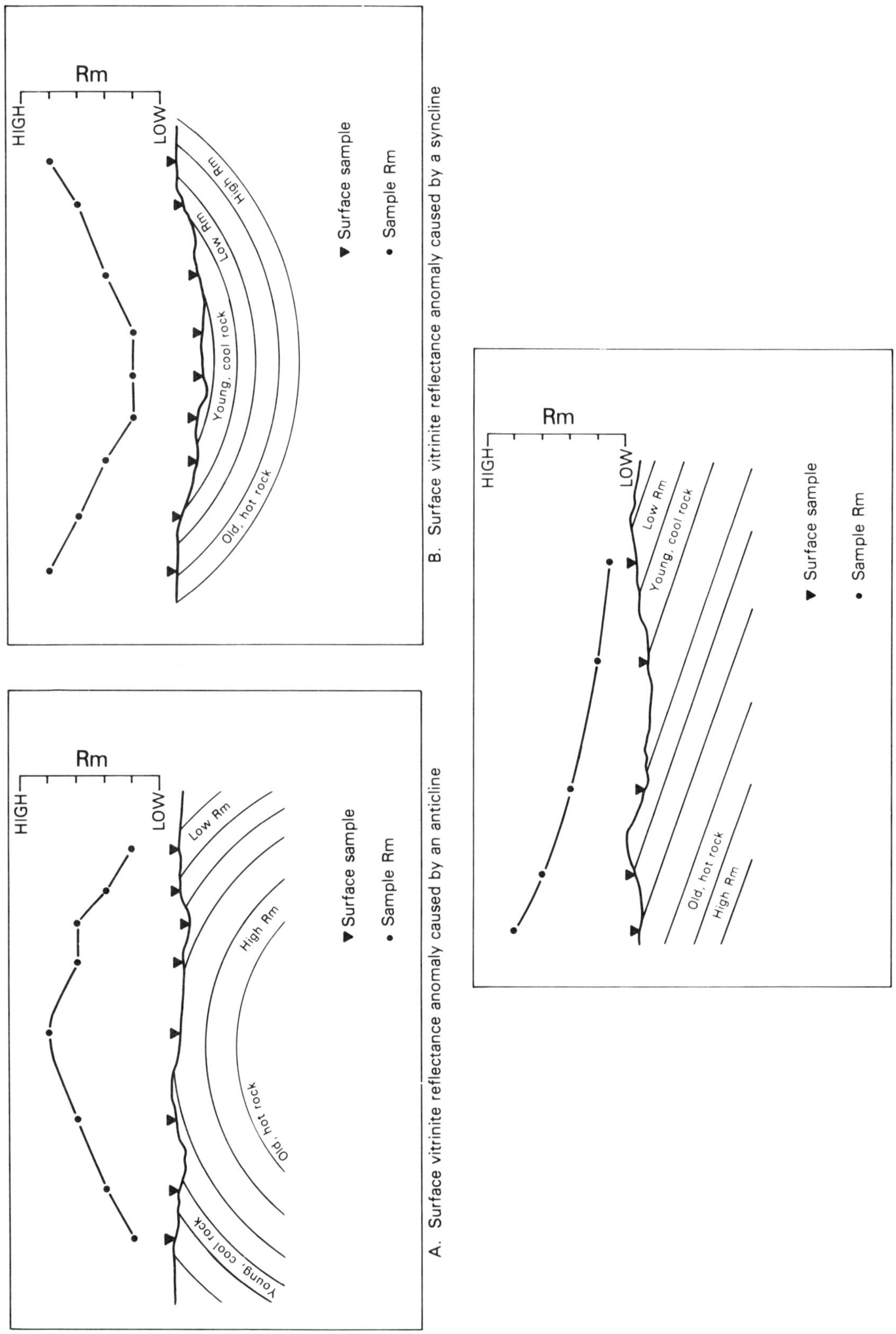

Fig. 7. Hypothetical structural effects on surface surveys of vitrinite reflectance.

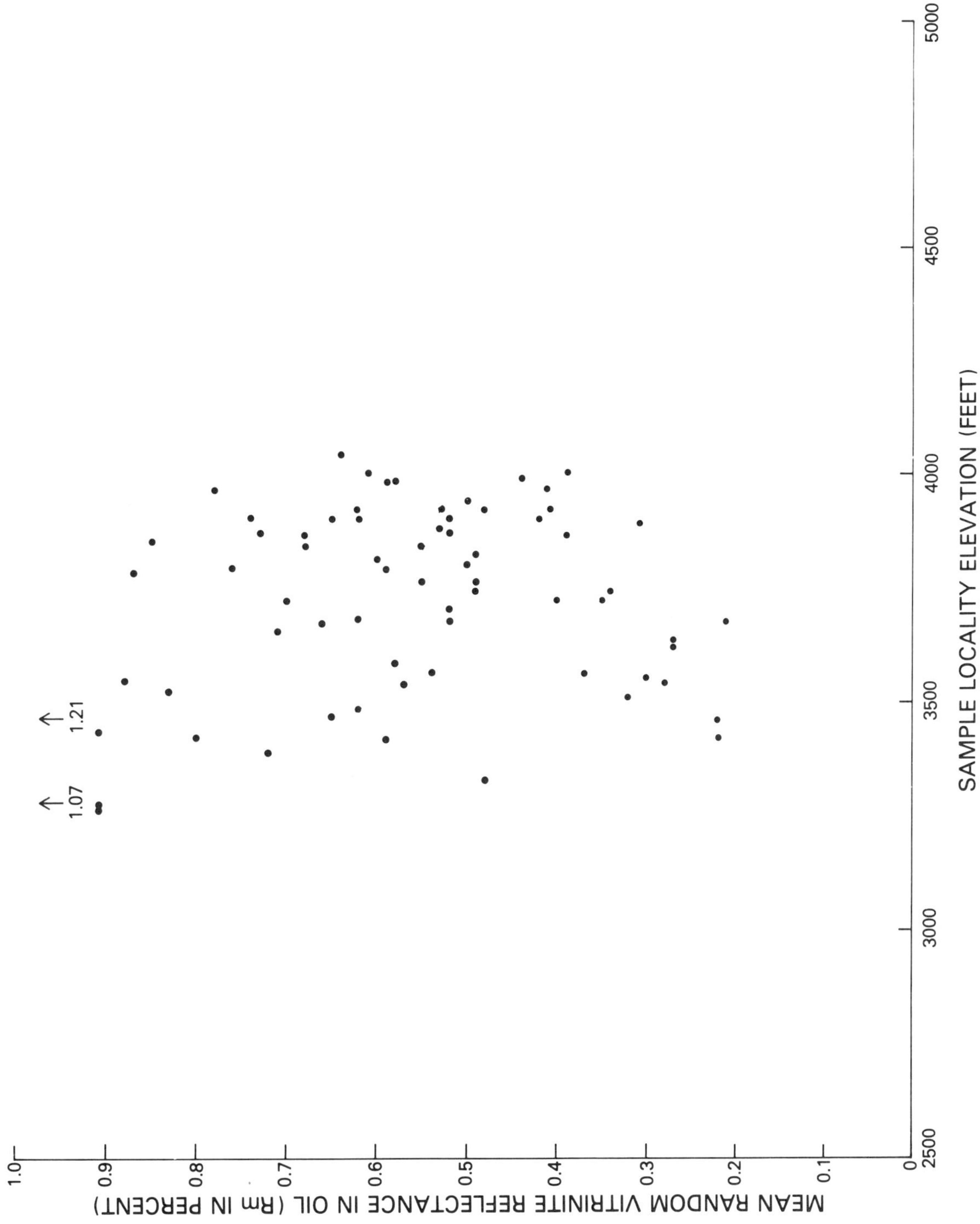

Fig. 8. Vitrinite reflectance versus sample locality elevation, Bell Creek Oil Field, Montana.

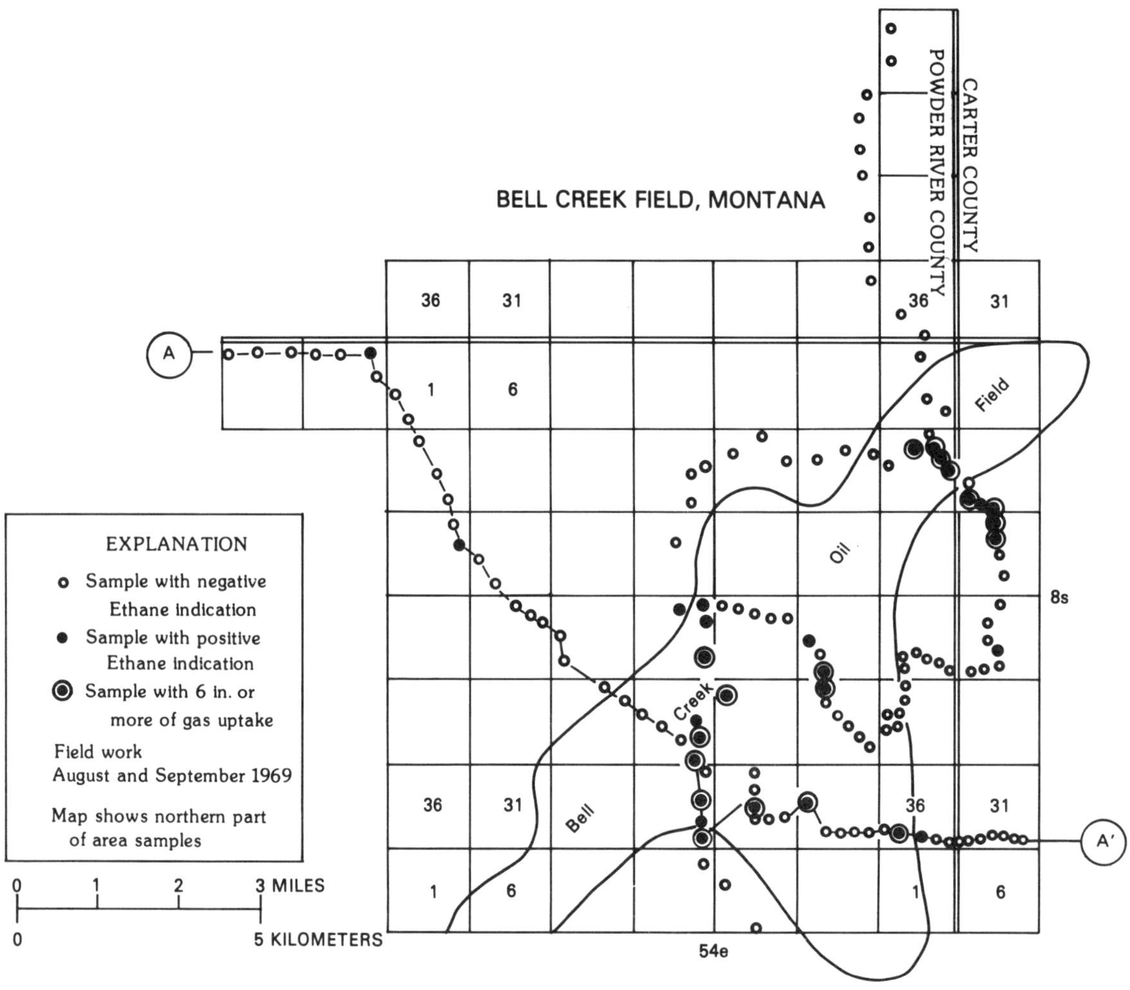

Fig. 9. Survey of microbial activity over the Bell Creek Oil Field, Montana. Modified from Miller (1976).

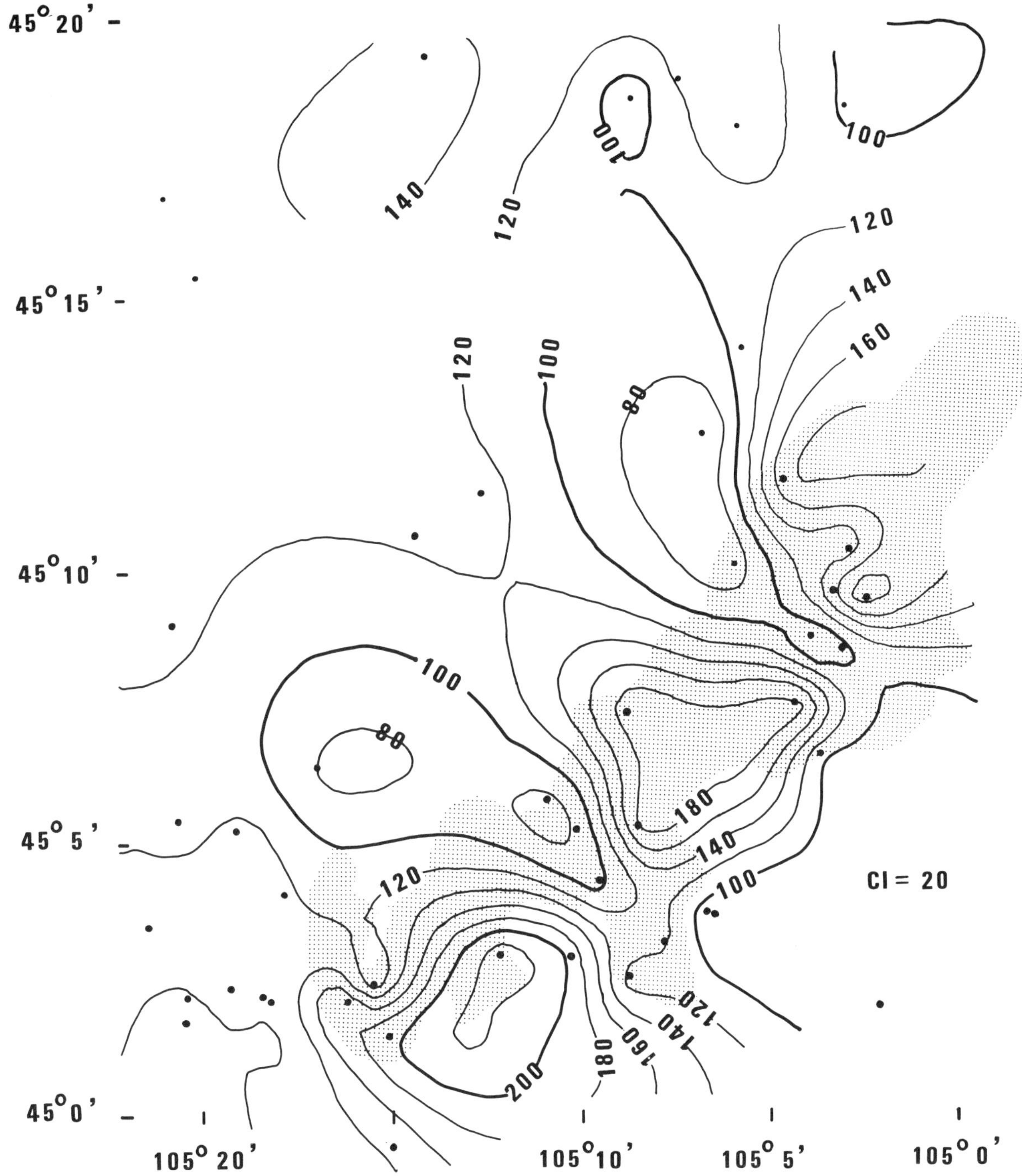

Fig. 10. Oxygen index of coal, Bell Creek Oil Field, Montana. Data from Rock-Eval pyrolysis. Solid circles represent sample localities. The stippled area is the Bell Creek production field. Contour interval is 20 mg CO_2/g organic carbon.

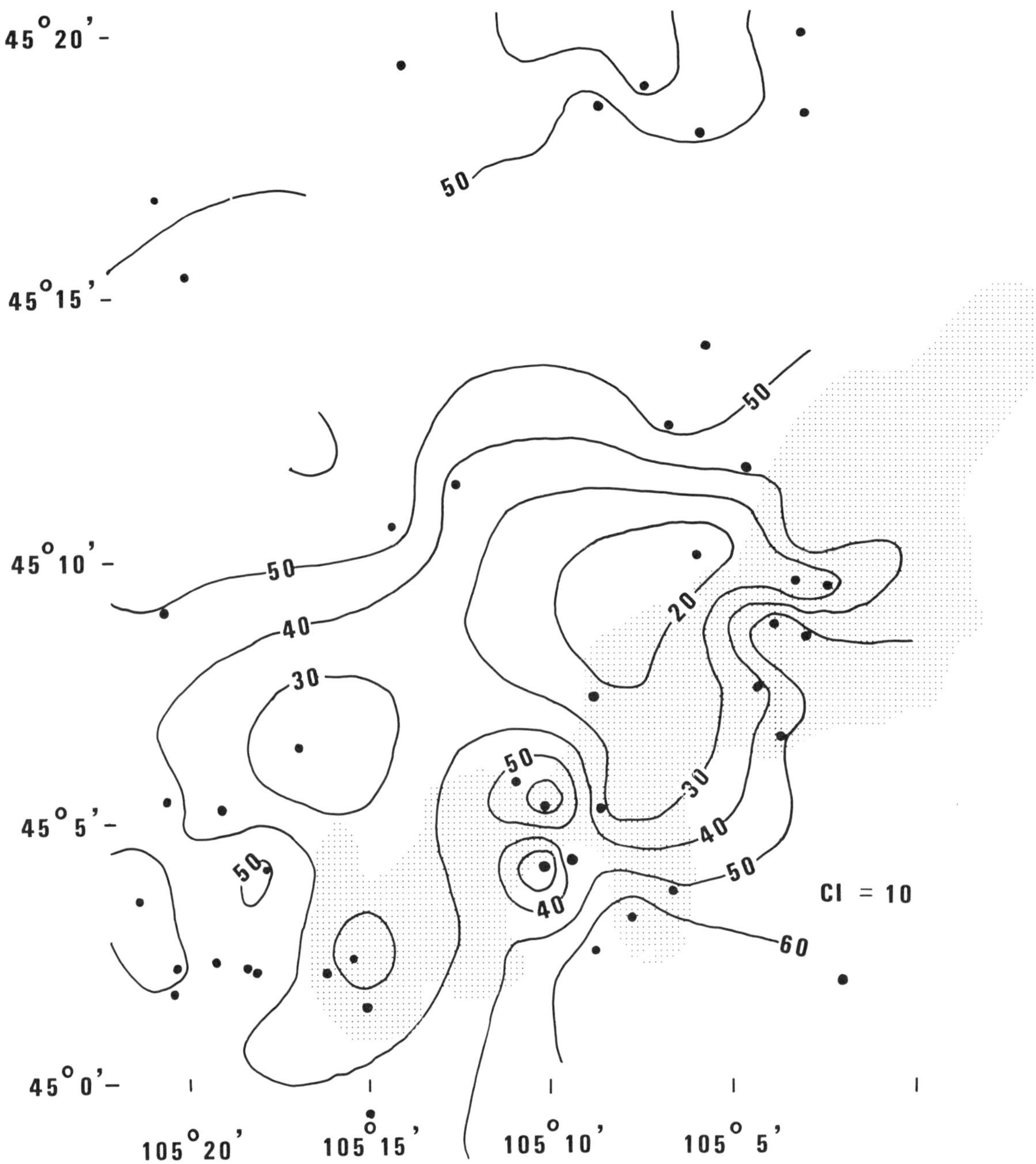

Fig. 11. Total organic carbon content of coal, Bell Creek Oil Field, Montana. Solid circles represent sample localities. The stippled area is the Bell Creek production field. Contour interval is 10 mg organic C/g rock.

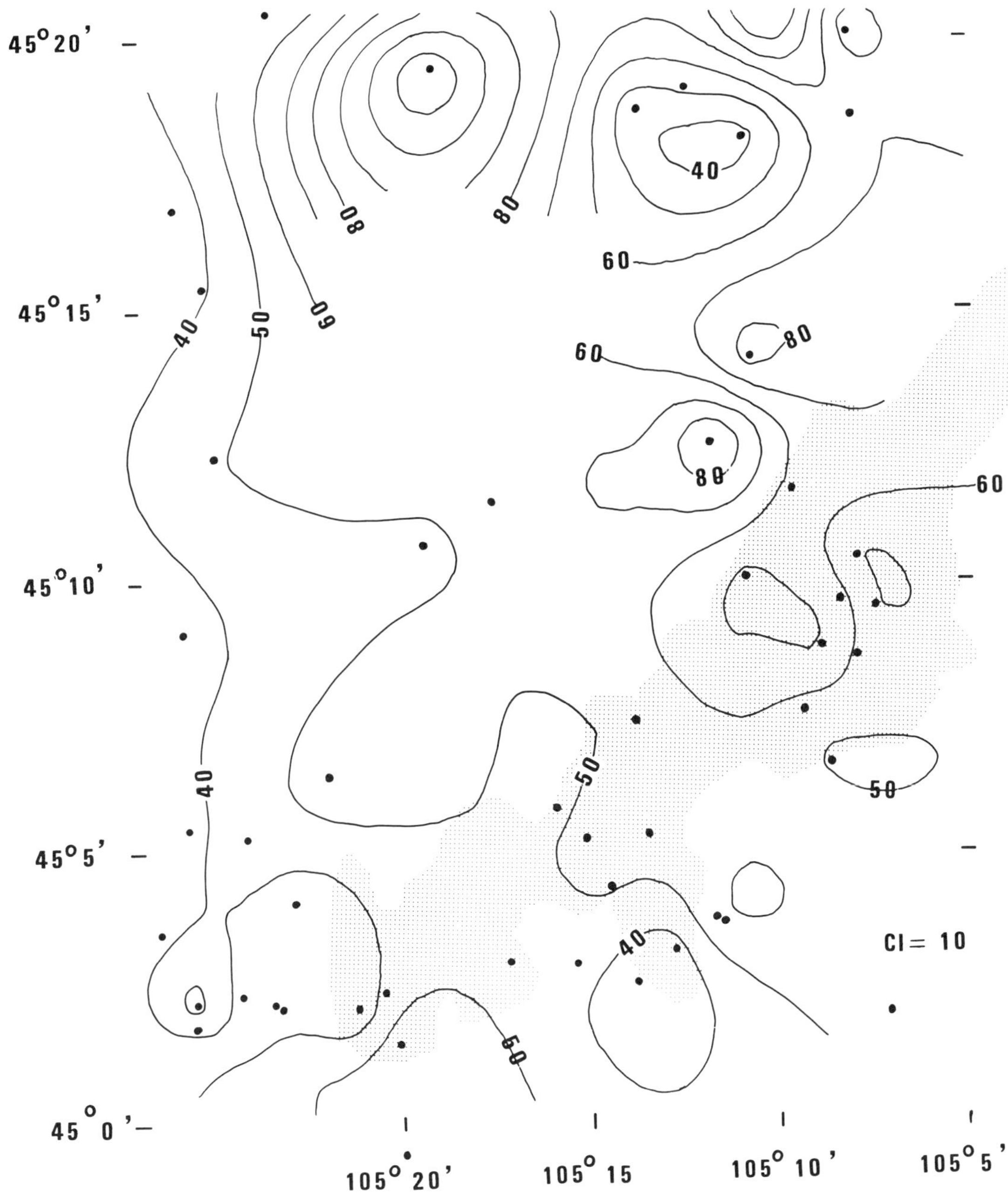

Fig. 12. Hydrogen index of coal, Bell Creek Oil Field, Montana. Data from Rock-Eval pyrolysis. Solid circles represent sample localities. The stippled area is the Bell Creek production field. Contour interval is 10 mg HC/g organic carbon.

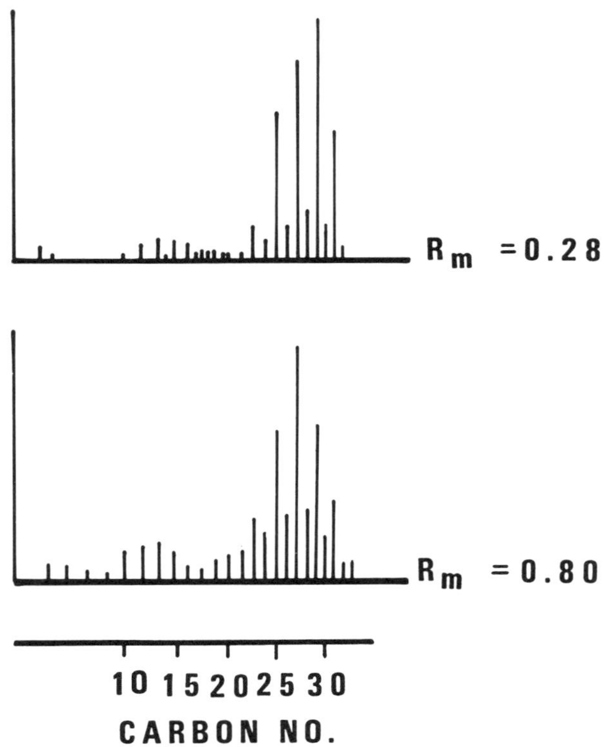

Fig. 13. Gas chromatography analysis of soxhlet extracts from coal as function of R_m, Bell Creek Oil Field, Montana.

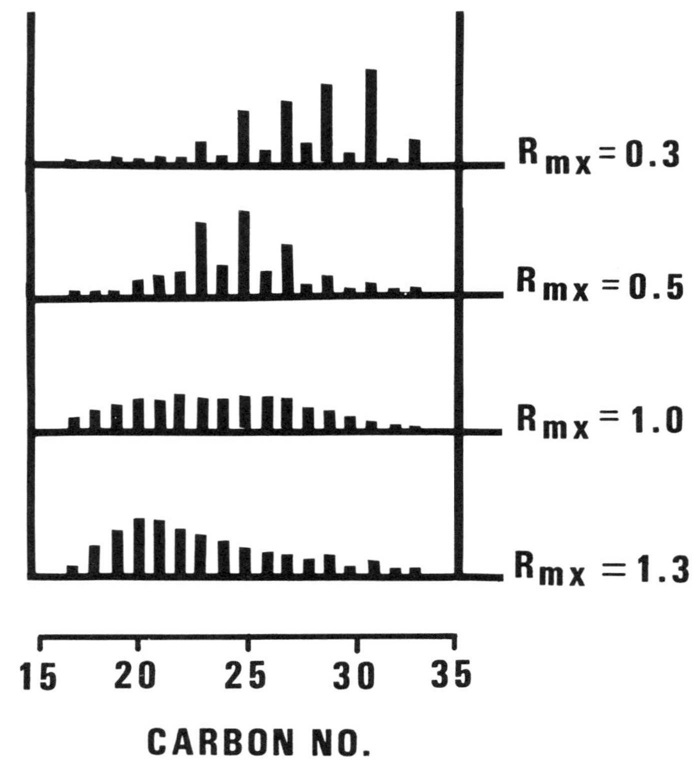

Fig. 14. Gas chromatography analysis of soxhlet extracts from coal as function of R_m. Modified from Hollerbach and Hagemann (1981).

HYDROCARBON GEOCHEMICAL EXPLORATION AFTER FIFTY YEARS

Leo Horvitz

HORVITZ RESEARCH LABORATORIES, INC.

ABSTRACT

The first hydrocarbon geochemical survey of record was conducted in Germany and involved determinations of methane in near-surface samples of soil air that were collected over and beyond the productive limits of a known gas field. The results of this survey were reported in 1933. Two years later, Russian investigators published their experiments with soil air. In 1936, American scientists entered the field of geochemical exploration, using the soil as the sample medium. Their initial experimental data were published in 1938 and 1939.

In the early years, geochemical exploration was controversial. This is surprising since the first oil field and many subsequent discoveries resulted from drilling in the vicinity of surface oil and gas macroseepages. The controversy surrounding geochemistry started to subside in the early 1970s and now, in the mid 1980s, near-surface geochemistry is an accepted exploration technique.

In order to evaluate prospective areas geochemically, the light saturated hydrocarbons (methane through pentane) should be searched for, both onshore and offshore, in whatever media they may occur. Heavier hydrocarbons, especially of the aromatic series, are also found in near-surface soils and sediments. The presence of anomalous amounts of aromatic hydrocarbons is indicative of an oil-prone area.

An important parameter that has been introduced in recent years to geochemical exploration is the determination of stable carbon isotope ratios of near-surface methane. These ratios are useful in predicting the contents of subsurface reservoirs.

INTRODUCTION

The publication by Laubmeyer (1933) of the results of a soil gas survey over a gas field in Germany marked the beginning of near-surface geochemistry. After Laubmeyer, Russian investigators entered the field (Sokolov, 1935). Fifty years have already passed since the first geochemical experiments were conducted in this country using soil as the sample medium (Rosaire, 1938, Horvitz, 1939). After struggling amidst controversy, for about half a century, near-surface geochemistry has finally been accepted by the oil industry as an important exploration technique. The achievements that account for this acceptance include the following:

1. Confirmation, by drilling, of many geochemical anomalies, especially in offshore areas (for an example of an offshore geochemical confirmation, see Horvitz, 1980).

2. Recognition of similarities in compositions of near-surface hydrocarbons and hydrocarbon gases in subsurface reservoirs (Jones et al., 1983).

3. Proof of the presence of thermogenic methane in near-surface sediments by stable carbon isotope determination (Horvitz, 1981; Stahl et al., 1981).

4. The discovery that indigenous biogenic methane can be removed from near-surface soils and sediments, while leaving thermogenic methane tightly bound to the fine-grained particles (Horvitz, 1980, 1985).

Many years of experience in the field of hydrocarbon geochemical exploration have demonstrated the importance of searching for near-surface saturated hydrocarbons in whatever media they occur. Some of the light hydrocarbons are present in the interstitial spaces of the soil, some are loosely held by soils and sediments, and some are dissolved in offshore bottom waters. However, most of the light hydrocarbons are usually tightly bound to the fine-grained portions of soils and sediments. Aromatic hydrocarbons, such as naphthalenes and phenanthrenes that are present in most crude oils, are also often present in the earth's near-surface in the vicinity of oil fields. Stable carbon isotope ratios of methane have been proven useful in geochemical exploration and, therefore, they too should be included in evaluating prospective areas.

The results of an experimental survey conducted in 1980 in the Santa Barbara Channel, offshore California, illustrate the value of using multiple parameters in geochemical exploration (Horvitz, 1982, 1984).

SANTA BARBARA CHANNEL SURVEY

The survey area (Fig. 1) includes three known oil fields, Hondo (H), Pescado (P), and Sacate (S), that were not producing at the time of the survey but were known to contain large reserves in the fractured Monterey shale section. To the north of these fields, outside of the survey area, several relatively small fields are located. The most important of these is the

Molino (M) gas field that has produced about 250 billion cu ft of gas and 4,500,000 barrels of condensate. Just beyond the northwest corner of the survey, the important Conception (CO) Oil Field is located. The other, smaller fields are Cuarta (CU), Alegria (A), Caliente (C), and Gaviota (G).

The survey involved the collection of bottom water and sediment samples from 189 stations located on approximately a 1 mi grid. Sediment samples were collected below the sea bottom with a piston-type corer fitted with a removable plastic liner. The sediment cores ranged from 1 to 6 ft in length; 162 were 4 to 6 ft long and 27 were shorter. The sampling procedure permitted some of the bottom water to collect in the liner immediately above the sediment. At each station, 500 cc of the sea water was sealed in a 950 cc plastic jar with approximately 250 cc of sediment that was taken from the bottom part of the core. Approximately 10 drops of 17% zepherin chloride solution (a bactericide) were added to the jar prior to sealing and inverting it. This sample was used to determine the sum of the free hydrocarbons contributed by the water and sediment.

A second sediment sample, approximately 1,000 g, was removed from the lower part of each core and securely wrapped in a plastic bag after pressing out the enclosed air. Samples can be stored for months without losing hydrocarbons when they are packaged in this manner. Portions of these sediment samples were used for adsorbed light hydrocarbon and hydrocarbon analyses and for determining carbon isotope ratios. At some locations, about 300 cc of sediment was sealed in a 600 cc can. These samples were stored at $-20°C$ for future use.

DISCUSSION OF RESULTS

Bottom Water and Interstitial Hydrocarbons

The data in Figure 1 represent the sum of the methane extracted from the bottom water and from the interstitial spaces of the sediment. These data were obtained by analyzing samples of headspace gas that were withdrawn from the jars containing sediments and sea water. The values are expressed as 10^{-6} cc CH_4/cc sediment (parts per million by volume of sediment).

The most outstanding anomalous feature produced by the free methane is in the southeastern part of the survey where methane attained values up to 67,500 parts per million. Such high methane concentrations are indicative of a biogenic source. The free methane in the northern part of the survey is probably thermogenic. This is suggested by the close relationships between the methane patterns and some of the productive fields. The halo-type methane anomaly that is associated with the Hondo field is of particular interest.

Adsorbed Hydrocarbons

Before conducting the analyses for the adsorbed hydrocarbons, biogenic methane was removed from each sediment sample. The biogenic methane removal process (also referred to as the sand removal process, Horvitz, 1980) involves the following procedure: about 250 g of sediment is stirred in a blender with distilled water for five minutes. The resulting mixture is then wet-sieved and the fine-grained material smaller than 63 microns is collected together with the water that is later removed by filtration. The thermogenic methane and other heavier hydrocarbons remain adsorbed or occluded to the remaining fine-grained fraction (Horvitz, 1980).

The hydrocarbons were extracted from approximately 25 g of the undried, fine-grained sample by acid and heat treatment (Horvitz, 1972), after which the extracted gases were analyzed by gas chromatography. The resulting hydrocarbon values are expressed in parts per billion by weight of dry sample (g hydrocarbon per 10^9 g dry sediment). The moisture content needed for calculating the results on a dry sample basis was determined on a separate portion of each sample.

Adsorbed Methane

The adsorbed methane values are shown in Figure 2 and range from 93 to 382 parts per billion by weight. The values that are considered anomalous, 175 parts per billion and above, were selected from histograms. Although no striking anomalies were found over or around the Hondo (H), Pescado (P), or Sacate (S) fields, some anomalous values are clearly associated with these fields. The relatively weak, incomplete near-surface methane anomalies reflect the low gas-oil ratios in the Monterey reservoirs.

Anomalous adsorbed methane values are associated with the Molino (M), Caliente (C), and Alegria (A) fields in the northern part of the survey. The group of anomalous methane values in the southeastern part of the survey may be pointing to a subsurface accumulation farther to the southeast.

Adsorbed Ethane and Heavier Hydrocarbons

Although the ethane and heavier hydrocarbon values (Fig. 3) are relatively high, ranging from 54–179 parts per billion by weight, no outstanding anomalies developed. The values are, however, in the range that is encountered in petroliferous areas.

Fluorescence Data

A group of hydrocarbons that often provides useful information is classified as aromatic hydrocarbons. These compounds are present in crude oils, and one of their characteristics is that they fluoresce when exposed to ultraviolet light. Aliphatic hydrocarbons, which are also present in crude oils, do not possess this property. When dilute solutions of crude oil in solvents such as hexane are exposed to ultraviolet light of a specific wavelength, the aromatic compounds emit light at certain higher wavelengths. The intensities of the emitted light can be measured with sensitive spectrofluorometers and recorded as fluorograms. The substances that fluoresce are probably 2-ring and 3-ring compounds such as naphthalenes and phenanthrenes that are present in most crude oils in relatively large quantities. The aromatic compounds are often present in measurable amounts in near-surface sediments over or adjacent to oil fields. When used together with the light hydrocarbon data, fluorescence determinations are useful in distinguishing a geochemical anomaly that is reflecting an oil accumulation from one that is produced by a subsurface gas deposit.

Fluorograms of two crude oil samples and a sediment sample from the survey area are shown in Figure 4. The fluorograms of the oil samples were obtained by exposing dilute solutions of the oils, which were dissolved in spectroscopically pure hexane, to ultraviolet light having a wavelength of 265 nanometers (nm). Two emission peaks appear on the fluorogram of the crude oil that has an API gravity of 35.9°; one of the peaks is at a wavelength of 320 nm and the other is at 365 nm. The middle fluorogram, that of crude oil having an API gravity of 29.2°, has a shoulder at 320 nm and a well-defined peak at 365 nm. The fluorogram of sediment sample 0-3 also has a distinct peak at 365 nm but only a slight shoulder at 320 nm. However, all three samples produced emissions at the same wavelengths.

Sediment sample 0-3 and those of the other samples collected in the survey were prepared for fluorescence measurements by the following procedure: 10 cc of spectroscopically pure hexane was added to 1 g of fine-grained sediment that had been dried at 40°C. The mixture was allowed to stand at room temperature for fifteen minutes with intermittent shaking. The clear liquid above the sediment was then poured into a quartz cuvette and the latter was introduced into the fluorescence spectrophotometer. The solution was exposed to ultraviolet light at 265 nm and the fluorescence emission intensity versus wavelength was recorded on chart paper for the range of wavelengths from 300 nm to approximately 600 nm. The relative fluorescence values are expressed in peak height units after taking into account sensitivity and attenuation factors. Other fluorescence techniques have also been described (Campbell, 1946; Brooks, 1983; Hebert, 1984).

Fluorescence 320 nm

The fluorescence values for the 320 nm emissions are shown in Figure 5. As in the case of the lighter adsorbed hydrocarbons, anomalous fluorescence values occur in the northern part of the survey where the productive areas are located.

Fluorescence 365 nm

The 365 nm emission values shown in Figure 6 represent one of the most important sets of data produced by the survey. The main feature is in the northern part of the survey where unusually high fluorescence values were found. Anomalies developed around the Hondo (H) field as well as over the Pescado (P) and Sacate (S) fields.

The high 365 nm values, to the south of the Molino gas field, suggest the presence of oil in the Molino structure. This possibility may be tested by a major oil company that plans to drill a series of wells within the present limits of the Molino structure in search of deep gas and to test the Monterey section for oil (Anonymous, 1984).

FLUORESCENCE RATIOS

The 365 nm and 320 nm fluorescence data are of value in determining the oil potential of an area. Therefore, their appearance in anomalous quantities in the vicinity of, or over, oil fields is to be expected. Additional information of value, however, can be determined by considering their ratios. For example, the ratio of the 365 nm to the 320 nm peak heights (Fig. 4) is 1.2 for crude oil having a gravity of 35.9° API and 1.5 for the lower gravity (29.2° API) crude. The highest ratio, 4.0, was attained by sediment sample 0-3, located near the northern edge of the Hondo field. During drill stem tests of wells of this field, oil was produced with gravities ranging from 14°–26° API (USGS 1974). These fluorograms suggest that the ratios of the 365 nm to the 320 nm values of near-surface sediment samples are useful in predicting the relative gravities of subsurface oil deposits. The higher the ratio, the lower the gravity of oil to be expected in subsurface reservoirs.

CARBON ISOTOPE RATIOS

Isotope ratios of carbon in methane that is extracted from near-surface soils and sediments provide data that are useful in predicting the contents of underlying reservoirs (Horvitz, 1981; Stahl et al., 1981). This capability is based on the fact that carbon isotope ratios (C^{13}/C^{12}) of methane in petroleum reservoirs depend upon the types of hydrocarbons in the reservoirs (Schoell, 1980).

To obtain carbon isotope ratios of thermogenic methane in near-surface soils, it is first necessary to remove any indigenous biogenic methane that may be present. This is accomplished by the previously described biogenic methane removal process. After the hydrocarbons have been extracted by acid treatment of about 150 g of the processed sediment, the methane is separated from the other hydrocarbons and any carbon dioxide that may be present. The methane is then converted to carbon dioxide and water, and after removing the water, the purified carbon dioxide is introduced into a Varian MAT 250 isotope mass spectrometer for the carbon isotope ratio determination relative to the PDB (Peedee Belemnite) standard.

Figure 7 shows the carbon isotope ratios that were obtained in the Santa Barbara Channel survey. The $\delta^{13}C$ values range from -38 to -55 per mil for the 180 samples analyzed. Ninety-two percent of the isotope values are in the -40 to -49 per mil range, values that are usually obtained for carbon in methane that is associated with oil. Isotopically lighter methane (-44 to -47 per mi) was found in the vicinity of the productive areas rather than in the background areas; the isotopically lighter methane may be pointing to zones of active upward migration of methane from subsurface petroleum reservoirs.

Of particular interest is the close resemblance of the patterns that were produced by the carbon isotope ratios of methane (the lightest constituent of the near-surface sedimentary hydrocarbons) to those patterns produced by the fluorescence of 365 data (Fig. 6) which represent the heaviest compounds observed in the survey. Anomalous isotope patterns are frequently associated with productive areas and reinforce geochemical interpretations. Unlike offshore, onshore isotope ratio determinations should be made on deeper samples, substantially below the near-surface oxidized zone.

The importance of removing biogenic methane from soil and sediment samples prior to hydrocarbon analysis and carbon isotope ratio determination is illustrated in Figure 8. The upper chromatogram was produced from headspace gas that was removed from the canned sample taken at station N-9. This sample was stored at $-20°C$ for eighteen months prior to analysis. The chromatogram indicated that the headspace gas contained about 9,000 parts per million of methane and the $\delta^{13}C$ value of this methane was found to be -104 per mil. Both the high methane value and the correspondingly low methane isotope ratio indicate that the methane present in the headspace is biogenic. It may also be observed that the large quantity of biogenic methane obscures any ethane or propane that may be present.

The lower chromatogram (Fig. 8) resulted from analysis of a 20 g portion of the fine-grained sediment that remained after the contents of the can were subjected to the biogenic methane process. The complete suite of hydrocarbons, methane through n-pentane, is clearly seen, and traces of three hexane isomers are also visible. The adsorbed hydrocarbon values obtained in this analysis differed by less than 5% from those of the original hydrocarbon analysis of the sediment taken from the same core at station N-9 (Figs. 2 and 3). The carbon isotope ratio of −45 per mil that was obtained indicated that the biogenic methane had been eliminated.

It is of interest to compare the data of Figure 7 to carbon isotope data previously published by Faber et al. (1983) for samples taken from the same area in the Santa Barbara Channel. The samples used by these authors were kept immersed in liquid nitrogen (−196°C) from the time of collection to the time that they were subjected to isotopic measurement. Only bulk samples were used and $\delta^{13}C$ ratios ranging from −95 to −30 per mil were reported. Distribution curves prepared from these carbon isotope ratios have a broad peak at approximately −90 per mil, an indication of methane of a biogenic source. There is also a second peak between −50 to −38 per mil that is characteristic of thermogenic methane. When such mixtures of biogenic and thermogenic methane are present, confusing areal distribution patterns are produced by the isotope data. The results shown in Figure 7 also suggest that storing sediment samples in liquid nitrogen prior to determining carbon isotope ratios is not necessary.

RELATIONSHIP OF NEAR-SURFACE GEOCHEMICAL DATA TO SUBSURFACE GEOLOGY

A map that was prepared from deep seismic, core hole and exploratory well data available in the area of the Santa Barbara Channel Survey is shown in Figure 9 (U.S.G.S., 1974). Close relationships between the near-surface geochemical data and the subsurface structural information are apparent. A fault, upthrown to the south and located along the northern edge of the Hondo field, is of interest. Anomalous fluorescence 365 values (Fig. 6) and carbon isotope ratios (Fig. 7) in the vicinity of this fault suggest that it is the conduit that transports hydrocarbons from the Hondo reservoir to the near-surface.

Anomalous hydrocarbon values also occur on the upthrown side of the fault that parallels the Molino, Caliente, and Algeria fields. Here, too, the fault may be involved in the process of transferring hydrocarbons from the subsurface accumulations to the surface. However, anomalous geochemical values over the Pescado and Sacate fields, where no faulting is indicated on the subsurface map, suggest that hydrocarbons may reach the surface by routes other than those provided by known faults. Lineaments (which may be fractures rather than faults) that extend from the subsurface petroleum accumulation to the surface of the earth may be an alternative route (Richers et al., 1982).

CONCLUSIONS

Near-surface geochemical exploration involves the determination of hydrocarbons in several different media. Hydrocarbons may occur in the free state in bottom waters and in near-surface soils and sediments. They may be loosely bound to the soil or sediment but, generally, the largest quantities are adsorbed tightly to the fine-grained particles. In addition to the aliphatic hydrocarbons, in the range from methane through pentane, heavier straight chain hydrocarbons, particularly hexanes and heptanes, may be detected.

Aromatic compounds are also found in the near-surface. When present in anomalous amounts they often indicate oil in subsurface reservoirs.

The elimination of biogenic methane from soils and sediments is essential in order to detect methane that has migrated from petroleum reservoirs. Stable carbon isotope ratio determinations made on migrated methane are helpful in confirming anomalies based on heavier hydrocarbon data. The isotope ratios are also useful in predicting the contents of subsurface reservoirs.

Striking relationships are often observed between geochemical anomalies and studies based on subsurface geology and geophysics. Therefore, integration of near-surface geochemistry with these and other valid exploration methods should lead to improved success in the search for petroleum.

ACKNOWLEDGMENTS

The author is indebted to Tenneco Oil Company for underwriting the sampling phase of the Santa Barbara Survey, and to Exxon Company U.S.A. for permission to take samples from the Exxon leases within this area. Thanks are also due to E. Philip Horvitz for his invaluable assistance in the preparation of this paper.

REFERENCES

Anonymous, 1984, Shell to resume deep pay probe off California: Oil and Gas Journal, v. 82, no. 7, p. 38.

Brooks, J. M., Kennicutt, M. C., II, Barnard, L. A., Denoux, G. J., and Carey, B. D., Jr., 1983, Applications of total scanning fluorescence to exploration geochemistry: OTC 4624, 15th Offshore Technology Conference: Houston, Texas, May 2–5, 1983, p. 393–400.

Campbell, O. E., 1946, The fluorographic method of petroleum exploration: World Petroleum, March, 1946.

Faber, E., Rehder S., and Stahl, W., 1983, Surface sediment gas data from petroliferous regions of the North Sea and Offshore California: Erdol und Kohle—Erdgas—Petrochemie, v. 36, no. 8, p. 357–361.

Hebert, C. F., 1984, Geochemical prospecting for oil and gas, using hydrocarbon fluorescence techniques, *in* Davidson, M. J., and Gottlieb, B. M., eds., Unconventional methods in exploration for petroleum and natural gas III: Dallas, Southern Methodist University Press, p. 40–58.

Horvitz, L., 1939, On geochemical prospecting: Geophysics, v. 4, p. 210–225.

———, 1972, Vegetation and geochemical prospecting for petroleum: American Association of Petroleum Geologists Bulletin, v. 56, p. 925–940.

———, 1980, Near-surface evidence of hydrocarbon movement from depth, *in* Roberts, W. H., III, and Cordell, R. J., eds, Problems of petroleum migration: American Association of Petroleum Geologists Studies in Geology, no. 10, p. 241–269.

———, 1981, Hydrocarbon geochemical prospecting after forty years, *in* Gottlieb, B. M., ed., Unconventional methods in exploration for petroleum and natural gas II: Dallas, Southern Methodist University Press, p. 83–95.

———, 1982, Upward migration of hydrocarbons from gas and oil deposits: Presented before the Division of Geochemistry of the American Chemical Society at the 183rd American Chemical Society National Meeting, Las Vegas, Nevada, March 28–April 2, 1982.

———, 1984, Near-surface geochemistry in exploration for low gravity petroleum: Read before the American Association of Petroleum Geologists Research Conference on Exploration for Heavy Crude Oil and Bitumen, Santa Maria, California, October 29–Novermber 2, 1984. To be published in forthcoming AAPG Special Volume.

———, 1985, Geochemical exploration for petroleum: Science, v. 229, no. 4716, p. 821–827.

Jones, V. T., and Drozd, R. J., 1983, Predictions of oil and gas potential by near-surface geochemistry: American Association of Petroleum Geologists Bulletin, v. 67, p. 932–952.

Laubmeyer, G., 1933, A new geophysical prospecting method, especially for deposits of hydrocarbons: Petroleum, v. 29, no. 18, p. 1–4.

Richers, D. M., Reed, R. J., Horstman, K. C., Michels, G. D., Baker, R. N., Lundell, L., and Marrs, R. W., 1982, LANDSAT and soil-gas geochemical study of Patrick Draw Oil Field, Sweetwater Country, Wyoming: American Association of Petroleum Geologists Bulletin, v. 66, p. 903–922.

Rosaire, E. E., 1938, Shallow stratigraphic variations over Gulf Coast structures: Geophysics, v. 3, p. 96–115.

Schoell, M., 1980, The hydrogen and carbon isotopic composition of methane from natural gases of various origins: Geochimica et Cosmochimica Acta, v. 44, p. 649–661.

Sokolov, V. A., 1935, Summary of the experimental work of the gas survey: Neftyanoye Khozyaystvo, v. 27, no. 5, p. 28–34.

Stahl, W., Faber, E., Carey, B. D., and Kirksey, D. L., 1981, Near-surface evidence of migration of natural gas from deep reservoirs and source rocks: American Association of Petroleum Geologists Bulletin, v. 65, p. 1543–1550.

U.S. Geological Survey, 1974, Proposed plan of development, Santa Ynez Unit, Santa Barbara Channel, Offshore California—Final Environmental Statement: U.S. Geological Survey FES 74-20, v. 1, p. II-118 to II-174. Figure II-20 is a structure map furnished by the Exxon Co., USA, and slightly modified by the Conservation Division, U.S. Geological Survey. Deep seismic data and all available core hole and exploratory well data utilized in preparing this map.

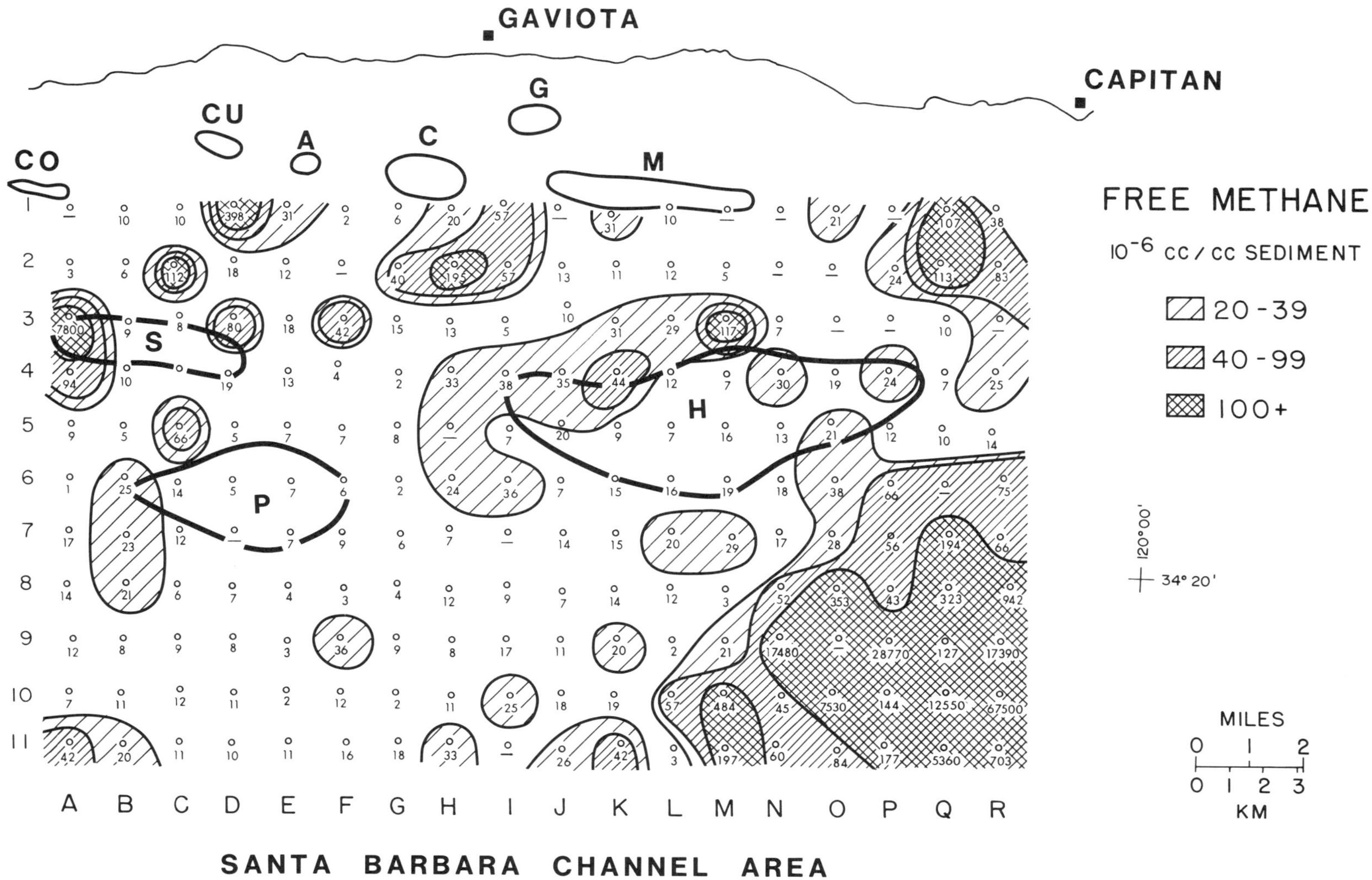

Fig. 1. High concentrations of free methane in southeastern part of survey is of biogenic origin. Lower, but anomalous, concentrations in northern part are apparently related to productive areas. Halo type pattern around Hondo (H) field is of particular interest. From Horvitz, 1984.

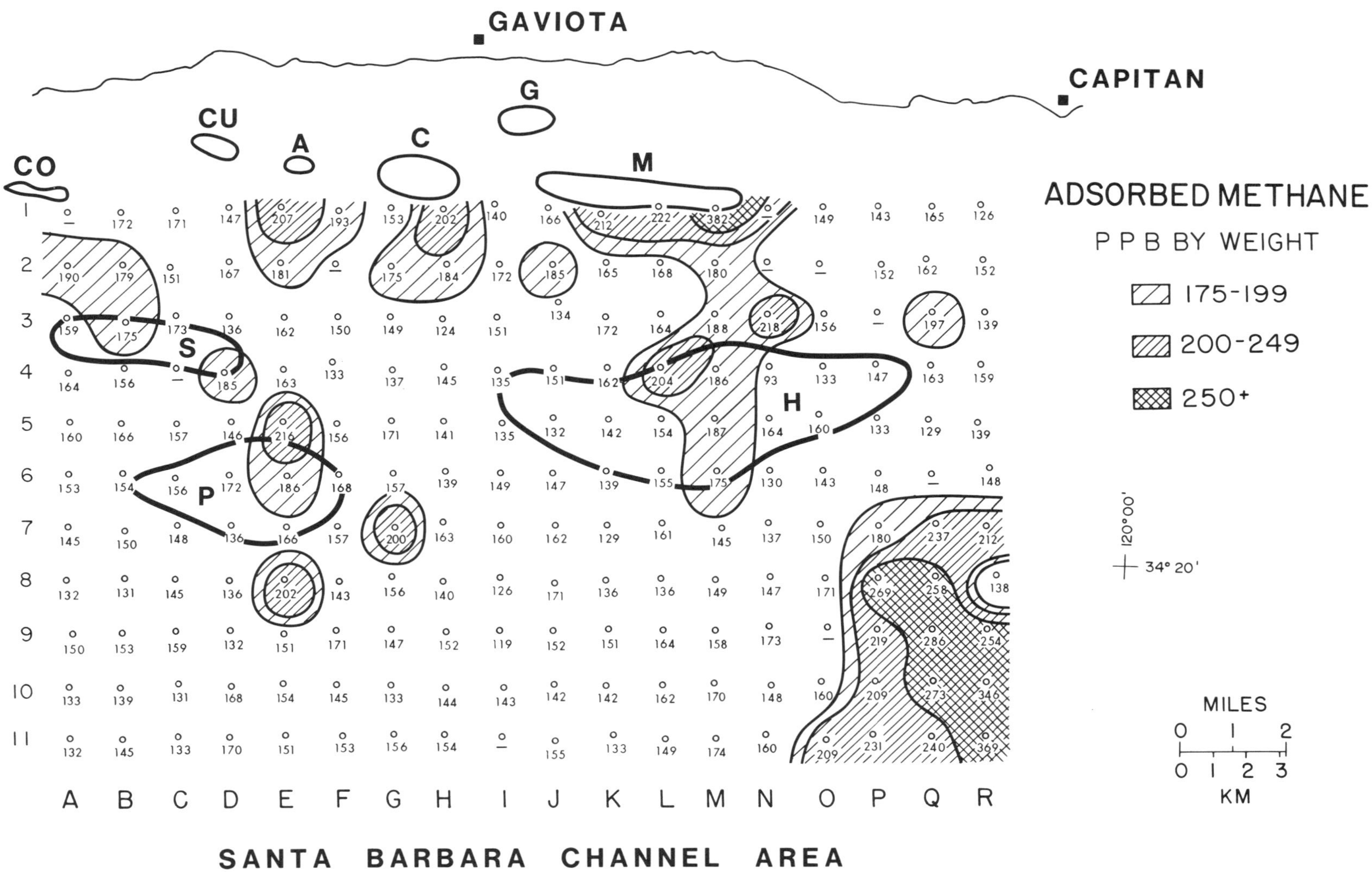

Fig. 2. Adsorbed methane from fine-grained portions of sediment (freed of indigenous biogenic methane) produced anomalous values that are related to productive areas. Incomplete anomalies over Hondo (H), Pescado (P), and Sacate (S) oil fields are expected because of low gas-oil ratios in Monterey reservoirs. From Horvitz, 1984.

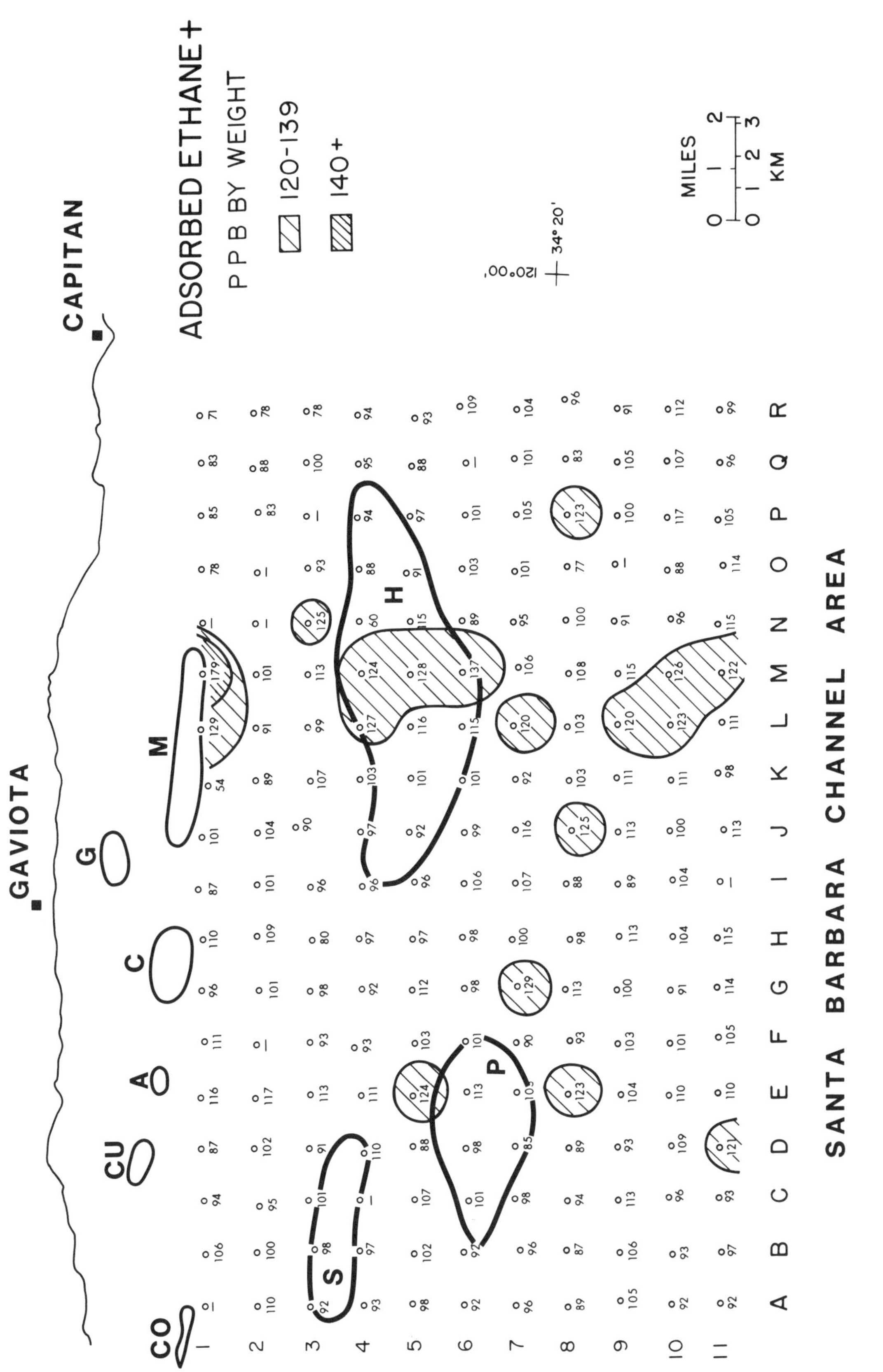

Fig. 3. Weak ethane through pentane (ethane+) anomalies are due to the low gas content of subsurface heavy oil deposits. From Horvitz, 1984.

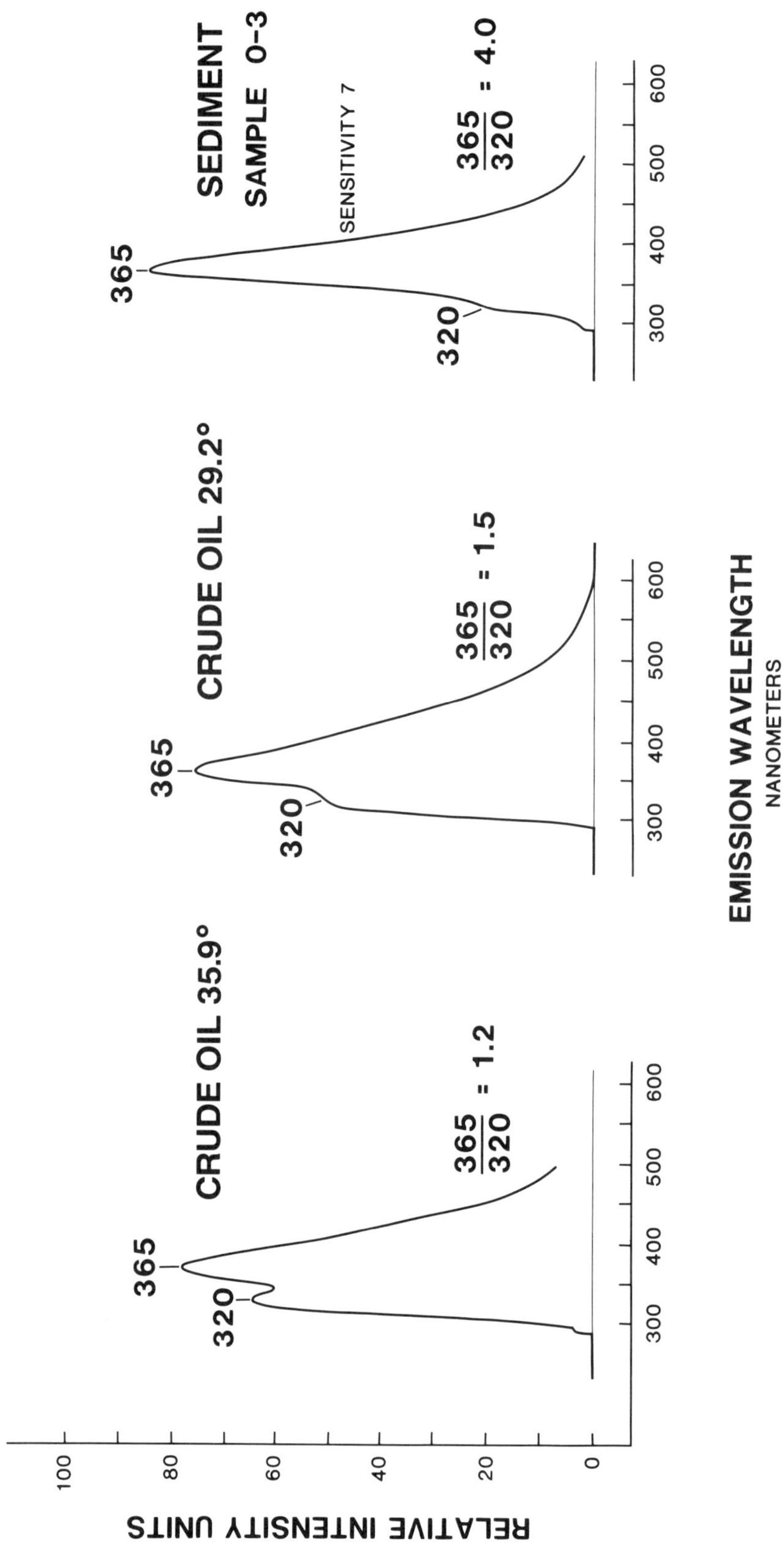

Fig. 4. Fluorogram of sediment sample near Hondo field is similar to those of crude oil samples. Ratios of 365 nm peak heights to 320 nm peak heights are related to gravity of oil; the higher the ratios, the lower the API gravity. From Horvitz, 1984.

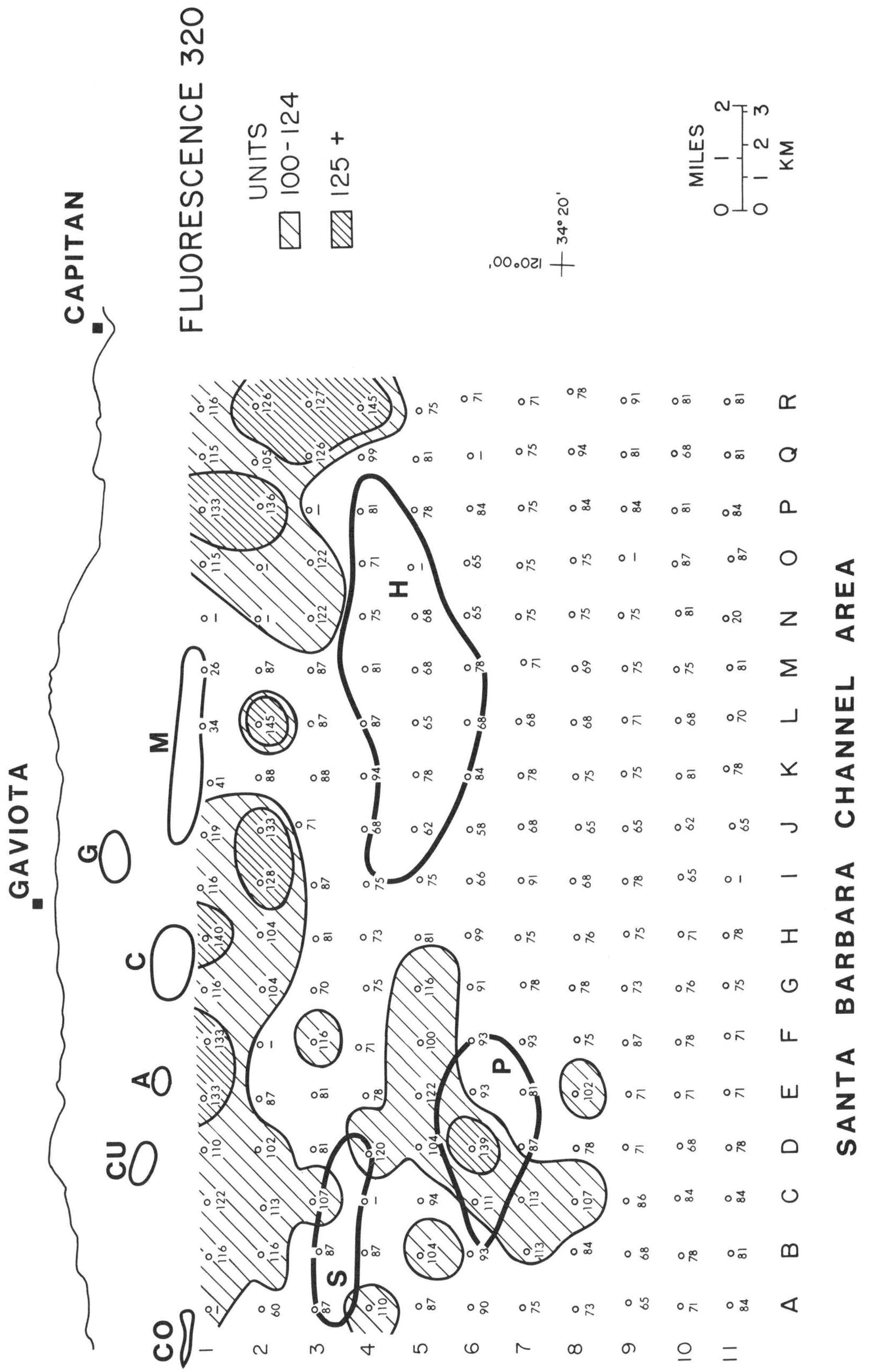

Fig. 5. Anomalous fluorescence 365 values appear in northern part of survey in vicinity of productive areas. From Horvitz, 1984.

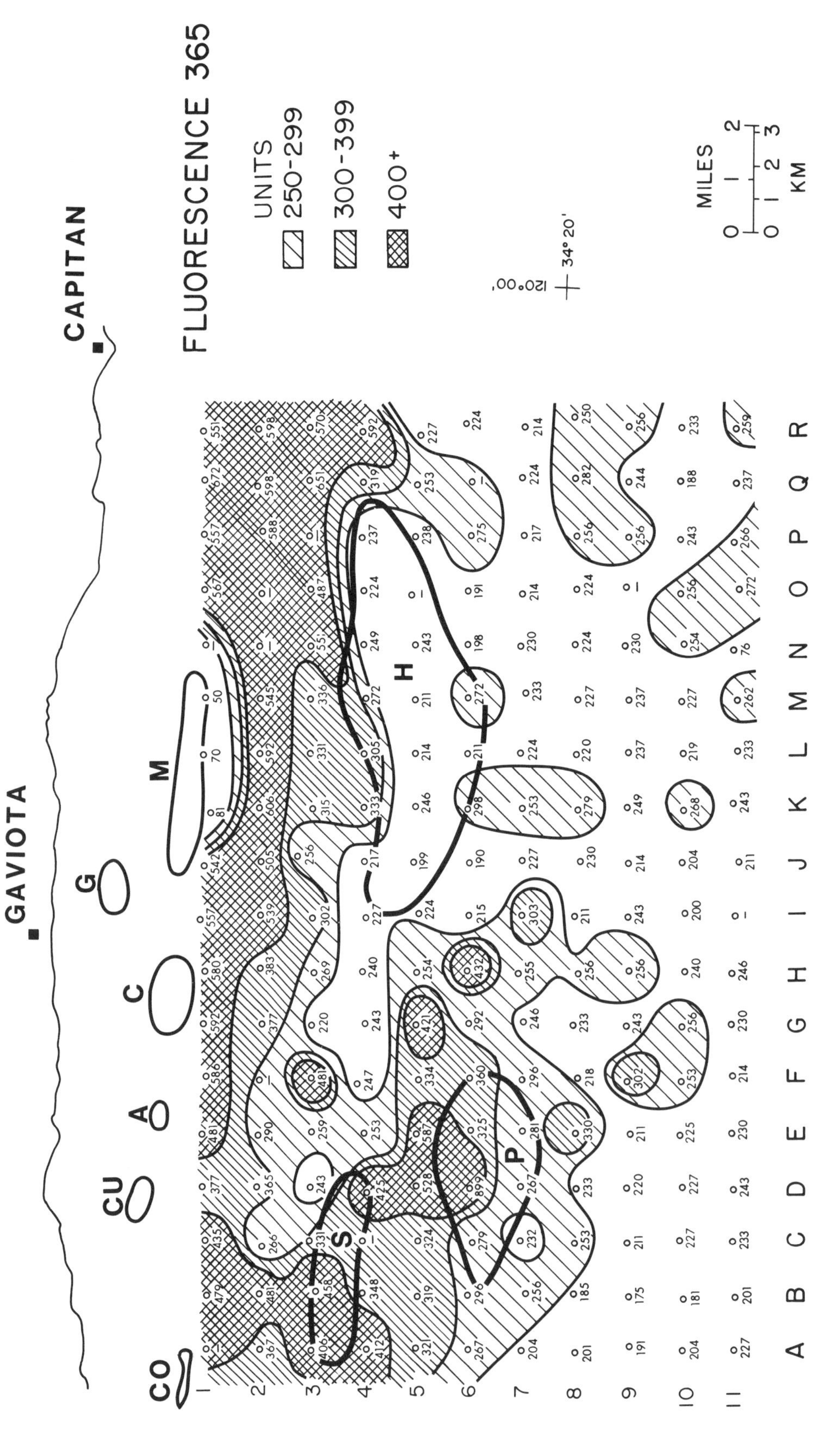

Fig. 6. Anomalously high fluorescence 365 values reflect low gravity subsurface oil accumulations. High concentrations south of Molino (M) gas field indicates possible presence of heavy oil in Molino structure. From Horvitz, 1984.

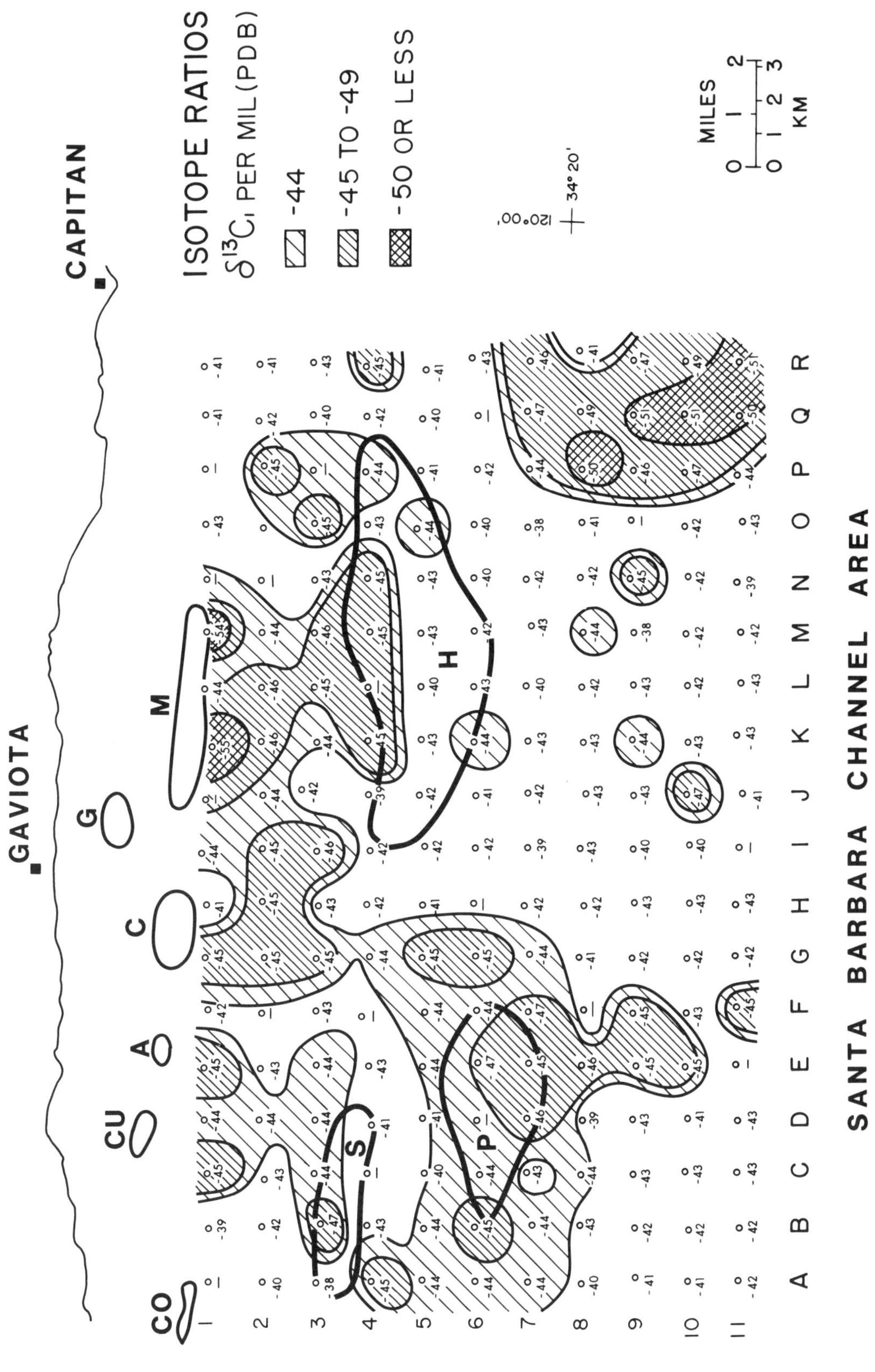

Fig. 7. Carbon isotope ratios of adsorbed methane extracted from fine-grained portions of sediment samples. Most values are in the −40 to −50 per mil range that is usually found in oil prone areas. Anomalous isotope ratio patterns developed that resemble closely those produced by the fluorescence 365 nm determinations (Figure 6). From Horvitz, 1984.

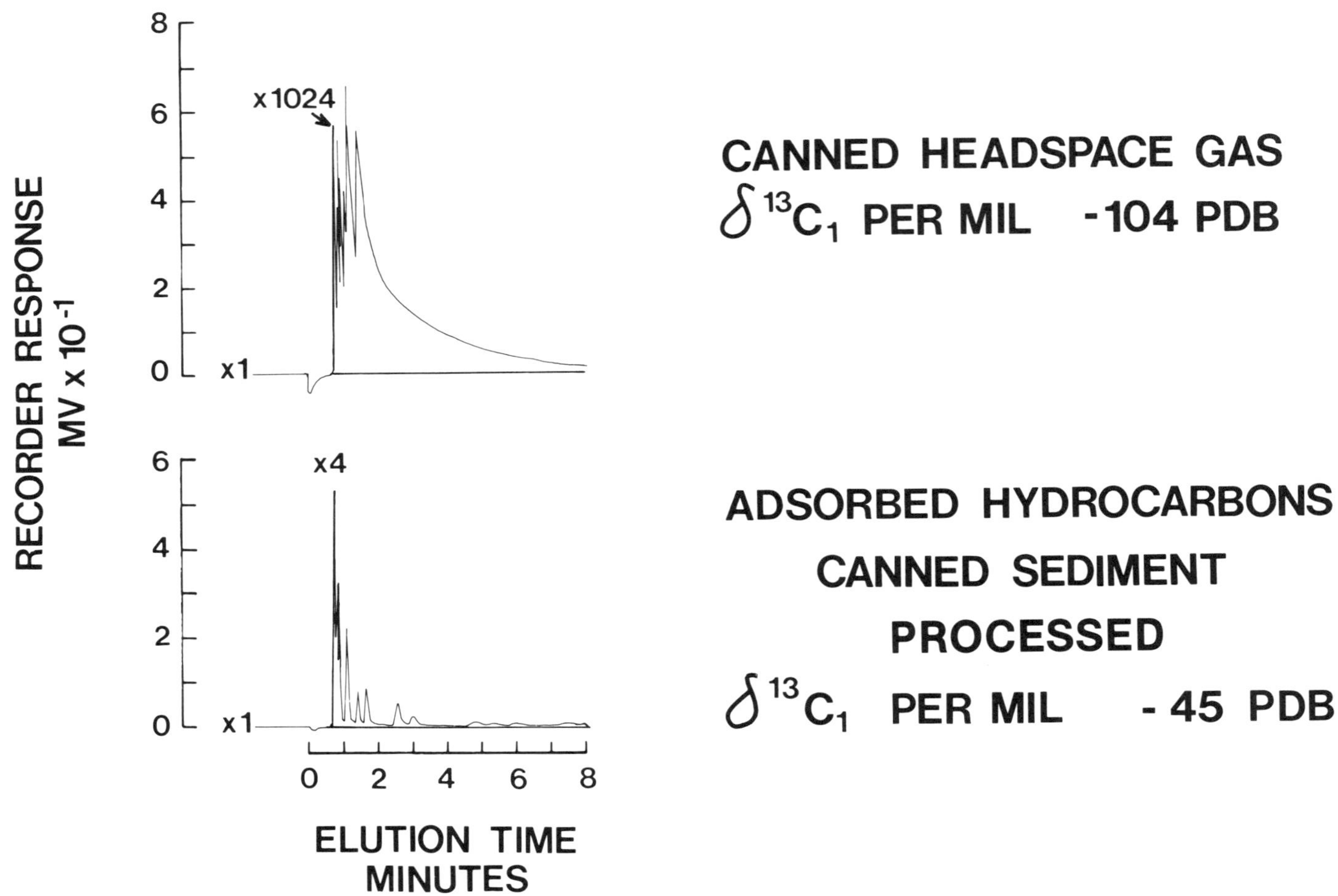

Fig. 8. Importance of determining hydrocarbons in fine-grained portions of sediment rather than in headspace gas, especially in areas containing indigenous methane, is demonstrated. Canned headspace gas (upper chromatogram) is composed mainly of biogenic methane ($\delta^{13}C$ of -104 per mil). Sediment sample (from same can), after being subjected to biogenic methane removal process, contains the whole suite of hydrocarbons including hexane isomers (lower chromatogram). Isotope value of -45 per mil indicates the methane is thermogenic.

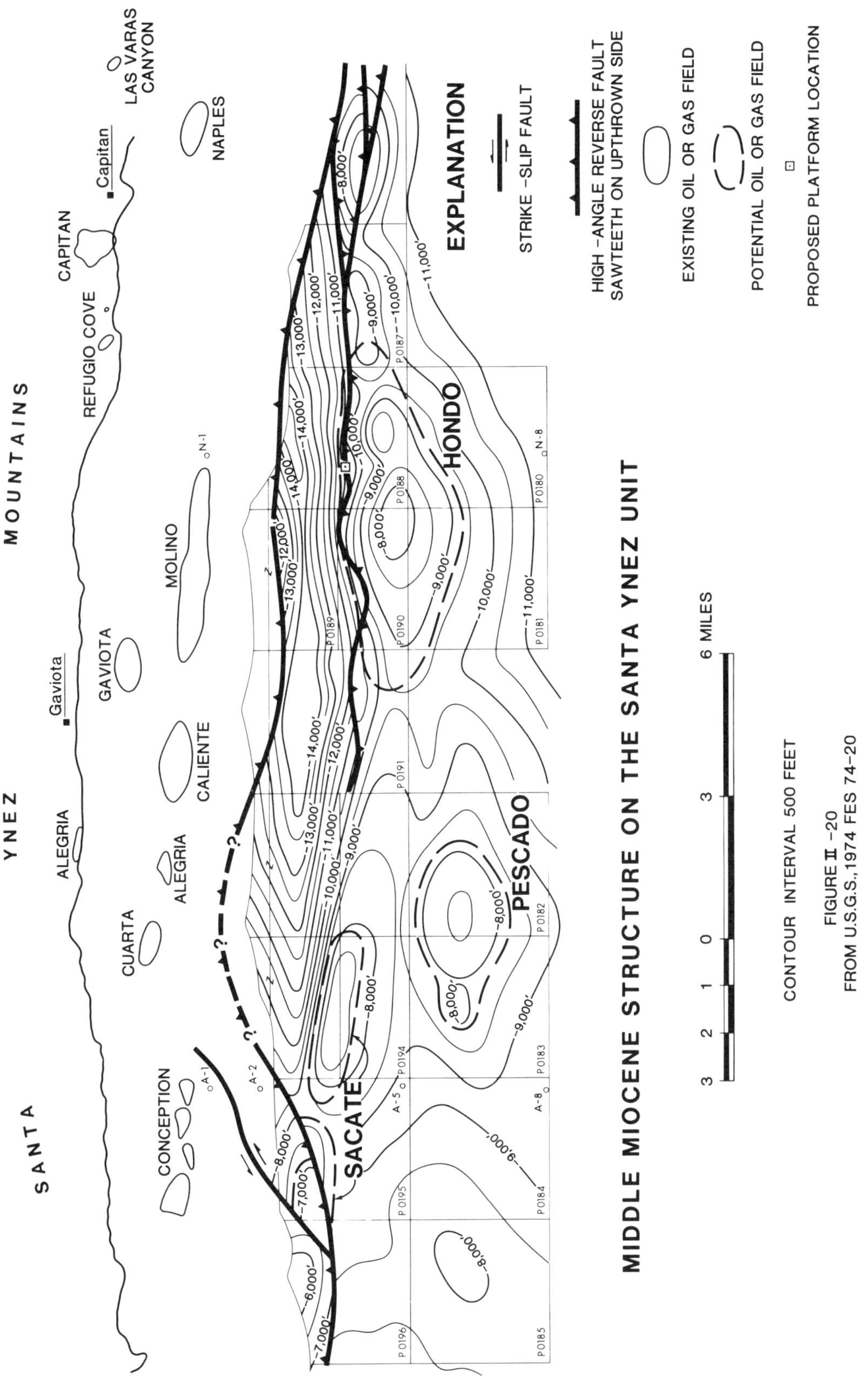

Fig. 9. Close relationships of geochemical data to Miocene structure are observed in Santa Barbara Channel area when Figures 2, 6, and 7 are compared with above map (U.S. Geological Survey, 1974).

REMOTE DETECTION AND CORRELATION OF DISTRIBUTION OF NATURAL TREE SPECIES AND SOIL CONCENTRATIONS OF LOW MOLECULAR WEIGHT HYDROCARBONS

B. N. Rock, H. R. Lang, J. B. Parrish, and C. J. LaVine

JET PROPULSION LABORATORY
CALIFORNIA INSTITUTE OF TECHNOLOGY

ABSTRACT

As part of the joint NASA/Geosat Test Case Study, scientists from the Jet Propulsion Laboratory produced vegetation classification images of a heavily forested area in the Ridge and Valley Province of West Virginia. These images revealed vegetation anomalies that were later shown to be associated with high soil hydrocarbon gas concentrations. The multispectral data used to produce these images were acquired by an airborne LANDSAT Thematic Mapper simulator flown in October, 1980, at the time of peak fall foilage display. The site was selected because it was a totally vegetated area overlying a known gas field at a depth of approximately 6,000 ft. It provided the opportunity to test remote sensing as a tool for discovering vegetation anomalies which might characterize surface zones affected by hydrocarbon seepage from a subsurface natural gas reservoir.

Detailed study of the vegetation classification images, coupled with extensive field assessment in 1981, led to recognition of subtle vegetation distributional anomalies consisting of areas of unusual maple concentrations. Maples are normally replaced by species of oak in the process of plant succession in this region. In 1982 it was proposed that these "maple anomalies" represented the forest community's response to anaerobic soil conditions that could be caused by natural gas seepage from the underlying gas field. Greenhouse studies indicate that above normal soil methane concentrations may have an adverse affect on the distribution of oak seedlings.

As part of the continuing study of the site, Gulf Science and Technology Company conducted a soil gas survey of the area in 1983. The survey identified order of magnitude higher soil concentrations of methane, ethane, propane, and butane associated with one of the maple anomalies. Soil methane concentrations from this site rank above the 97th percentile when compared with all other samples analyzed. Soil samples from two other maple anomalies were also analyzed, and yielded methane values ranked above the 75th percentile. The fact that soil samples from the maple anomaly sites contain significant amounts of ethane, propane, and butane suggest that the source of gaseous hydrocarbons at all three sites is nonbiogenic in origin.

These results demonstrate the value of employing remote sensing systems, such as the Thematic Mapper on board LANDSAT 5, to map vegetation communities as a part of hydrocarbon exploration in vegetated terrain. For the first time, remote sensing has led to imaged vegetation community maps that, in turn, have located sites of anomalous distribution of native tree species that are tolerant of high soil gas conditions. In the future, vegetation remote sensing may play a leading role in detecting hydrocarbon seepage in heavily forested areas.

This research was carried out at the Jet Propulsion Laboratory, California Insitute of Technology, under contract with the National Aeronautics and Space Administration.

INTRODUCTION

The Lost River, West Virginia, area was selected for study as part of the joint NASA/Geosat Test Case Project (Abrams et al., 1985) because it represents a heavily forested site occurring over a natural gas (methane) field. The reservoir, a Lower Devonian Oriskany sandstone, is associated with the Whip Cove Anticline at a depth of approximately 6,000 ft. The Oriskany and associated hydrocarbon-bearing units crop out at the surface approximately 4 mi to the east of the Whip Cove Anticline. Such a heavily forested site, occurring over a potential source of microseepage of methane, provides an ideal setting in which to identify vegetation responses to methane-rich soil gas conditions, using remote sensing systems such as the Thematic Mapper Simulator (TMS).

As stated above, the Lost River site is forested (approximately 80%), representing a deciduous-evergreen (oak-hickory-pine) cover type. The non-forested areas are under cultivation (lawns, pastures, and field crops such as corn and hay)and thus, surface reflectance data are nearly totally dominated by various types of vegetation.

At the beginning of the study it was assumed that any potential vegetation anomalies associated with methane microseepage would be either distributional, phenological (seasonal), or physiological (see Goetz et al., 1983, for a discussion of various vegetation responses to geochemical stress). Such anomalies might be the result of either the presence of methane in the soil, or the alteration of the soil environment by methane. In order to recognize any type of vegetational anomaly, it was necessary to determine the normal physiological state, phenological condition, and distribution pattern(s) characteristic of the local site variables (slope, aspect, elevation, soils, etc.) within the study area. The supervised vegetation classification images were produced so that such normal conditions would be recognized.

METHODS

An initial phase of the study involved development of preliminary field vegetation surveys (Rock, 1982) to establish a total of thirty-five training areas for use in a supervised classification of airborne TMS (the NASA NS-001) multispectral data acquired at peak fall foliage display (Mission 431, line 1, October 21, 1980). See Table 1 for a comparison of NS-001 TMS and LANDSAT Thematic Mapper (TM) bandpasses. Two supervised vegetation classification images were produced using a hybrid parallelepiped-Bayesian maximum likelihood classifier. The raw NS-001 band data used in these classifications were: band 2 (0.52–0.60μm), band 3 (0.63–0.69μm), band 4 (0.76–0.90μm), band 5 (1.00–1.30μm), band 6 (1.55–1.75μm), and band 7 (2.08–2.35μm). These data were acquired at an altitude which delivered an average 15 m pixel at nadir. Details regarding production of these two images are found in Abrams et al., 1985.

During the summer of 1981 detailed vegetation analyses were conducted over the entire test site, using a modified 0.1-acre circular plot method as suggested by Lindsey et al. (1958). This method has been shown to be an efficient and accurate means of woody habitat assessment (Lindsey et al., 1958; James and Shugart, 1970) and provides an accurate determination of woody plant dominance as expressed in basal area (Kulow, 1965). Field assessment of the accuracy of both supervised vegetation classification images was conducted at this time by random selection of test sites in which 0.1-acre plot analyses were compared with assigned image vegetation classes.

An extensive soil gas survey of the entire Lost River study area was conducted during July and August, 1983, by Gulf Science and Technology Company, Houston, Texas (Mathews et al., 1984). The method used employed a probe inserted into the soil to a depth of 3 ft, and an evacuated bottle to sample soil gases located at the bottom of the hole. Soil gas analyses were then conducted using a field portable gas chromatography unit, using techniques developed by Gulf.

RESULTS

As the result of work at the Jet Propulsion Laboratory, two supervised vegetation classification images were produced, one utilizing data from NS-001 bands 2,3, and 4 and band ratios of 4/5 and 5/7, and the other utilizing NS-001 bands and band ratios of 6, 3/5, 4/5, 5 and 3/6. Both of the supervised vegetation classification images recognize nine vegetation classes and one bare soil class. Based on an evaluation of interband correlation matrices, principal component analyses, and stepwise, linear discriminant analyses for the supervised classification images produced, a determination was made of the relative utility of bands for vegetation discrimination. The four most useful NS-001 bands in order of decreasing utility were: 6 (1.55–1.75μm), 3 (0.63–0.69μm), 5 (1.00–1.30μm) and 4 (0.76–0.90μm). It should be noted that NS-001 band 5 (1.00–1.30μm) is not included on the LANDSAT 4 and 5 Thematic Mapper (Table 1).

TABLE 1
NS-001 THEMATIC MAPPER SIMULATOR BANDPASSES

NS-001 Band #	Wavelength μm	Equivalent LANDSAT 4, 5 TM Band #
1	.45 – .52	1
2	.52 – .60	2
3	.63 – .69	3
4	.76 – .90	4
5	1.00 – 1.30	Not Available
6	1.55 – 1.75	5
7	2.08 – 2.35	7
8	10.4 – 12.5	6

The selection of bands for production of the first supervised vegetation classification image (2, 3, 4, 4/5, and 5/7) was based on a favorable comparison of a number of initial images (a band 2, 3, 4 color composite image; a 4/3, 4/5, 5/7 color ratio composite image; and a band 2/3 ratio color density slice) with preliminary field vegetation maps. The selection of bands for production of the second supervised vegetation classification image (6, 3/5, 4/5, 5, and 3/6) was determined by use of the stepwise, linear discriminant analysis program.

As noted above, a total of nine vegetation classes were discriminated and mapped in both supervised vegetation classification images (Rock, 1982; 1984). The classes recognized are: Evergreen, Chestnut Oak, Chestnut Oak Shaded, Scarlet Oak, Other Oak, Maple, Field 1 (dry or senescent pastures), Field 2 (green, actively photosynthetic pastures), and Scrub. A single soil class is also mapped.

A comparison of assignment of vegetation classes on each of the images with actual ground conditions shows both classifications to be highly accurate. During the summer of 1981 both of the classification images were field checked. For the second vegetation classification image a total of 289 field checks were made (using the 0.1-acre plot analysis method) of selected broadleaf forest sites, and of these, 256 were correctly classified (88.6%). The high degree of accuracy of discrimination of forest communities is likely the result of acquisition of multispectral data in October, a time of maximum differences among various community members. Based on field evaluations, the chief source of error in both classification images is the result of errors in training area assignments.

As a result of the 1981 summer evaluation of the classification images produced, subtle distributional vegetation anomalies were recognized (Rock, 1982), in the form of concentrations of both red and sugar maples occurring at low elevation sites where normal plant community succession would be expected to result in oak-hickory climax communities. Sugar maples normally occur above 2,400 ft elevation in the Lost River area, and the anomalous maple concentrations all occur at or below 2,000 ft elevation.

The results of the field assessment of the supervised classification images were presented at the fourth Geosat workshop held in Flagstaff, Arizona, in the spring of 1983 (Henderson and Rock, 1983). In response to this presentation, Gulf Science and Technology agreed to conduct a soil gas survey. The Gulf soil gas survey has identified an order of magnitude methane, ethane, propane, and butane soil gas anomaly which precisely coincides with the linear maple anomaly identified on the NS-001 vegetation classification images (Matthews et al., 1984). Figure 1 offers a comparison of the locations of three maple anomalies recognized on the vegetation classification images and the relative methane values determined by the soil gas analysis. Figure 2 presents actual methane values for twenty-one soil samples taken in the immediate area of the linear maple anomaly. Comparison of methane concentration values in Figure 2 with the summary methane values for the entire study area (Fig. 3) indicates the anomalous nature of the gas samples found at the linear maple site. Table 2 provides typical soil gas values (methane through butane) from representative sites.

Based on a statistical analysis conducted by Matthews (private communication) on these soil gas data, background values of methane for the Lost River soils range from 0.4–7.0 ppm. Of the twenty-nine gas samples taken within the three maple anomalies tested, twenty are ranked above the 75th percentile, while five of the eight samples from within the linear anomaly are ranked above the 95th percentile ranking (90 ppm or above). It is significant to note that few soil methane values below the 75th percentile ranking are found within any of the maple anomalies sampled and none are at background levels, while background values are found for soil samples taken immediately outside all of the anomalies (see Figs. 1 and 2). At sites A and B (Fig. 1) methane highs are noted that are not associated with maple concentrations. Site A occurs in an open field (i.e., no trees at this point in time) while Site B is occupied by scrub growth which, in this case, is a mix of young red maples, pines, and black locusts, and represents a recently abandoned cultivated area.

The only hydrocarbon soil gas values higher than those found within the maple anomalies are found in association with the Oriskany and older outcroppings located to the east of the gas field. Within these outcrop areas the full range of soil gas values are seen from background to above the 95th percentile ranking. To date, no detailed correlation of soil gas values with the occurrence of maples has been made for this area, since it is considered to be outside of the original test case site. It should be noted, however, that maples do occur in concentrated units along these outcrops and their associated soils, especially those of the Marcellus shales and the Oriskany sandstones.

DISCUSSION

As a result of the Lost River geobotanical studies (both ground assessment and production of the supervised classification images) the major broadleaf deciduous and needle-leaf evergreen vegetation types have been well documented in terms of their repeated association with specific topographic, lithologic, and/or physiographic features characterizing the area. When compared with the various types of forest cover occurring in West Virginia (Core, 1966), the woody vegetation found at the test

TABLE 2
GULF SOIL PROBE SAMPLE ANALYSES FOR HYDROCARBON GASES
(Samples Taken at 3 ft Depth)

Site Number	Site Description	Soil Gas Concentrations (PPM) C_1	C_2	C_3	C_4
204	Typical background over gas field (younger than Oriskany)	6.61	.15	.07	.02
628	Maple anomaly over gas field (younger than Oriskany)	179.85	13.89	12.02	5.76
17	Oriskany and older to east of gas field	122.32	4.28	2.39	.96

site represents a typical oak-hickory-pine association, characteristic of the Ridge and Valley Province of the state. The present cover represented by each vegetation class as derived from pixel assignments from both images were compared (Rock, 1982) with data presented by Sturm (1977) for the Lost River State Park and were found to correlate very well with Sturm's assessment of vegetation cover in the area. The truly anomalous nature of concentrations of maple at the lower elevations at Lost River, West Virginia, has been recently demonstrated in a study of vegetation distribution patterns for a similar Ridge and Valley site located in the Harrisburg, Pennsylvania, area (Price et al., 1985).

Laboratory studies (Parrish and Rock 1983; Gilman et al., 1981), in which woody plant species have been grown in contact with methane under controlled conditions, indicate that certain species are likely to be more tolerant of such conditions than others. Leone et al. (1977) suggest that the influence of methane is indirect, generating an anaerobic soil condition. Since red maples are commonly found growing along streambanks in which soils are likely to be saturated with water at least part of a growing season (creating an anaerobic soil condition), it is reasonable to consider them as candidates for growth in methane-rich soils.

Normal plant succession for the Lost River area should result in a climax forest cover of oak-hickory, based on the distribution of species associated with methane-free soils (Rock, 1982, 1984; Sturm, 1977). Native species possessing a tolerance of the anaerobic soil conditions that might be associated with high soil gas conditions would then be at a competitive advantage in occupying such sites when compared with intolerant species such as oak and possibly hickory. Figure 4 graphically depicts the proposed competitive succession which followed timbering activities in the Lost River area during the early 1930s.

Maples have been shown to require a different type of mycorrhizal (root) fungus than do oaks (Parrish and Rock, 1983; Abrams et al., 1985), which suggests a possible explanation for the maple's tolerance of the anaerobic soil conditions. Maples require a form of fungal association known as Endomycorrhizal, which consist of lower (i.e., more primitive) forms of fungi, while the oaks require an Ectomycorrhizal association, consisting of a higher form of fungus. Metabolic differences between the two forms of required fungal associations may allow those plants possessing Endomycorrhizal fungi to be at a competitive advantage in terms of tolerating anaerobic soil conditions. If such is the case, use of this approach need not be site or species specific. Remote discrimination, identification, and mapping of anomalous distribution patterns of species known to require Endomycorrhizal associations, and thus, more likely to be tolerant of anaerobic soil conditions possibly associated with microseepage, will prove to be a valuable exploration tool. Advanced remote sensing systems such as the TM, with its improved vegetation discriminatory capability, and airborne imaging spectrometers, with their higher spectral resolution, will prove most useful in this approach (Rock, 1983; 1984).

CONCLUSIONS

The techniques employed at the Lost River test site for vegetation discrimination and mapping have been very successful, and highly accurate supervised vegetation classification images have been produced (approximately 90% accurate assignment of forest classes). The 15 m IFOV and bandwidth selection provided by the airborne NS-001 Thematic Mapper Simulator (TMS) are well suited for vegetation mapping of heterogeneous deciduous forest cover. Multispectral data acquired with the NS-001 TMS during October foliage display proved valuable in allowing successful species discrimination and mapping. Based on data from this study, NS-001 bands 2 (0.52–0.60μm), 3 (0.63–0.69μm), 4 (0.76–0.90μm), 5 (1.00–1.30μm), 6 (1.55–1.75μm) and 7 (2.08–2.35μm) are useful in providing accurate vegetation discrimination, classification, and mapping of eastern deciduous forest species and species associations, with bands 6, 3, 5, and 4 the most useful respectively.

Analysis of the supervised vegetation classification images has identified vegetational distributional anomalies in the form of high concentrations of maples at low elevation sites in areas where normal plant successional development would be expected to lead to oak/hickory climax communities. Soil gas analyses at three such maple anomalies has identified an order of magnitude soil methane, ethane, propane, and butane anomaly at the linear maple anomaly, and soil gas (all fractions) concentrations above the 75th percentile ranking at the other two.

Based on the results of the Lost River study, a useful remote-sensing-based geobotanical exploration approach has been demonstrated. Discrimination and mapping of anomalous distribution patterns of native species tolerant of anaerobic soil conditions will assist in focusing additional exploration activities on localized sites within an area of study. Heavily forested terrain may thus be used to advantage in the remote detection of hydrocarbon microseepage.

REFERENCES

Abrams, M. J., Conel, J. E., and Lang, H. R., 1985, The joint NASA/Geosat test case project final report: American Association of Petroleum Geologists, Tulsa, Oklahoma, vol. 2.

Core, E. L., 1966, Vegetation of West Virginia: Parsons, West Virginia, McClain Printing Co.

Gilman, E. F., Leone, I. A., and Flower, F. B., 1981, The adaptability of 19 woody species in vegetating a former sanitary landfill: Forest Science, v.27, no.1, p.13–18.

Goetz, A. F. H., Rock, B. N., and Rowan, L. C., 1983, Remote sensing for exploration: An overview: Economic Geology, v. 78, no. 4, p. 573–590.

Henderson, F. B., III, and Rock, B. N., eds., 1983, Frontiers for geological remote sensing from space: American Society of Photogrammetry, 80 p.

James, F. C., and Shugart, Jr., H. H., 1970, A quantitative method of habitat description, Audubon Field Notes, v. 24, no. 6, p. 727–736.

Kulow, D. L., 1965, Elementary point-sampling, Morgantown, West Virginia: West Virginia University Agricultural Experimental Station Circular, no. 116.

Leone, I. A., Flower, F. B., Arthur, J. J., and Gilman, E. F., 1977, Damage to woody species by anaerobic landfill gases: Journal of Arboriculture, v. 3, no. 12, p. 221–225.

Lindsey, A. A., Barton, J. D., and Miles, S. R., 1958, Field efficiencies of forest sampling methods: Ecology, v. 39, p. 428–444.

Matthews, M. D., Jones, V. T., and Richers, D. M., 1984, Remote sensing and surface hydrocarbon leakage: International Symposium on Remote Sensing of the Environment, Proceedings, p. 663–670.

Parrish, J. B., and Rock, B. N., 1983, The effect of soil methane on *in vitro* growth and vigor of *Quercus prinus* seedlings, Botanical Society of America, v. 70, no. 5, pt. 2, p. 52.

Parrish, J. B., 1985, The use of geobotanical and remotely sensed data in an Appalachian hydrocarbon exploration strategy: The Lost River, West Virginia, NASA/Geosat test site, University Park, Pennsylvania, Ph.D. diss., Pennsylvania State University, University Films International, Ann Arbor, Michigan.

Price, C. V., Birnie, R. W., Logan, T. L., Rock, B. N., and Parrish, J., 1985, Discrimination of lithologic units on the basis of botanical associations and LANDSAT TM spectral data in the Ridge and Valley Province, International Symposium on Remote Sensing of Environment, Proceedings, p. 125–140.

Rock, B. N., 1982, Mapping of deciduous forest cover using simulated LANDSAT-D TM data: International Geoscience and Remote Sensing Symposium (IGARSS '82), Digest, v. 1, WP-5, p. 3.1–3.5.

———, 1983, Preliminary airborne imaging spectrometer vegetation data: 1983 International Geoscience and Remote Sensing Symposium (IGARSS '83) Digest, v. 2, FA-1, p. 5.1–5.4.

———, 1984, Remote detection of geobotanical anomalies associated with hydrocarbon microseepage using Thematic Mapper Simulator (TMS) and Airborne Imaging Spectrometer (AIS) data: International Union of Geological Sciences and UNESCO, BRGM document no. 82, p. 299–309.

Sturm, R. L., 1977, Comparison of forest cover types in seven environmentally diverse areas in West Virginia, Morgantown, West Virginia: Master's thesis, West Virginia University.

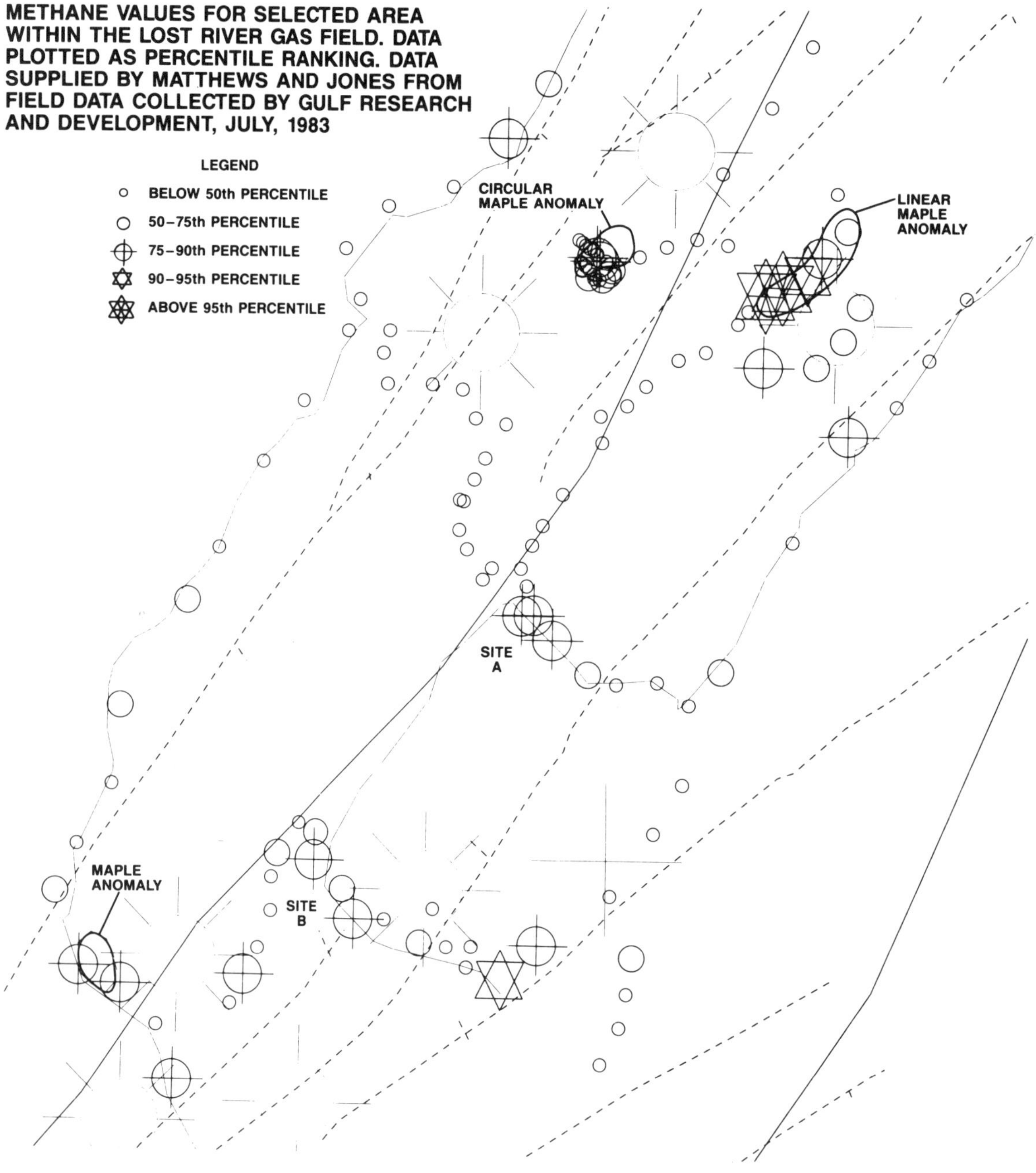

Fig. 1. Diagrammatic representation of results of 1983 soil gas analysis conducted by Gulf Science and Technology in the area of the three maple anomalies. Methane values are plotted as percentile ranking. Courtesy of M. Matthews.

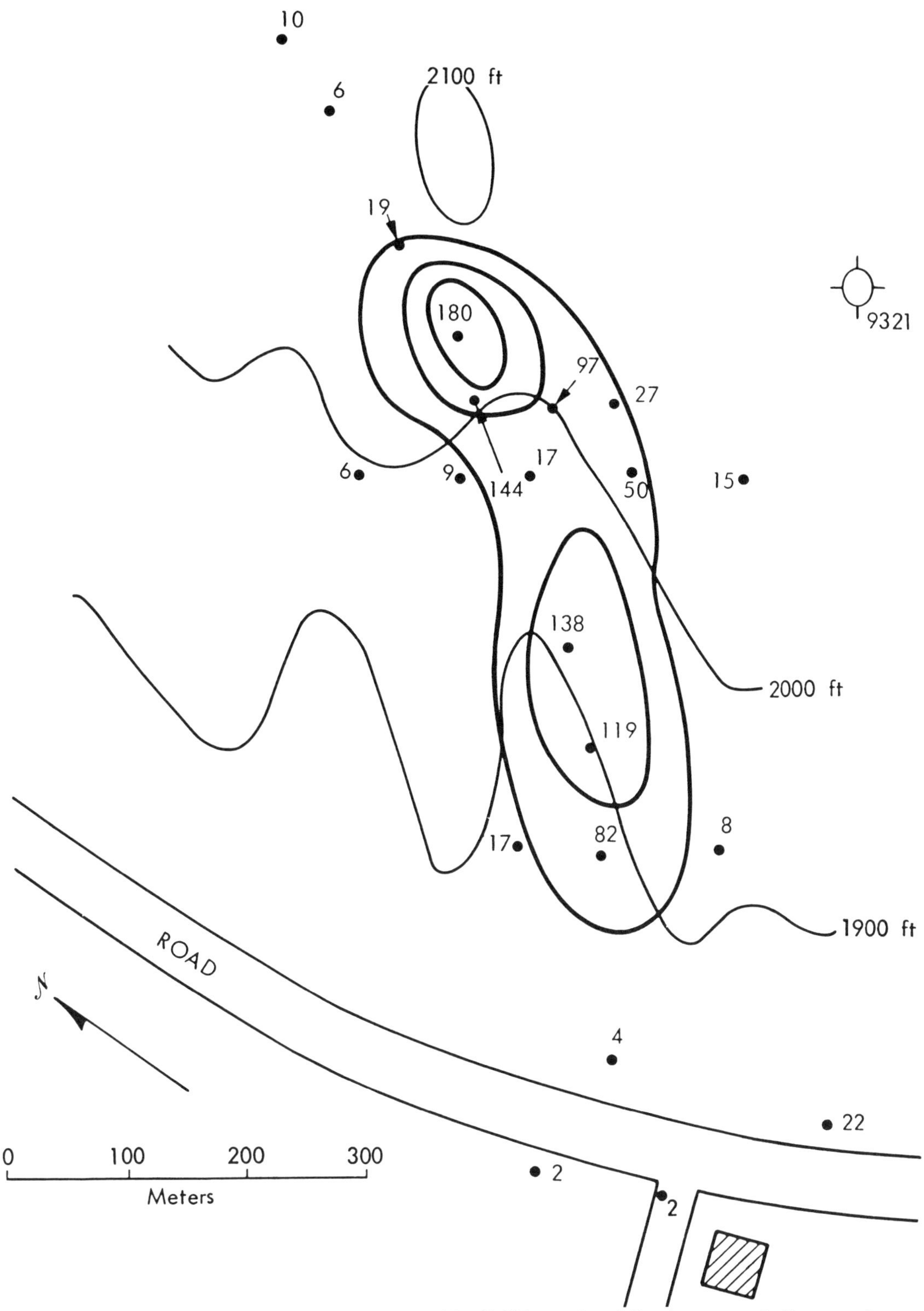

Fig. 2. Details of result of twenty-one gas samples collected by Gulf Research and Development at the linear maple anomaly. Numbers represent methane values in parts per million (ppm). Light lines represent topographic contours and cultural features. Heavy lines represent inferred concentration gradient for methane gas. The boundary of the linear maple anomaly corresponds precisely with the lowest gradient contour. Courtesy of J. Parrish and H. Lang (personal communication).

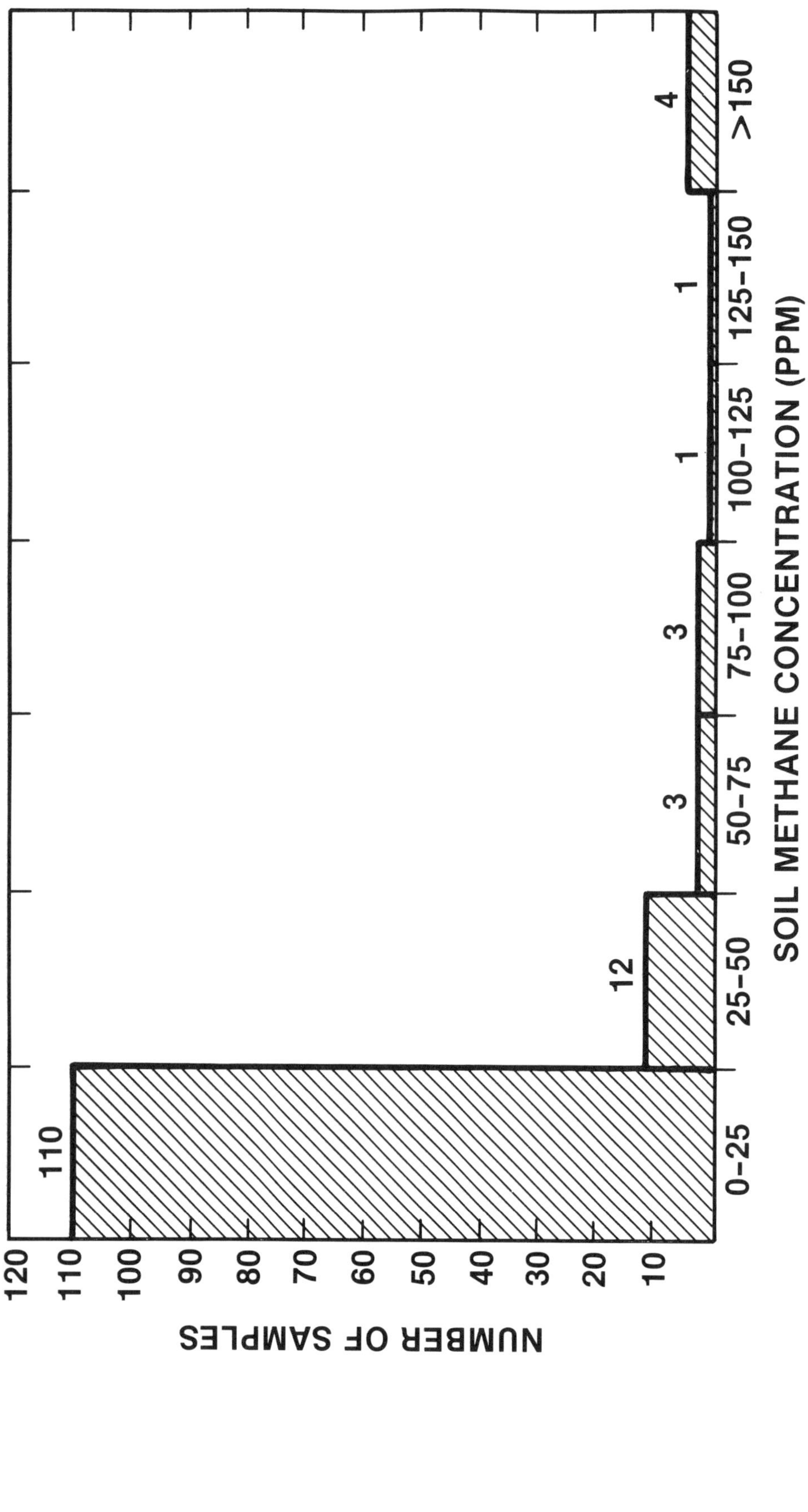

Fig. 3. Comparison of methane concentrations from randomly selected soil gas sample analyses provided by Gulf Science and Technology. Based on sample analyses for 20% of the total samples collected. According to Gulf statistical analysis, methane values of 90 ppm or more fall above the 95th percentile ranking.

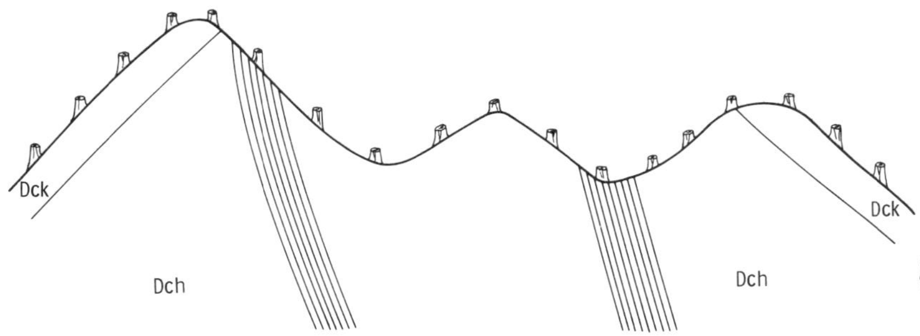

A. Area immediately following timbering activities.

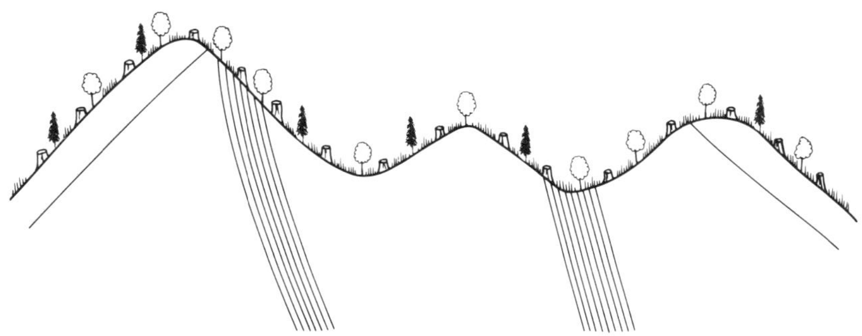

B. Pioneer regrowth community of grasses and scrub growth (a mix of broadleaf trees such as red maples, and evergreens such as pines).

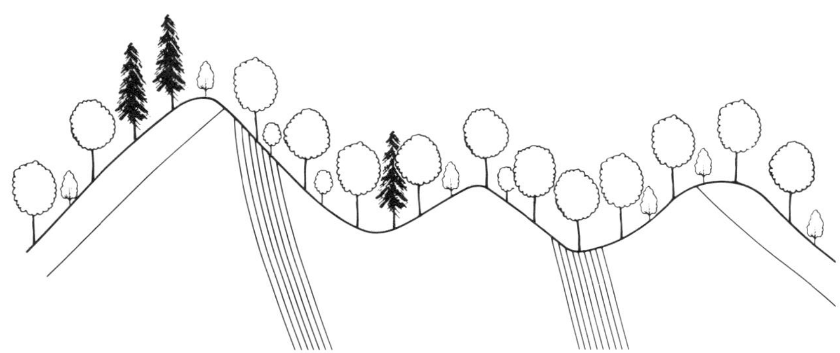

C. As pioneer community matures, shade created by its canopy prevents maple and pine seedlings from growing except where they are at a competitive advantage over more shade-tolerant species, such as oaks.

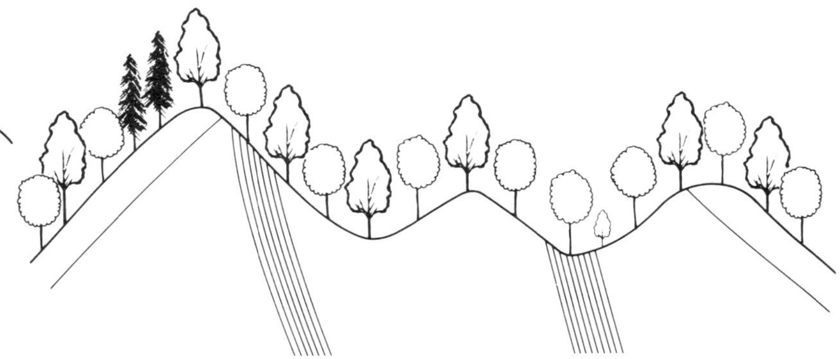

D. Oak and hickory climax species replace the pioneer species except where the climax species are excluded because of adverse conditions, such as anaerobic soils.

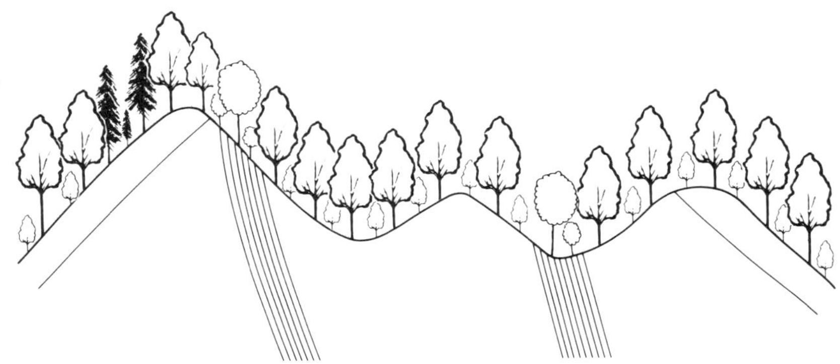

E. Climax community reproduces itself except at sites where certain species (maples) are at a competitive advantage.

Note: The composition of the forest community has changed in the intervening fifty years from a pioneer community dominated by red maples and pines to a climax community dominated by oaks. The residual maple concentrations persist because of their tolerance of anaerobic soil conditions (oaks are intolerant of the same conditions), while small patches of pines persist at higher elevations, especially at outcrop sites, characterized by thin soils.

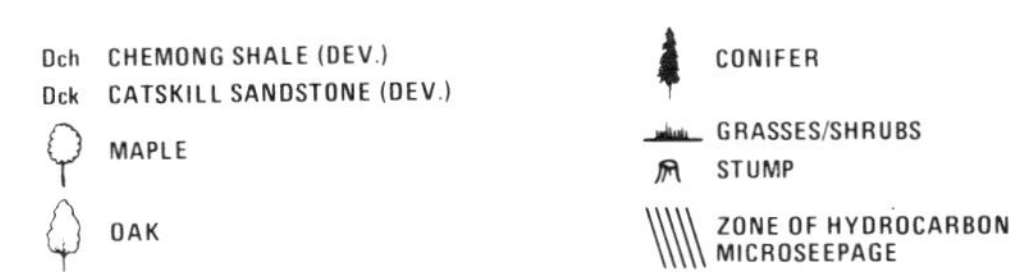

Fig. 4. Diagrammatic representation of forest successional sequence proposed to have occurred at the Lost River site after timbering activities in 1930–1935. Diagram courtesy of Parrish (1985).

DELTA CARBONATE GEOCHEMICAL HYDROCARBON PROSPECTING

William Duchscherer, Jr.

GEOCHEMICAL SURVEYS, INC.

ABSTRACT

Early in the history of petroleum exploration it was recognized that underground oil and gas accumulation were very often manifested through perceptible gas and oil seepages. Based upon that reality during the 1930s, Everett DeGolyer and Eugene McDermott, within Geophysical Service, Inc., systematically and at great expense developed a geochemical prospecting method now termed the Delta Carbonate method (ΔC). This specific ΔC method has been employed by Geochemical Surveys, Inc., since the separation of that company from G.S.I. in 1942. Hundreds of thousands of near-surface soil samples, taken from both onshore and offshore areas, have been analyzed for that specific carbonate which yields ΔC. The method depends upon measuring the yield of carbon-dioxide released from carbonate by thermal decomposition within a given temperature window.

What is presented here is an intellectual construct which attempts to explain the empirical results that have been used successfully for over fifty years. From January, 1942, to January, 1984, the ΔC method initiated the exploration process which resulted in the drilling of 184 prospects, ultimately resulting in 39 new oil and gas fields within which there were eventually drilled 1,084 field wells. The exploratory tests resulted in a 21% new field wildcat sucess ratio.

It has been observed that the ferroan nature of carbonates allows the incorporation of ferrous ions (Fe^{++})in the carbonate lattice structure, either adsorbed or absorbed, to produce ferroan calcite. In ΔC geochemical exploration the thermal oxidation (combustion) of ferroan calcite ($FeCaCO_3$) may produce the CO_2 in accordance with the reaction:

$$FeCa\,(CO_3)_2 \longrightarrow FeO + CaCO_3 + CO_2.$$

Nisle (1941) provides a plausible explanation for the progressive accumulation in the near-surface soils of such salts as carbonates. It is not inconceivable, however, that there may be additional sources of thermally released carbon dioxide within the specific temperature window; e.g., the metabolic action of microorganisms acting upon hydrocarbons.

The amount of carbon dioxide (dissociated in the thermal decomposition process) is measured statistically and plotted on a map at the location from which the soil sample was taken. The results are then contoured and interpreted. The resulting geochemical anomaly invariably manifests itself as a halo yielding high carbon dioxide values around the edges of oil and gas fields, and low values over the center of the field. The migration mechanism by which the low molecular weight hydrocarbons move from a petroleum accumulation into the near-surface may not yet be understood completely. However, the halo as an empirical phenomenon has been evidenced by geochemists using methods other than ΔC geochemistry. Apical anomalies, as opposed to halo anomalies, usually represent smaller or narrower fields.

Experience with the ΔC method suggests that it is an excellent supplementary tool when used with subsurface and seismic methods either as a reconnaissance tool or, more significantly, in upgrading the quality of conventional prospects.

INTRODUCTION

The numerous and very evident oil and gas seeps of earlier times are gone, but we now find geochemical prospectors searching for traces of mobile low molecular weight hydrocarbons (methane, ethane, propane, etc.) which have migrated into the near-surface and left their signatures in one measurable form or another. Once identified, these hydrocarbons, or their effects (e.g. ΔC), may be used to locate deeply buried subsurface petroleum accumulations.

The early scientists of modern geochemical hydrocarbon prospecting (such as Laubmeyer, Sokolov, Blau, Rosaire, Horvitz, McDermott, and Mogilevsky) searched for the causes of the geochemical effects they observed. These scientists were not always convincing when describing the mechanisms they felt accounted for the surface anomalies. However, the empirical relationship between petroleum accumulations at depth and surface geochemical anomalies was and is a valid one (Horvitz, 1980).

Scientific thinking involves the attempt to relate cause and effect. Deductive reasoning, as used in mathematics, begins with an established cause and then looks for the logical effects resulting from that cause. Inductive reasoning, as used for example in geochemical hydrocarbon prospecting, views material effects and attempts to determine, or reasonably to justify, the material causes of such observable effects. To denigrate the role of inductive reasoning is, literally, to deny both the history of science and the history of technology. Those who belittle geochemical hydrocarbon prospecting because all the t's are not yet crossed, nor all the i's dotted, effectively impede the achievement of its potential.

The last half-century has witnessed vast growth in all phases of geochemical knowledge and measuring techniques. Further research and continued scientific investigation will find superior answers to existing questions, which will in turn provide the confidence to rely more strongly on geochemically based prospecting decisions. Past geochemical prospecting successes are not isolated or random events. Rather, they are the consequences of geochemical processes in nature that have become increasingly evident.

Figure 1 is a conceptualized vertically migrating hydrocarbon pattern which gives rise to the ΔC halo phenomenon. The halo is empirically determined from a multitude of soil samples measured for ΔC and may result from either a structural or stratigraphic trap. W.H. Roberts III (1981) observes that deep-water discharge carries into the reservoir some oil and gas which rises to the surface. He states: "The discharge of chemically imprinted waters from depth shows up on air photos and satellite imagery in the form of anomalous soil and vegetation patterns."

Since the inception of petroleum geochemical exploration, a rather severe state of tension has existed between surface geochemical and other more widely used exploration methods. The accumulation of increased quantities of vertical migration data of many different kinds (of which surface geochemistry is only one kind) are now easing these past tensions. Even though theoretical advances are being achieved, it is the empirical case histories, both successes and failures, which account for the belated and rapidly growing acceptance of those surface methods which are dependent upon vertical migration.

The hydrocarbon geochemist evaluates his conceptual system by the faithfulness with which conclusions drawn from that system correspond to laboratory or field observations. This is an essential part of the scientific method (Duchscherer, 1984a). It seems apparent today that many petroleum geochemists tend to be reductionists; i.e., they have the tendency to reduce the multidimensionality of hydrocarbon geochemistry to a single dimension producing a one-sided view of reality. Each has his own narrow concept of the process of vertical migration which is taken to be the only valid one. There is much in geochemical hydrocarbon prospecting that needs to be more closely observed, better understood, and more effectively applied. In geochemical prospecting there are many "mansions" and there are many different roads to reach them.

THE ΔC METHOD

Duchscherer (1984a) provides evidence in the form of actual ΔC case histories of producing oil and gas fields. The soil alteration or indicator material yielding ΔC is a special form of carbonate mineralization in the soil overlying petroleum accumulations. This mineralization or soil alteration finds its origin in the oxidation of vertically migrating low-molecular-weight hydrocarbons upon their arrival in the near-surface soils (Duchscherer, 1979, 1980, 1981, 1982, 1984 a,b).

A perfectly satisfactory explanation of the emplacement mechanisms of this special carbonate in the near-surface soils does not exist. It is observed, however, that leaking hydrocarbons provide a locally reducing environment and thus contribute to the reduction, mobilization, and subsequent redistribution of various chemical elements and compounds distributed over and around underlying petroleum accumulations. It is also observed that the ΔC anomaly will remain in the near-surface soils and sediments as a fossil anomaly even after the petroleum accumulation is depleted by production or otherwise dissipated by normal geological processes.

A plausible but much overlooked mechanism for the origin of the observed geochemical anomalies over and around petroleum accumulations is given by the thermodynamic calculations of Nisle (1941) by means of which it is illustrated that gas migrating vertically through the near-surface soils is capable of enhancing the normal evaporation rate of the ground water, thus causing an alteration of the sedimentary overburden through which these gases migrate. Because of natural processes, such as osmotic pressure and capillary action, the evaporated water would be replaced by normal ground fluids carrying their quota of normal salts from the surrounding sediments. This overall evaporative effect would result in a progressive accumulation in the near-surface of such inorganic salts as carbonates, sulfates, halides, iodides, chlorides, and bromides. These materials may be determined easily and accurately through soil sampling collected from a depth which will insure against near-surface contaminants.

In general practice, ΔC soil samples are collected at a depth of 5–7 ft below the surface of the ground. About a pint of soil is placed in an ordinary sample sack. Since the soil alteration is a stable, cumulative, and indurated carbonate, no precautions or other extraneous effects need be guarded against.

Soils are defined herein as the products of weathering that remain in situ above weathered parent rocks, and their complex chemical composition consists largely of alumina, silica, and calcium compounds. Soil may be viewed as the reaction occurring between a parent material consisting of some rock formation, and a "dead end" product, usually extremely resistant, of the weathering process. Soils may be transported from this "in-situ" weathering location and are usually finely divided particles which present a physiochemically active surface upon which the low-molecular-weight hydrocarbons are retained by adsorption and/or absorption.

Geochemical Surveys, Inc., in its soil sample analysis methods, determines the special carbonate mineralization in its ΔC method by means of thermal decomposition. Briefly, the clay fraction of the soil sample is first preheated in an oxygen atmosphere at 500°C to decompose any low temperature carbonates or carbonaceous materials. The sample, after a critical time period, is then passed to a 600°C furnace, in a nitrogen atmosphere, where the rate of evolution of carbon dioxide evolved is measured by gas thermal conductivity cells. The quality of carbon dioxide has been given the name ΔC. Through forty-three years of stringent laboratory controls and research, including ongoing intensive literature search, it has been determined that there is no known massive carbonate compound, thus far investigated, that will thermally decompose in the temperature window of 500°C–600°C. The above noted procedure of treating soil samples will have removed any vegetable matter and/or extraneous organic materials not related to oil and gas accumulations prior to reaching the particular temperature window. The following chemical reactions are offered to explain how the carbon dioxide may be derived from some of the lighter hydrocarbons.

$$CH_4 + 2O_2 \rightarrow 2H_2O = CO_2 \quad \text{methane}$$

$$2C_2H_6 + 7O_2 \rightarrow 6H_2O + 4CO_2 \quad \text{ethane}$$

$$C_3H_8 + 5O_2 \rightarrow 4H_2O + 3CO_2 \quad \text{propane}$$

$$2C_4H_{10} + 13O_2 \rightarrow 10H_2O + 8CO_2 \quad \text{butane}$$

The carbon dioxide thus evolved dissolves and weathers the alumino-silicates in the soils to produce carbonates, silica, and alumina much in the following manner:

$$CaAl_2Si_2O_8 + 2CO_2 + H_2O \rightarrow Ca^{++} + 2HCO_3^- + Al_2O_3 + 2SiO_2$$

which yields,

$$Ca^{++} + 2HCO_3^- \rightleftharpoons CaCO_3 + H_2O + CO_2$$

The final dissolution products, $CaCO_3$ and CO_2, may incorporate with other elements to be precipitated as a diagenetic carbonate cement in the near-surface soils over and around underlying petroleum accumulations. Donovan (1974, 1977) explains these diagenetic features as near-surface evidence of buried hydrocarbon deposits and suggests that the observed color changes in the surface rocks were due to the reduction of iron by migrating hydrocarbons. This reduction may occur by at least two chemical mechanisms. In the first, reduced iron combines with hematite and water to form magnetite:

$$Fe^{++} + Fe_2O_3 + H_2O \rightarrow Fe_3O_4 + 2H^+$$

In the second, Fe^{++}, having been reduced at depth, migrates vertically into the oxidizing zone where it is oxidized, producing magnetite and maghemite and/or ferroan calcite cements directly. The surface soils are thus significantly altered and possess distinctive new physiochemical properties that are measureable.

Pirson (1973) states that the reducing environment associated with underlying petroleum accumulations can be of great magnitude. This would enable divalent iron to be organically complexed or adsorbed/absorbed, allowing the incorporation of Fe^{++} ions to form ferroan calcite cements.

To arrive at the special carbonate which yields ΔC in the temperature window of 500°C–600°C (Fig. 2), a type of siderite, $FeCO_3$, and/or ferroan calcite, $FeCa(CO_3)_2$, will evolve carbon dioxide.

Figure 3 illustrates the endothermic and exothermic phase transformations of the more common carbonates. Conventionally, the recorded curves are oriented so that the endothermic reaction (heat consuming) is shown as a downward deflection and the exothermic reaction (heat releasing) as an upward deflection (Kulp, 1951).

Lilburn and Zuhair (1983) explain the ferroan nature of carbonates by the incorporation of Fe^{++} ions in the carbonate lattice to produce ferroan calcite by the following reaction:

$$CaCO_3 + xFe^{++} \rightleftharpoons (Ca_{1-x}, xFe_x)CO_3 + xCa^{++}$$

Figure 4 illustrates how the CO_2, named ΔC, is arrived at through thermal decomposition. Carbon dioxide may also be formed by microbes in the near-surface oxidizing environment; however, in the ΔC method of analysis the preheating temperature destroys these extraneous effects.

Another diagenetic product found over and around petroleum accumulations is reported by Donovon, Forgey, and Roberts (1979). They interpret the magnetic anomaly over the Cement Oil Field "as reflecting abundant near-surface magnetite formed by the reduction of hydrated iron oxide and/or hematite as a direct result of petroleum microseepage." They also observed a strong petroleum influence on the isotope ratios of carbon in the carbonate cements of surface rocks over the oil and gas deposit.

It is these soil alterations that produce the halo pattern found around the perimeters of most underlying petroleum accumulations (Duchscherer, 1980). The halo type of anomalous pattern, resulting from soil gas or soil sample analyses, has been reported in many different types of surveys, e.g., soil hydrocarbons, soil fluorescence, soil wax, trace metals, iodine and radiometrics (Horvitz, 1945; Ransone, 1958; Debnam, 1969).

EXPERIENCE WITH ΔC GEOCHEMICAL PROSPECTING

Early geochemical prospecting for hydrocarbons utilized methods that, empirically, were not adequately tested at the time. Nearly all the early surveys were limited to small areas with insufficient background coverage, and mainly over producing fields. Analytical errors were found to be large; in many cases this was due to the very nature of the indicator used. Sparse sampling led to arbitrary contouring of the data, which on close scrutiny proved to have several alternative interpretations. There was also a tendency of many early geochemical prospectors to place too much emphasis on geochemical methods alone in order to outline drillable prospects.

Effective ΔC geochemical prospecting requires extensive surface sampling coverage in order to arrive at a reliable value for the geochemical background. All ΔC values are subjected to frequency curve studies and evaluated statistically for a contour value. The final ΔC values are then recorded at the sample location on base maps at scales of 1 in or 2 in to the mile. Once a contour value is determined, anomalies are contoured and the resulting patterns are interpreted as edge leakage forms or halos and are graded in terms of size, intensity, and completeness of form. The usual test of quality is the "contrast ratio," R, for each geochemical anomaly. This ratio shows the degree of contrast between the anomalous pattern and the geochemical background as follows:

$$R = \frac{\text{Average } \Delta C \text{ values inside leakage pattern}}{\text{Mode}}$$

It has been demonstrated that contrast ratios generally greater than 1.5 show greater success than those with lower contrast ratios. A good general correlation between ΔC and known fields exists in all areas investigated (Duchscherer, 1984a). One thousand eighty-four wells were drilled in the thirty-nine fields listed in Table 1. Twenty-five of the field discoveries were stratigraphic traps which, it is reasonable to assert, could not have been found at the time of discovery by any conventional exploration method. Thirty-five case histories are presented in Duchscherer (1984a). From January, 1942, to January, 1984, the ΔC method was employed in the exploration and eventual drilling of 184 oil and gas fields for an overall new field wildcat success ratio of 21%.

CONCLUSIONS

The costs of reflection seismic exploration, notwithstanding the significant improvements in that method, have been increasing with respect to the value of the oil and gas discovered. Judging from the results of geochemical exploration, it is possible to achieve a marked decrease in the costs of exploration by judicious use of selected unconventional methods presently available to the industry.

TABLE 1
FIELDS POSSESSING ΔC ANOMALIES PRIOR TO DRILLING

Field	Location	Trap	Disc. Date	No. Wells	Cuml. Bbls.	Cuml. MCF
1. Reddin	Taylor County, TX	Pennsylvania Reef	1942	309	10MM	
2. Adamson	Jones County, TX	Structural closure	1950	1	150M	
3. Hardy	Jones County, TX	Sand pinch out	1943	30	3MM	
4. Willman-Maedgen	San Patricio County, TX	Sand pinch out	1945	62	2.8MM	1.8MM
5. Spiekerman	San Patricio County, TX	Structural closure	1957	6		174M
6. Ingram	Taylor County, TX	Structural closure	1946	9	240M	
7. Green Lake	Calhoun County, TX	Structural closure	1947	9	180M	124M
8. Ferrell	Taylor County, TX	Structural closure	1947	5	141M	
9. Long Mott	Calhoun County, TX	Structural closure	1948	31	2.8MM	
10. Akens	Taylor County, TX	Structural closure	1948	2	21M	
11. Bird	Taylor County, TX	Structural closure	1948	12	713M	
12. Gibson	Taylor County, TX	Pennsylvania Reef	1948	35	1.7MM	
13. Caldwell Ranch	Taylor County, TX	Sand lens	1949	18	757M	
14. Sojourner	Haskell County, TX	Sand pinch out	1950	63	15MM	
15. South Crews Fry	Runnels County, TX	Sand Lens	1952	30	1.3MM	
16. Traylor	Runnels County, TX	Sand lens	1952	1	75M	
17. Gillespie	Haskell County, TX	Sand pinch out	1950	5	96M	
18. Ballinger Palo Pinto	Runnels County, TX	Porosity trap	1951,55	63	1.1MM	
19. North Weinert	Haskell County, TX	Sand pinch out	1954	8	1.9MM	
20. Shafter	Kern County, CA	Structural closure	1954	2		1MM
21. Section Seven	Jones Cunty, TX	Structural closure	1954	8	280M	
22. Plains (Blk-D)	Yoakum County, TX	Structural closure	1957	2	385M	
23. Bittern Lake	Alberta, Canada	Sand lens	1957	8	S.I.	
24. Seymour	Baylor County, TX	Sand lens	1950-52	44	3MM	
25. S. Crews Cardner	Runnels County, TX	Sand lens	1952	57	2.2MM	
26. West Hatchel	Runnels County, TX	Structural closure	1956	16	393M	
27. O'Brien	Haskell County, TX	Structural closure	1951	66	3.6MM	
28. Russell	Bottineau County, ND	Sand pinch out	1963	2	205M	
29. Sixto	Starr County, TX	Sand pinch out	1964	22	640M	
30. Harris	Irion County, TX	Sand pinch out	1965	3	250M	
31. Salinas	Starr County, TX	Sand pinch out	1964	1		S.I.
32. Arden South	Irion County, TX	Sand pinch out	1965	17	430M	112M
33. Davis	Shackelford County, TX	Sand pinch out	1957	1		112M
34. Berthelote	Toole County, MT	Sand lens	1966	2	150M	
35. N.E. Foothills	Burke County, ND	Porosity trap	1959	38	3MM	5.5MM
36. South Glen Burn	Renville County, ND	Structural closure	1958	6	1MM	79M
37. Barstow	Ward County, TX	Structural closure	1969	13		188MM
38. Cajoh (extension)	Rawlins County, KN	Stratigraphic	1977	1		
39. KWB	Tom Green County, TX	Stratigraphic	1981	76	927M	416M[1]
	Total through 9/31/84			1084	58MM	197MM

[1]Liquids = 14,000,587 gals.

REFERENCES

Debnam, A. H., 1969, Geochemical prospecting for petroleum and natural gas in Canada: Geological Survey of Canada Bulletin 177, 26 p.

Donovan, T. J., 1974, Petroleum microseepage of Cement, Oklahoma, evidence and mechanism: American Association of Petroleum Geologists Bulletin, v. 58, n. 3, p. 429–446.

———, and Dalziel, M. C., 1977, Late diagenetic indicators of buried oil and gas: U.S. Geological Survey Open-File Report 77-817, p. 38.

———, Forgey, R. L., and Roberts, A. A., 1979, Aeromagnetic detection of diagenetic magnetite over oil fields: American Association of Petroleum Geologists Bulletin, v. 63, p. 245–248.

Duchscherer, Wm., Jr., 1981, Carbonates and isotope ratios from surface rocks: A geochemical guide to underlying petroleum accumulations: *in* Gottlieb, B. M., ed., Unconventional Methods in Exploration for Petroleum and Natural Gas II: Dallas, Southern Methodist University Press., p. 201–218.

———, 1980, Geochemical methods of prospecting for hydrocarbons: Oil and Gas Journal, December 1, 1980, p. 194–208.

———, 1981, Nongasometric geochemical prospecting for hydrocarbons with case histories: Oil and Gas Journal, October 19, 1981, p. 312–327.

———, 1982, Geochemical exploration for hydrocarbons: No new tricks but an old dog: Oil and Gas Journal, July 5, 1982, p. 163–176.

———, 1984a, Geochemical Hydrocarbon Prospecting: Tulsa, PennWell Press, p. 196.

———, 1984b, Discovery and development of KWB field, 1: Oil and Gas Journal, May 21, 1984, p. 175–182.

———, 1984b, Discovery and development of KWB field, 2: Oil and Gas Journal, May 28, 1984, p. 143–151.

Horvitz, L., 1939, On geochemical prospecting: Geophysics, v. 4, p. 210–228.

———, 1945, Recent developments in geochemical prospecting for petroleum: Geophysics, v. 10, p. 487–493.

———, 1969, Hydrocarbon geochemical prospecting after thirty years: *in* Heroy, W. B., Jr., ed., Unconventional Methods in Exploration for Petroleum and Natural Gas: Dallas, Southern Methodist University Press, p. 205–208.

———, 1980, Non-surface evidence of hydrocarbon movement from depth: *in* Roberts, W. H., III, and Cordell, R. J., eds., Problems of petroleum migration: American Association of Petroleum Geologists Studies in Geology, n. 10, p. 241–269.

Kulp, J. L., Kent, P., and Kerr, P. F., 1951, Thermal study of the Ca-Mg-Fe carbonate materials: American Mineralogist, v. 36, nos. 9 and 10, 643 p.

Laubmeyer, G., 1933, A new geophysical prospecting method, especially for deposits of hydrocarbons: Petroleum, v. 29, p. 1–4.

Lilburn, R. A., and Zuhair, Al-Shaieb, 1983, Geochemistry and isotopic composition of hydrocarbon-induced diagenetic aureole (HIDA), Cement Field, Oklahoma: Shale Shaker, v. 34 p. 40–56.

Nisle, R. G., 1941, Considerations on the vertical migration of gases: Geophysics, v. 6, p. 449–454.

Pirson, S. J., 1973, Direct oil and gas finding methods: An answer to the energy crisis in the United States?: Texas Business Review, v. XLVII, n. 7, July 1973.

Ransone, W. R., 1958, Case history of the Sojourner oil field, Haskell County, Texas: A geochemical confirmation: Geophysics, v. 23, n. 3, p. 547–578.

Roberts, W. H., III, 1981, Design and function of oil and gas traps: Problems of petroleum migration: American Association of Petroleum Geologists Studies in Geology, n. 10, p. 217–240.

Rosaire, E. E., 1938, Shallow stratigraphic variations over Gulf Coast structures: Geophysics, v. 3, p. 96–115.

Sokolov, V. A., 1933, A new method of surveying oil and gas formations: Tekhn Nefti, v. 16, 163 p.

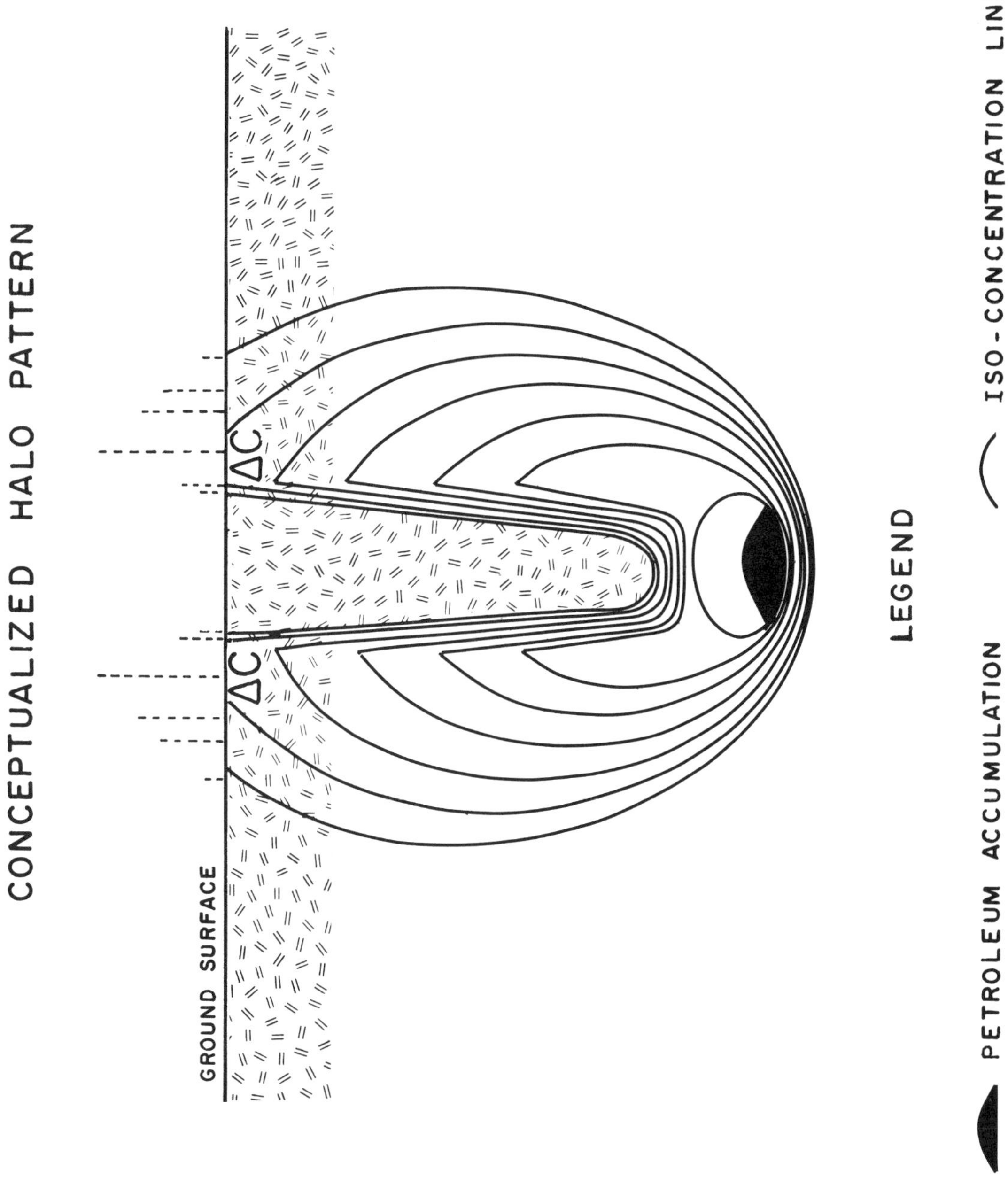

Fig. 1. Conceptualized halo pattern.

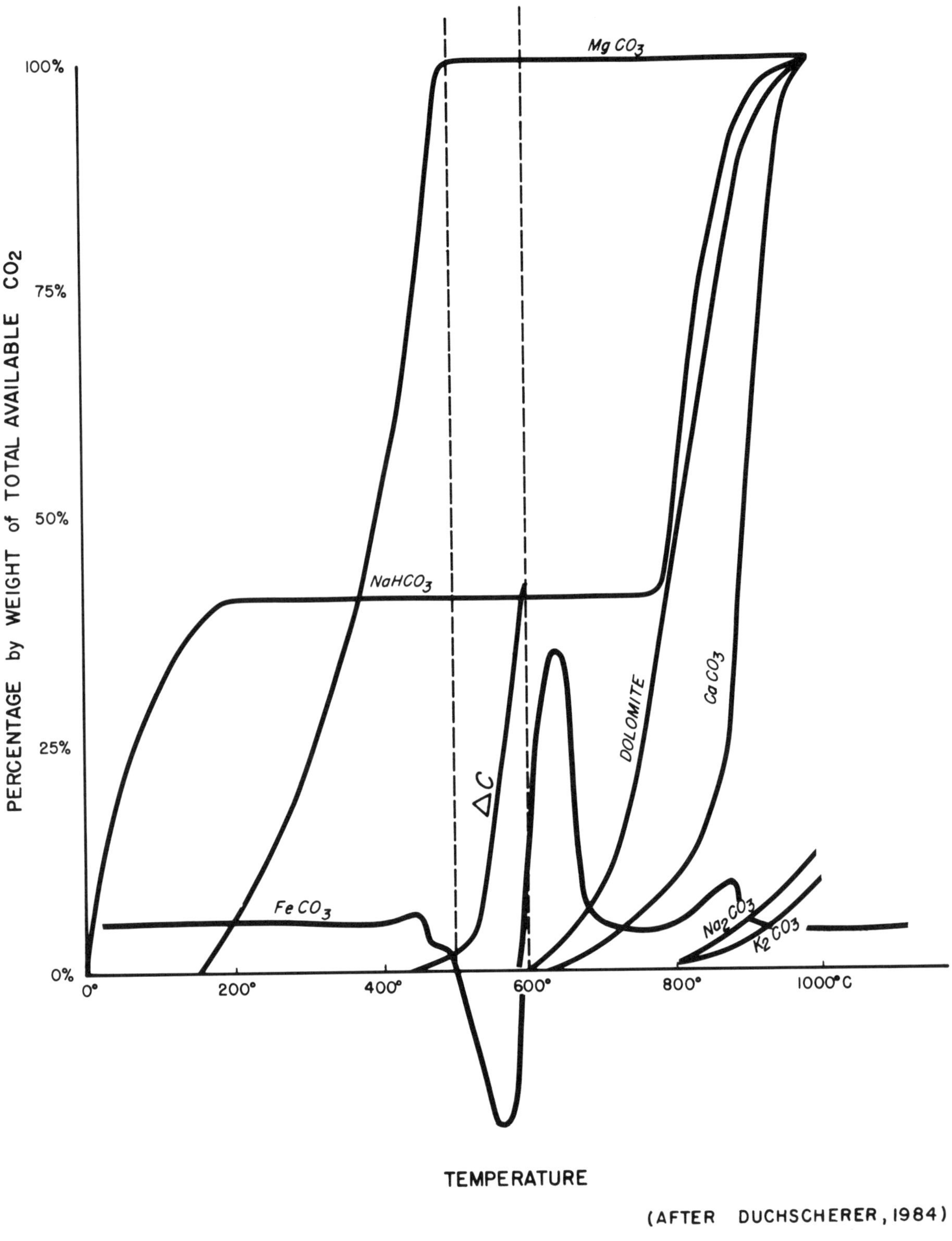

Fig. 2. Thermal dissociation curves.

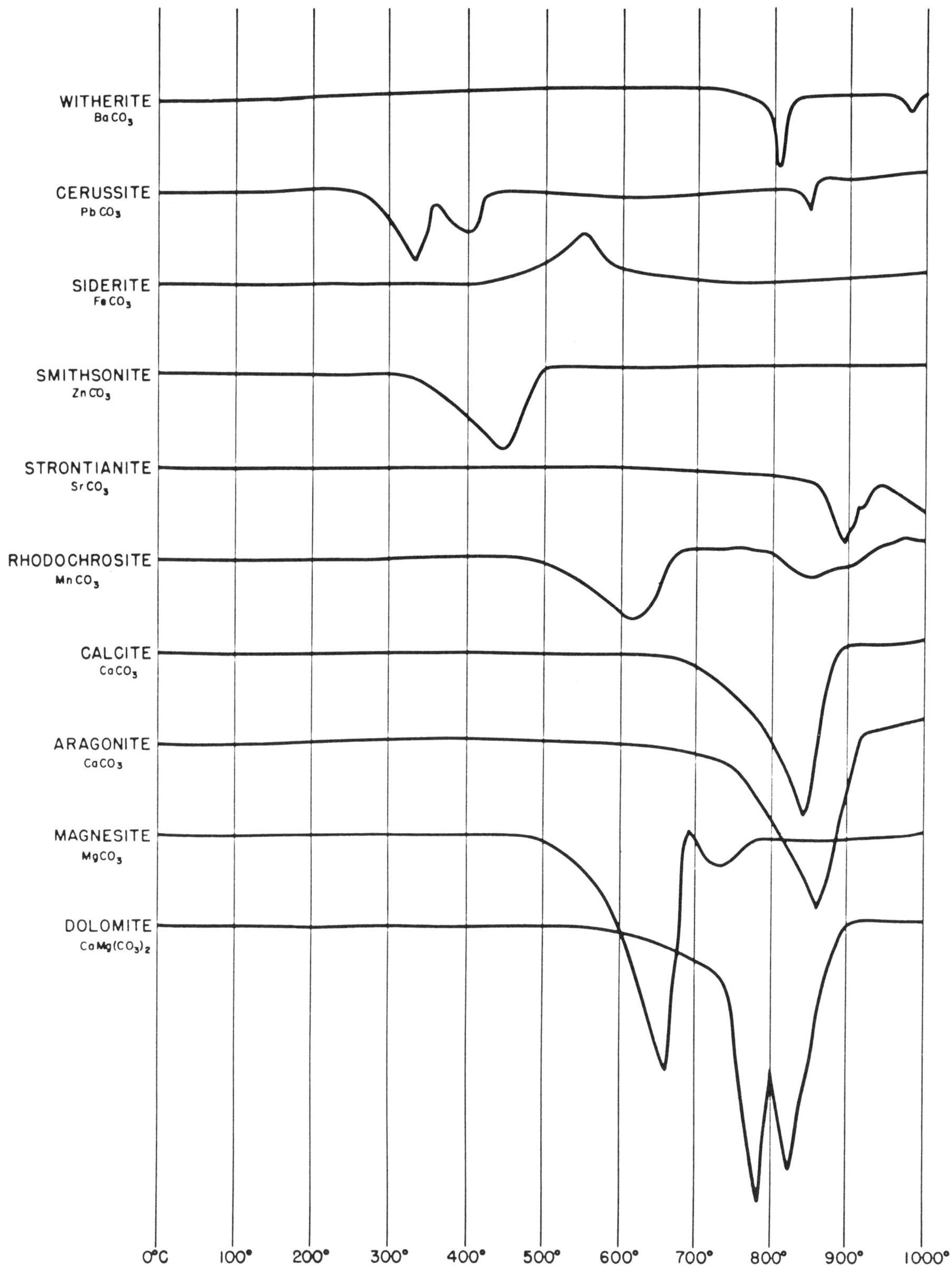

Fig. 3. Differential thermal curves.

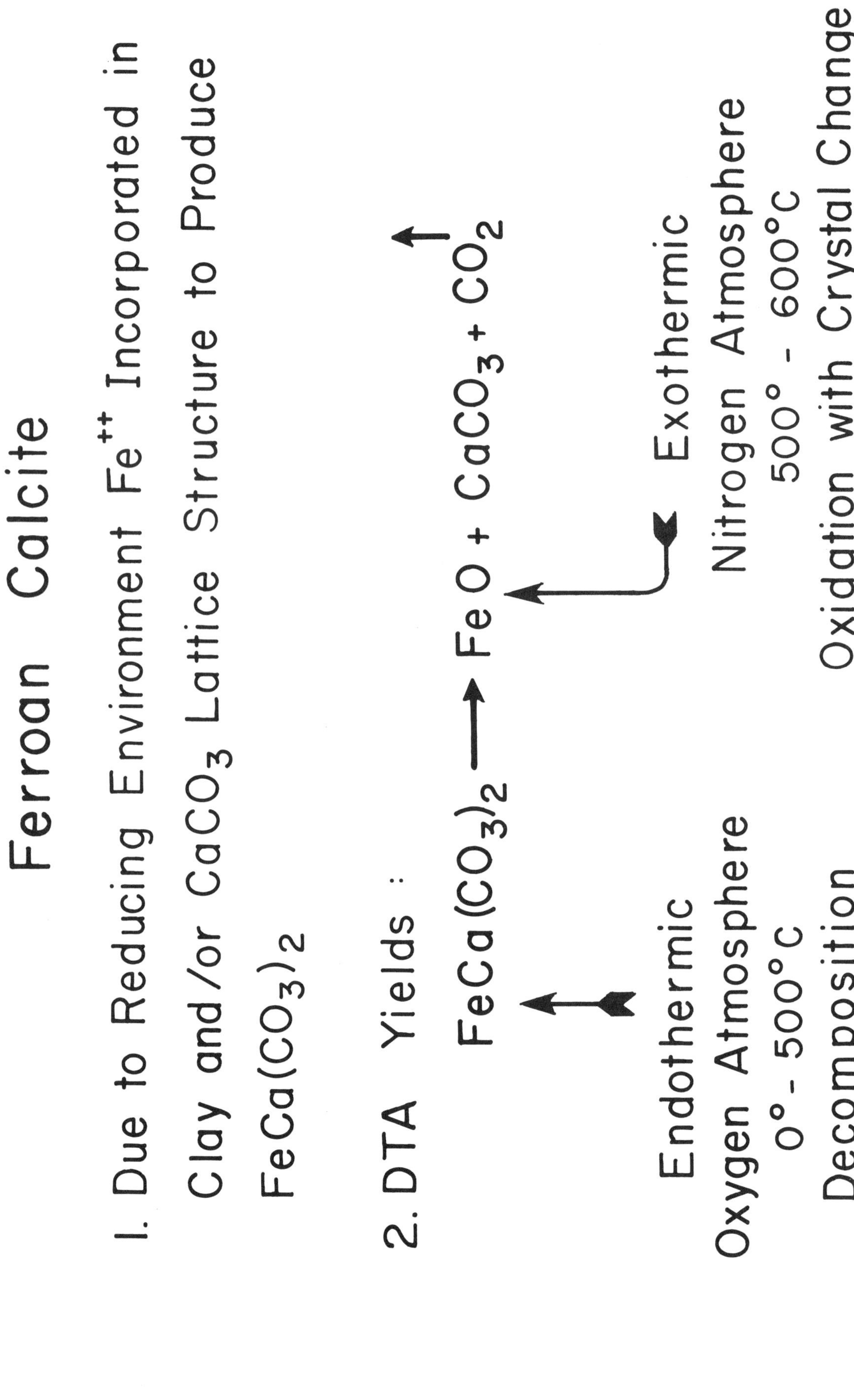

Fig. 4. Ferroan calcite.

NEAR-SURFACE HYDROCARBON GAS MEASUREMENT OF VERTICAL MIGRATION

Gary K. Rice

GEOPHYSICAL SERVICE, INC.

ABSTRACT

Recent advances in near-surface interstitial gas sampling technology have resulted in improved resolution of low-molecular-weight hydrocarbon (LMWHC) data used in geochemical exploration for petroleum. These advances, coupled with new data processing technology, are key factors in the current transition of soil gas geochemistry from a reconnaissance technique to a valuable petroleum exploration tool which is capable of locating and delineating petroleum reservoirs. Development of this technology requires continuous investigation of vertical migration pathways and deviations of vertical migration of the LMWHC gases by joints, faults, and fractures.

Gas data collected from near-surface horizons at dense grid spacings exhibit patterns that are in good agreement with the location and extent of petroleum reservoirs established by well data. The locations of near-surface gas concentration gradients relative to known reservoir boundaries provide a means to measure lateral displacement during gas migration to the surface. Cases have been studied where data show that, in the absence of faults, fractures, and large bedding plane dip, migration from a reservoir to the surface can be very nearly vertical.

INTRODUCTION

The earliest recorded literature on the measuring of nonvisible near-surface effects of vertical fluid migration can be attributed to Laubmeyer in 1933. Literature during the last fifty years has recorded the development of many techniques based on vertical migration since the pioneering efforts of Sokolov (1937), Horvitz (1939), and Rosaire (1939). During this time, vertical migration methods have contributed to a large number of successes and have taken the blame for some failures in the search for oil and gas. Successes can be attributed to technical advances in data collection, data processing, and interpretation by these pioneers, with significant contributions from later workers. At the same time, the cases of failure point to limitations in the technology. Philp and Crisp (1982) have provided a useful review of the literature in this field.

The concept of vertical migration of petroleum gases or other fluids from a reservoir to the near-surface through conduits formed by faults and fractures is becoming widely accepted in the industry. "Micro seepages" of this type are widely known and can be viewed as extensions of the larger visible seeps described by DeGolyer (1940) and Link (1952). Extension of these principles to measurable migration from a reservoir through several thousand feet of overburden in the absence of known faults or fractures has gained acceptance among exploration geologists at a much slower rate.

Technical advances during the last ten years have contributed greatly to the data collection process, including especially compact, reliable flame ionization gas chromatographs and computing integrators for measuring hydrocarbon gases. Data processing has also advanced markedly due to the availability of powerful and economical digital computers with sophisticated processing and display algorithms. Interpretation has advanced at a much slower rate. This can be attributed to little real progress during the last fifty years in understanding the detailed physics and chemistry of the vertical migration process. Progress has been limited, at least in part, by a lack of data from the area between the reservoir and the near-surface where this process takes place.

Even with limited information about the fundamental process, the success of vertical migration methods cannot be denied. It is because of noteworthy contributions to petroleum exploration and the perseverance of workers in the field that vertical migration methods currently play important roles in many exploration programs. However, wide application of vertical migration methods is currently limited by a lack of understanding of the fundamental migration mechanism. As more data are acquired to define accurately the migration process, these limitations can be removed and vertical migration methods will be able to provide increasingly detailed and precise exploration information.

VERTICAL MIGRATION

Fluid migration from depth has been applied directly in petroleum exploration by analysis of near-surface hydrocarbons (Horvitz, 1954), and indirectly by a variety of techniques, most of which depend upon chemical alterations, including visible mineral changes (Donovan, 1974). There have been few efforts to bring together the many observations from the large number of vertical migration techniques into a unified theory of the migration process. Davidson (1982) reviewed results from a number of vertical migration methods and explained the data in terms of modern theories of subsurface fluid migration. But even with this effort and with fifty years of measurements of various effects of the migration process, there is still an incomplete understanding of the vertical migration mechanism.

Importance of Vertical Migration

A large number of methods are currently in use to measure end results from vertical migration processes. Remote sensing by LANDSAT and aerial photography has been used successfully to detect alteration zones caused by hydrocarbon migration to the near-surface (Donovan et al., 1979; Saunders 1979). Biogeochemical prospecting has been suggested to exploit alterations in plant growth caused by vertical migration induced chemical changes affecting micronutrient availability to plants (Dalziel and Donovan, 1980). Radiometric data, which depend upon near-surface radioactive nuclide redistribution caused by vertical migration, have been shown to correlate well with subsurface reservoir location (Weart and Heimberg, 1979; Curry, 1984). Techniques based upon pathfinder elements, such as iodine (Gallagher, 1984) and various trace metals, all depend upon vertical migration from depth and chemical alteration at the sample collection depth. Pirson (1969) suggested a model for the formation of several varieties of alteration anomalies due to vertically migrating fluids.

Another class of vertical migration methods depends upon chemical alterations below the surface, but at depths considerably less than that of most petroleum reservoirs, e.g., upper 1,000 ft (300 m). Diagenetic magnetite, formed from the reduction of hematite by vertically escaping fluids, can be detected magnetically and used as an indicator of subsurface petroleum deposits (Donovan et al., 1979). Induced polarization methods may detect other reaction products of chemical alteration by these fluids, e.g., pyrite (Sternberg and Oehler, 1984). Ferguson (1979) observed visible geochemical alterations in well cuttings.

Direct methods, which measure one of the vertically migrating fluids, are in widespread use. Aromatic hydrocarbons are measured in shallow soil samples by fluorometry (Hebert, 1984). Examples of direct measurement of vertically escaping, low-molecular-weight hydrocarbons include the acid extraction technique (Horvitz, 1969) and several interstitial gas hydrocarbon measurement methods (Richers et al., 1982; Jones and Drozd, 1983).

These widely different techniques are representative of the variety and number of methods developed during the last fifty years for measuring effects from the vertical migration process. Many of these are now in use as petroleum exploration tools.

Vertical Migration Measurement

A review of the literature shows that even though many references describe techniques for measuring effects caused by vertical migration, very few investigate the migration process itself. All vertical migration methods are therefore limited in application by such questions as: (1) How does the near-surface data relate to reservoir location? (2) Why does the data indicate oil (or gas)? (3) When did the gas measured at the near-surface leave the reservoir and will it reflect recent development activities?

The size and lateral displacement of gas patterns in the near-surface have been related to known reservoir limits and have shown that vertical migration can occur with only relatively small displacements from true vertical (Horvitz, 1965; McIver et al., 1985). The composition of light hydrocarbon gases migrating from depth to the near-surface has been investigated, and many near-surface samples were found to be similar to reservoir composition (Jones and Drozd, 1983). However, vertical profiles made during drilling showed specific cases of composition changes occurring during migration from reservoir depths to the near-surface (Reitsema, 1985). There have been notably few reports on measurements of vertical migration rate, partly because this parameter requires either introduction of a tracer into the reservoir and detection at the surface or artificial perturbation of the reservoir system and intensive monitoring at the near-surface over long periods of time. Some data have been obtained from a coal burn project where changes in shallow subsurface pressure produced detectable concentration changes at the near-surface in three to six days (Jones and Thune, 1982). These studies are all important to the changing emphasis in recent years toward determining spatial resolution, composition changes, and rates of vertical migration.

SPATIAL RESOLUTION

Determining the spatial relationships between near-surface data and a petroleum reservoir requires detailed knowledge of each. In the case of the reservoir, information about location and extent may be available from detailed 2-D or 3-D seismic data. A highly developed oil field may yield sufficient information from subsurface data. Equally important are the detail and precision of the near-surface measurements.

Data Collection

Figure 1 shows a sampling and analysis system designed by GSI and used to obtain interstitial gas samples at 8 ft (2.5 m) depth by extracting soil gas from a tube extending to the bottom of an augered sample hole while the auger remained in place. This method has proved to be reproducible in a variety of soil types and conditions and has been used to obtain concentration gradient information over very small, e.g., 10 ft (3 m), horizontal distances.

Samples were analyzed by gas chromatography, using flame ionization detectors. Peak areas, which are proportional to concentration, were calculated with a computing integrator, and the results were recorded on magnetic tape for computer analysis. All the analytical equipment was contained in the doghouse shown in Figure 1, and analysis was carried out near the sampling point with minimum sample storage time.

Sampling and analysis resolution are of fundamental importance in any measurement process. A basic measure of resolution is reproducibility; i.e., the same measurement repeated under the same conditions within an acceptable tolerance. This is a particularly difficult parameter to establish in soil sampling because soil is a nonhomogeneous system and concentration gradients of the light hydrocarbon gases can be extremely high. Resolution also includes the sensitivity of the sampling and analysis methods to small changes in sample location. Sampling and analysis reproducibility are normally determined by taking replicate samples. This has to be done in an area where concentration gradients were known to be small. Figure 2 shows one example of sampling reproducibility taken near a producing field where ethane concentration at three sample points spaced 50 to 70 ft (15 to 20 m) were in good agreement. Concentration reproducibility across areas as large as that in Figure 2 is very unusual.

Knox County Example

A small area in Knox County, Texas, was sampled on a 500 ft (150 m) spacing. A total of ninety-six samples were collected over an area covering approximately 1 sq m (2.6 sq km). These data were collected in a producing oil field which had been extensively developed. Production was at approximately 2,000 ft (600 m) in the Tannehill Sand, which is a complex stratigraphic channel system of Permian age. A contour map of the propane concentration data is shown in Figure 3. An area of high concentrations trending east-west can be seen in the lower half of the mapped data. Another trend extending southwest to northeast is apparent in the top portion of the concentration map.

Mapped geochemical data that have been contoured can be interpreted by examining images formed in the mapped data in a manner similar to techniques used in geomorphology (Rice and Jackson, 1984). This method is illustrated in Figure 3, where producing wells and dry holes were also mapped. It can be readily seen that the highest concentrations of propane gas in the near-surface occurred over the producing field. In addition, the general shape of the contour lines tended to conform roughly to the outline of production. This highly developed field with extensive well data met the criterion of dense subsurface control that was needed for an effective test of the spatial resolution of near-surface data.

Possible sample contamination always must be considered when working with near-surface data in a producing field. There was no evidence of surface spills contributing to the near-surface data; compositions did not indicate patterns expected from weathered oil. The possibility of casing leakage contributing to the near-surface data could not be entirely eliminated, but there was no indication in the data that the highest concentrations occurred at well locations. In fact, this possibility was partially controlled by collecting samples away from the existing wells and tanks. Therefore, it was concluded that the near-surface data in Figure 3 represented vertical migration from the reservoir and was not caused by field production activity.

When tops of the perforated zone were added to the map, as shown in Figure 4, even more details became apparent. The field was outlined by propane concentration data where contour lines became closely spaced. The spacing of contour lines is representative of the concentration gradient. Therefore, images projected in contoured near-surface data could be used to determine reservoir limits at the transition between high and low concentration gradients.

The concentration gradient concept can be visualized, as shown in Figure 5. If the concentration data are represented as a solid surface with concentration displayed as distance in the Z direction and mapped coordinates in the X and Y directions, a concentration gradient may be represented by a plane tangent to that surface at the sample point location. The gradient plane can be represented by two vectors, one in the X direction and one in the Y direction. The lengths of these vectors, or the scalar components, represent concentration changes along the X and Y mapped coordinates respectively. When these vectors were summed the result was a vector which represented a combination of the partial derivatives in the X and Y directions with a scalar component representative of the magnitude of the gradient and at an angle representative of the direction of the gradient. These vectors were depicted as arrows and mapped as shown in Figure 6.

The data shown in Figure 6 were regridded from the original 500 ft sample spacing to approximately 125 ft (38 m) grid cell spacings, and gradients were calculated at each of the regridded cell values. The regridding algorithm was critical only with respect to being a reasonable interpolation among the measured data points. The additional point density accomplished by regridding was done to help illustrate the gradient values covering the mapped area. The concentration gradient, located at the point of each arrow symbol in Figure 6, is represented by the size and orientation of each arrow. It can readily be seen by comparing Figure 6 with Figure 4 that the oil reservoir areas were covered by relatively high concentration gradients in the near-surface data.

The gradient data include the representation of spatial relationships among the data points as well as their absolute magnitudes. This can be illustrated by reference to the histograms in Figures 7 and 8. The concentration data distribution in Figure 7 is typical of many types of near-surface geochemical data. There is no clear distinction between background data and data collected over the reservoir. The histogram of the scalar component of the concentration gradient data shown in Figure 8 has a more distinct multimodal distribution. In fact, the scalar quantity of 0.22 may be selected as a threshold between background and nonbackground data for the data shown in Figure 8. A single contour at this gradient threshold shown in Figure 9 illustrates the spatial relationship to the producing area. A good outline of the producing reservoir is apparent in near-surface data displayed in Figure 9. Using concentration gradients it can be seen that the spatial resolution of the near-surface propane concentration data versus the subsurface reservoir data is correct to within the 500 ft (150 m) sampling interval of the near-surface data.

The gradients were calculated and displayed in this example to illustrate the gradient concept. In many interpretation applications this calculation effort is not carried out. Approximate gradients may be determined from computer contoured data by merely observing the spacing of contour lines. By understanding the gradient concept and using the contour line spacings in Figure 3, an outline of the Tannehill reservoir can be visualized.

EXPLORATION APPLICATIONS

Producing oil fields such as the example cited above are important for testing vertical migration concepts and may be the only way of obtaining detailed subsurface information for determining spatial resolution of near-surface data. The real test comes, however, when a method is applied successfully and results in a new discovery.

South Texas Example

The data shown in Figure 10 were collected over an area of approximately 1.5 sq mi (3.9 sq km). At the time the data were acquired, two dry holes had been drilled in this area. The concentration data show a series of high values in the northwest corner and along the western edge of the surveyed area.

Drainage patterns were analyzed and showed no geomorphic evidence of faults or fractures that could affect the gas migration patterns (Rice and Jackson, 1984). Stratigraphy in the area was known to be characterized by virtually horizontal bedding planes. Therefore, it was concluded that there were no structural reasons for significant lateral migration of the hydrocarbon gases.

Using concentration gradients depicted by contour spacing as described above, a boundary was determined between the concentration high area in the northwest part of the map and the concentration low area approximately along the 1.0 ppm bold contour line. Possible well locations were restricted by lease boundaries, which were at the edges of the survey in the northwest corner, and the small size of the area of interest. An offset of the near-surface data by 500 ft (150 m) could have made the difference between success and failure. Without identifiable migration pathways, there was no reason to expect a lateral offset of this amount if the hydrocarbon gases were migrating from a shallow reservoir, e.g., less than 5,000 ft (1500 m).

From information based largely on this interpretation, three wells were drilled at the locations shown in Figure 11, and all are currently in production. Two zones were encountered at approximately 3,000 ft (900 m) in Fry Sand (Strawn series, Pennsylvanian age) and in Goen Lime (also Strawn). Since these formations had not been encountered in the other wells, it is believed these are stratigraphic traps that pinch out abruptly in the vicinity of the concentration gradient change.

Near-surface geochemistry and other vertical migration methods give no information about depth, reservoir thickness, or permeability. But spatial resolving power of near-surface geochemical exploration in the case of a shallow reservoir is illustrated with this discovery which occurred in the very corner of a lease in a small area. This case illustrates migration from depth to the near-surface with a registration of near-surface patterns defined by the concentration gradients within a few hundred feet of the actual reservoir location. This South Texas example illustrates the power of modern near-surface geochemical exploration with its rapid, economical, areal coverage and its ability to define spatially the size, shape, and location of petroleum reservoirs.

CONCLUSIONS

Vertical migration is a fundamental mechanism of many exploration methods, and it needs to be understood in order to utilize fully the strengths and to realize the limitations of these methods. To understand better the process of vertical migration, measurements are needed of migration rates, composition changes, and spatial relationships between near-surface data and reservoirs. In measurement processes, and especially in those measuring vertical migration, it is important to determine the resolution of the measurement in terms of its precision and sensitivity.

When low-molecular-weight hydrocarbon gas concentrations are determined with precise near-surface measurement at a sampling density sufficient to give high resolution, concentration gradients provide a simple method for analyzing data in terms of spatial relationships among adjacent data points. Concentration gradients are valuable aids for (1) measuring spatial resolution of vertical migration and (2) delineating reservoirs. In cases without intervening structural perturbations, methods dependent on vertical migration can give accurate information about size, shape, and location of subsurface reservoirs.

By taking advantage of opportunities where detailed subsurface information is known, and with detailed near-surface sampling, it is possible to measure spatial resolution of vertical migration. In cases where vertical migration gives precise near-surface images of reservoirs, such as the example in Figure 4, it is possible to obtain detailed information about reservoir location which can contribute significantly to new discoveries.

During the last fifty years, much has been learned about vertical migration, and many new methods have been developed for measuring effects resulting from the migration process. During recent years more information is being gathered to measure the vertical migration mechanism. As this process is better understood and the technology continues to develop, application of vertical migration methods will accelerate, and these techniques will assume increasingly important roles in exploration programs, becoming second only to seismic methods in scope and frequency of application.

ACKNOWLEDGMENTS

The author is grateful for early contributions to the concepts presented in this paper by Vernon Jackson, who recently retired after more than thirty years of service to the petroleum and minerals industries. This study became a reality with the diligent assistance of Judy Smith, who completed word processing, and Tom Kebr, who programmed computer calculations and created the computer-generated displays.

Constructive reviews of this writing by Dr. Don Saunders and by Dr. Mark Potts are gratefully acknowledged.

Any technical contributions this study may have made should be credited to the oil companies which supported development of the concepts reported here, and especially to those companies which permitted use of their data, information of vital importance to this paper.

REFERENCES

Curry, William H., III, 1984, Evaluation of surface gamma radiation surveys for petroleum exploration in the deep powder river basin, Wyoming, *in* Davidson, M. J., and Gottlieb, B. M., eds., Unconventional methods in exploration for petroleum and natural gas III: Dallas, Southern Methodist University Press, p. 25–39.

Dalziel, Mary C., and Donovan, Terrence J., 1980, Biogeochemical evidence for subsurface hydrocarbon occurrence, Recluse Oil Field, Wyoming: Preliminary results: Geological Survey Circular 837.

Davidson, Martin J., 1982, Toward a general theory of vertical migration: Oil and Gas Journal, v. 80, no. 25, p. 288–300.

DeGolyer, E., 1940, Direct indication of the occurrance of oil and gas, *in* Elements of the petroleum industry: New York, American Institute of Mining Engineers, p. 21–25.

Donovan, Terrence J., 1974, Petroleum microseepage at Cement, Oklahoma: Evidence and mechanism: American Association of Petroleum Geologists Bulletin, v. 58, no. 3, p. 429–446.

———, Forgey, Randal L., and Roberts, Alan A., 1979, Aeromagnetic detection of diagenetic magnetite over oil fields: Reprinted for private circulation from the American Association of Petroleum Geologists Bulletin, v. 63, no. 2.

———, Termain, Patricia A., and Henry, Mitchell E., 1979, Late diagenetic indicators of buried oil and gas II: Direct detection experiment at Cement and Garza Oil Fields, Oklahoma and Texas, using enhanced LANDSAT I and II images: United States Department of the Interior Geological Survey Open-File Report 79-243.

Ferguson, Jerry D., 1979, The subsurface alteration and mineralization of Permian red beds overlying several oil fields in southern Oklahoma: Shaleshaker, v. 29, p. 172–208.

Gallagher, Alton V., 1984, Iodine: A pathfinder for petroleum deposits, *in* Davidson, M. J., and Gottlieb, B. M., eds., Unconventional methods in exploration for petroleum and natural gas III: Dallas, Southern Methodist University Press, p. 148–159.

Hebert, Carroll F., 1984, Geochemical prospecting for oil and gas, using hydrocarbon fluorescence techniques, *in* Davidson, M. J., and Gottlieb, B. M., eds., Unconventional methods in exploration for petroleum and natural gas III: Dallas, Southern Methodist University Press, p. 40–58.

Horvitz, Leo, 1939, On geochemical prospecting I: Geophysics, v. 4, no. 3, p. 210–228.

———, 1954, Near-surface hydrocarbons and petroleum accumulation at depth: Mining Engineering, December, p. 1205–1209.

———, 1965, Case histories show results of drillng geochemical prospects: World Oil, November, p. 161–165.

———, 1969, Hydrocarbon geochemical prospecting after thirty years, *in* Heroy, W. B., Jr., ed., Unconventional methods in exploration for petroleum and natural gas: Dallas, Southern Methodist University Press, p. 205–218.

Jones, Victor T., and Drozd, R.J., 1983, Predictions of oil or gas potential by near-surface geochemistry: American Association of Petroleum Geologists Bulletin, v. 67, no. 6, p. 932–952.

———, and Thune, Howard W., 1982, Surface detection of retort gases from an underground coal gasification reactor in steeply dipping beds near Rawlins, Wyoming, *in* 57th annual fall technical conference and exhibition of the Society of Petroleum Engineers of AIME.

Laubmeyer, G., 1933, A new geophysical prospecting method, especially for deposits of hydrocarbons: Petroleum, v. 29, no. 18, p. 1–4.

Link, Walter K., 1952, Significance of oil and gas seeps in world oil exploration: American Association of Petroleum Geologists Bulletin, v. 36, no. 8, p. 1505–1540.

McIver, Richard D., McIver, Richard W., and McIver, David W., 1985, Near-surface methane anomaly over shallow Foley gas field, Baldwin County, Alabama: American Association of Petroleum Geologists 1985 annual convention, New Orleans, Louisiana.

Philp, R. P., and Crisp, P.T., 1982, Surface geochemical methods used for oil and gas prospecting, a review: Journal of Geochemical Exploration, v. 17, p. 1–34.

Pirson, Sylvain J., 1969, Geological, geophysical, and chemical modifications of sediments in the environment of oil fields, *in* Heroy, W. B., Jr., ed., Unconventional methods in exploration for petroleum and natural gas: Dallas, Southern Methodist University Press, p. 159–186.

Reitsema, R. H., 1985, Gas compositions in a well column: American Association of Petroleum Geologists 1985 annual convention, New Orleans, Louisiana.

Rice, Gary K., and Jackson, Vernon N., 1984, Geochemical

techniques in exploration: 1984 American Society of Photogrammetry Fall Convention Technical Papers, p. 620–623.

Richers, D. M., Reed, R. J., Horstman, D. C., Michels, G. D., Baker, R. N., Lundell, L., and Marrs, R. W., 1982, LANDSAT and soil-gas geochemical study of Patrick Draw Oil Field, Sweetwater County, Wyoming: American Association of Petroleum Geologists Bulletin, v. 66, no. 7, p. 903–922.

Rosaire, E. E., 1939, The handbook of geochemical prospecting, 1st ed., published by the author.

Sternberg, Ben K., and Oehler, Dorothy Z., 1984, Electrical methods for hydrocarbon exploration I, Induced polarization (indepth) method, *in* Davidson, M. J., and Gottlieb, B. M., eds., Unconventional methods in exploration for petroleum and natural gas III: Dallas, Southern Methodist University Press, p. 188–201.

Saunders, Donald F., 1981, Use of LANDSAT geomorphic and tonal anomalies in petroleum prospecting, *in* Gottlieb, B. M., ed., Unconventional methods in exploration for petroleum and natural gas II: Dallas, Southern Methodist University Press, p. 63–82.

Sokolov, V. A., 1937, The gas survey as a method of prospecting for oil and gas formation: Technika NGR, no. 1.

Weart, Richard C., and Heimberg, George, 1981, Exploration radiometics: postsurvey drilling results, *in* Gottlieb, B. M., ed., Unconventional methods in exploration for petroleum and natural gas II: Dallas, Southern Methodist University Press, p. 116–123.

Fig. 1. Sampling and analysis vehicle.

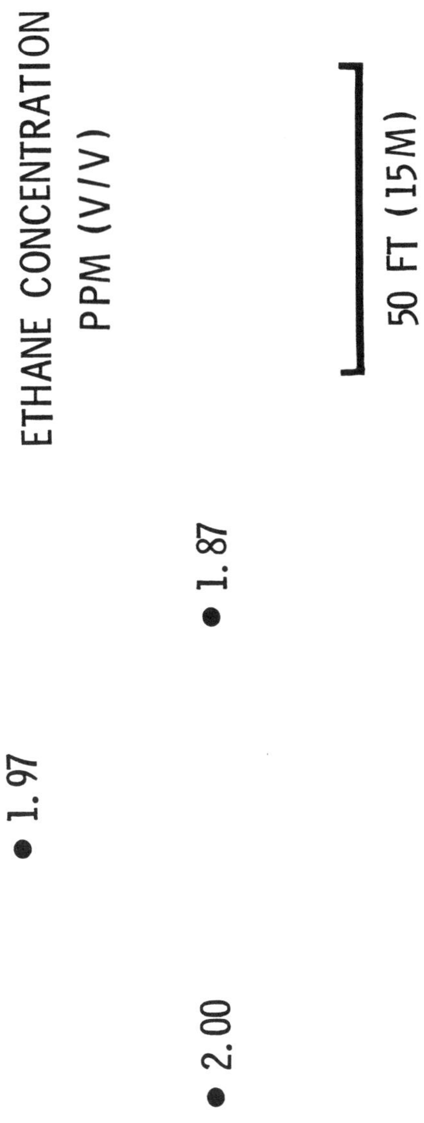

Fig. 2. Sampling resolution test. Sampling and analysis are reproducible in areas with unusually low concentration gradients.

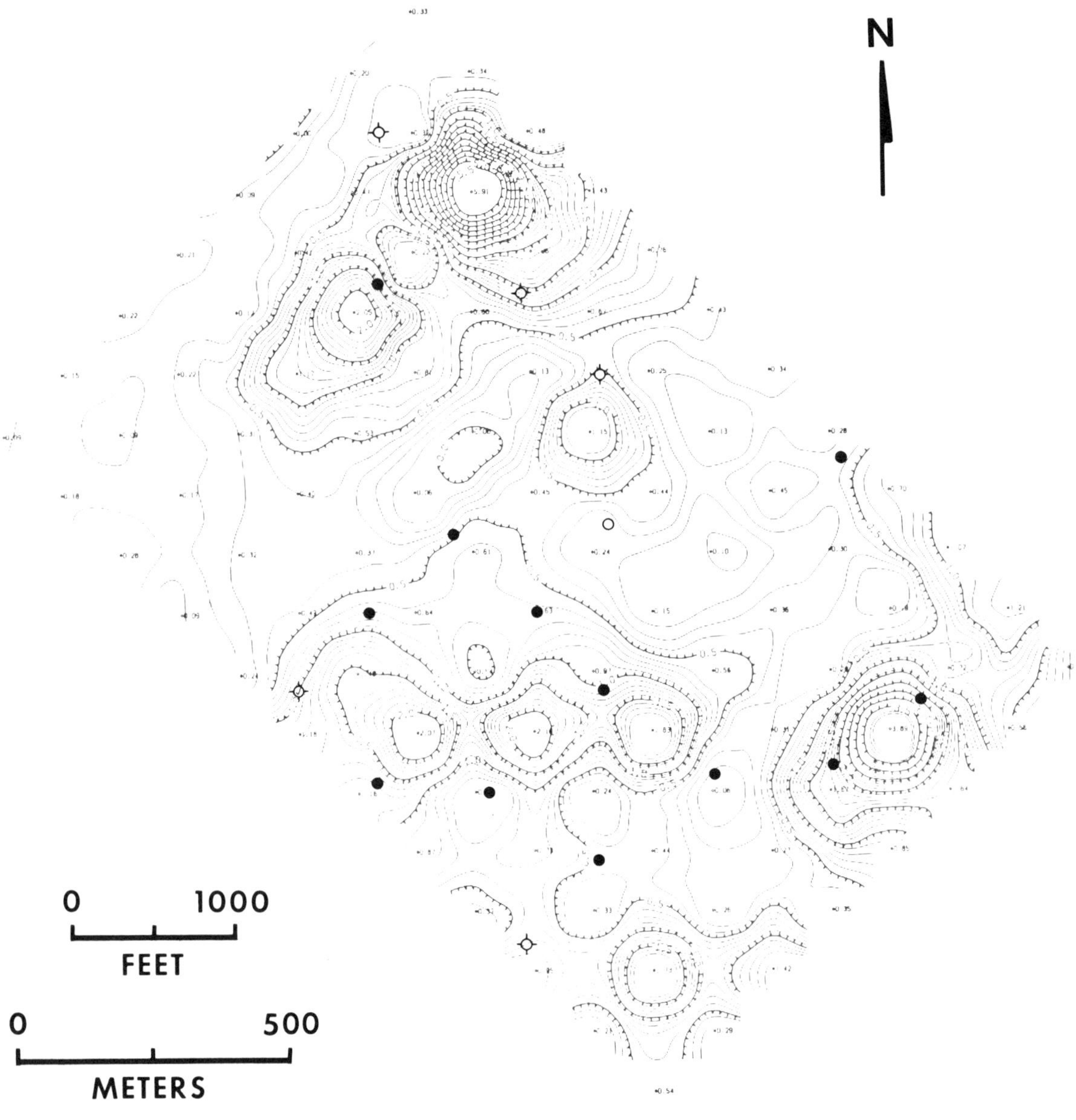

Fig. 3. Tannehill Sand propane concentration map, Knox County, Texas.

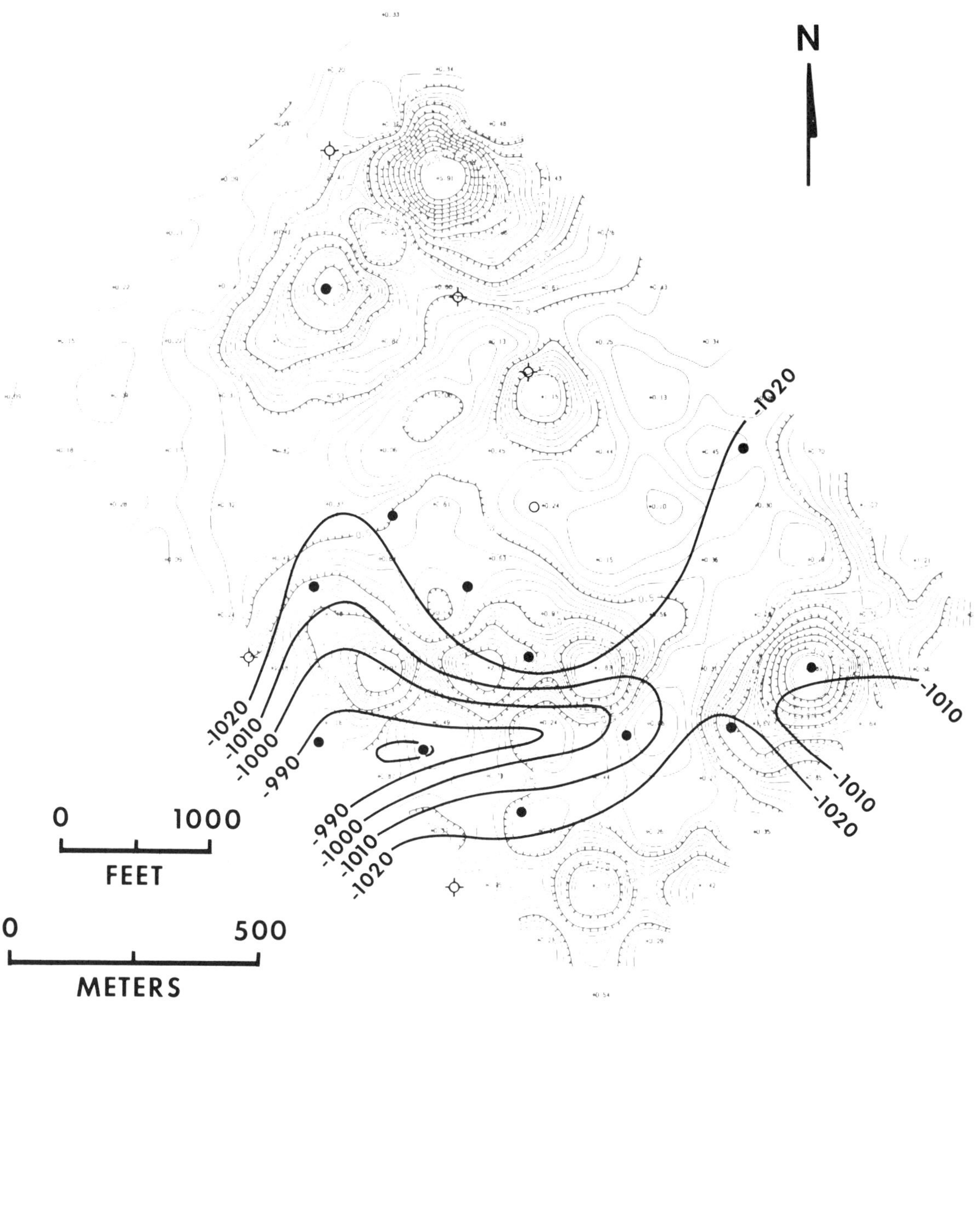

Fig. 4. Subsurface map with near-surface propane concentration superimposed, Knox County, Texas.

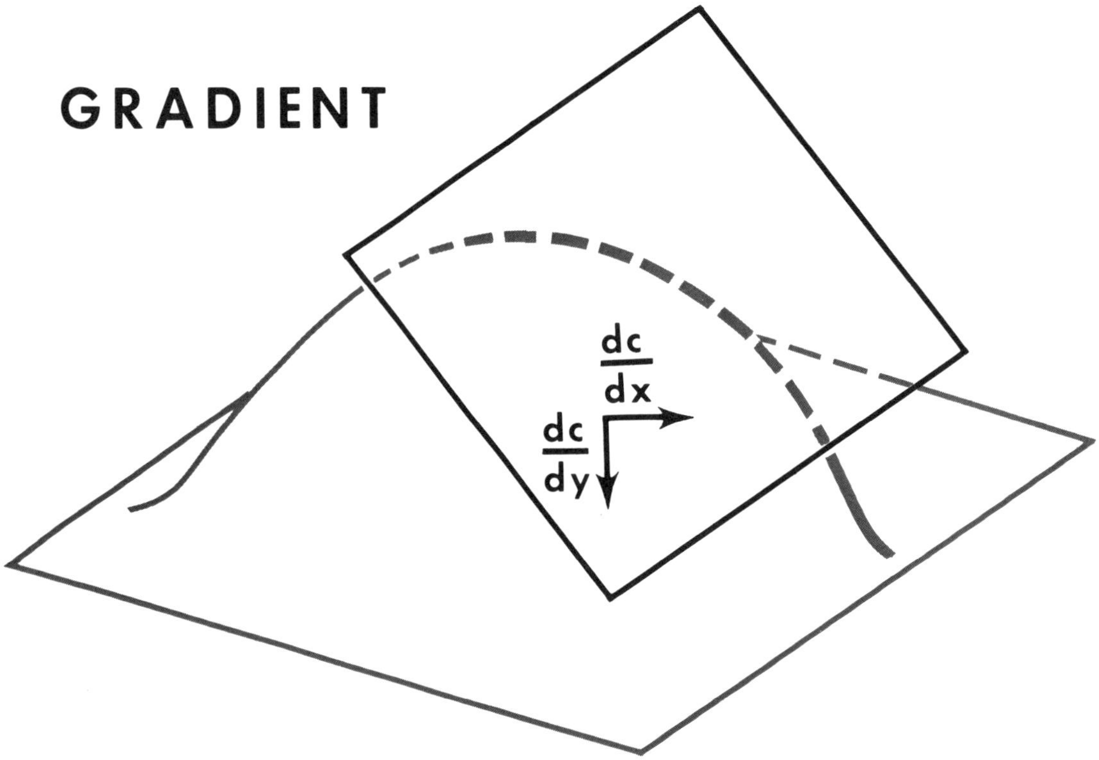

Fig. 5. Illustration of concentration gradient.

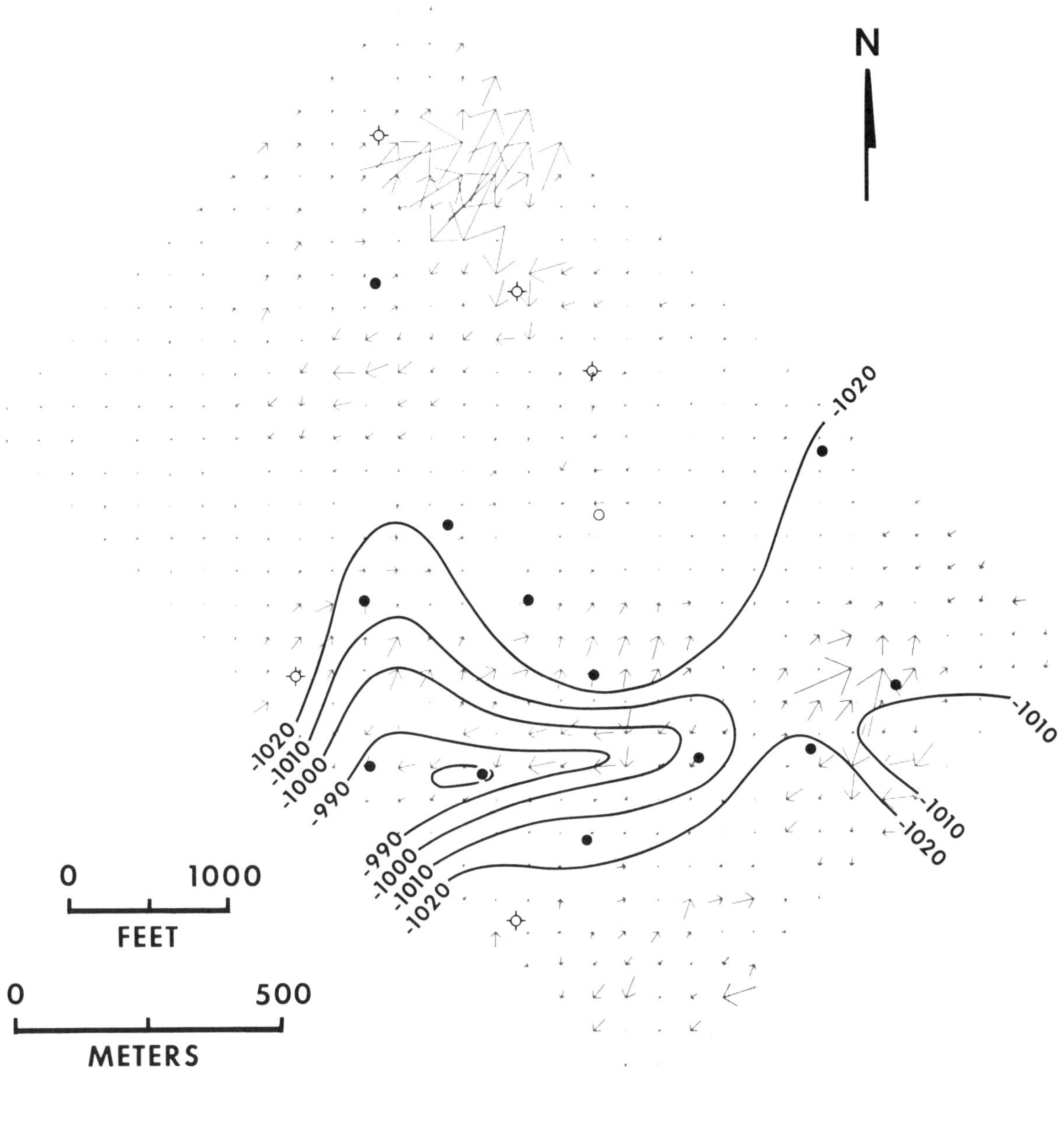

Fig. 6. Gradient vector map, Knox County, Texas.

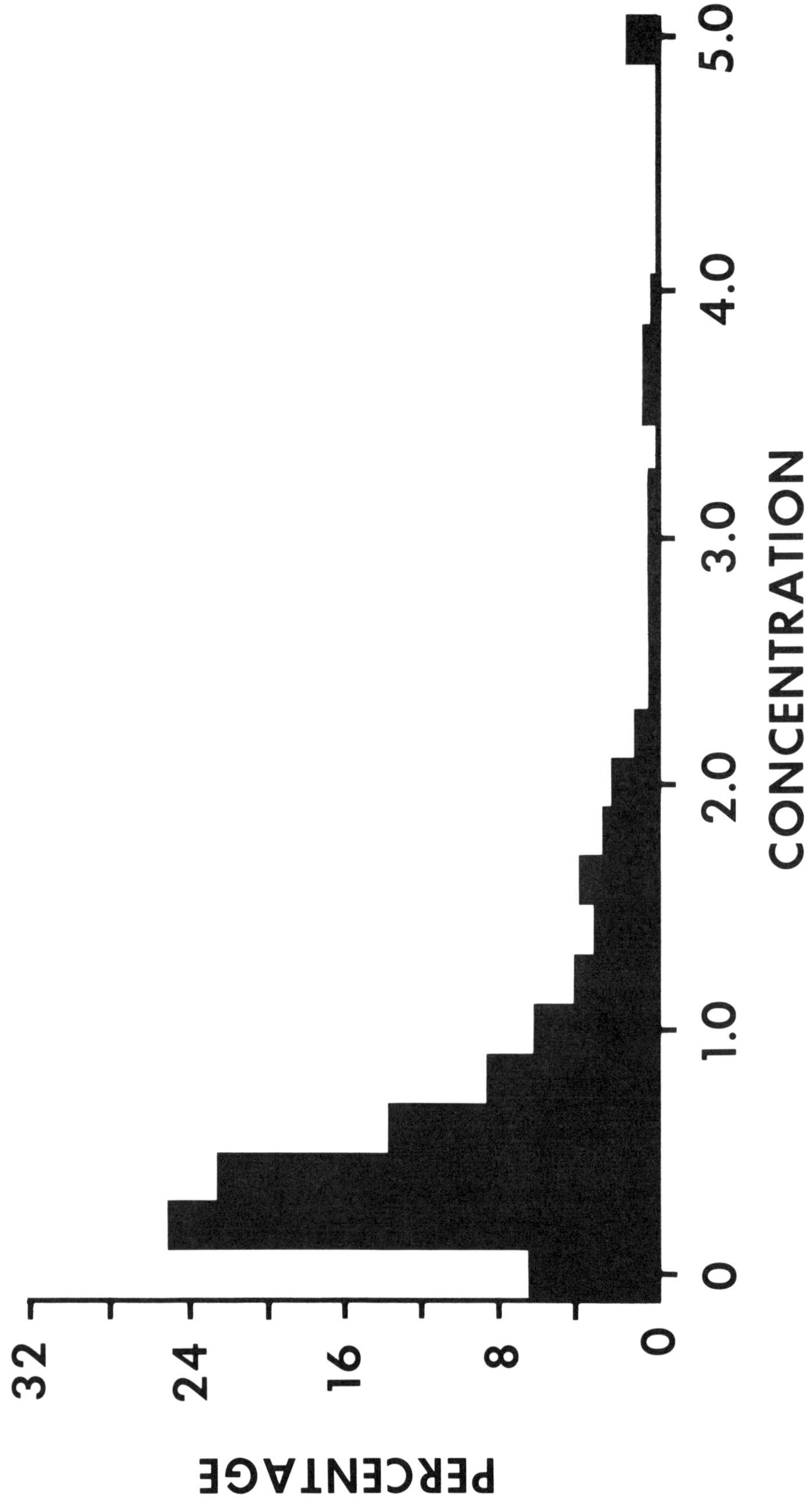

Fig. 7. Propane concentration data histogram.

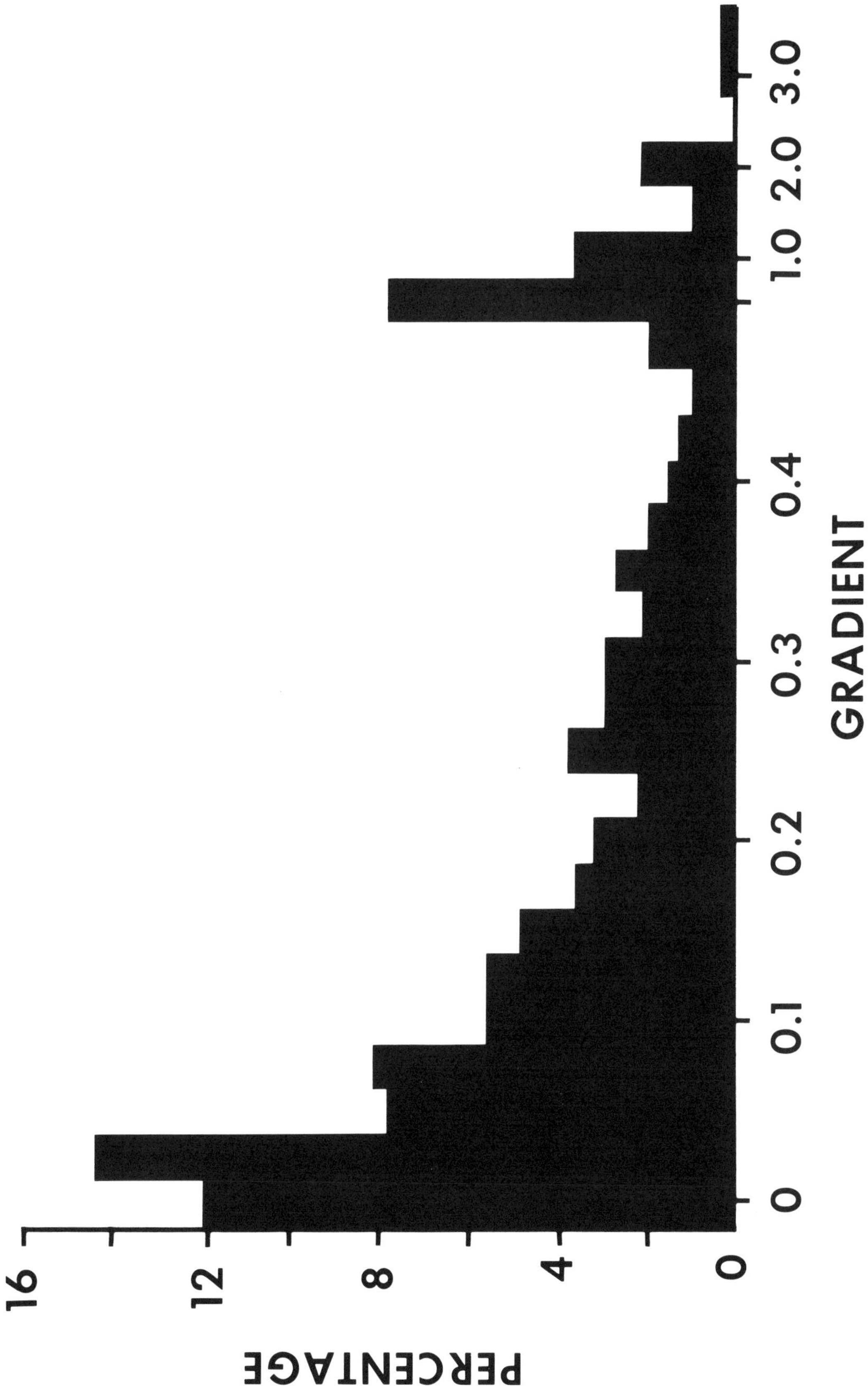

Fig. 8. Propane concentration gradient histogram.

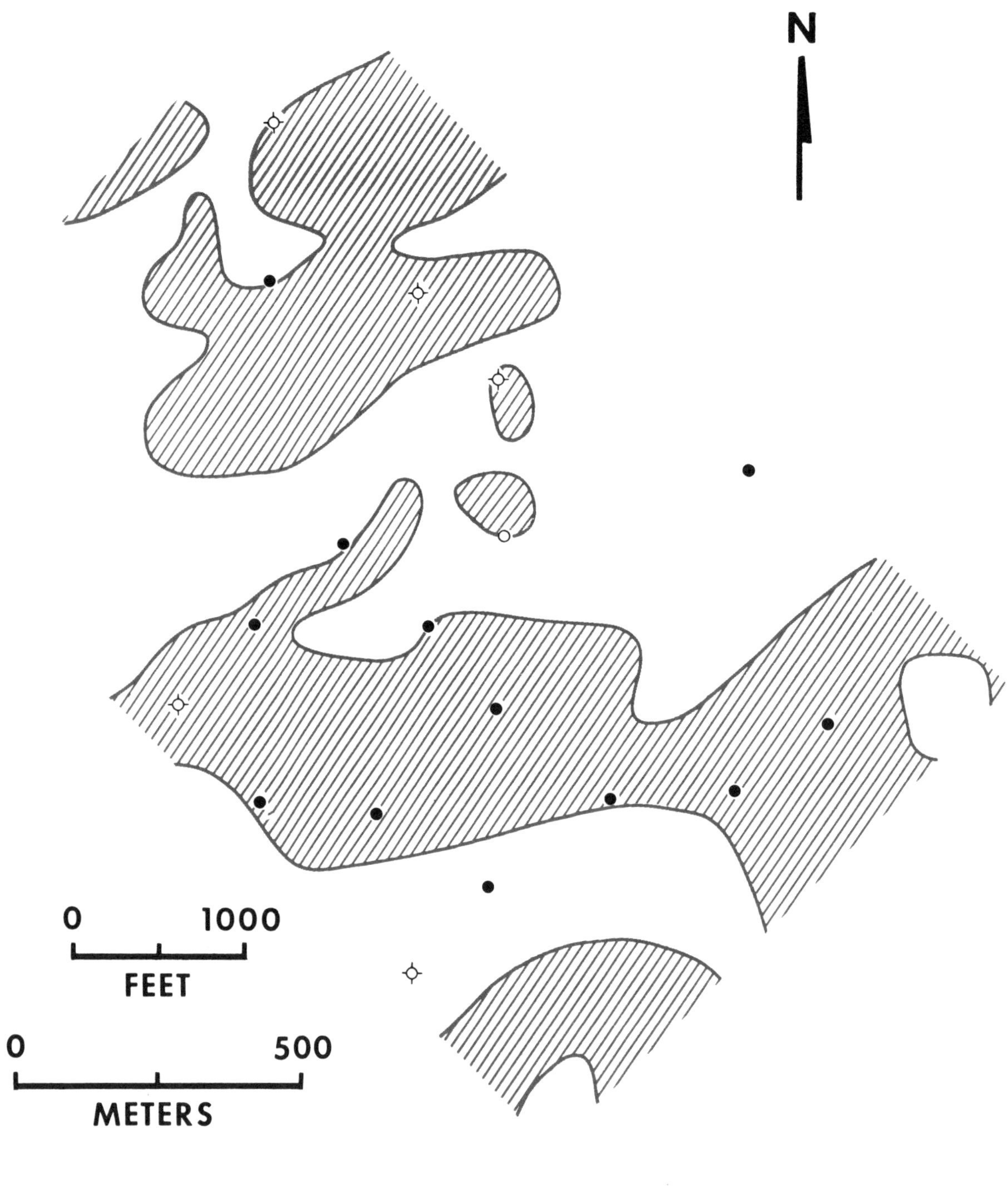

Fig. 9. Propane gradient map with well location, Knox County, Texas.

Fig. 10. South Texas propane concentration map.

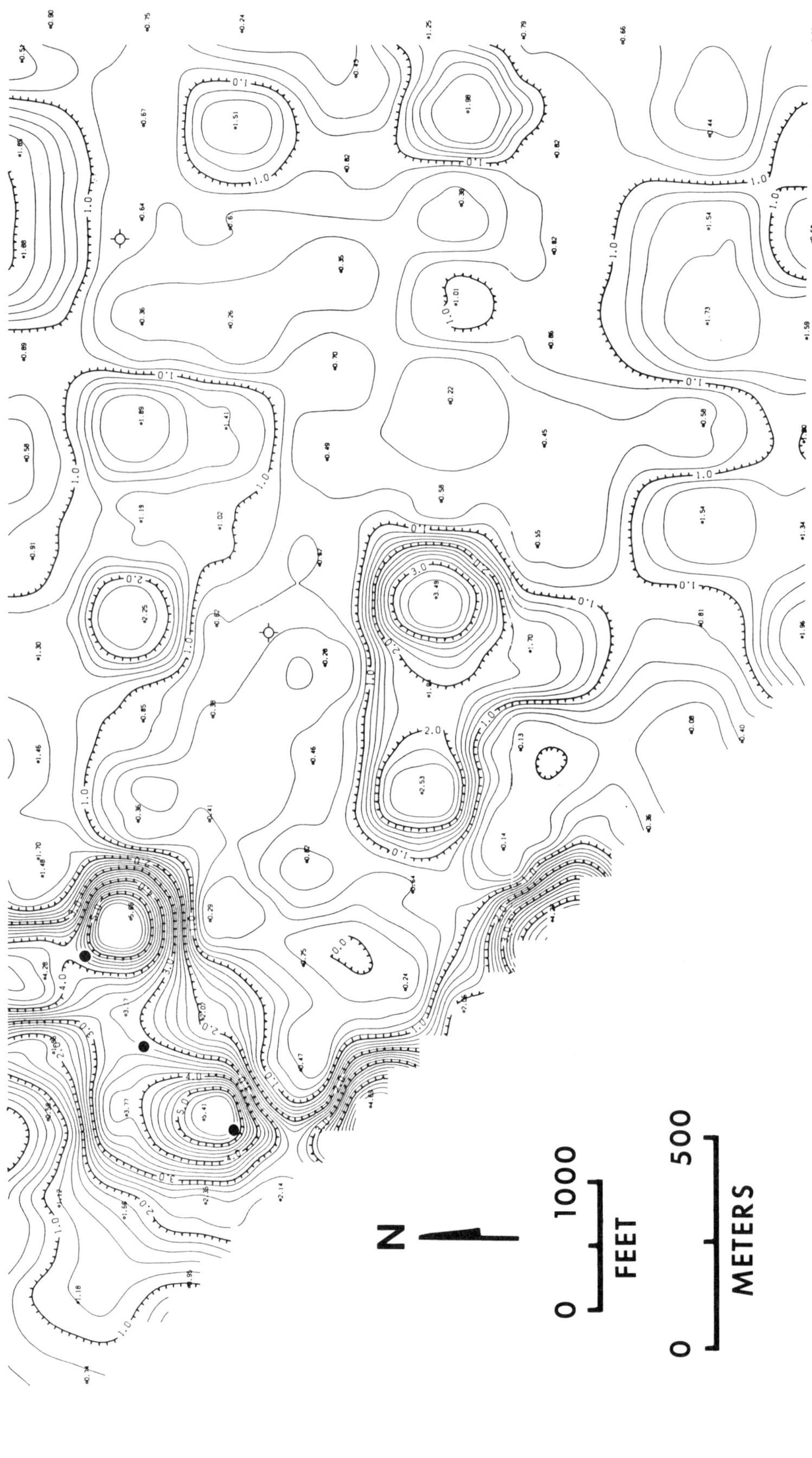

Fig. 11. South Texas propane concentration with post-survey wells.

MULTIPLE CORRELATION GEOCHEMICAL PROSPECTING

A. R. Barringer and J. S. Lovell

BARRINGER RESOURCES, INC.

INTRODUCTION

One of the ongoing problems of surface exploration geochemsitry for oil and gas is the variability of the two-dimensional surface expressions of underlying oil and gas fields. Surface anomalies can be continuous or patchy halos around the fields, apical anomalies, flanking fault-controlled anomalies, fracture elongated zones, etc. Some geochemical techniques tend to provide apical anomalies directly over the accumulation, while other techniques provide halo anomalies. Even with a single technique, the proper determination of the valid anomaly is never wholly dependable. The resulting uncertainties in interpretations have created a serious impediment to the general acceptance of surface geochemical methods.

The purpose of this paper is two-fold. Primarily, it is to suggest that by stacking layers of data it is possible to reduce exploration costs by first applying remote sensing and airborne geochemical and geophysical methods that have a very low cost per acre. If these are followed by inexpensive near-surface geochemical or geophysical methods, targets for the much more costly tools of seismic and drilling can be more accurately defined.

The second purpose of this paper is to demonstrate the collection and integration of a number of near-surface geochemical parameters.

SATELLITE IMAGERY TECHNIQUES

Examples of oil exploration applications of satellite imagery are given below:

1. Structural and fracture pattern analysis can be used for the location of traps, since these are frequently controlled by structures that have expressions on LANDSAT imagery. Such features include faulting, fault intersections, curvilinear and circular features associated with domes, uplifts, anticlines, etc.

2. Tonal features in soils and rocks relating to reducing zones associated with hydrocarbon leakage. Such leakage stimulates the activity of hydrocarbon oxidizing bacteria that deplete the near-surface soils of oxygen. Local changes in the oxidation state of iron minerals may alter the soil coloration (Donovan, 1974).

Geochemical leakage zones are also often associated with the hydrodynamic discharge of formation waters giving rise to the surface accumulation of salts. In semiarid regions, this can cause pronounced whitening of the soils—sometimes in an apical fashion, at other times in flanking or halo patterns.

3. Vegetation fertilization or stress features that can give rise to increases or decreases of plant vigor. In some climatic and soil environments, small amounts of hydrocarbon leakage increase the organic or nitrogen content of soils and stimulate the growth of vegetation (Harper, 1939; Adamse et al., 1972). In other situations such leakage can cause hydrocarbon bacteria to deplete the soils of oxygen and create anoxic conditions. In extreme cases, this can heavily suppress growth and inhibit the growth of certain plant species, allowing more tolerant types to take over. Thus, localized changes in ecology or plant vigor are clues to possible geochemical anomalies for ground follow-up. (Richers et al., 1982).

RAPID RECONNAISSANCE AIRBORNE SURVEYS

In many areas of the world that may be either remote or highly developed, there are major opportunities for an effective airborne survey approach. Land access may be physically difficult or costly, and the rapid coverage of areas either large or small leading to the prior selection of targets by remote sensing methods is attractive. Costly land acquisition is minimized and the expenditure of exploration money in the field is more directly focused.

An important phenomenon that makes possible the airborne hydrocarbon leakage detection methods described in this paper is the fact that surface microseepage of liquid phase hydrocarbons is ubiquitously associated with source rocks in oil fields. Such seepages are seen over a wide range of geological situations varying from source rock outcrops to both commercial and noneconomic oil and gas accumulations lying at depths of thousands of feet. They are also associated with strong fracturing and pervasive microfracturing that may be invisible to normal mapping or remote sensing methods.

Details of the mechanisms of basin dewatering, the vertical ascent of formation waters, and polar and nonpolar products of maturation will not be discussed here. It has been observed, however, that a whole range of thermally generated compounds occur at the surface of oil-rich sedimentary basins in both onshore and offshore environments. These occurrences appear to be related to vertical migration, an observation supported by examples such as the close correlation that exists between materials extracted from piston cores sampled in the Gulf of Mexico and oil fields lying at depths of many thousands of feet below (Brooks et al., 1983). The same situation can be reasonably expected onshore, with the qualification that near-surface bacterial oxidation, ground water movement, and the absence of a continuous water saturated column to the surface can be anticipated as complicating factors. It is fortunate, however, that many constituents of crude oil are quite resistant to bacterial destruction. It is these compounds that are observed in trace amounts at the surface of sedimentary basins in a wide range of soil and climatic environments.

Twomey and McMaster (1955) and Twomey (1960) reported that the evaporation of saline solutions from salt or soils generates large numbers of particulates that can act as condensation nuclei in cloud formation. Twomey's experiments have been repeated and quantitative measurements have been made of particulate and elemental release rates under standardized laboratory conditions. The apparatus diagrammatically illustrated in Figure 1 has been employed for this work. Similar studies of the inorganic particulates generated by plants were carried out by Beauford et al. (1975). These studies have now been extended to cover the joint release of organic and inorganic materials in particulate form by both soils and plants. The apparatus used for the plant studies is shown in Figures 2a and 2b. The same equipment that is shown in Figure 1 was also used for the measurement of organic particulates that are released from soils. The results can be summarized as follows:

1. Soils that have been saturated with saline water carrying traces of crude oil in true and micellar solution spontaneously liberate particulates, during evaporation, that carry both the salts and the crude oil originally contained in the oil solution. Similar results are obtained when the salts are replaced by a percentage of clay minerals. The fluorescent characteristics of the oil released from the soil in particulate form are very close to the characteristics of the oil in solution and do not exhibit the very heavy bias toward the light ends that occur in vapor phase release.

2. Plants fed with water containing salts and traces of crude oil in true and micellar solution also exhibit an anomalous release of organic particulates when compared with plants that are fed with the same saline solution but without the addition of oil. The composition of the organics released, however, is modified as compared with composition of the oil carried in the water fed to the plants. Preliminary data indicates that this is associated with the preferential uptake by plants of the components of crude oil that are carried in true solution, thereby accentuating, for example, the more soluable light aromatics. Some anomalous release of organics may also be related to plant stress when components in the nutrient are toxic (Fig. 3).

Details of this work will be documented in a separate publication, but it is mentioned here since it forms the basis of the airborne exploration system to be described. The system known as AIRTRACE, described by Barringer (1981), has now been updated. In earlier work, there was insufficient understanding of the basic phenomena involved to allow full interpretation and exploitation of the data obtained. However, recent results acquired over heavily vegetated areas such as those illustrated over the Colchester Field in southwestern Ontario (Fig. 4) have provided the basis for a complete reappraisal and further development of the technology. More advanced equipment is currently going into operation.

AEROMAGNETIC METHODS

Aeromagnetic equipment has also been developed that provides for compensation down to better than 1 gamma when flying in turbulent air at low altitudes. Low level precision magnetic surveys of this type allow the simultaneous mapping of not only the magnetic effects of the basement, but also near-surface magnetic anomalies that are sometimes associated with hydrocarbon leakage (Donovan et al., 1979).

GROUND GEOCHEMICAL SURVEYS

The use of a multiparameter ground geochemical approach has been developed not only for the follow-up of satellite and airborne targets, but also as a general ground reconnaissance tool. The basic principles involved relate to the demonstrable fact that when a variety of geochemical surveys are carried out over a known oil field, there is often a considerable variability in the shape and position of the anomalies obtained by different methods. In particular, methods that rely upon the collection and analysis of soil-gases for helium, hydrocarbons, and other gases often yield anomalies that are dependent upon diurnal, meteorological, seasonal, or cultural variables and are therefore generally difficult to repeat. Other near-surface methods may include the desorption of hydrocarbons from subsurface soil samples, extractions from surface soils for fluorescent analyses, and the analysis of ground waters.

Additional analyses may determine indirect hydrocarbon indicators such as iodine, carbonate cements, and a variety of inorganic species. Such surveys may lead to results that include the following anomaly types: (a) continuous or discontinuous halos surrounding the field; (b) apical anomalies on top of the field; and (c) structurally biased anomalies that are strongly influenced by the geometry of fracturing or other structures associated with the field.

Experience has shown that no one class of sampling and analysis consistently provides either apical or halo anomalies, a fact that makes interpretation difficult and can lead to completely erroneous predictions. The problem is further compounded by the tendency for different components to show modifications in their distribution even for a single sampling and extraction technique.

Preliminary orientation surveys over known fields do not necessarily solve the problem, since the response between techniques can vary considerably within a very local area. Futhermore, seasonally changing field conditions severely restrict the extent to which data may be extrapolated from survey to survey. The approach that is described here tends to minimize this problem by the use of multiple methods in the detailed phase of geochemical surveys. Since this technique adds to survey costs, it is not necessarily used during the reconnaissance phase. A single surface or near-surface method that first identifies general areas of interest may be preferred. Ideally, this should be a satellite and/or airborne, remote sensing, geophysical or geochemical method that can be applied economically on a reconnaissance scale to define targets efficiently for more expensive detailed follow-up.

As an example of the effectiveness of this technique, a variety of surface geochemical methods were applied to a study over a field in the Denver Basin. It must be stressed that the objective of using a known producing area as an example is, principally, to demonstrate that many geochemical methods frequently yield data that are even unable to distinguish between background areas and a producing oil field. When a variety of methods are applied, however, they yield a consensus of data over the area of leakage while simultaneously suppressing the random effects of geochemical noise in the background areas. The final data set is therefore a more specific indication of true leakage.

The Wigwam Field lies in the eastern portion of the Denver-Julesberg Basin. Oil is produced from a D-sand reservoir at a depth of 7,000 ft (Fig. 5). The trapping mechanism is thinning, and facies changes in D-sand channels. The field produced a total of 269,000 barrels of oil and 147,000 mcf of gas by December, 1981. It was discovered in 1974.

The methods that were applied here include soil analysis for C_{10+} hydrocarbon, methane, ethane, propane, total volatile hydrocarbons, and volatile sulfur. In addition, helium was determined in soil-gas samples collected from a depth of 1 m.

Helium is an indirect indicator of leakage that may coincidentally reflect anomalous concentrations of this gas in the reservoir. Over the Wigwam Field (Fig. 6) helium reveals a series of well-defined anomalies that tend to concentrate over the main portions of the production. However, there are values that could still be considered anomalous that are adjacent to dry holes, particularly over the eastern portion of the survey area.

Volatile hydrocarbon analysis determines the total hydrocarbons that are released from the sample after heating. As the detector response is species dependent, and as the volatiles are not speciated, it is not possible to calibrate the data. The results are therefore presented in parts per million methane equivalent (Lovell, 1979). The data reveal a well-defined anomaly lying along the central portion of a producing channel with the highest values lying over the best developed reservoir (Fig. 7).

Volatile sulfur is determined simultaneously with the volatile hydrocarbons. The results are presented in parts per million sulfur dioxide equivalent. The results (Fig. 8) do not reflect the trend of the channel with the precision of the volatile hydrocarbons. There are anomalously low values lying over some portions of the field and anomalously high values in background or dry-hole areas. There is, however, a well-defined anomaly over the thickest portion of the reservoir. Although this method does not even distinguish between background and production, it is central to the philosophy of this approach that it should be applied, since it is not possible in a prospective situation to predict which method is likely to be definitive and which is not.

Methane in soils is determined after a prolonged extraction. This gas generally forms the bulk of the hydrocarbon gases even from thermogenic sources. However, it is also generated almost ubiquitously in the near-surface by anaerobic bacteria. In the Wigwam area the strongest anomalies are formed to the west of the field and the field itself is overlain by some anomalously low values (Fig. 9), although methane constitutes more than 60% of the gas from this field.

The propane data show a very well-defined east-west trend through the field that does not conform to the north-south channel trend (Fig. 10). There are highly anomalous values overlying the production, but equally anomalous values extend into dry-hole areas to the east, west, and northwest.

The C_{10+} analyses determine the concentrations of a number of the more biologically resistant contituents of crude oils that reveal the location of strong fluid leakage. The analyses were carried out on extractions from soil samples. The data from the Wigwam Field (Fig. 11) for one of the C_{10+} components reveal a series of discontinuous anomalies that tend to concentrate over the central portion of the field. There are, however, a number of values beyond the field in background or dry-hole areas.

Having generated a series of different geochemical maps from a series of variables, it is now necessary to stack or process the data into a coherent picture which will indicate areas showing consistent evidence of leakage. There are a number of methods that can be used to stack the data, including factor analysis, discriminant analysis, and characteristic analysis. Each of these approaches has limitations. Factor analysis requires predetermination of the significant factors, characteristic analysis is inherently unsatisfactory as it relies on the use of models from other areas, and discriminant analysis relies on the subjective selection of a group of samples as being either anomalous or background, which are then used to model the rest of the area.

The data cannot be simply summed, as there is a wide numeric range within the different sets. The data are therefore reduced to a common scale by standard normalization either with or without logarithmic transformation.

Figure 12 shows the results of stacking the data shown in the previous figures. There is a very well-defined series of anomalies that all lie within the confines of the field and are especially concentrated in the areas of major production. The most significant feature of this approach is that there are no strong anomalies beyond the known production limits. The method does not necessarily reproduce the isopachs of the reservoir at the surface; however, the principal objective is to site the discovery hole. A hole sited on any of the geochemical anomalies would have intersected production.

This approach has been applied in prospective situations, and it has been convincingly demonstrated that there is detectable leakage at the surface from unexploited reservoirs that cannot be ascribed to contamination of surface disturbance.

An item of considerable importance that has to be added to the geochemical coverage is that of subsurface geology. Where there is extensive drilling control, the subsurface geology is a prime item in the stacking of data. Information on the existence of multiple pay zones, the types of hydrocarbons present, etc., is also very important for use in the interpretation.

An additional factor to be taken into account in the interpretation of ground geochemistry is considered to be that of basin dewatering. There is evidence, occurring in surface soils of sedimentary basins, of the widespread existence of hydrodynamic discharge zones containing anomalous amounts of elements such as iodine, manganese, calcium, strontium, etc. Space limits discussion on this topic, but it is suggested that large scale upward movement of deep formation waters provides a major carrier for solutions and micelles of both gaseous and liquid phase hydrocarbons. The inorganic associations of hydrocarbon anomalies therefore provide clues on their mode of emplacement.

MULTIPLE OVERLAY TECHNIQUES

The computer integration employed to stack the various classes of information calls for the use of a geographic data base with an appropriate scaling being adopted according to whether the information is regional or localized in nature. Data types stacked include the following: aeromagnetics, remote sensing imagery, gravity, airborne and ground geochemistry, and even digitized topographic data. The stacking may take into account both positive and negative features in the data, as well as various ratios.

CONCLUSIONS

The shortcomings of surface geochemical methods in the interpretation of complex leakage patterns can be reduced by the use of multiple correlation methods. Different geochemical techniques have a tendency to produce anomalies that show, on comparison, significant discrepancies. The statistical reliability of predictions can be greatly improved, however, by stacking several types of geochemical measurements along with geological and geophysical data. Such information can be obtained from satellite imagery as well as from airborne geophysical and geochemical surveys that are coupled with ground investigations.

Improvements in the accuracy of predictions and in success ratios can be anticipated when multiple correlation methods are used.

Stacking various scales of data from regional to a detailed scale can also substantially reduce the costs of an exploration program by using remote sensing and airborne geochemical and geophysical techniques to define targets for more expensive detailed investigation.

REFERENCES

Adamse, A. D., Hocks, J., DeBont, J. A. M., and Van Kessel, J. F., 1972, Microbial activities in soil near natural gas leaks: Arch. Mikrobiol. v. 83, p. 32–51.

Barringer, A. R., 1981, Airborne geochemical reconnaissance methods for oil and gas exploration, *in* Gottlieb, B. M., ed., Unconventional Methods in Exploration for Petroleum and Natural Gas II: Dallas, Southern Methodist University Press, p. 219–239.

Beauford, W., Barber, J., and Barringer, A. R., 1975, Heavy metal release from plants into the atmosphere: Nature (London) 256(5512), p. 35–37.

——, 1977, Release of particles containing metals from vegetation into the atmosphere: Science 195(4278), p. 571–573.

Brooks, J. M., Kennicutt, M. C., Barnard, L. A., Denoux, G. J., and Carey, B. B., 1983, Application of total scanning fluorescence to exploration geochemistry, 15th Annual Offshore Technology Conference, Proceedings, OTC 4624, Houston: Offshore Technology Conference, p. 393–400.

Donovan, T. J., 1974, Petroleum microseepage at Cement, Oklahoma: Evidence and mechanism: American Association of Petroleum Geologists Bulletin, v. 58, no. 3, p. 429–446.

——, Forgey, R. C., and Roberts, A. A., 1979, Aeromagnetic detection of diagenetic magnetite over oil fields: American Association of Petroleum Geologists Bulletin, v. 63, no. 2, p. 245–248.

Harper, H. J., 1939, The effect of natural gas on the growth of microorganisms and the accumulation of nitrogen and organic matter in the soil: Soil Science, v. 48, p. 461–466.

Lovell, J. S. 1979, Applications of vapour geochemistry to mineral exploration: Ph.D. dissertation, University of London.

Richers, D. M., Reed, R. J., Horstman, K. C., Michels, G. D., and Baker, R. M., 1982, LANDSAT and soil-gas geochemical study of Patrick Draw Oil Field, Sweetwater County, Wyoming: American Association of Petroleum Geologists Bulletin, v. 66,no. 7, p. 903–922.

Twomey, S. A., and McMaster, K. N., 1955, Production of condensation nucleii by crystallizing salt particles: Tellus, v. 7, p. 458–461.

Twomey, S. A., 1960, On the nature and origin of natural cloud nuclei: Bulletin de l'Observatoire due Pui de Dome, Jan–Mar 1960, p. 1–19.

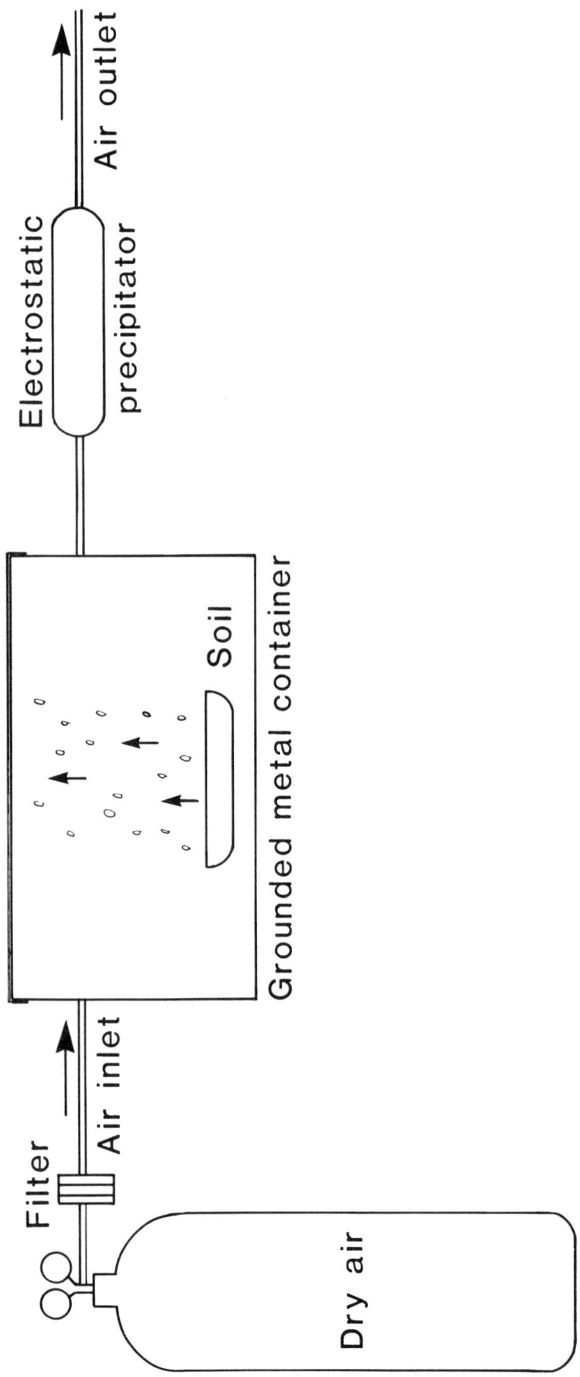

Fig. 1. Apparatus for the study of particulates released during evaporation from a moist soil.

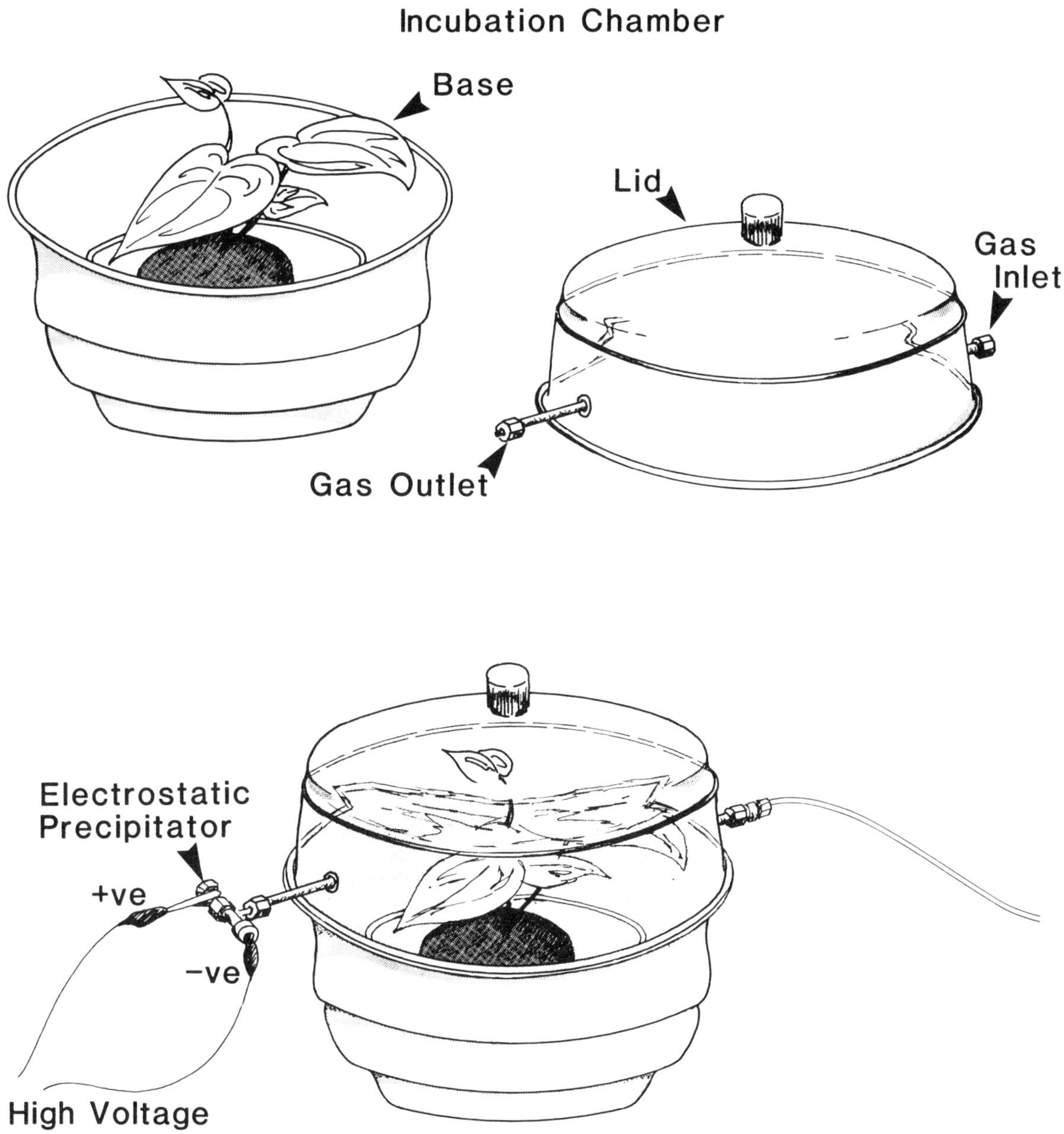

Fig. 2. Apparatus for collecting particulates released during transpiration from plants.

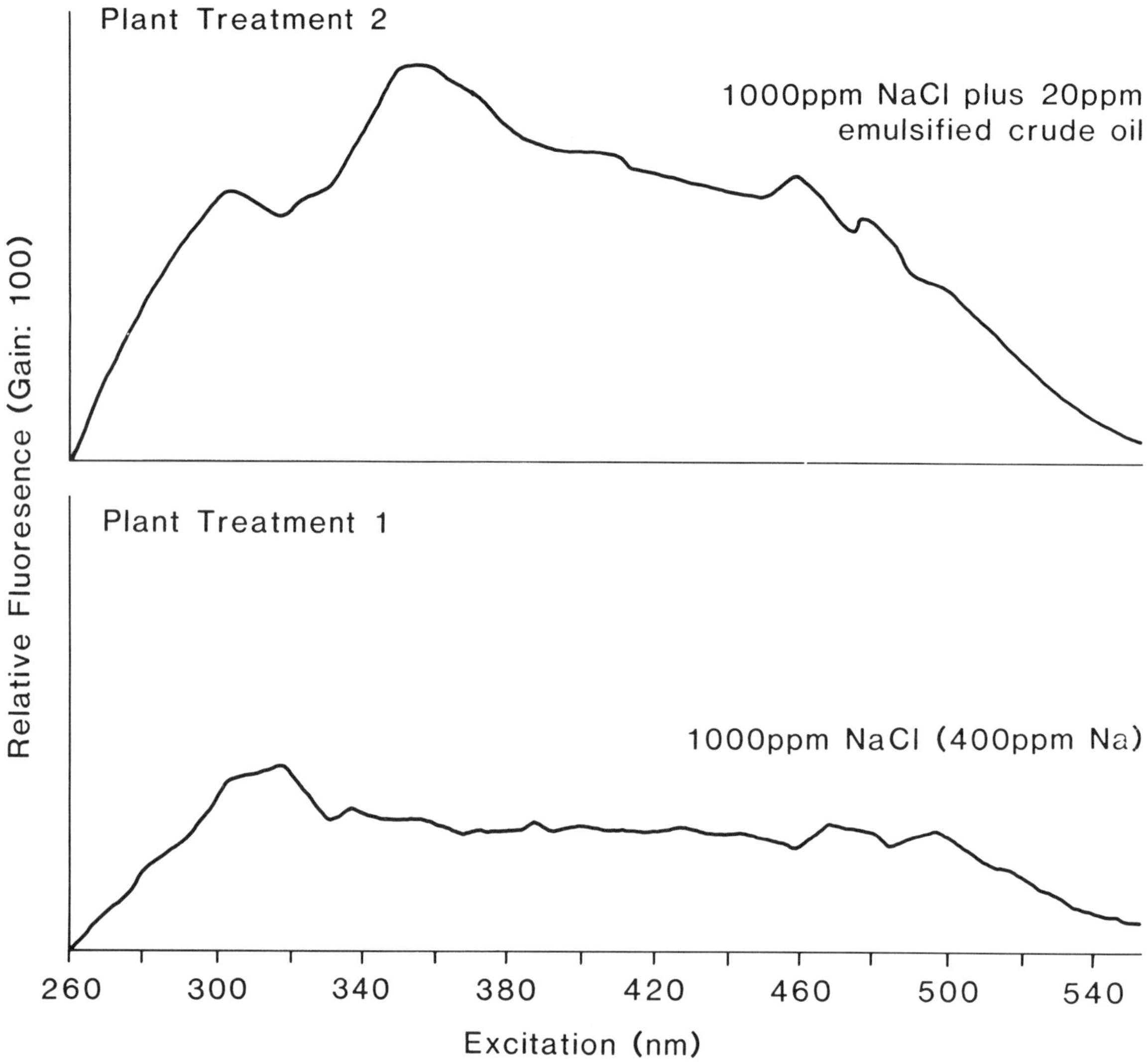

Fig. 3. Synchronous fluorescent scans of solvent extracts of particulates released from plants grown in oil-free and oil-spiked nutrient solutions. Synchronous scanning employs simultaneous scanning of both excitation and emission monochrometers with a 25nm separation. 1,000 ppm of NaCl was added to the nutrient solution of both plants.

Fig. 4. AIRTRACE® results, Colchester Field, Ontario, Canada.

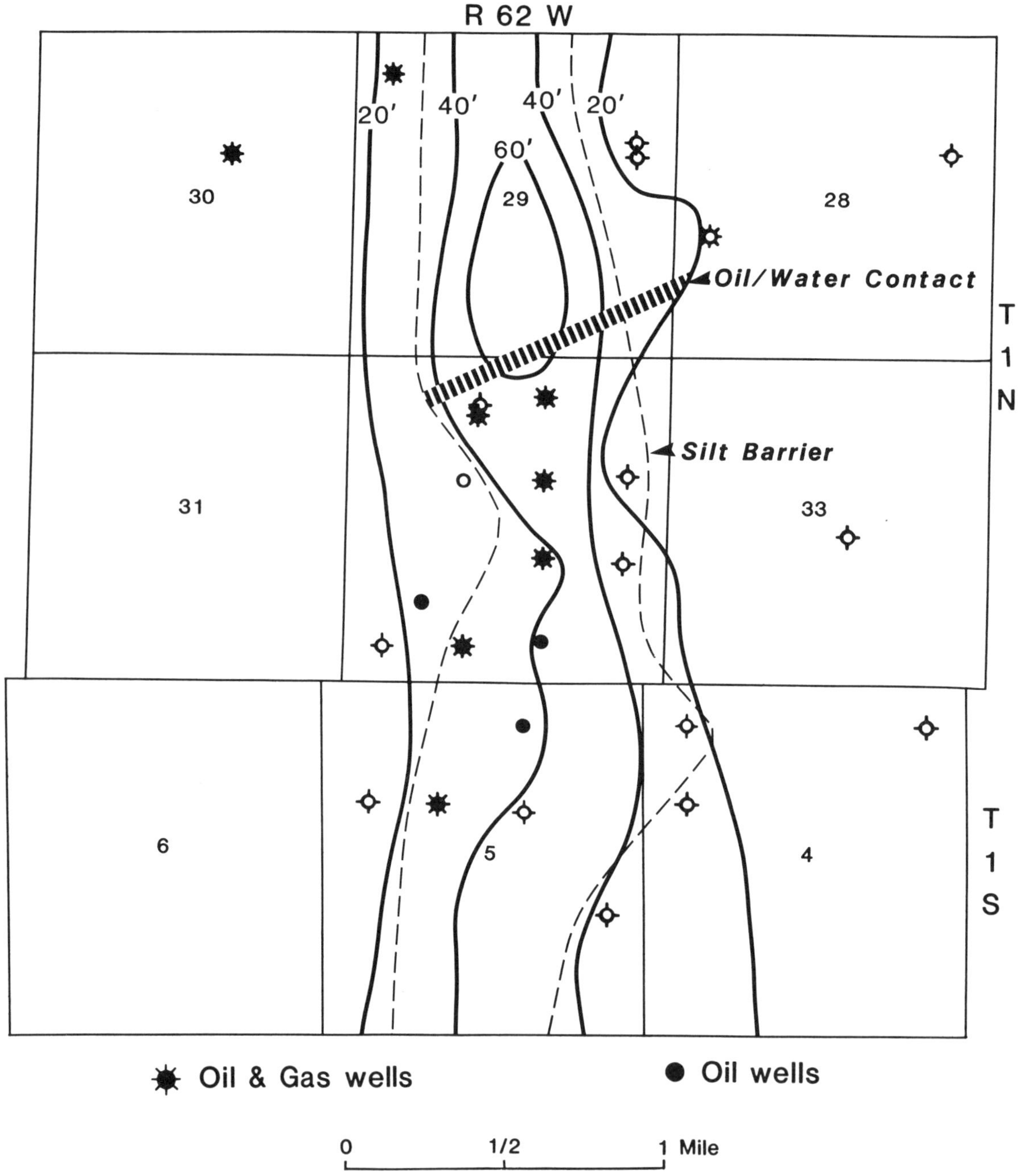

Fig. 5. D-sand isopach map, Wigwam Field, D-J Basin, Colorado.

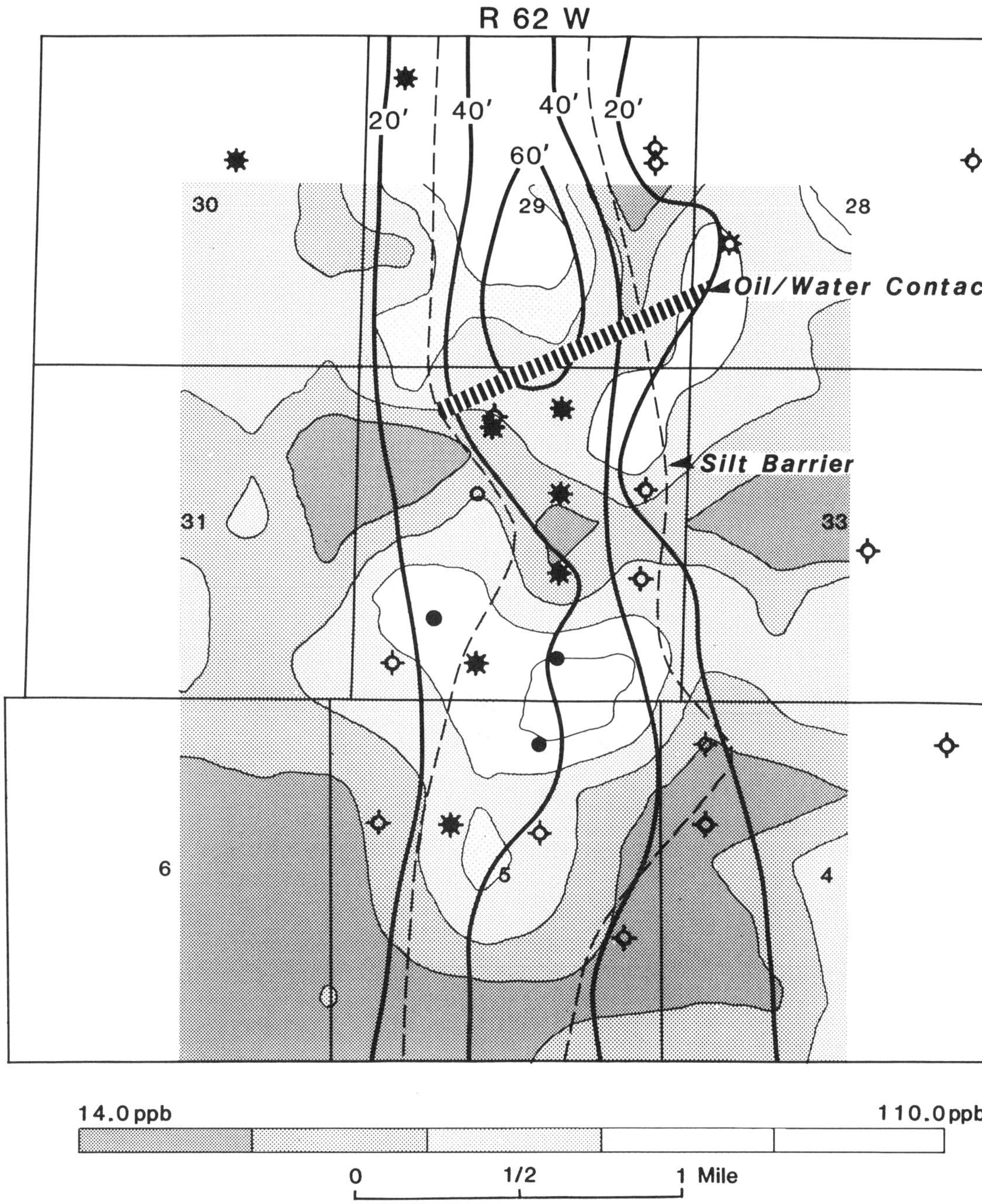

Fig. 6. Helium in soil gas, Wigwam Field, Colorado.

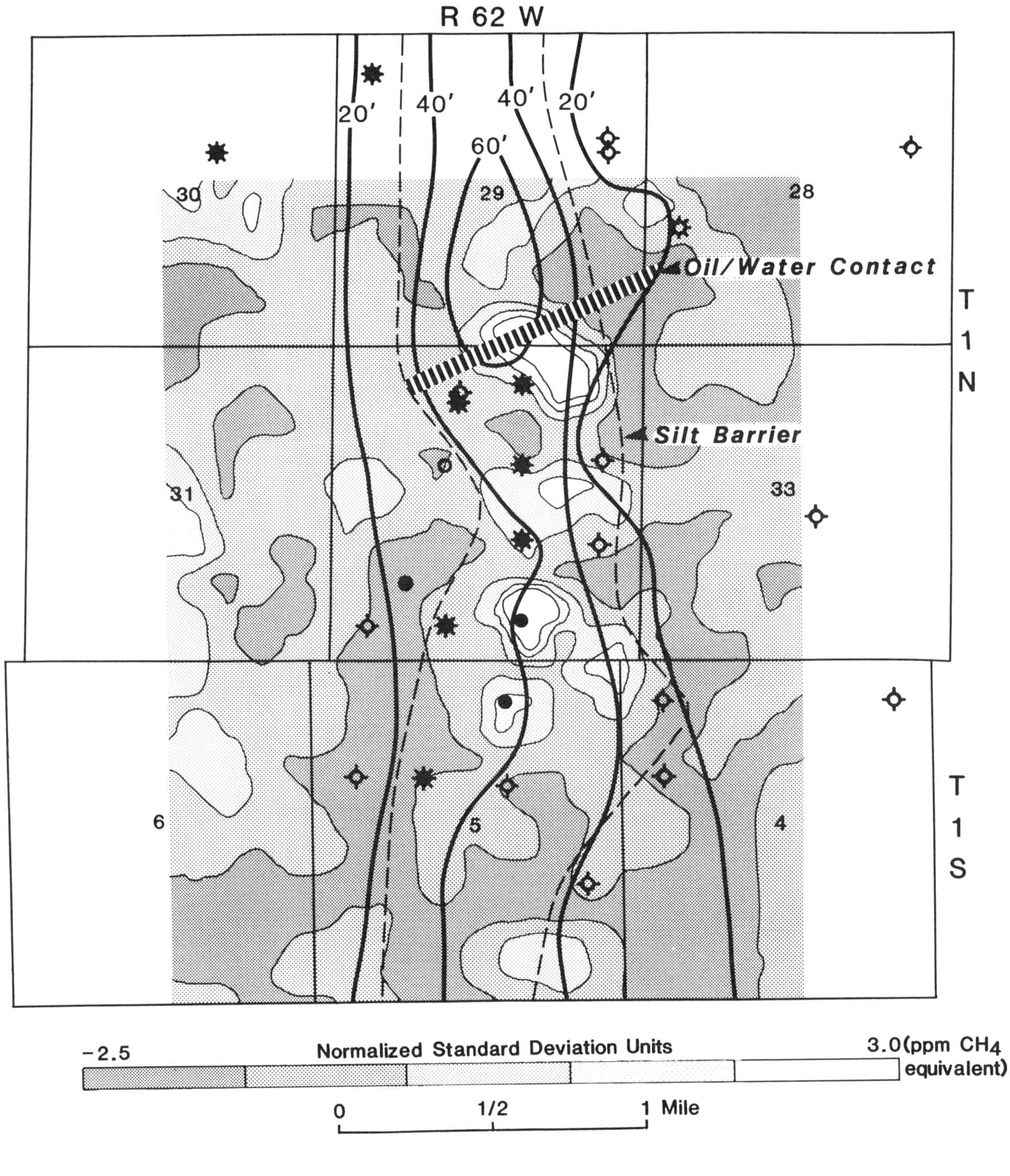

Fig. 7. Volatile hydrocarbons in soils, Wigwam Field, Colorado.

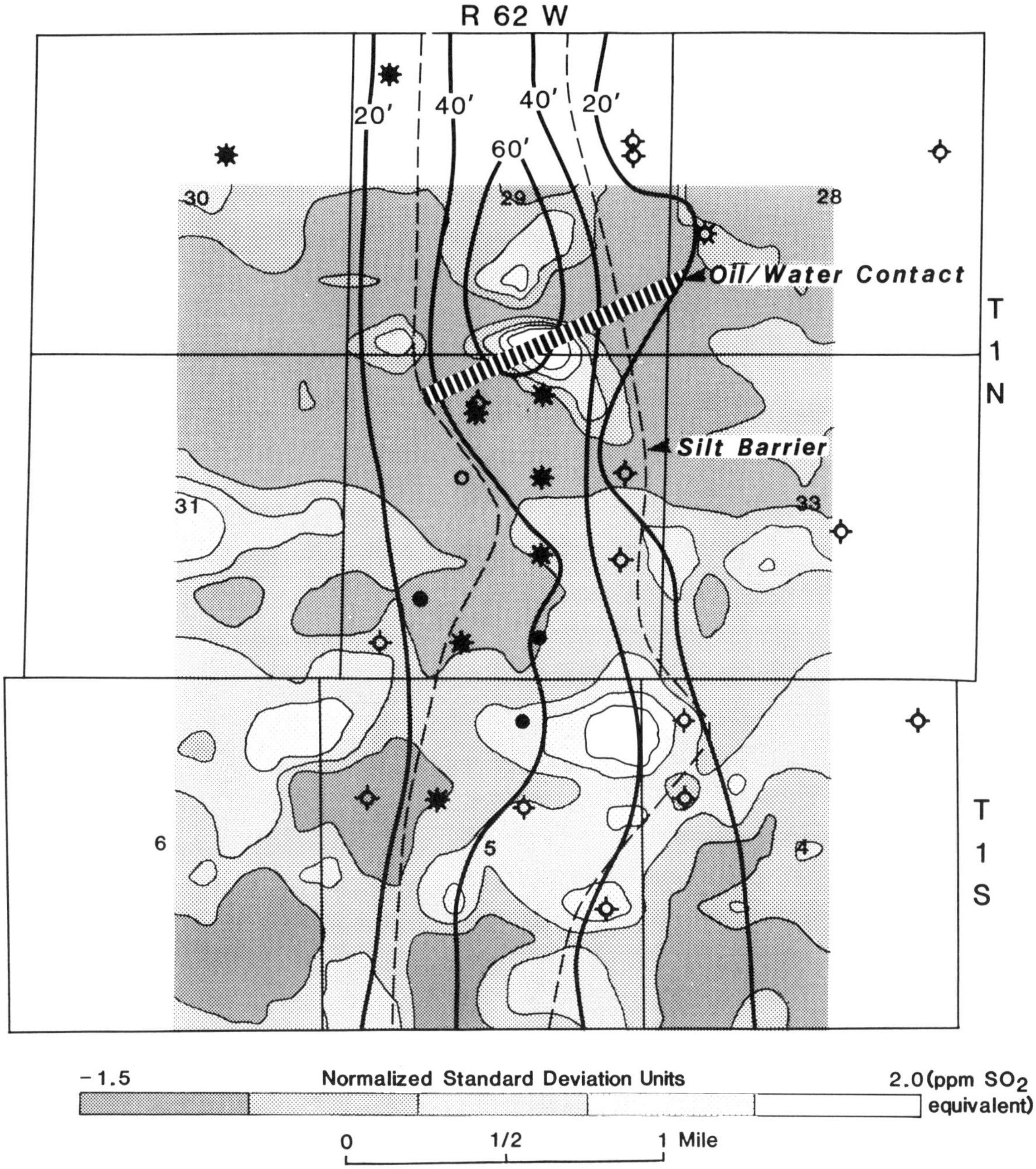

Fig. 8. Volatile sulfur in soils, Wigwam Field, Colorado.

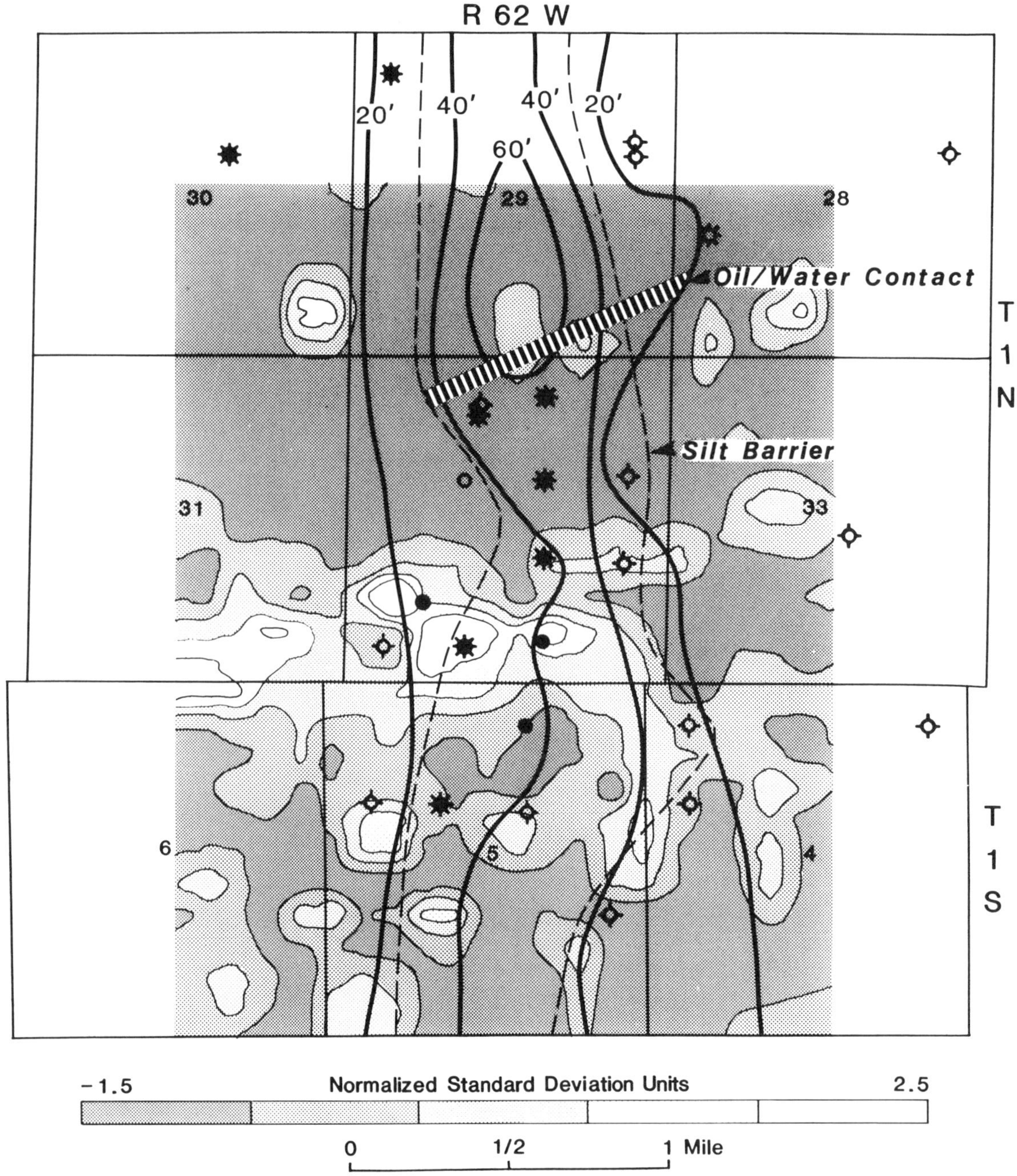

Fig. 9. Methane in soils, Wigwam Field, Colorado.

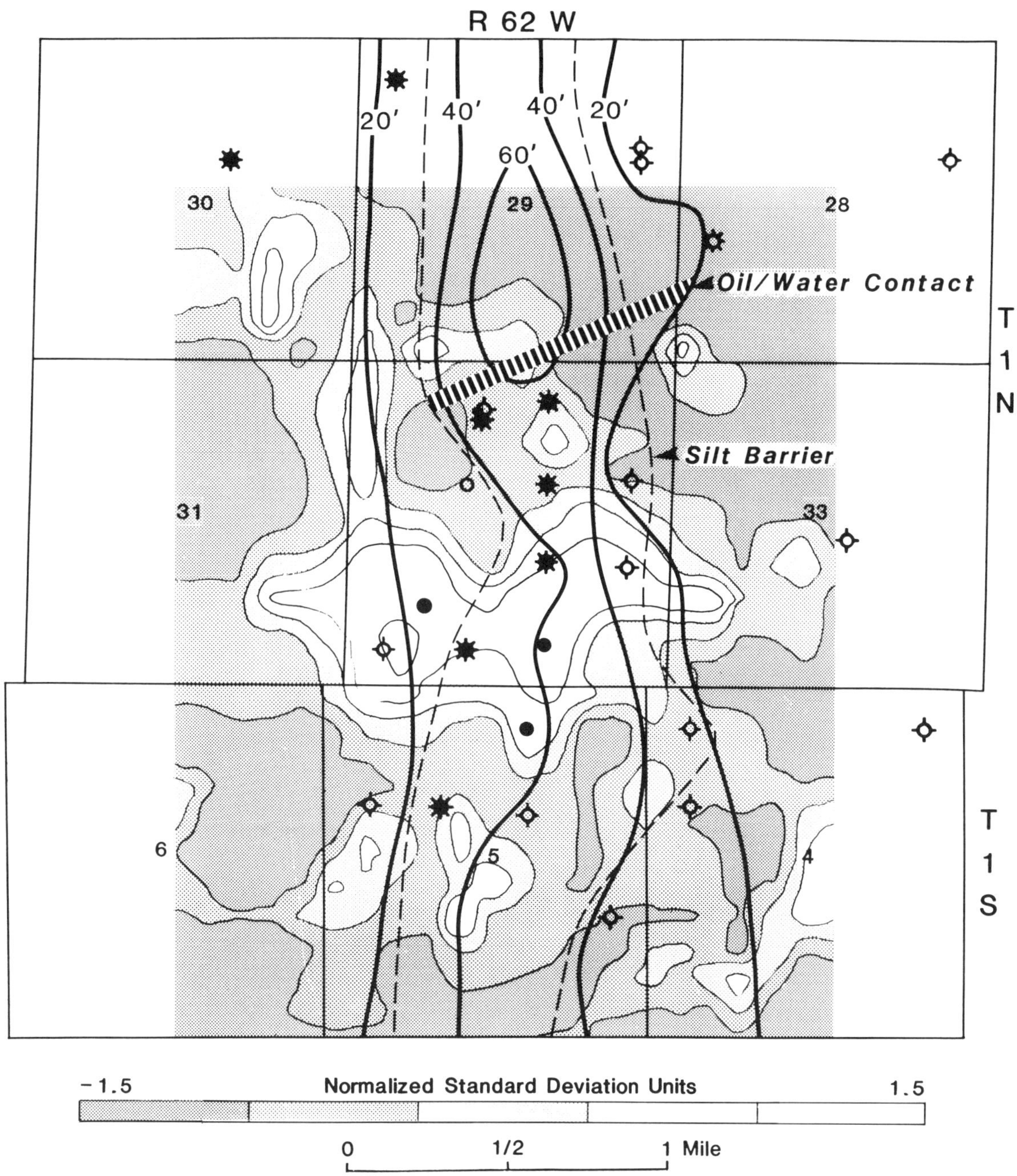

Fig. 10. Propane in soils, Wigwam Field, Colorado.

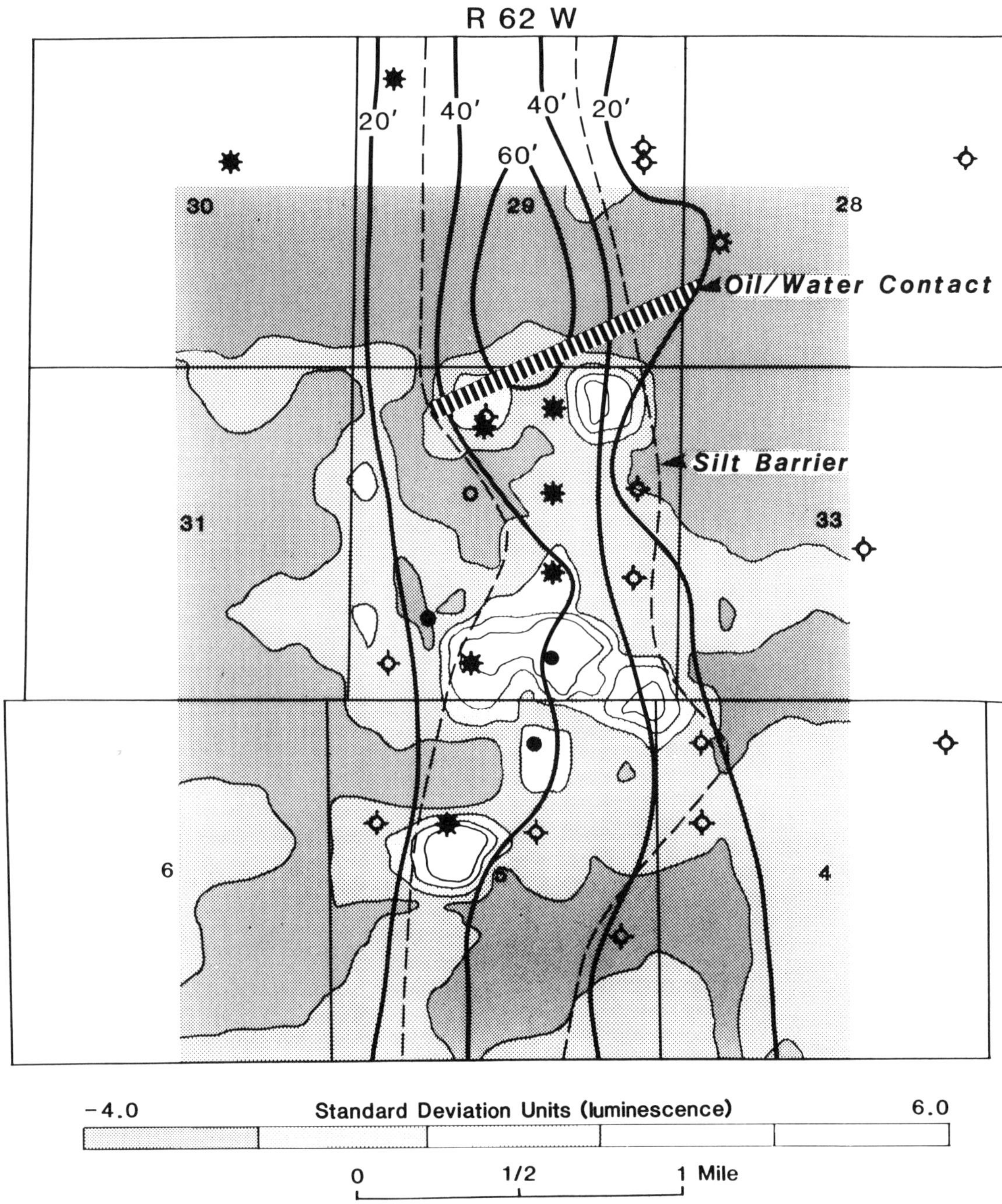

Fig. 11. C_{10+} hydrocarbons in soils, Wigwam Field, Colorado.

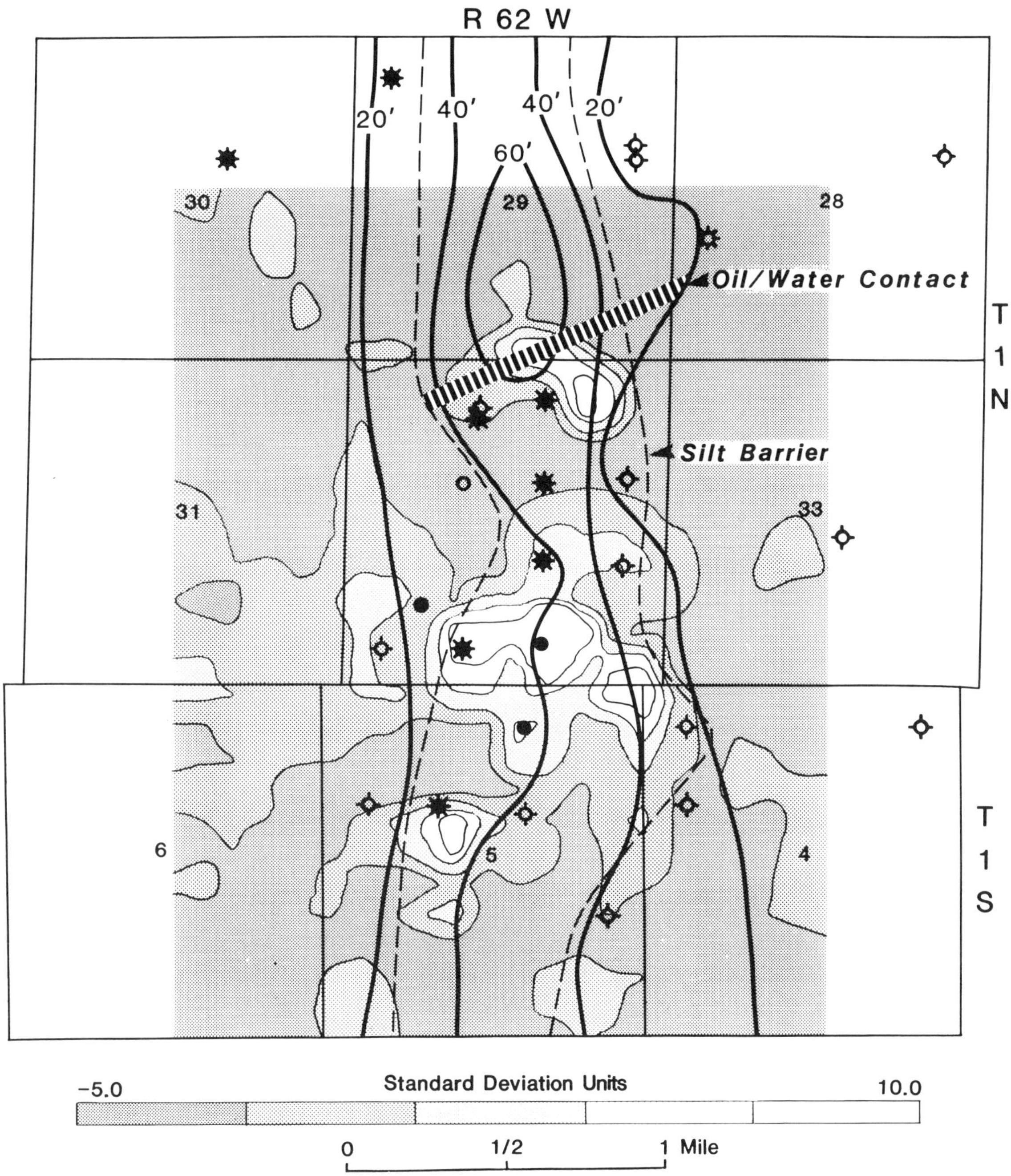

Fig. 12. Sum of geochemical parameters, Wigwam Field, Colorado.

APPLICATION OF THE K-V FINGERPRINT TECHNIQUE FOR PETROLEUM EXPLORATION

Ronald W. Klusman, Kent J. Voorhees, James C. Hickey, and Michael J. Malley

COLORADO SCHOOL OF MINES

ABSTRACT

The K-V Fingerprint Technique couples integrated soil gas collection with mass spectrometry and pattern recognition in the direct detection of hydrocarbons. A ferromagnetic wire coated with activated charcoal is buried at shallow depth to adsorb volatile compounds for a period of one to two weeks. The adsorbed compounds are analyzed following thermal desorption, using a quadrupole mass spectrometer. Pattern recognition using samples collected over known production is used to characterize survey samples.

Case studies from the Denver-Julesburg Basin, Uinta Basin and northwestern Colorado are presented. The Denver-Julesburg case study demonstrates trends parallel to that shown by well control. The Uinta Basin example demonstrates the vertical migration effect and resolution confirmed by drilling. The northwestern Colorado case study demonstrates the ability to characterize and define high potential for difficult-to-explore fault/fracture reservoirs.

INTRODUCTION

The use of soil gases from the earth as an exploration tool for petroleum extends back to at least the 1930s, (Horvitz, 1939; Rosaire, 1940). Gases are highly mobile and are indicative of their source due to composition and, to a lesser extent, their amount. The early studies resulted in mixed interpretations, primarily due to sampling and analytical reasons. Prior to the development of the gas chromatograph and mass spectrometer, direct measurement of soil gases was difficult. Combustion and manometry methods did not have sufficient sensitivity.

Other problems associated with use of gas fluxes are related to migration processes, multiple sources, and possible modification of composition during transport because of differential mobility. There is considerable chemical and physical interaction possible on the way to the surface. This interaction can be with solid materials, other fluids—such as groundwater—or with the atmosphere in the near-surface. Biological consumption and production in the near-surface can significantly alter the original composition and amount of migrating gas. There is potentially considerable interference with the original signal, making discrimination difficult. Noise of this type cannot be overcome by improvement in analytical techniques or sensitivity.

The interaction of gases with the overlying sedimentary column during vertical migration can result in chemical and mineralogic changes characteristic of indirect indicators, such as iron reduction or botanical effects (Donovan, 1974; Donovan et al., 1975, 1979, 1981; Dalziel and Donovan, 1980, 1984; Duchscherer, 1981, 1982; Richers et al., 1982; Rock, 1984; Matthews et al., 1984; Roeming and Donovan, 1985). Other gases, such as helium and hydrogen, may have an indirect association with petroleum and natural gas, migrate with hydrocarbons to traps, but may not have the same origin (Hawkes, 1972; Roberts et al., 1976; Roberts, 1981; Pogorski and Quirt, 1981).

The surface direct measurement of hydrocarbons in soil gas or marine sediments has been practiced as an indicator of subsurface oil and gas (Brooks et al., 1979; Kvenholden et al., 1981; Kvenholden and Redden, 1980; Reed and Kaplan, 1977; Sackett, 1977; Jones and Drozd, 1983; Simoneit and Galimore, 1984; Sigalove, 1985). Because of the high mobility of soil gases, and to a lesser extent in marine sediments, near-surface measurements of the quantity of gases exhibit considerable spatial and temporal variations. This results in extreme contrasts over short distances and in the lack of reproducibility in flux or emanation rate measurements.

The integration of gas flux measurements over longer periods of time reduces short term temporal variations. The integration or continuous collection of soil gas may be experimentally difficult and requires two visits to the field site. Soil or marine sediment has been used as an integrating medium for soil/marine sediment gas (Horvitz, 1972; Richers et al., 1982; Vankatesan and Kaplan, 1982). Soil and, to a lesser extent, marine sediment is a complex material varying substantially in mineralogy, organic material character and content, temperature, and other factors which affect the ability to retain adsorbed gases. Temperatures of marine sediments are more constant than soils, but may degas as a result of pressure changes during transportation to the surface. These characteristics make soil or bottom sediment less than ideal as a medium for integration of volatiles. Separation of a particular size fraction of the soil or sediment is a step toward improving the ability to interpret the signal (Horvitz, 1980).

K-V FINGERPRINT TECHNIQUE

A new technique was developed in the Department of Chemistry and Geochemistry at the Colorado School of Mines and Petrex for the direct surface detection of and characterization of petroleum and natural gas using hydrocarbons (Klusman and Voorhees, 1983; Hickey, 1985; Hickey et al., 1985). The objectives of the original method development were: (1) to develop an integrative device that provides more reproducibility than soils or marine sediments, (2) a completely portable integrative collector that could be placed in the field without power equipment or environmental impact, (3) an analytical technique which allows rapid analysis of a broad range of components which might be collected during a significant period of gas flux integration, (4) computer-aided data interpretation system that objectively relates the broad range of data to subsurface accumulations.

This method of integrative collection uses a ferromagnetic wire partially coated with charcoal using an inorganic cement. This gives nearly identical collectors, which is not possible using soils. A 2.5×15 cm culture tube is presently used for both transportation and field placement of the wire (Fig. 1). A hole approximately 30 cm deep is dug or augered, the cap removed, and the tube with the wire inside is placed in an inverted position in the hole. The hole is backfilled and the location marked to allow retrieval of the wire in one to two weeks. The caps and tubes are retained in a clean, sealed container until wire retrieval, cap replacement, and sample numbering. This method provides the advantage of integrative collection and better reproducibility than the soil or marine sediment samples.

The samples are analyzed by computerized Curie point mass spectrometry. The system allows the introduction of the sample without chemical preparation directly into the mass spectrometer and extends the analysis into a mass range encompassing only slightly volatile compounds. The higher molecular weight compounds are more difficult and time-consuming to analyze by gas chromatography and are usually not included in a soil gas analysis. The ferromagnetic wire is heated by a radio frequency source to its Curie point. The adsorbed gases are desorbed without pyrolysis and separated by a quadrupole mass spectrometer. The mass spectrum, including a mass range of 15–240, is stored on a hard disk by a minicomputer. The computer control and data capture results in a rapid reliable collection of the large amount of data generated by mass spectrometry. The mass spectrum is a representation of the composition and is proportional to the flux of a complex mixture of volatile compounds and molecular fragments. The flux of individual compounds is a function of a number of uncontrolled factors, such as season, underlying permeability, and biological activity for some gases. Interpretation of the flux of a few gases is confounded by these unknown parameters.

Because interpretation of flux is difficult, a more powerful and useful method was developed. This approach uses the entire mass spectrum or the pattern of gas fluxes which can be compared to a fingerprint. The mass spectrum of soil gas samples is complex, often containing in excess of one hundred peaks of varying intensity (Fig. 2). An underlying accumulation, such as a petroleum trap, has a pattern or fingerprint at the surface which is different and distinguishable from the fingerprint observed in the absence of petroleum. The interpretation and characterization is accomplished by using pattern recognition to discover and group data by the fingerprint. The mass spectral pattern or fingerprint allows computer recognition of the distinguishable data structures.

One means of grouping volatile emissions is by cluster analysis. The interpretation of multivariate cluster analysis data is aided by a reduction of dimensionality. Table 1 is a hypothetical analysis of four samples containing three components each. A three-component mixture can be graphically represented by a three-dimensional diagram such as Figure 3a. The three-dimensional diagram can be projected to a two-dimensional diagram in a way

TABLE 1

HYPOTHETICAL HYDROCARBON GAS MIXTURE

Sample	Percent Ethane	Percent Propane	Percent Butane
1	30	60	10
2	29	61	10
3	14	50	36
4	70	20	10

that the spatial characteristics of the data are preserved, as shown by Figure 3b. A similar procedure can be used for the complex mass spectra of Figure 2. Although these two spectra are obviously different, the classification of large numbers of spectra with more subtle variations is not as obvious. This type of projection allows a quick determination of sample clusters and of samples which are similar or dissimilar in pattern or fingerprint. For example, where a single subsurface hydrocarbon accumulation will yield a single cluster, often multiple but different hydrocarbon accumulations yield different fingerprints and distinguishable clusters (Fig. 4).

The pattern or the fingerprint of a sample can be increased in usefulness by the quantitative determination of the similarity to a reference set of samples. This reference set of samples is termed a training set. If the subsurface accumulation has a consistent composition, the volatile emissions should be consistent and should be possible to characterize. A group of samples collected over a known petroleum accumulation can be used to characterize or fingerprint the emanations. The grid or unknown survey samples can then be compared by calculating a multidimensional Euclidean distance for each sample from the centroid of the training set samples. The difference between the sample at the greatest distance from the centroid and the sample at the minimum distance is linearly scaled from 0–100. All samples are scaled to this distance and result in a parameter called the similarity index with values from zero to one hundred. A similarity index is plotted in map form for interpretation. This parameter is a quantitative, dimensionless comparison of the pattern or fingerprint rather than a measure of the quantity of individual volatile species.

Another type of emanation observed in the mass spectral data is the fault/fracture signature. Areas of increased permeability exhibit high relative proportions of high molecular weight species. These are readily recognized by the presence of these high molecular weight components, and the commonly observed high flux rate does not mislead the interpretation (Fig. 5).

AREAS OF STUDY

The application of the K-V Fingerprint Technique to three areas will be presented. One of these is a stratigraphic trap in the Uinta Basin. The second area is in northwestern Colorado, where the target is production from fault/fracture reservoirs. The third area, in the Denver-Julesburg Basin, has a number of stratigraphic traps and demonstrates the effectiveness of the K-V Fingerprint Technique in the use of detailed surveys following a reconnaissance survey.

UINTA BASIN

The first example demonstrates the potential resolution of the method and strongly supports the vertical migration effect. A detailed grid of samples was established over a hypothesized channel in the Dakota formation which produced from a depth of approximately 2000 ft in the southern Uinta basin (Fig. 6). The overlying unit is the Cretaceous age Mancos shale, which does not have significant sand lenses or channels that might contain hydrocarbons that were being detected at the surface. A shut-in well was used to develop a geochemical model. Calculated similarity indices indicated a high degree of similarity extending to the south, which the client interpreted as the extension of the channel (Fig. 6).

A well drilled on the high similarity zone near the south end of the survey area was dry. Subsequent downhole measurements indicated the bit had drifted and the bottom of the hole was not in the area of high similarity. A second well was drilled at a point where the similarity zone was judged to be wider. The result was a commercial well. The results of these two wells strongly suggest that vertical migration with minimal dispersion does occur. The fingerprinting of a narrow compositional model using heavier gases probably enhances the effect of vertical migration. The lighter hydrocarbons are likely to be more dispersed during migration than heavier hydrocarbons.

Northwestern Colorado Fault/Fracture Reservoir

The presence of faults and fractures in the sedimentary column creates high permeability zones for fluid migration and possible accumulation. The Cretaceous Niobrara formation of northwestern Colorado is a petroleum source rock where reservoirs occur only in areas of extensive faulting and fracturing.

Preliminary selection of areas of interest is usually made through the use of LANDSAT data or aerial photography. In some cases faults can be mapped on the surface. The difficulty in exploration for this type of reservoir is the areal discontinuity frequently observed in production. Production might be obtained from many of the faults and fracture systems in an oil-prone area, but not from intervening blocks that do not have the porosity and permeability provided by the faults and fracture systems.

A grid covering an area approximately 1/2× 1-1/2 mi with a 200 ft sample interval was established over a prospective area (Fig. 7). Three wells with fracture production from approximately 2,700 ft were used to develop a composite geochemical model. The use of more than one well for development of a geochemical model can add stability to the calculated similarity index and average out heterogeneities which can be reasonably expected in this environment.

The objective of the survey was to map areas with high similarity to those samples collected around the training set wells and map areas with fault/fracture characteristics. Figure 5 has a typical mass spectrum observed for this area where fault/fractures are not present and where they are present. Those samples with fault/fracture characteristics typically have significant fluxes of compounds with molecular weights greater than 140.

Figure 8 has the areas of high similarity and the areas of fault/fracture characteristics as determined by mass spectra superimposed over the survey area. The faults mapped at the surface using photography are also shown. A well drilled in the northwest part of the area where both production and fault/fracture characteristics occur proved to be a producer. Figure 9 is a schematic cross section of the prospect. A directional well is drilled into the Niobrara formation at a point directly below the surface observation of high similarity and fault/fracture characteristic. These data and other similar situations in the same environment support the concept of vertical migration.

Denver-Julesburg Basin Reconnaissance and Follow-up Surveys

The northern D-J Basin has numerous stratigraphic traps and the D sand is the target in the area of a reconnaissance survey. A sample spacing of 1/2 mi was used in the initial survey covering approximately one township (Fig. 10). The objectives of the survey were to (1) fingerprint the hydrocarbon emanations over known D-sand production at a location approximately 10 mi north of the survey area, (2) determine the existence of emanating hydrocarbons in the prospect area and map regions having a similar fingerprint to the known D-sand production, (3) correlate the K-V fingerprint and hydrocarbon flux data with LANDSAT data, (4) conduct detailed surveys in prospective areas identified in the initial survey, and (5) aid in well-site selection.

The reconnaissance survey indicated broad regions of high similarity to the D-Sand model over the survey area, as indicated in Figure 11a. The hydrocarbon flux for C2–C10 alkanes is shown in Figure 11b and has a distribution different from that of the similarity zones map of Figure 11a. Four existing dry holes fell outside of the high similarity zones. Two of these dry holes fell into high flux zones. Two others fell into zones indicating low flux. The superimposed maps are illustrated in Figure 12 and show the similarity zones map to be more indicative of both production and existing well control in the prospect area. An initial discovery well, currently in production, was then drilled in the high similarity zone located in the southwestern part of the survey area (Fig. 12).

Detailed surveys were carried out in three separate stages over a period from September through November. The surveys were in different parts of the reconnaissance area and extended to the east and south. In addition to the original dry holes, the discovery well was used to develop further the geochemical model for the area. Sample density increased to 16 to 32 per sq mi as a result of the successive stages, and the surveys were used to resolve and define the high similarity zones more accurately. The first detailed surveys covered portions of the northern and eastern part of the reconnaissance area (Fig. 13a), the second covered portions of the western and southern part of the reconnaissance area (Fig. 13b), and the third was used to infill in the northeast and to define more precisely the area around the discovery well (Fig. 13c).

The combined similarity zones data for the three detailed surveys are shown in Figure 14. A northwest–southeast trend is suggested, which may be due to higher porosity zones within a deltaic complex. The high flux areas from the detailed surveys are superimposed and, as in the initial reconnaissance survey, there is little correspondence with similarity zones and well control. The high flux appears to have a northeast–southwest trend at 90° to the trend in the similarity data.

Figure 15 shows the areas with the fault/fracture characteristic as defined by the distribution fluxes of high molecular weight components in the mass spectra, which are superimposed over LANDSAT defined lineaments. Both methods of defining fault/fracture characteristics and zones of high flux have a northeast-southwest trend. It has been hypothesized by the client that the Niobrara formation which overlies the D-Sand is fractured and may be the reason for the high observed fluxes and fracture character. Use of flux data alone in this environment will be misleading.

SUMMARY AND CONCLUSIONS

Previous work has shown that measurement of soil gas quantities from the earth are difficult to reproduce and interpret. Integrative collection of gases and interpretation based on composition allow a reduction in the temporal noise observed and a collection of lower volatility components.

A new method of soil gas collection and analysis, utilizing activated charcoal samples analyzed by mass spectrometry and pattern recognition, overcomes many of these problems. The K-V Fingerprint Technique characterizes volatile emissions around existing production and computes a quantitative degree of similarity for the survey or grid samples.

The K-V Fingerprint Technique allows recognition of leakage characteristic of faults and fractures, which is demonstrated to be effective in exploration for petroleum contained in this type of reservoir. The K-V Fingerprint Technique maintains its effectiveness in different seasons and allows integration of multiple surveys. The observed geochemical trends related to production are maintained in overlapping surveys. Flux data appear to be more closely related to structural trends which may not be associated wtih production. Based on the results of these surveys, the conclusion may be drawn that vertical migration, with little dispersion, exists.

REFERENCES

Brooks, J. R., Bernard, B. B., and Sackett, W. M., 1979, Characterization of gases in marine waters and sediments, *in* Watterson, J. R., and Theobald, P. K., eds., International Geochemical Exploration Symposium: Golden, Colorado, Association of Exploration Geochemists, Rexdale, Ontario, Proceedings, p. 337–345.

Dalziel, M. C., and Donovan, T. J., 1980, Biogeochemical evidence for subsurface hydrocarbon occurrence, Recluse Oil Field, Wyoming: United States Geological Survey Circular 837, 11 p.

———, and Donovan, T. J., 1984, Correlations of suspected petroleum-generated biogeochemical and aeromagnetic anomalies, Bell Creek Oil Field, Montana, *in* Davidson, M. J., and Gottlieb, B. M., eds., Unconventional methods in exploration for petroleum and natural gas III: Dallas, Southern Methodist University Press, p. 59–69.

Donovan, T. J., 1974, Petroleum microseepage at Cement, Oklahoma: Evidence and mechanism: American Association of Petroleum Geologists Bulletin, v. 58, p. 429–446.

———, Forgey, R. L., and Roberts, A. A., 1979, Aeromagnetic detection of diagenetic magnetite over oil fields: American Association of Petroleum Geologists Bulletin, v. 63, p. 245–248.

———, Noble, R. L., Friedman, I., and Gleason, J. D., 1975, A possible petroleum-related geochemical anomaly in surface rocks, Boulder and Weld counties, Colorado: United States Geological Survey, Open-file Report, 75-47, 11 p.

———, Roberts, A. A., and Dalziel, M. C., 1981, Epigenetic zoning in surface and near-surface rocks resulting from seepage-induced redox gradients, Velma Oil Field, Oklahoma: A synopsis: Shale Shaker, v. 32, p. 1–7.

Duchscherer, W., Jr., 1981, Carbonates and isotope ratios from surface rocks: A geochemical guide to underlying petroleum accumulations, *in* Gottlieb, B. M., ed., Unconventional methods in exploration for petroleum and natural gas II: Dallas, Southern Methodist University Press, p. 201–218.

———, 1982, Geochemical exploration for hydrocarbons: No new tricks, but an old dog: Oil and Gas Journal, v. 80, no. 27, p. 163–176.

Hawkes, H. E., 1972, Free hydrogen in genesis of petroleum: American Association of Petroleum Geologists Bulletin, v. 56, p. 2268–2270.

Hickey, J. C., 1985, Development of an integrative gas geochemical technique for petroleum exploration: Department of Chemistry and Geochemistry, Colorado School of Mines, Golden, Colorado, Master's thesis.

———, Voorhees, K. J., and Klusman, R. W., 1985, A new integrative gas geochemical technique for petroleum exploration: American Association of Petroleum Geologists Bulletin, in review.

Horvitz, L., 1939, On geochemical prospecting: Geophysics, v. 4, p. 210–228.

———, 1972, Vegetation and geochemical prospecting for petroleum: American Association of Petroleum Geologists Bulletin, v. 56, p. 925–940.

———, 1980, Near-surface evidence of hydrocarbon movement from depth, *in* Roberts, W. H., III, and Cordell, R. J., eds., Problems in petroleum migration: American Association of Petroleum Geologists Studies in Geology, no. 10, p. 241–269.

Hunt, J. M., Miller, R. J., and Whelan, J. K., 1980, Formation of C4-C7 hydrocarbons from naturally occurring terpenoids: Nature, v. 288, p. 577–578.

Jones, V. T., and Drozd, R. J., 1983, Prediction of oil and gas potential by near-surface geochemistry: American Association of Petroleum Geologists Bulletin, v. 67, p. 932–952.

Klusman, R. W., and Voorhees, K. J., 1983, A new development in petroleum exploration technology: Mines Magazine, v. 73, no. 3, p. 6–10.

Kvenvolden, K. A., and Redden, G. O., 1980, Hydrocarbon gas in sediment from the shelf, slope and basin of the Bering Sea: Geochimica and Cosmochimica Acta, v. 44, p. 1145–1150.

———, Bogel, T. M., and Gardner, J. V., 1981, Geochemical prospecting for hydrocarbons in the outer continental shelf, southern Bering Sea, Alaska: Journal of Geochemical Exploration, v. 14, p. 209–219.

Matthews, M. D., Jones, V. T., and Richers, D. M., 1984, Remote sensing and surface hydrocarbon leakage: International Symposium on Remote Sensing of the Environment, Third Thematic Conference, Colorado Springs, April 16–19, 1984.

Pogorski, L. A., and Quirt, G. S., 1981, Helium emanometry in exploring for hydrocarbons I, *in* Gottlieb, B. M., ed., Unconventional methods in exploration for petroleum and natural gas II: Dallas, Southern Methodist University Press, p. 124–135.

Reed, W. E., and Kaplan, I. R., 1977, The chemistry of marine petroleum seeps: Journal of Geochemical Exploration, v. 7, no. 2, p. 255–292.

Richers, D. M., Reed, R. J., Horstman, K. C., Michels, G. D., Baker, R. N., Lundell, L., and Marrs, R. W., 1982, LANDSAT and soil-gas geochemical study of Patrick Draw Oil Field, Sweetwater County, Wyoming: American Association of Petroleum Geologists Bulletin, v. 66, no. 7, p. 903–922.

Roberts, A. A., 1981, Helium emanometry in exploring for hydrocarbons II, *in* Gottlieb, B. M., ed., Unconventional methods in exploration for petroleum and natural gas II: Dallas, Southern Methodist University Press, p. 136–149.

———, Dalziel, M. C., Pogorski, L. A., and Quirt, G. S., 1976, A possible petroleum-related helium anomaly in soil gas, Boulder and Weld counties, Colorado. United States Geological Survey, Open-file Report, 76-544, 7 p.

Rock, B., 1984, Remote detection of geobotanical anomalies associated with hydrocarbon microseepage: Third International Symposium on Remote Sensing of the Environment, Colorado Springs, Colorado, Proceedings, April, 1984.

Roeming, S. S., and Donovan, T. J., 1985, Correlations among hydrocarbon microseepage, soil chemistry, and uptake of micronutrients by plants, Bell Creek Oil Field, Montana: Journal of Geochemical Exploration, v. 23, p. 139–162.

Rosaire, E. E., 1940, Geochemical prospecting for petroleum: American Association of Petroleum Geologists Bulletin, v. 24, p. 1400–1433.

Sackett, W. B., 1977, Use of hydrocarbon sniffing in offshore exploration: Journal of Geochemical Exploration, v. 7, p. 243–254.

Sigalove, J. J., 1985, Geochemical identificaiton of resource potential: Oil and Gas Journal, v. 83, p. 164–166.

Simoneit, B. R. T., and Galimov, E. M., 1984, Geochemistry of interstitial gases in quaternary sediments in the Gulf of California: Chemical Geology, v. 43, p. 151–166.

Venkatesan, M. I., and Kaplan, I. R., 1982, Distribution and transport of hydrocabons in surface sediments of the Alaskan outer continental shelf: Geochimica Cosmochimica Acta, v. 46, p. 2135–2150.

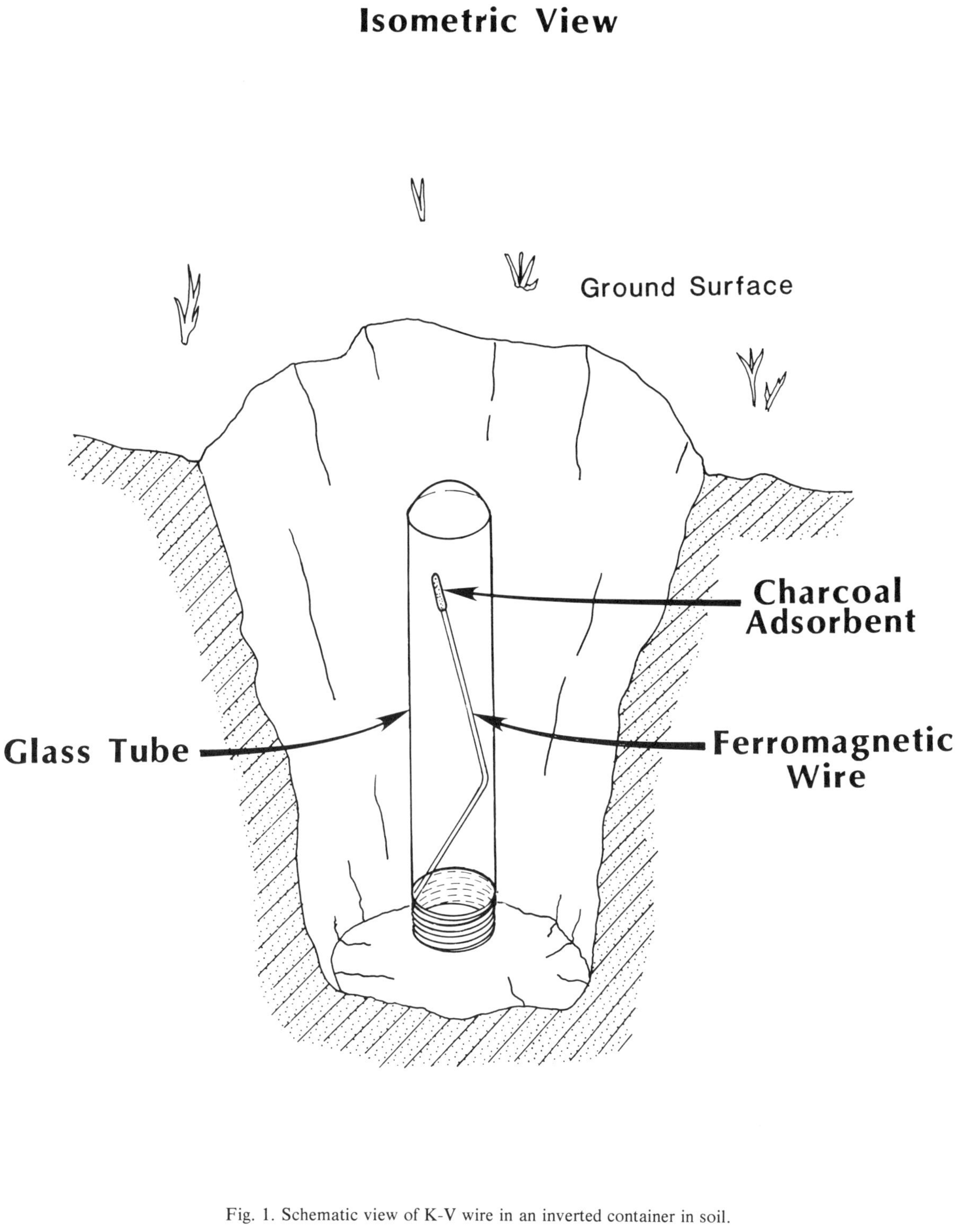

Fig. 1. Schematic view of K-V wire in an inverted container in soil.

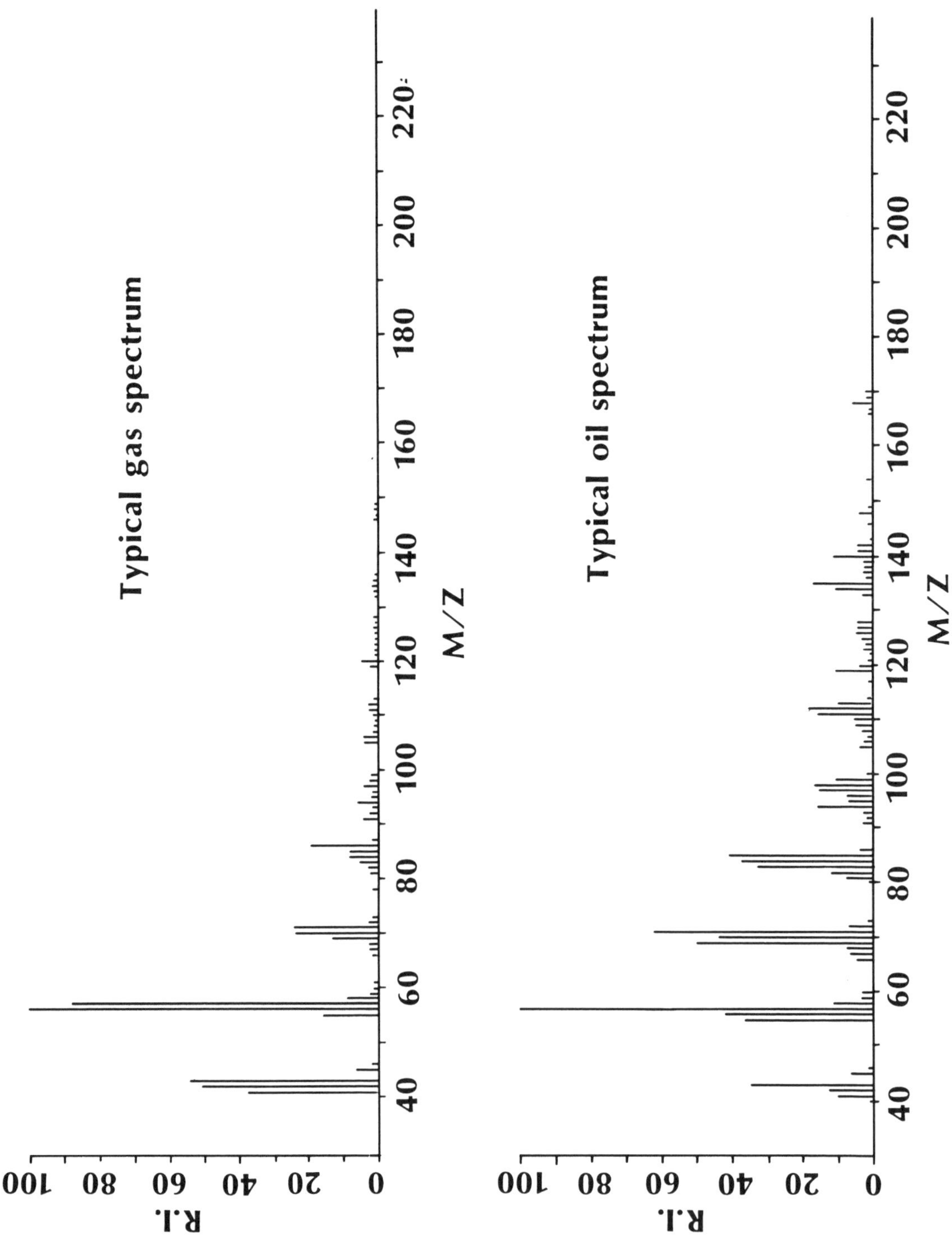

Fig. 2. Mass spectra of samples typical of petroleum emanation and gas emanation.

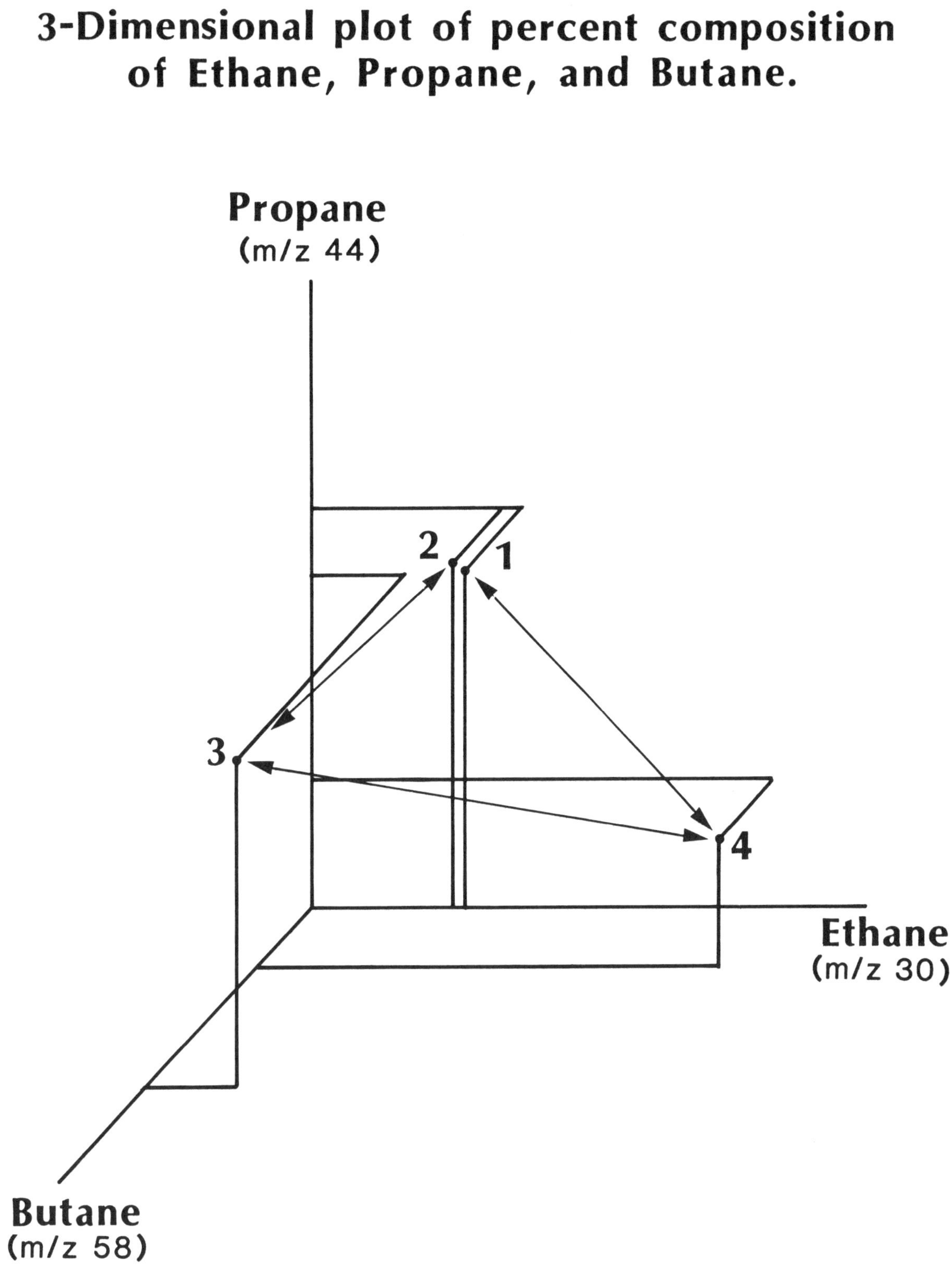

Fig. 3a. Three-dimensional plot of four samples of a three-component mixture.

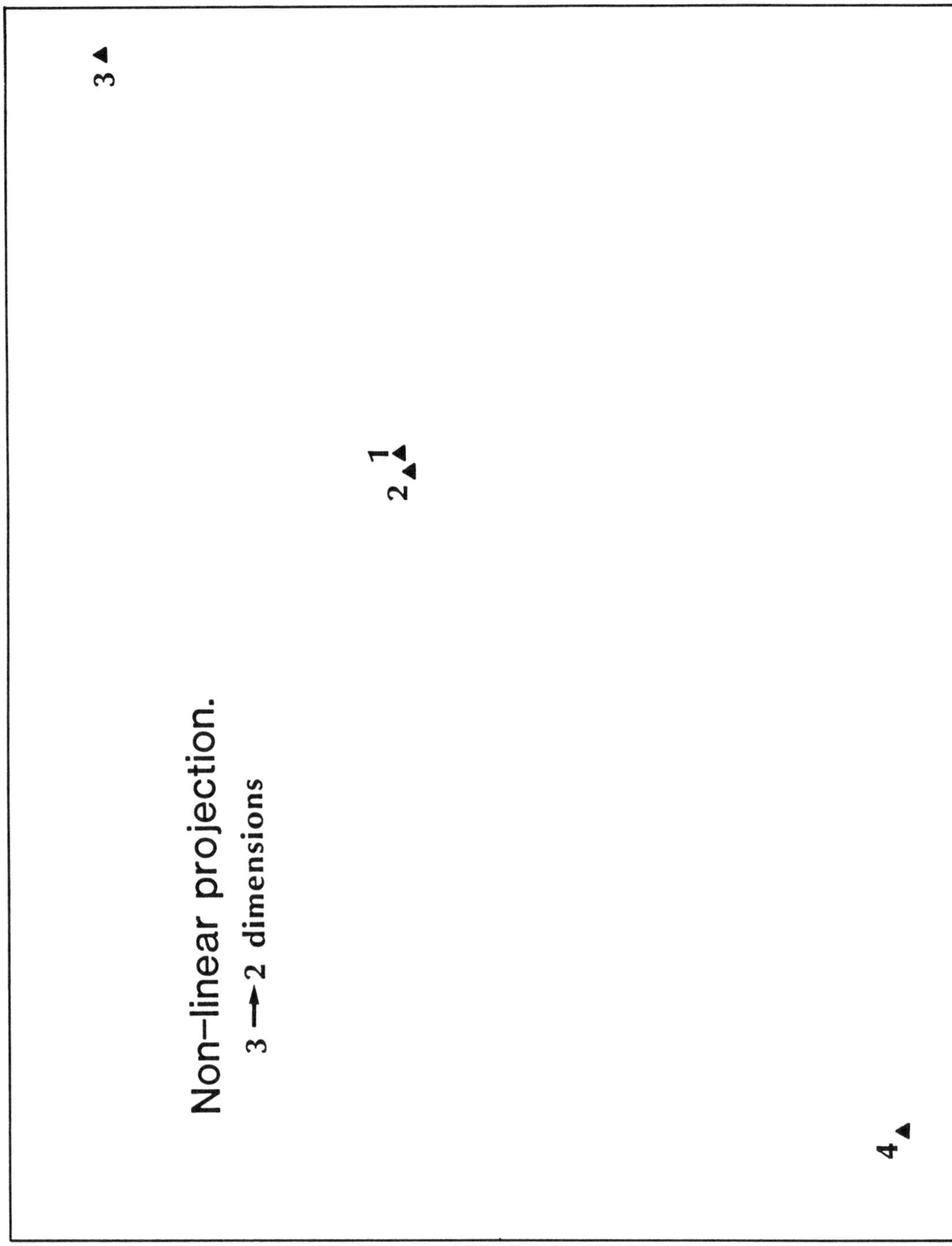

Fig. 3b. Two-dimensional projection of four samples of a three-component mixture.

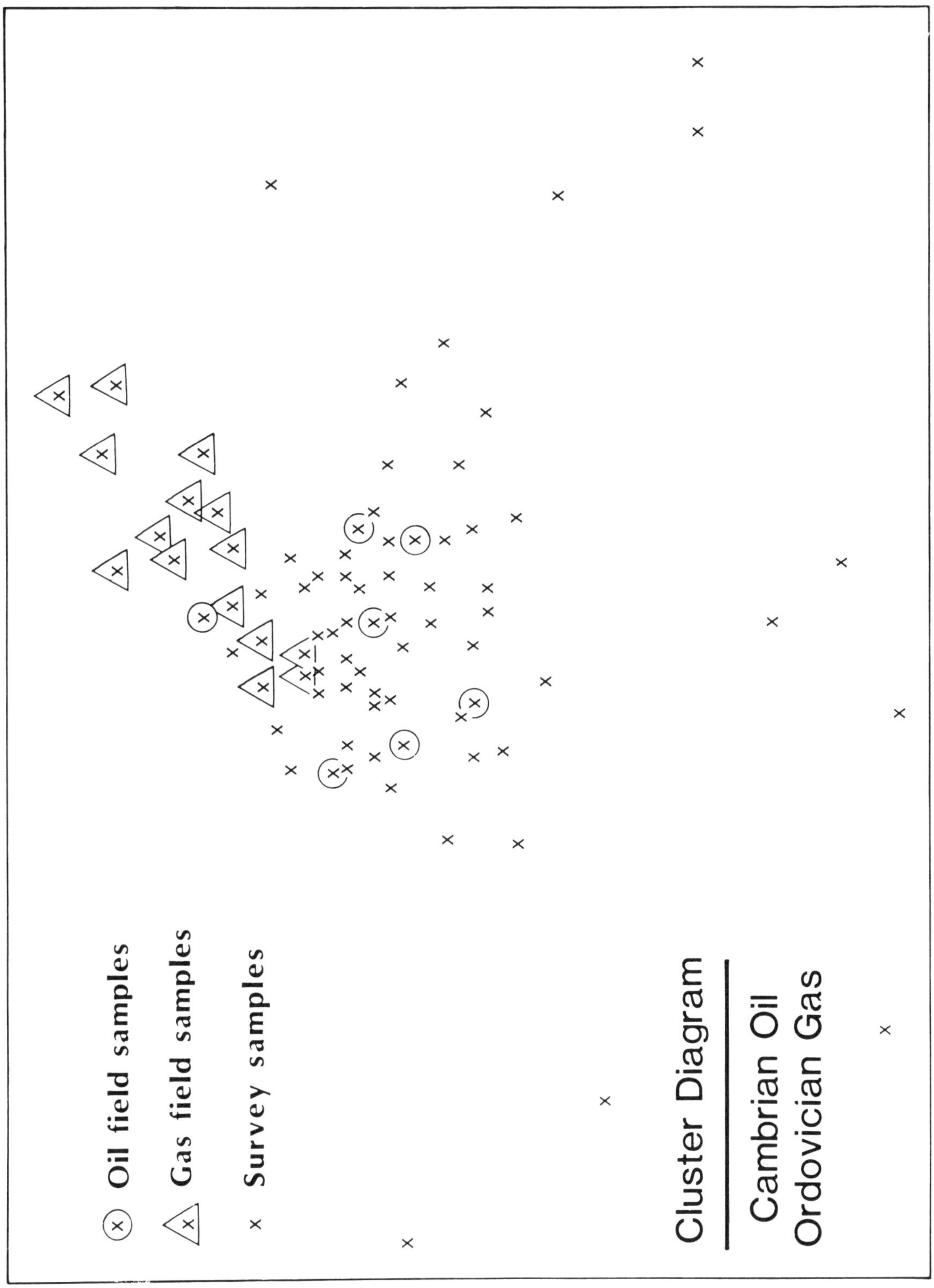

Fig. 4. Cluster diagram of mass spectra from an oil-prone area.

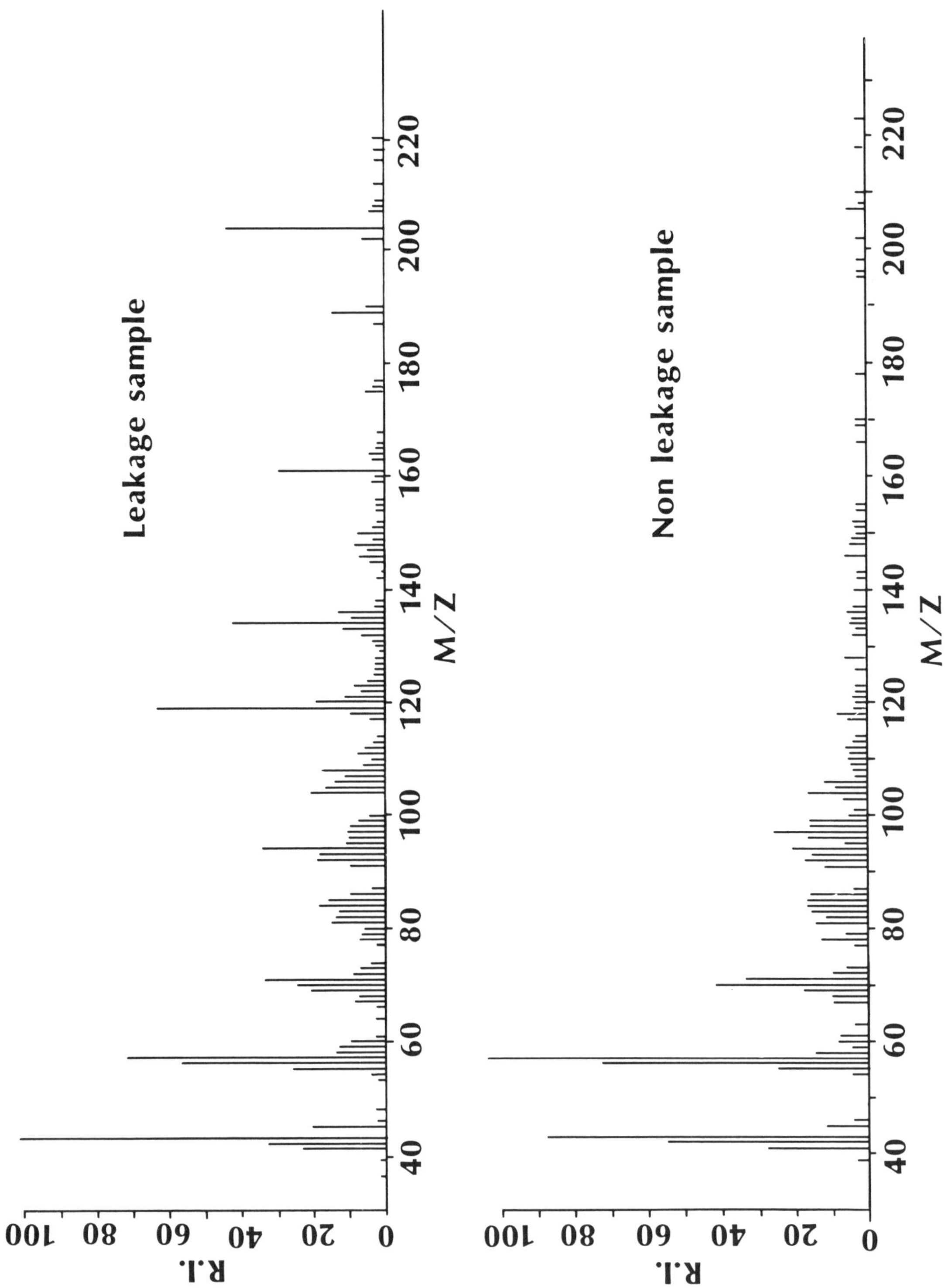

Fig. 5. Mass spectra of samples typical of an oil-prone area with and without fault/fracture characteristics.

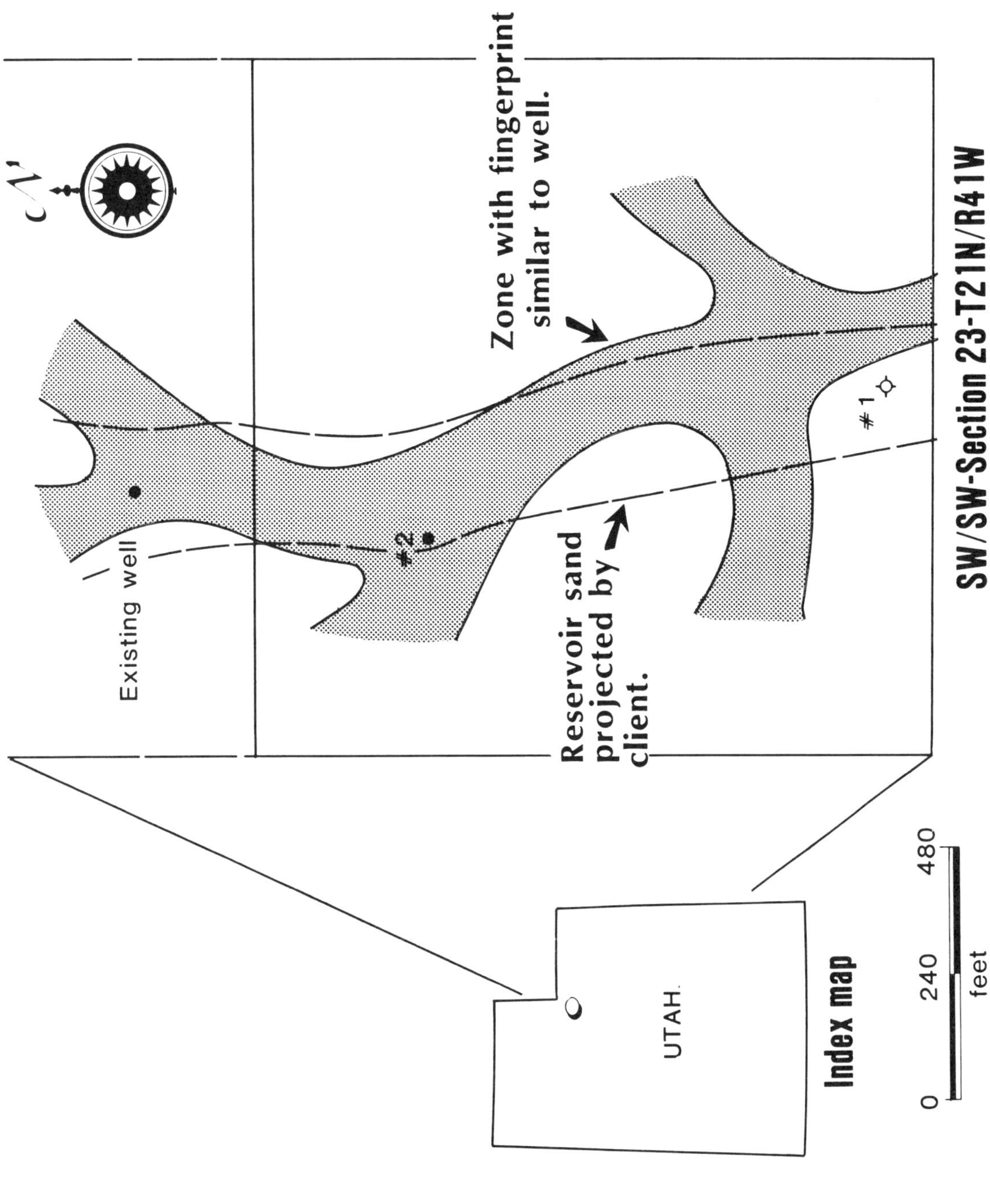

Fig. 6. Outline of hypothesized channel in Dakota formation of southern Uinta basin, as projected by client; together with outline of a zone determined by calculated similarity indices.

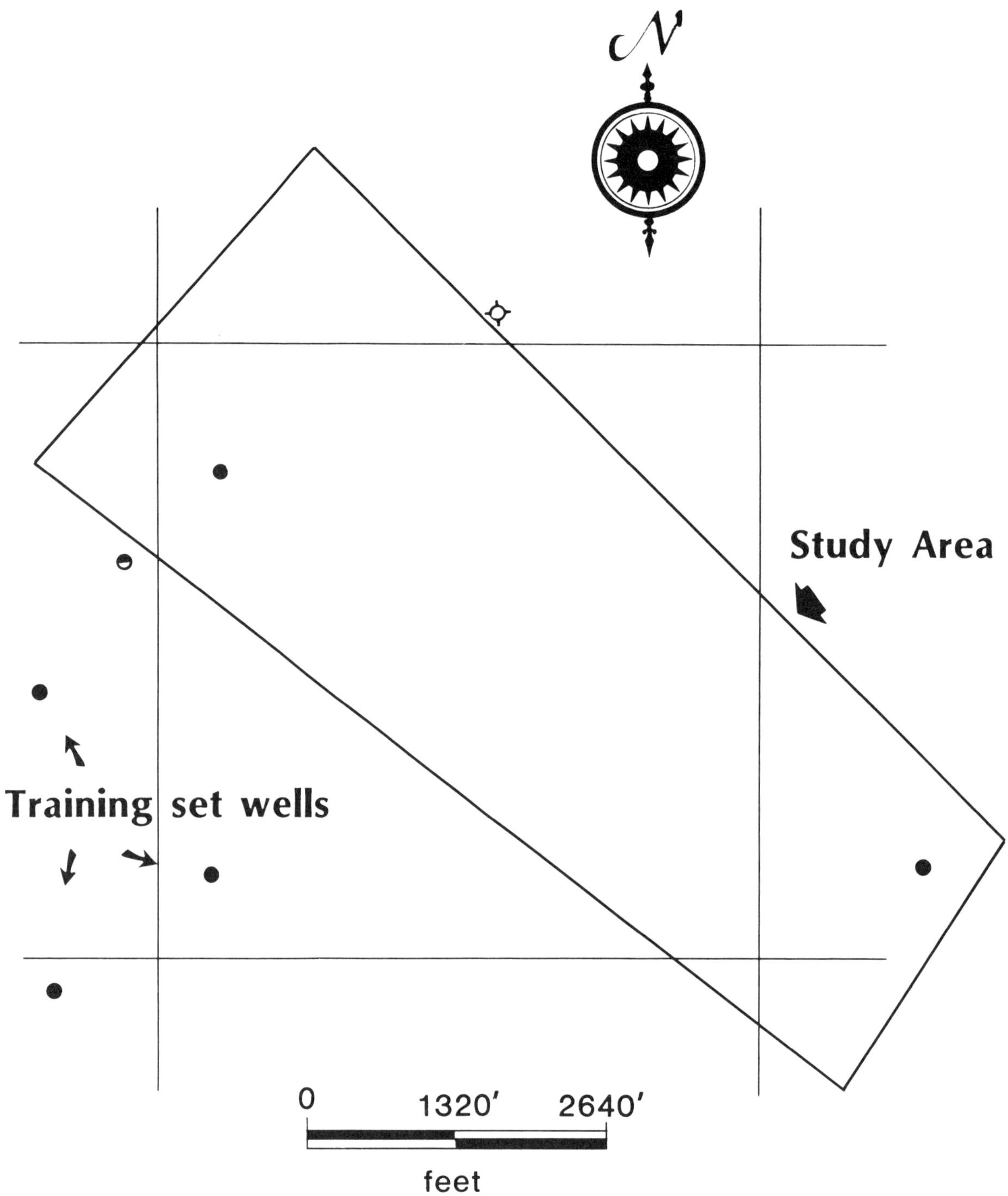

Fig. 7. Survey design for fault/fracture survey in area of possible Niobrara formation fracture production of northwest Colorado.

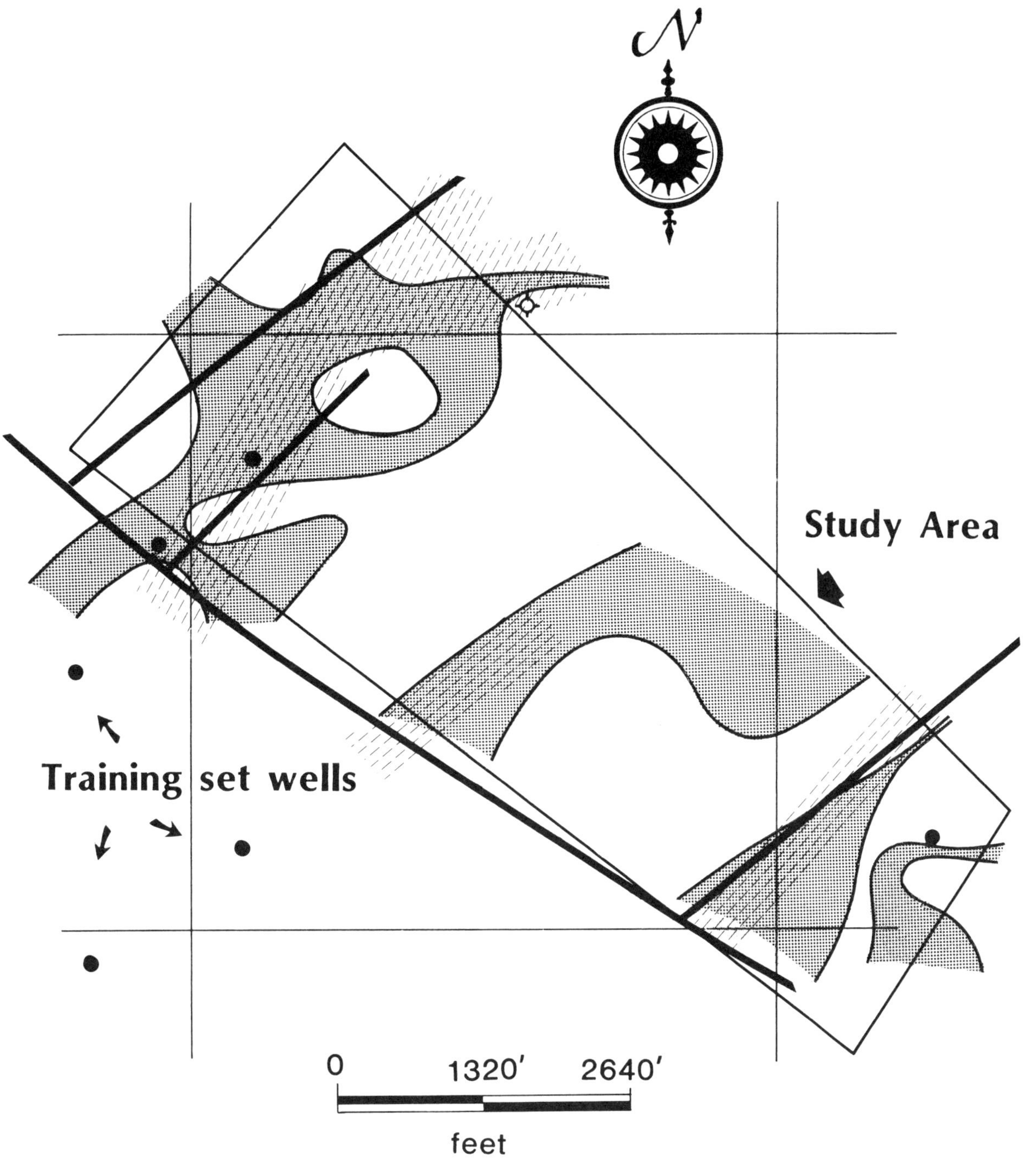

Fig. 8. Results of survey over Niobrara formation fracture production of northwest Colorado.

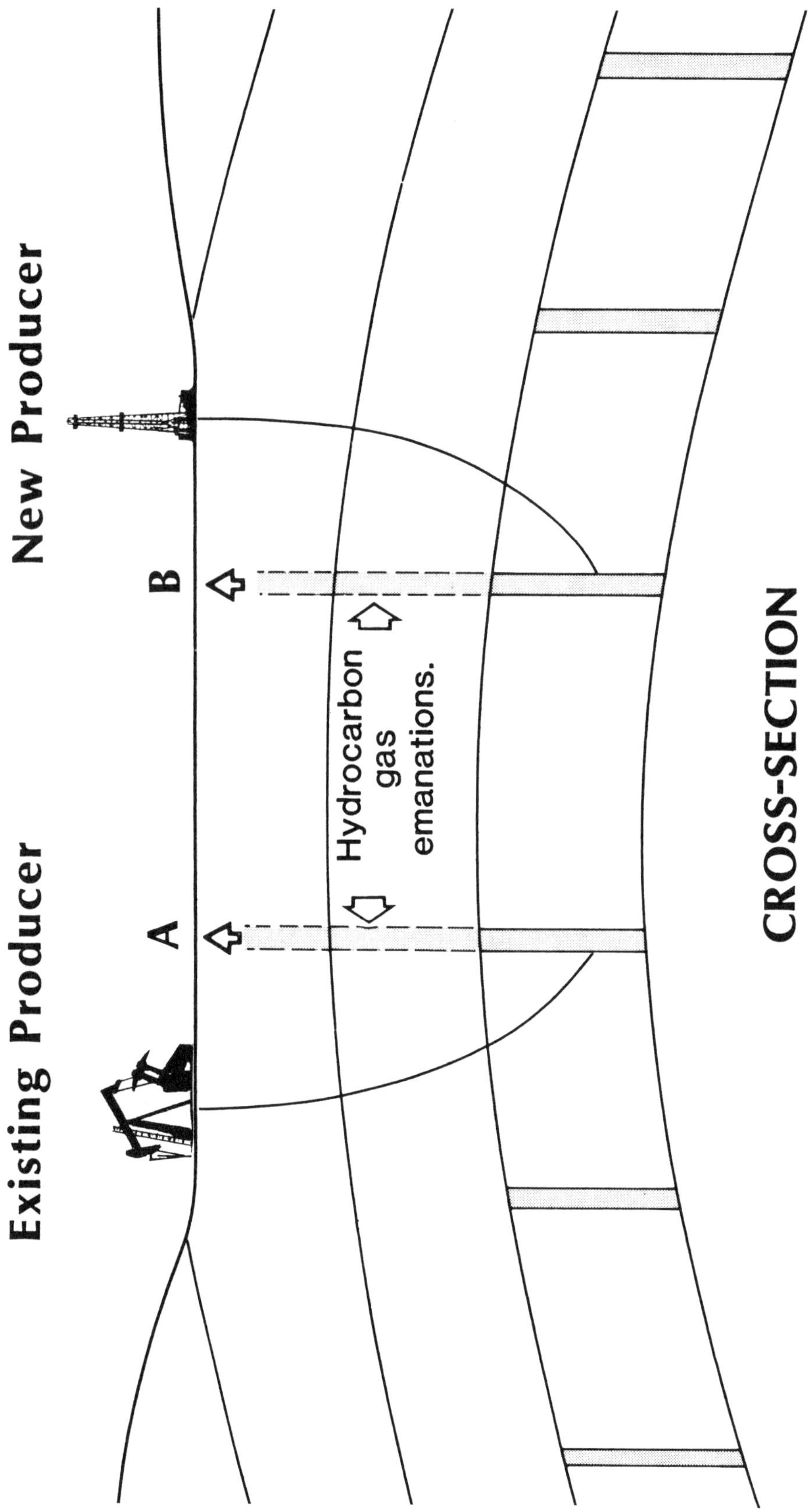

Fig. 9. Schematic cross section of Niobrara formation fracture production of northwest Colorado.

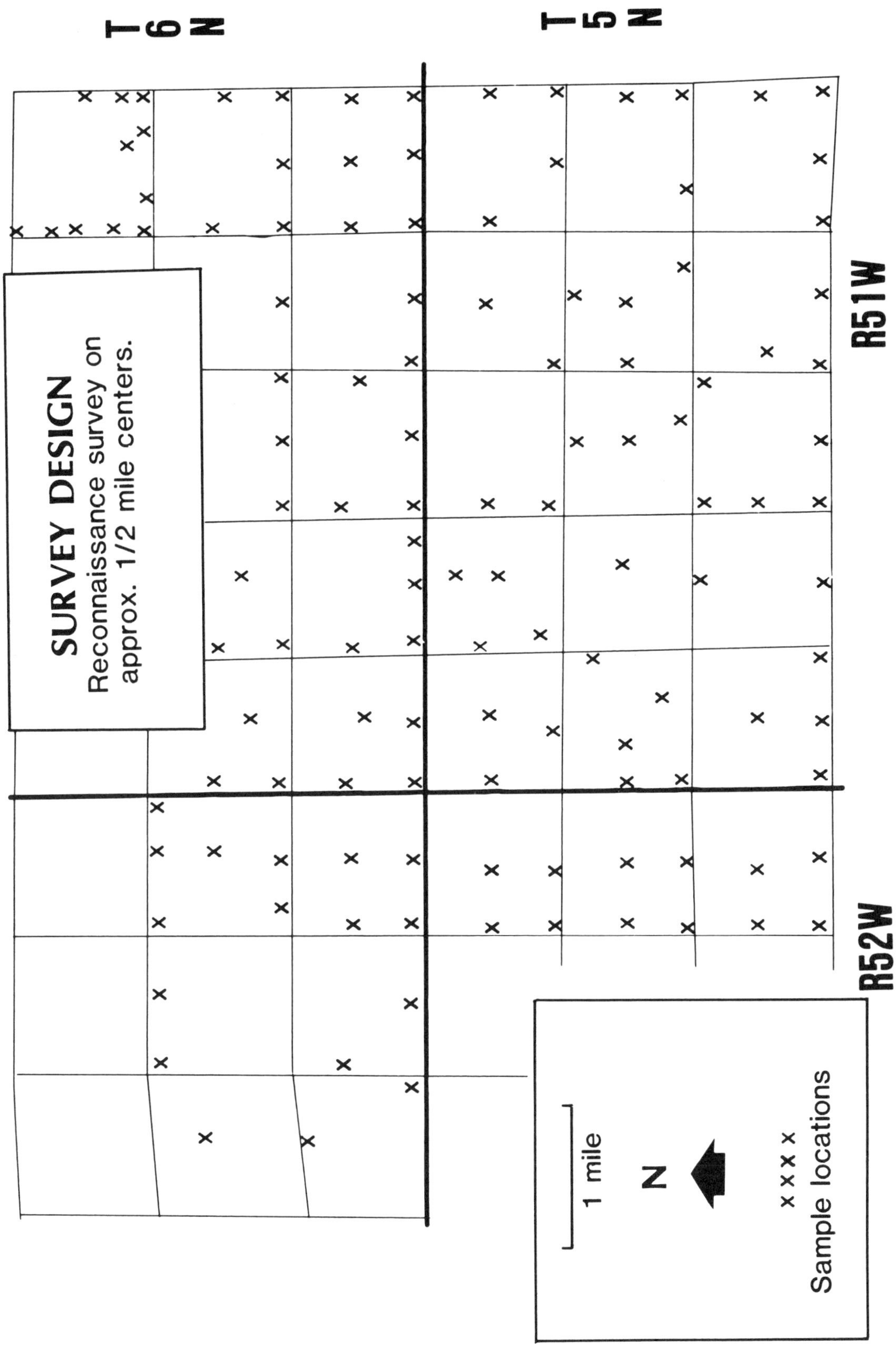

Fig. 10. Survey design for reconnaissance survey in northern Denver-Julesburg basin.

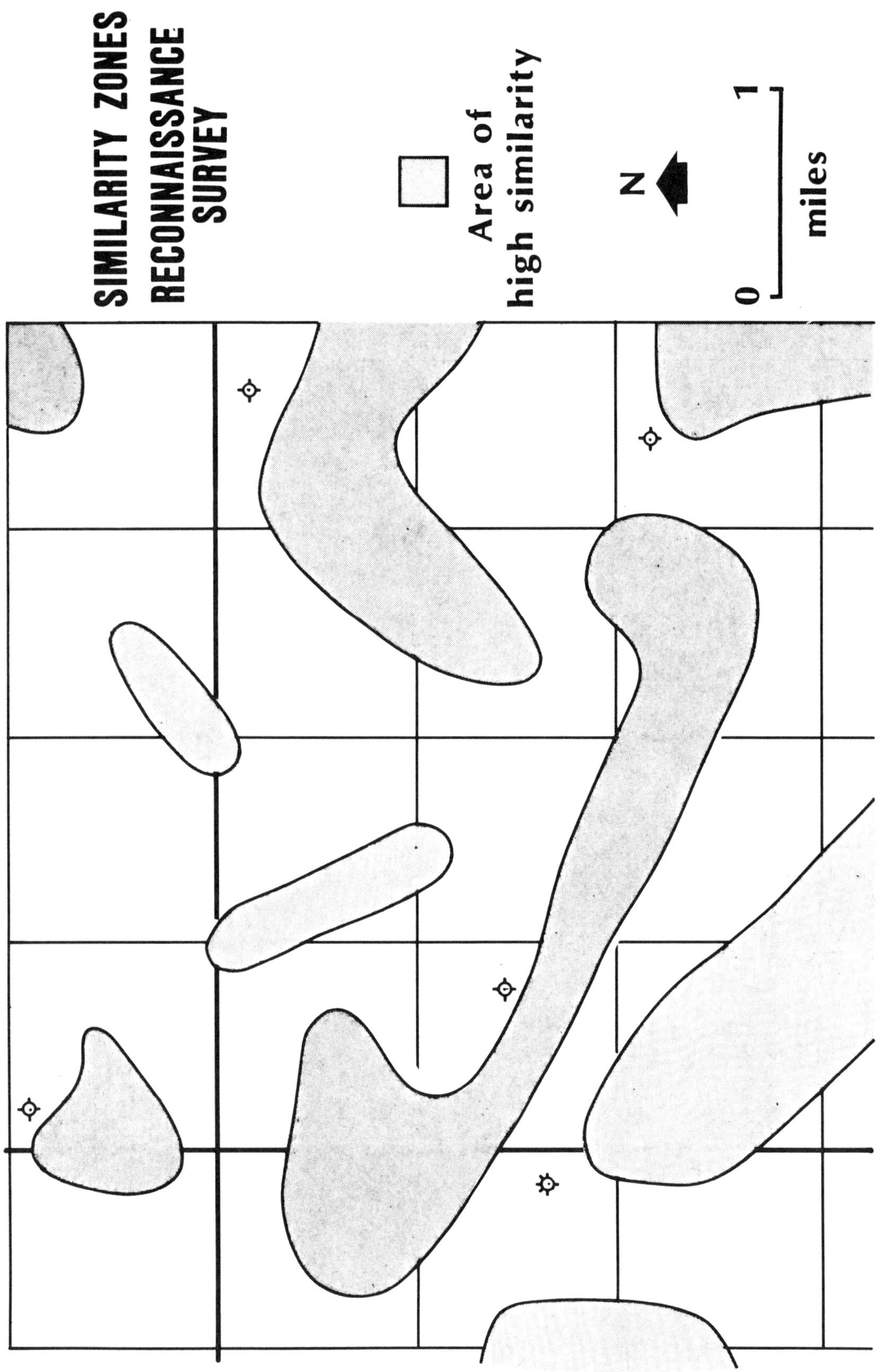

Fig. 11a. Petroleum production similarity zone for reconnaissance survey in northern Denver-Julesburg basin.

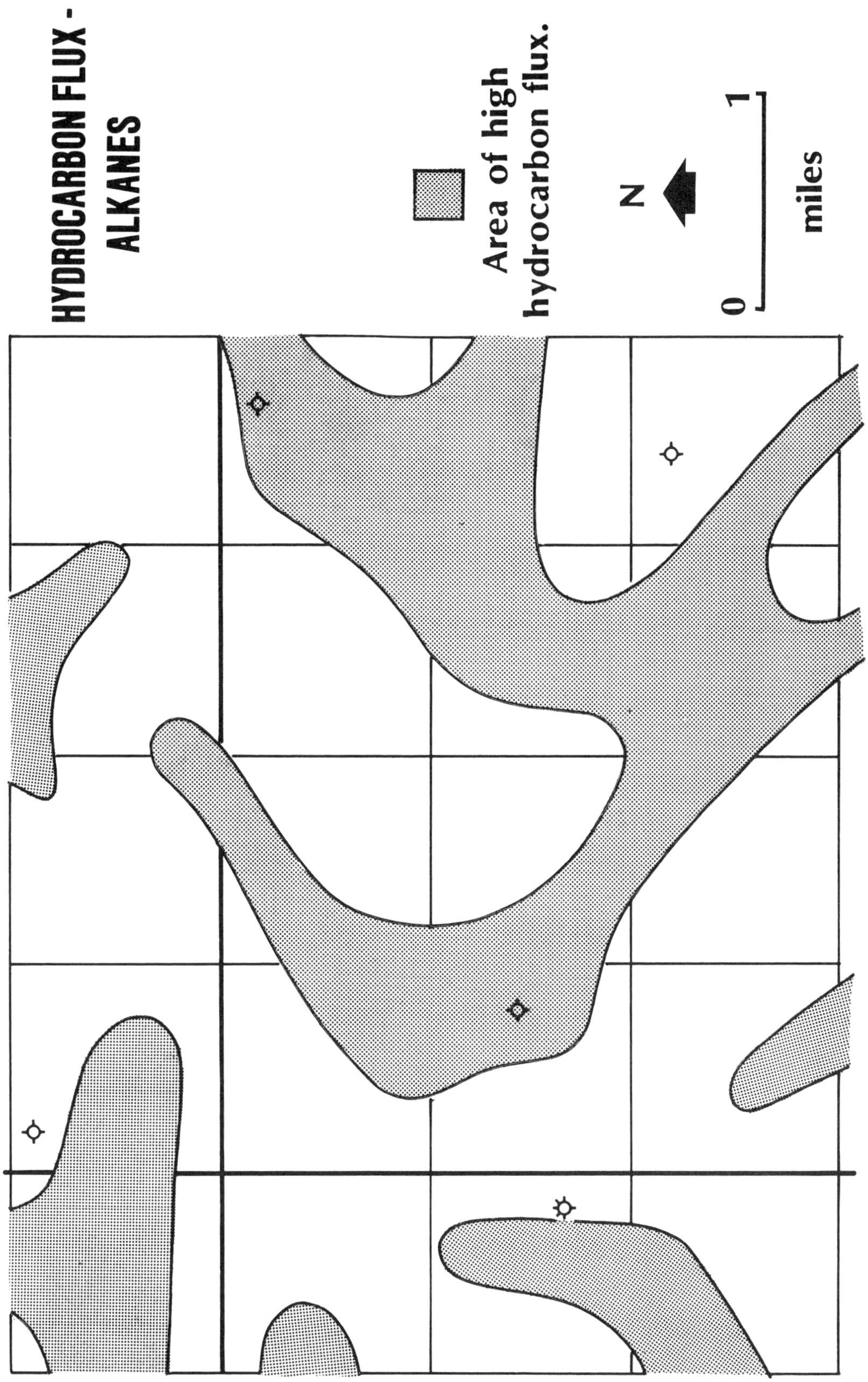

Fig. 11b. C2-C10 alkane flux for reconnaissance survey in northern Denver-Julesburg basin.

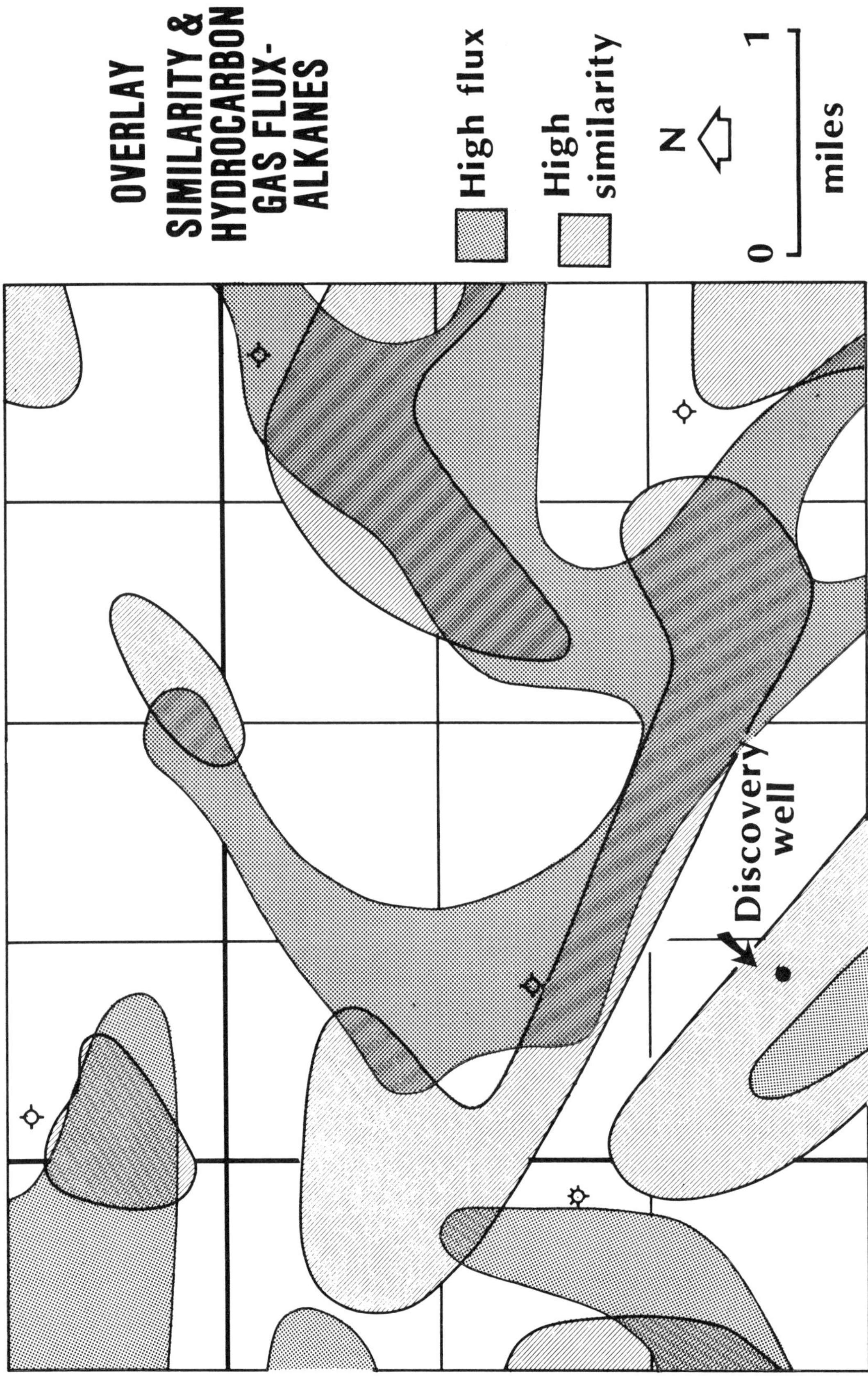

Fig. 12. Superimposed petroleum production similarity zone and alkane flux for reconnaissance survey in northern Denver-Julesburg basin.

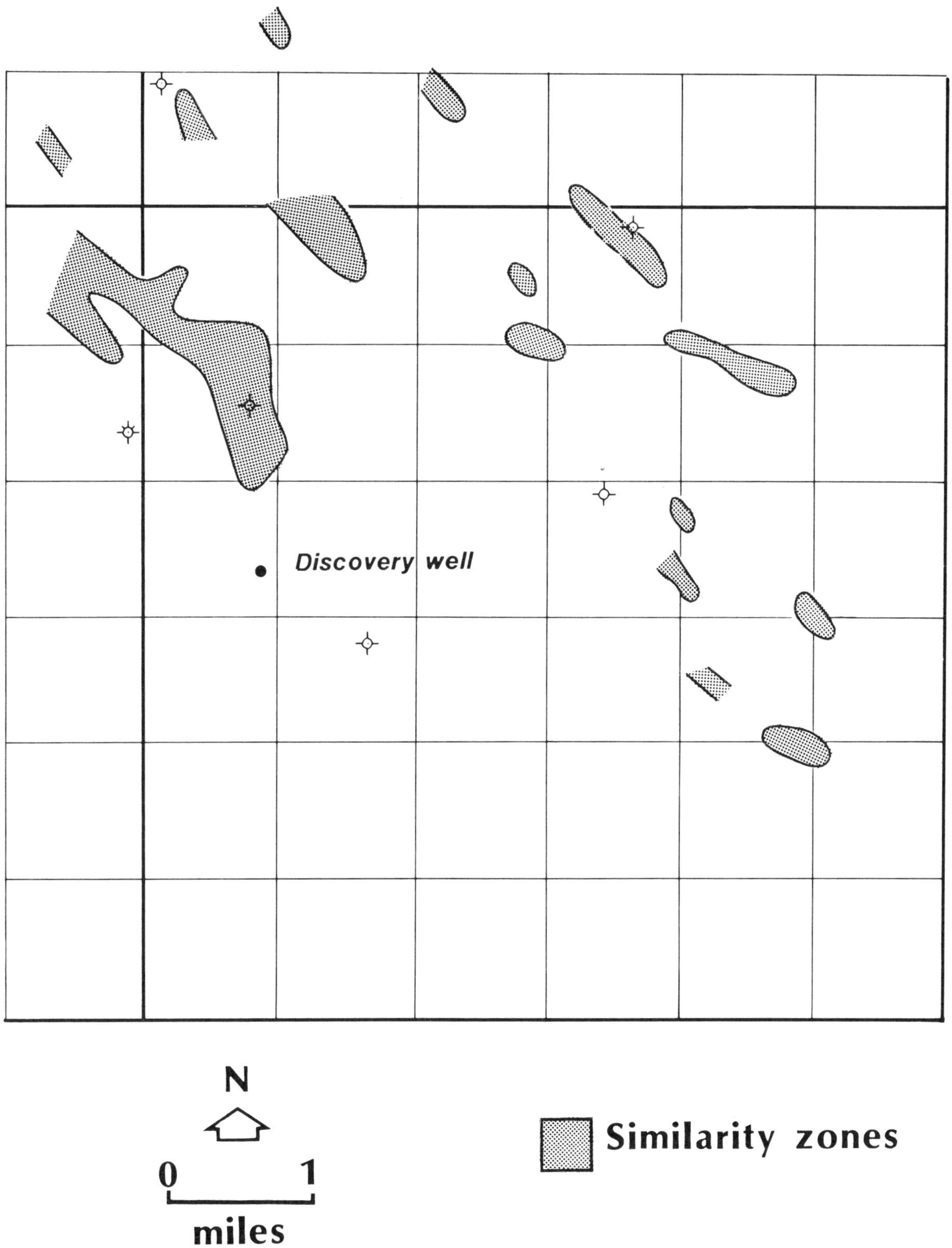

Fig. 13a. Detail survey no. 1 in northern Denver-Julesburg basin.

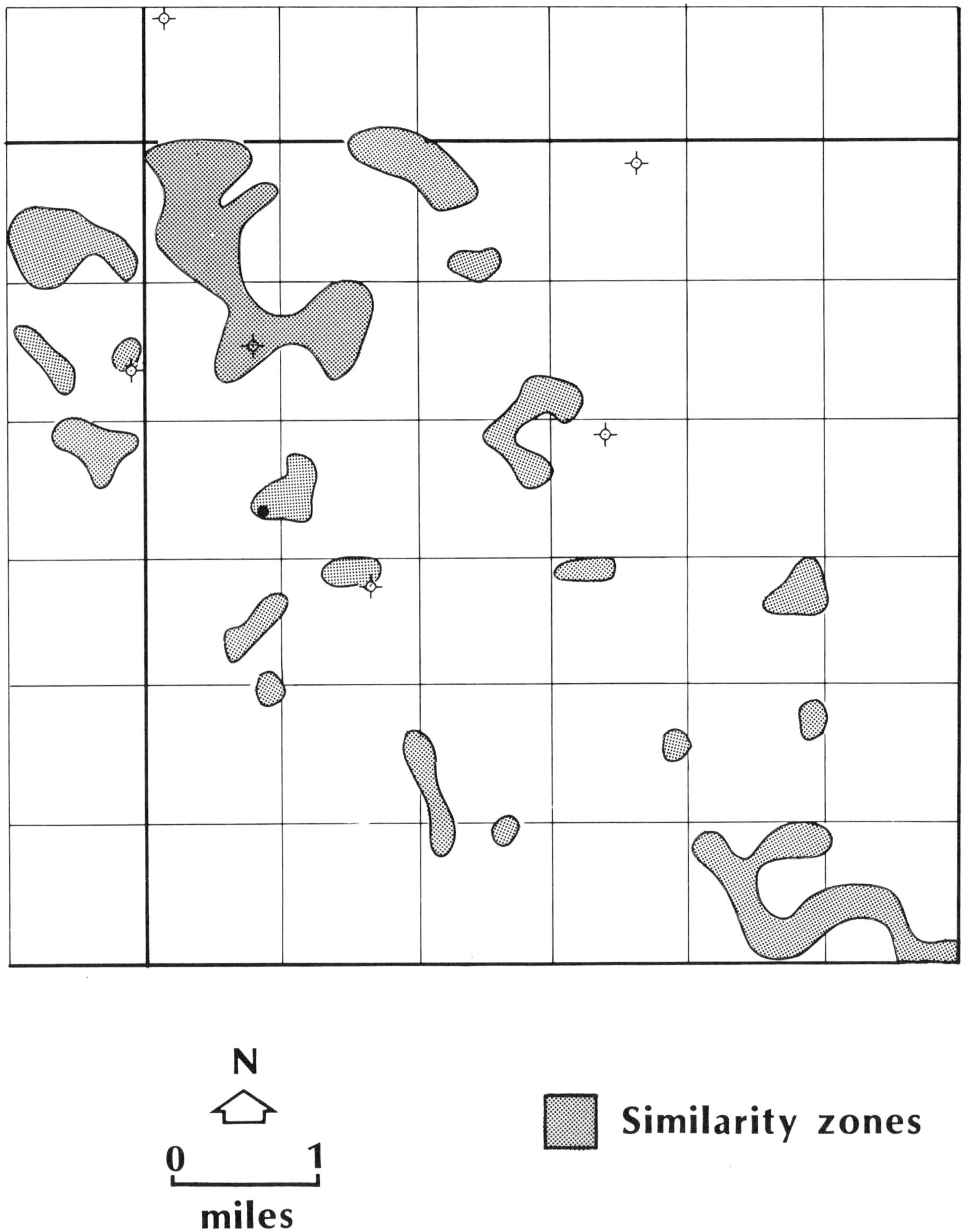

Fig. 13b. Detail survey no. 2 in northern Denver-Julesburg basin.

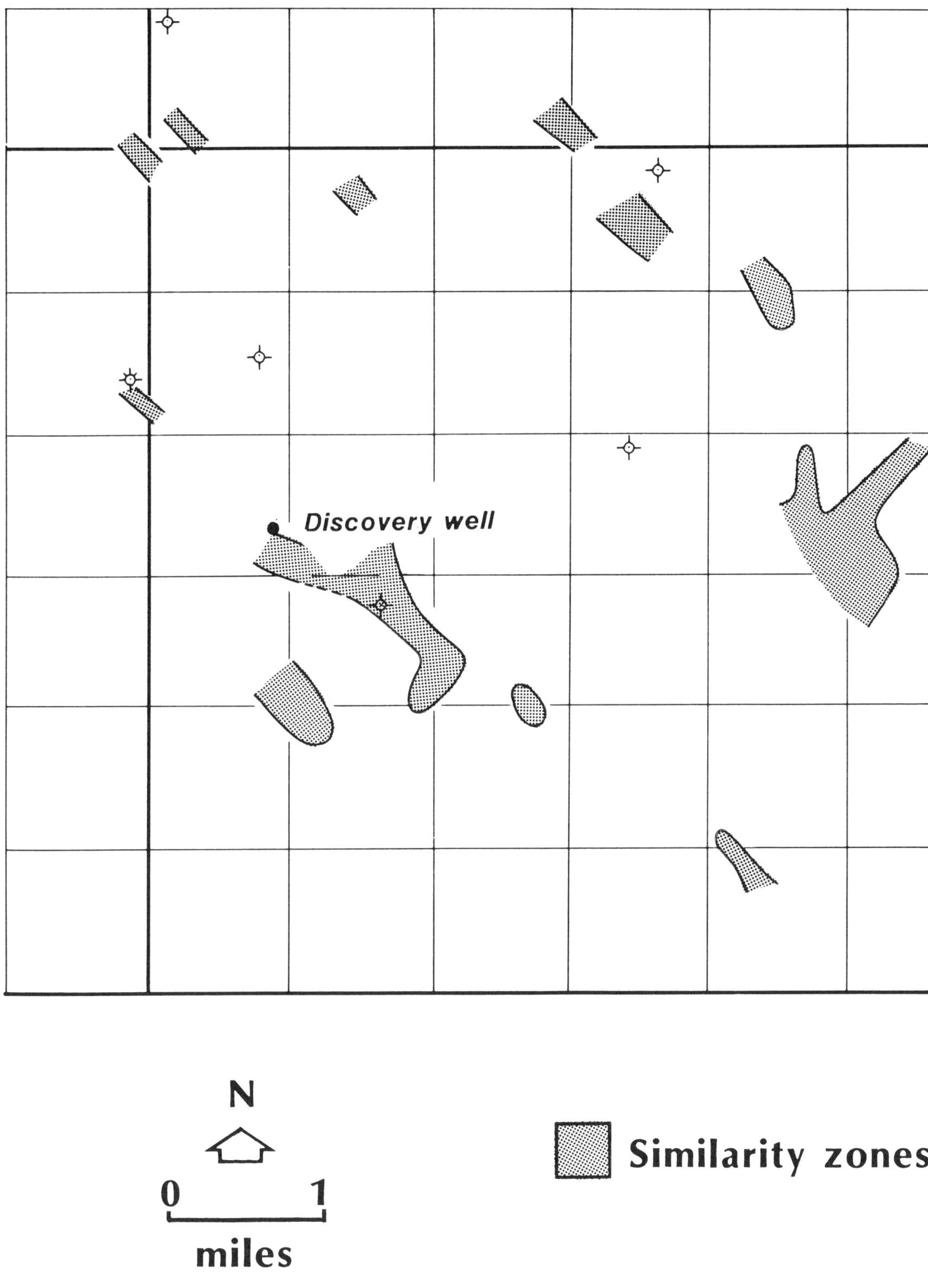

Fig. 13c. Detail survey no. 3 in northern Denver-Julesburg basin.

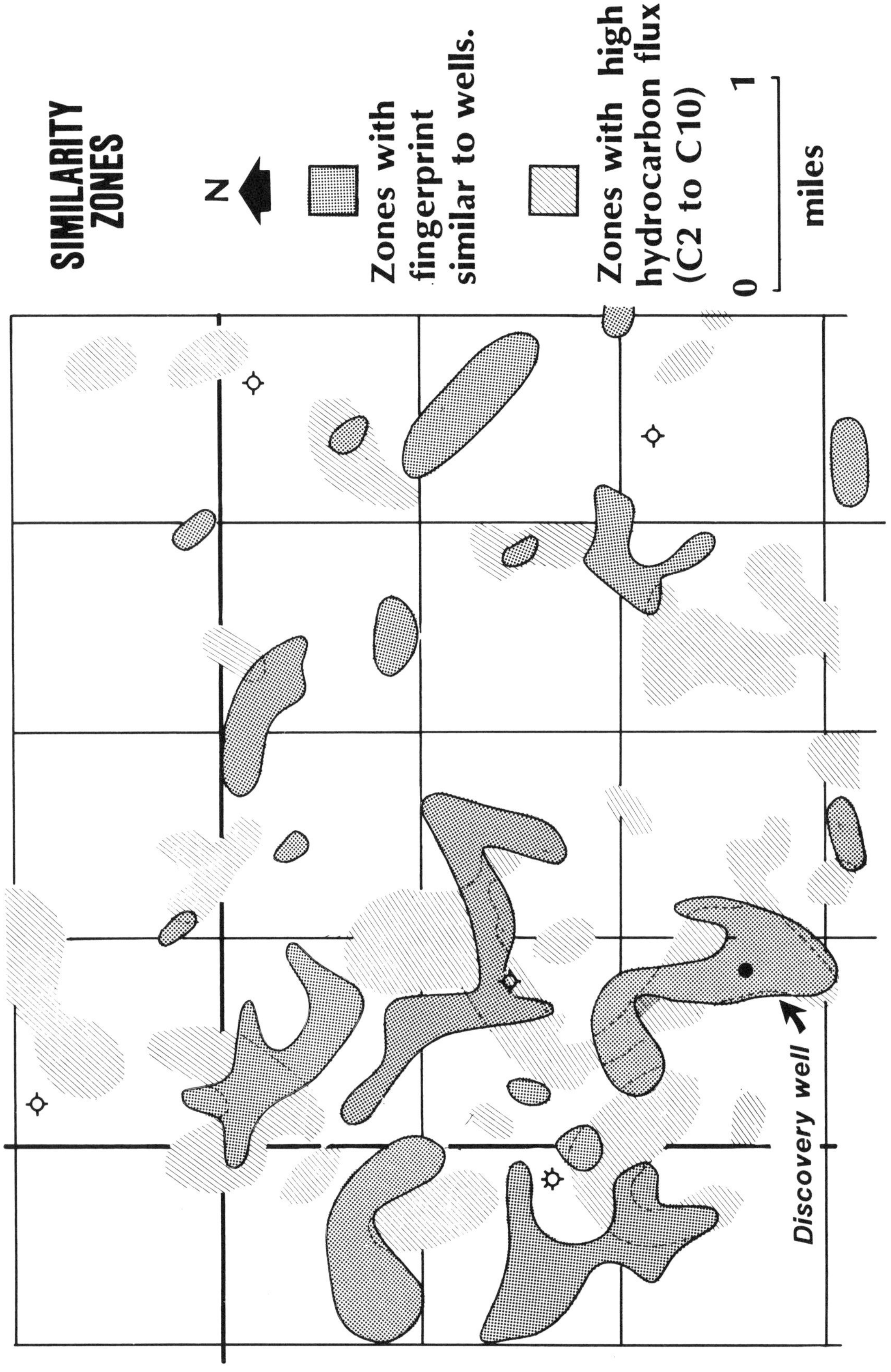

Fig. 14. Combined petroleum production similarity zone and alkane flux for three detailed surveys in northern Denver-Julesburg basin.

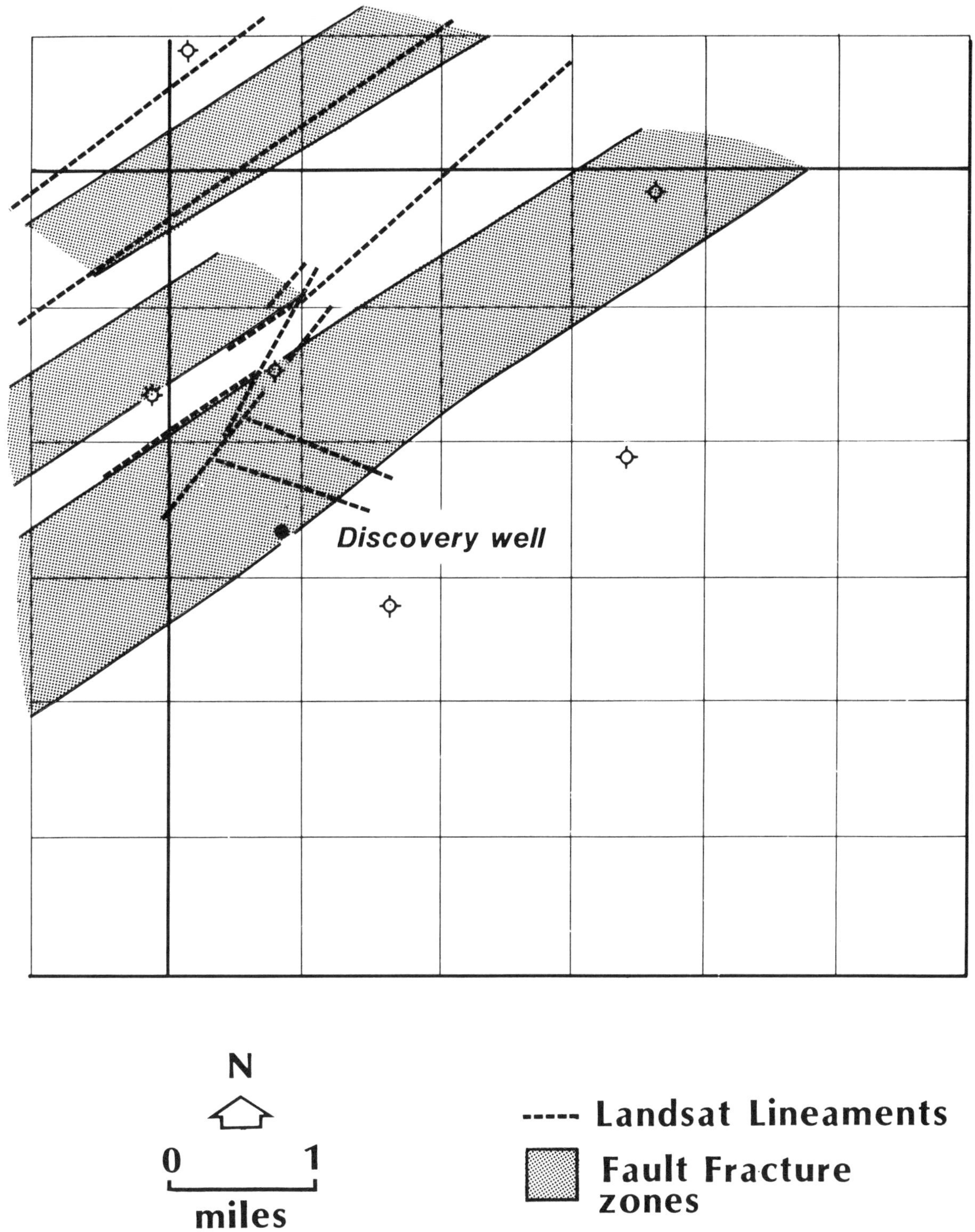

Fig. 15. Superimposed mass spectral fault/fracture characteristic and LANDSAT lineaments.

A CRITICAL OVERVIEW AND PROPOSED WORKING MODEL OF SURFACE GEOCHEMICAL EXPLORATION

Leigh C. Price

U.S. GEOLOGICAL SURVEY

ABSTRACT

Since the concept of surface geochemical exploration for oil and gas was first proposed by Laubmeyer and Sokolov over fifty years ago, the major Western effort has been one of creating a "service contract" market for these methods. Except for Soviet efforts, little research has been conducted on the causes, mechanism(s), consequences, and correct utilization of surface geochemical exploration or hydrocarbon microseepage. No area of science may achieve its ultimate commercial application until the basic principles of that scientific area have been fully delineated. This has not occurred with hydrocarbon microseepage, the basis of surface geochemical exploration. Answers are needed to explain (1) the hydrocarbon microseepage transport mechanism, (2) the cause of the "halo" anomaly, (3) the degree and nature of the interaction of microseeping hydrocarbons with their environment, and (4) the characteristics and causes of "chimneys" over hydrocarbon deposits. Especially important is the need of superior methods for the detection, measurement, and interpretation of hydrocarbon microseepage anomalies.

Although minimal fundamental research has been carried out on hydrocarbon microseepage, the existing data, when combined with established principles in petroleum geology and geochemistry, are sufficient to construct a *realistic* working hypothesis for hydrocarbon microseepage. The only transport mechanism in agreement with observed data is the vertical migration of microbubbles of natural gas through microfracture systems over hydrocarbon deposits. The role of aerobic and anaerobic bacteria in the sediments at all levels above hydrocarbon deposits, but especially in surface soils, has been all but unrecognized in the published literature. Various bacteria, by attacking vertically migrating hydrocarbons, including microseeping hydrocarbons, utilize free or complexed oxygen and produce CO_2 (and, by sulfate reduction, S^{-2}). Thus the pH and eH of the leaking hydrocarbon environment are dramatically changed as are consequently the stability fields of the different mineral species in that environment. These changes result in the precipitation, and/or solution and mobilization, of different mineral species and elements, such that the rock column above a leaking hydrocarbon deposit becomes measurably different from laterally equivalent rocks. In this manner a "chimney" is created. "Halo" anomalies from some analyses are due to near-surface oxidation of microseeping (or other vertically migrating) hydrocarbons by aerobic bacteria and other microbes.

Surface geochemical exploration continues to be controversial. Some investigators deny the existence of hydrocarbon microseepage. Others maintain that even if it does exist, it cannot outline surface traces of hydrocarbon deposits. Published data clearly establish the existence of hydrocarbon microseepage, and further establish that microseepage does outline distinct surface traces of hydrocarbon deposits. The Soviets have experienced marked success using surface geochemical exploration. There are published examples of fields discovered solely or largely on the basis of surface geochemical exploration, and it is probable that a much greater number of unpublished examples exist. There is no question but that surface geochemical exploration has been a viable exploration tool. The larger question is whether surface geochemical exploration techniques can dramatically contribute to the future search for hydrocarbons in stratigraphic and/or low-relief traps.

Direct geochemical detection techniques have not taken advantage of the revolution in analytical chemistry over the last twenty-five years. Conventional soil-air and soil-survey techniques require deep sampling (at least 6 ft, 2 m). The resulting need for truck-mounted augers drastically decreases field mobility and increases field time and expense. Moreover, the analytical methods for these techniques are tedious and expensive. The optimum method for the detection of C_2-C_5 microseeping hydrocarbons has not yet been devised. Whatever method is used, its optimum employment will depend on full cognizance of the seasonal (climatological) and geographic variance of the microbiologic filter through which microseeping hydrocarbons must pass.

INTRODUCTION

Surface and near-surface geochemical exploration for hydrocarbon deposits can be divided into two broad categories: direct geochemical detection techniques and indirect geochemical detection techniques. Direct geochemical detection techniques are based on the substantiated assumption that C_1 to C_5 hydrocarbons migrate to the earth's surface from thermogenic hydrocarbon deposits to form anomalously high, measurable surface concentrations of C_1 to C_5 hydrocarbons. Methane, however, is not uniquely indicative of thermogenic hydrocarbon deposits (Horvitz, 1939, 1954, p. 1206–1207; Kartsev et al., 1959; Sokolov et al., 1959, 1971; Davis, 1967, p. 203; Bernard, 1981; and Philp and Crisp, 1982, p. 3) and has only limited applicability as an indicator of hydrocarbon deposits. On the other hand, C_2-C_5 hydrocarbons are uniquely indicative of thermogenic hydrocarbon deposits (Sokolov et al., 1971; Davis, 1976, p. 203; Horvitz, 1980, p. 243; Bernard, 1981; and Philp and Crisp, 1982, p. 3) and anomalous surface concentrations of C_2-C_5 hydrocarbons are considered excellent exploration tools. *In this paper hydrocarbon microseepage is defined as the migration of principally the C_1 to C_5 hydrocarbons from hydrocarbon deposits to the earth's surface to form hydrocarbon and other anomalies.* C_{6+} hydrocarbons also migrate to the earth's surface by physical forces other than those responsible for hydrocarbon microseepage and also form anomalies. A direct geochemical detection method is any technique which measures either a relative or absolute concentration of C_2-C_5 hydrocarbons (or a microbial response thereof) in the surface or near-surface environment. Such methods include analyses for C_2-C_5 hydrocarbons either sorbed onto or occluded within soils, soil-air analyses, microbiologic analyses, and integrative absorption.

Indirect geochemical detection methods can be divided into two classes: (1) those related to or caused by vertically migrating hydrocarbons, and (2) those unrelated to vertically migrating hydrocarbons but thought possibly to be diagnostic of hydrocarbon deposits. Indirect geochemical detection methods of the first class are based on the premise that vertically migrating hydrocarbons from hydrocarbon deposits somehow change the physical characteristics of surface and near-surface soils above the deposits (or even of the entire rock column above a hydrocarbon pool). The reasons for and the mechanisms of these changes are never specified or explained. Techniques which measure such changes are termed indirect geochemical detection methods for two reasons: (1) Such changes may be caused by factors other than hydrocarbon microseepage from thermogenic hydrocarbon deposits. (2) When changes are due to hydrocarbon microseepage, such changes are a secondary, or indirect, result of hydrocarbon microseepage. Kartsev et al. (1959, p. 43), Davidson (1967), Debnam (1969, p. 4), and Duchscherer (1981A, p. 203) all listed different indirect geochemical detection methods of the first class, which include radiometric analyses, redox potential, bitumen analyses (luminescence or fluorescence), soil "salt" (mainly carbonates) analyses, trace metal or trace element analyses, geobotanical analyses, soil mineralogy (lithological), hydrochemical, and magnetic analyses. Many of these techniques can result from hydrocarbon microseepage. An indirect geochemical detection method characteristic of the second class is helium surveying.

Surface geochemical exploration for petroleum is controversial. As McCrossan et al. (1972, p. 4) noted:

> Surface geochemical prospecting for petroleum has had a long and checkered history going back almost 40 years. Much of the work done in this field has been of a rather superficial nature without proper regard for the many variables involved. Consequently, many misleading conclusions have been based on overly simplified assumptions, resulting in periods of interest followed by periods of disillusionment. Very little of the published work is sufficiently broad in scope or comprehensive enough to have clearly settled the issue one way or the other. Much that has been written on this subject has been published in the trade journals and is either poorly documented or not documented at all . . .

Philp and Crisp (1982, p. 25) have noted:

> The literature on geochemical prospecting is frequently contradictory and confused. There is a lack of vigour in many descriptions, much pertinent data from the major oil companies have not been published and some data (usually the most optimistic) are provided by companies offering geochemical prospecting services.

The author's exhaustive review of the literature on the subject substantiates these statements. Since the inception of surface geochemical exploration, due to the initial efforts of Laubmeyer (1933) and Sokolov (1936), the Western emphasis, with few exceptions, has been (and continues to be) placed on the merchandising of insufficiently documented and researched surface geochemical exploration methods. Unfortunately, there has been little fundamental research into the causes, mechanism(s), consequences, and optimum utilization of surface geochemical exploration. Although a voluminous amount of literature exists on surface geochemical exploration, this literature is remarkably uninformative and at times offers more contradiction than fact. Surface geochemical exploration is an area where speculation and unsubstantiated claims far outweigh hard data. The *definitive* literature on the subject is largely from the intense Soviet effort in this area and from Western microbiologists.

The established Western trend of assertions of the exploration proficiency of various surface geochemical exploration methods, unsubstantiated by published rigorous scientific research, continues to the present. The poor Western record in this area is somewhat understandable. The Soviet efforts have been government centralized and controlled, resulting in full scientific disclosure. On the other hand, Western commercial surface geochemical exploration venturers, if successful, would gain nothing by full disclosure of their method. Nonetheless, articles in both scientific and trade journals, which largely serve to advertise various surface geochemical exploration methods with little or no meaningful scientific documentation, are suggestive of "95% sure" stock or commodity market trading techniques. It would be most surprising if an inexpensive, simple, easily employed, and accurate method, which finally fulfilled the early promises of surface geochemical exploration was offered as a service for general use. It is more likely that a true "breakthrough" in this field would not be widely publicized, but rather would be employed by its developer as the effective exploration tool it would be.

Different investigators have offered reasons why surface geochemical exploration has not been accepted to the degree that it should be (Davidson, 1963, p. 98–100, 1967; Debnam, 1969, p. 1; Hunt, 1979, p. 428; Duchscherer, 1981B, p. 312; Dennison, 1983, p. 161–162; and Kuzmin, 1983, p. 1089). There is probably some validity to many of their reasons. No area of science, however, can achieve its ultimate commercial application until the basic principles and controlling parameters of that scientific area are delineated. Such delineation certainly has not occurred with surface geochemical exploration methods based on hydrocarbon microseepage. Explanations are needed for (1) the microseepage migration mechanism, (2) the "halo" anomaly, (3) the degree and consequences of the interaction of microseeping (and other vertically migrating) hydrocarbons with their environment, (4) the characteristics and causes of "chimneys," and (5) how (or even if) microseeping hydrocarbons create indirect anomalies (soil salt, iodine etc.). It is quite likely that the optimum method to detect and to measure hydrocarbon microseepage has not yet been developed.

Although this paper is not intended as a review of the field, at times it probably approaches that status. Reviews of surface geochemical exploration have been given by Kartsev et al. (1959, by far the most definitive work), Debnam (1969, p. 1–6), Boyle and Garret (1970, p. 63–65), McCrossan et al. (1972, p. 4–8), Siegel (1974), Hunt (1979, p. 425–433, somewhat negatively oriented), and Philp and Crisp (1982, the most recent comprehensive review). The models used in this paper have been developed on the basis of what reliable data exist in the literature, established principles from related fields, and extrapolations, when necessary, based on the author's best scientific judgement. No new data or research are presented here. Minimal fundamental research has been carried out on hydrocarbon microseepage in the last fifty years; nevertheless, enough basic data does exist which can be combined with established geologic/geochemical principles to construct a *realistic* working model for hydrocarbon microseepage upon which future rigorous and proper scientific Western research can be based. Perhaps the publication of new data from such research will reveal that many of the concepts of this paper are invalid and will have to be rejected. If so, this paper will have served its purpose. In any case, the author believes that the models developed in this paper are much closer to reality than other published models and/or speculative hypotheses concerning hydrocarbon microseepage or the vertical migration of hydrocarbons to the earth's surface in general.

MIGRATION MECHANISM

MacElvain (1963, p. 133) stated that knowledge of the actual mechanism by which C_1 to C_5 hydrocarbons migrate to the earth's surface is unimportant. He contended that it is of greater importance to recognize that such migration happens in the first place. In the author's opinion, this attitude is not entirely correct. Knowledge of the hydrocarbon microseepage mechanism (or mechanisms) would allow better utilization of surface geochemical exploration methods and give them greater credibility.

The principal migration mechanisms proposed to explain hydrocarbon microseepage are (1) diffusion (Rosaire, 1940; Kartsev et al., 1959; Baijal, 1962; Siegel, 1974, p. 231; Donovan and Dalziel, 1977; Duchscherer, 1980, 1981A,B, 1983; and Dennison, 1983), (2) effusion (Rosaire, 1940; Kartsev et al., 1959; Baijal, 1962; Donovan and Dalziel, 1977; Dennison, 1983; and Duchscherer, 1981B, 1983), (3) vertical transport of hydrocarbons by deep basinal waters (Pirson, 1960, 1962; Donovan and Dalziel, 1977; Davidson, 1981; and Duchscherer, 1981B, 1983), and (4) "permeation" (Horvitz, 1950, p. 939; Baijal, 1962, p. 8–10), a mechanism unexplained by those authors.

The American Geological Institute (1976, p. 120) defines diffusion as "the spreading out of molecules, atoms, or ions into a vacuum, a fluid, or a porous medium, in a direction tending to equalize concentrations in all parts of a system." Hunt (1979, p. 425–428) gave an elucidating discussion of the improbability of diffusion controlling hydrocarbon microseepage. (Also see MacElvain, 1969.) Hunt (1979) pointed out that appreciable vertical diffusion distances through shales of near zero permeability can be achieved only over hundreds of millions of years, times far too long for relevance to hydrocarbon microseepage. Hunt (1979, p. 428) concluded and the author agrees: "Clearly vertical diffusion from petroleum accumulations buried at depths such as 1,700 m (5,600 ft) is not the mechanism causing surface geochemical anomalies." Jones and Drozd (1983), on the basis of their studies, also rejected diffusion as a possible migration mechanism for hydrocarbon microseepage.

The strongest arguments against diffusion are (1) because it is a *spherically dispersive* process with no enhanced vertical component, it cannot possibly account for the sharp surface outlines of hydrocarbon deposits; (2) accounting for "halo" anomalies by diffusion is impossible; (3) diffusion, as Hunt (1979) established, is far too slow to account for the rapid anomaly developments and/or disappearances caused by changes in reservoir pressure; and (4) if diffusion were the migration mechanism responsible for hydrocarbon microseepage, surface anomalies would involve much greater concentrations of C_{6+} hydrocarbons than they do. Rosaire (1940) recognized the dispersive nature of diffusion and attempted to reconcile the serious contradiction of the observed data with diffusion by an improbable scenario of vertically migrating gases evaporating ground water and triggering lateral ground water flow to cause a halo anomaly. Some investigators, including Dennison (1983), have called for an "enhanced vertical component" of diffusion. This is not possible, since diffusion by definition is spherically dispersive. Further, as MacElvain (1969, p. 18) noted: ". . . pure diffusion-mechanics control of a methane concentration in a sedimentary column would expand and dilute the gas two feet laterally for each foot of vertical ascent." Diffusion is the process, however, which is largely responsible for establishing the "background" levels of surface hydrocarbon concentrations in sedimentary basins. This is not to be confused with the hydrocarbons which microseep from hydrocarbon deposits and cause the anomalies that are measured against that "background."

Effusion is a term directly contradictory to hydrocarbon microseepage. The American Geological Institute (1976, p. 139) defines effusion as: "The act or process of effusing or pouring out . . ." In a geologic sense, effusion refers to unrestrained flow, or ejection of fluids, especially volcanic fluids (lava) and thus contradicts the concept of losses of small amounts of hydrocarbons (microseepage). In such usage, effusion would refer to intermittent or continuous monophasic flow of crude oil or natural gas along faults or fractures according to Darcy's Law, combined with the principles of two phase fluid flow; in other words, *macroseepage* of crude oil and/or natural gas. Microseepage by effusion would result in surface-anomaly hydrocarbon concentrations several orders of magnitude higher than they are with significant concentrations of C_{6+} hydrocarbons.

Different investigators contend that hydrocarbon microseepage anomalies are due, at least partly, to vertical movement of deep-basinal compaction waters, or meteoric-recharge waters, moving through hydrocarbon deposits to the earth's surface (Pirson, 1960, 1963, 1964, 1969; Donovan and Dalziel, 1977; Roberts, 1980; and Davidson, 1982, 1984). Such waters are believed to carry hydrocarbons and dissolved mineral species which result in the formation of different types of anomalies. Pirson (1960, 1963, 1964B, 1969) and Roberts (1980, 1981) have constructed elaborate "forced draft" or "deep water discharge" theories, wherein primary petroleum migration and accumulation is achieved by hydrocarbon traps "filtering out" junk molecules or solubilized oil from water, allowing the water to pass through the hydrocarbon deposit to the earth's surface. Based on past and ongoing personal studies of the origin, migration, and accumulation of oil, the accumulation scenarios of Pirson (1960, 1963, 1964B, and 1969) and Roberts (1980 and 1981) are most improbable. Evidence of significant flow of deep-basinal compaction waters up major fault zones does exist from different petroleum basins (Price, 1976, p. 229–239; Roberts, 1981; and Jones, 1981, 1984), as well as a few examples of flow of deep-basinal compaction waters up major faults cutting hydrocarbon deposits. There is no published evidence, however, of significant vertical flow of compaction (or meteoric-recharge) waters through hydrocarbon deposits in traps of slight to moderate structural complexity, and especially through stratigraphic traps. Further, the principles of two-phase fluid flow and Darcy's Law dictate that such vertical water flow through the oil-charged porosity of a hydrocarbon deposit is impossible. (See Chapman, 1982, for a more complete discussion of this point.) The creation of hydrocarbon microseepage (or other) anomalies from vertical flow of deep-basin compaction or meteoric-recharge waters through hydrocarbon deposits to the earth's surface is not possible. (1) Darcy's Law dictates that significant fluid flow vertically through shales, a direction of effective zero permeability, is not possible; (2) there is no evidence of such flow, which should exist if it were occurring; (3) significant vertical fluid flow through shales would result in much higher surface hydrocarbon concentrations than those observed, and would also result in significant concentrations of C_{6+} hydrocarbons. There is no question but that significant flows of deep-basin compaction waters do occur up faults and fractures in sedimentary basins. Hydrocarbons in aqueous solution in these waters do create surface anomalies. These anomalies, however, are not from hydrocarbon microseepage and therefore do not necessarily indicate or relate to a hydrocarbon deposit.

Data from nature provide evidence of and put rigid restraints on the hydrocarbon microseepage mechanism. The speed with which surface hydrocarbon anomalies appear or disappear (a matter of weeks to months), because of changes in reservoir pressure thousands of feet deep, testifies to the rapid vertical movement of hydrocarbon gases. For example, Hunt (1979, p. 431) cited a Soviet study where a few months after underground gas storage, the gas content of a water sand located 300 m (984 ft) above the storage reservoir increased ten times over its initial value. The concentration of hydrocarbon-oxidizing bacteria also increased significantly in this sand. Sivaborvon (1974, p. 67) noted that a repressurization of the Hilbig Oil Field, Bastrop County, Texas, during secondary recovery attempts, caused a surface hydrocarbon anomaly over the field to increase in intensity. The study of Coleman et al. (1977) also provided evidence of the speed of hydrocarbon microseepage.

Sokolov et al. (1971) provided other data which point to the migration mechanism. The Soviets, in their strong commitment to surface geochemical exploration, routinely carry out coring programs to depths of 6 to 20 m (19.7 to 65.6 ft) as well as exploration/research coring programs to 500 m (1,640 ft) or deeper. Quantitative analyses of soil-air and soil-sorbed hydrocarbons are carried out along the entire length of the core in these programs ("hydrocarbon gas logging"). Microseeping hydrocarbons at the earth's surface have negligible or no amounts of C_{6+} hydrocarbons (Sokolov, 1970; Sokolov et al., 1971; and Philp and Crisp, 1982, p. 3). However, Sokolov et al. (1971) noted that in the deeper coring programs, a chromatographic effect is noticed over oil deposits, with a progressive increase in the appearance and concentration of higher molecular weight hydrocarbons with increase in depth. Rosaire (1940) also described this peculiarity (Fig. 8). Kartsev et al. (1959, p. 315) noted a similar feature: below the earth's surface, with decreasing distance above oil deposits, increasing populations of bacterial species which consume only higher molecular weight hydrocarbons are observed. Horvitz (1939) and Rosaire (1940) describe similar cases down successful exploration wells.

Other evidence of the hydrocarbon microseepage migration mechanism is in the sharp surface boundaries of apical or halo anomalies formed directly over hydrocarbon deposits. Reports of slight lateral offsets of hydrocarbon anomalies, even in cases of strong lateral ground-water flow in shallow sands, are unusual, and most investigators contend that ground-water flow does not laterally displace hydrocarbon anomalies (MacElvain, 1963; Siegel, 1974; Horvitz, 1980; and Jones and Drozd, 1983). Thus, a strong vertical (gravitational) component must be present in the migration mechanism. Such a component can only be supplied by the buoyancy differences from monophasic hydrocarbon movement through water. However, the lack of significant concentrations of C_{6+} hydrocarbons in hydrocarbon microseepage surface anomalies dictates that a gas phase must be migrating as opposed to a crude-oil phase. The speed with which hydrocarbon anomalies appear and disappear in response to reservoir pressure changes, and the very low hydrocarbon concentrations in these anomalies, further dictate that "bulk" gas flow (effusion) cannot be occurring. Instead, a colloidal size or microbubble of hydrocarbon gas must be migrating.

MacElvain (1969), in a well-documented, elegant, and simple presentation, first proposed the formation of hydrocarbon microseepage from the vertical ascent of ultra-small (colloidal size) gas bubbles. Their small size allows frictional "clinging" effects to be overcome by Brownian motion. As MacElvain (1969, p. 25) noted:

> In this manner, colloidal-size gas bubbles are readily displaced upward by the surrounding water at rates up to several millimeters per second regardless of any sedimentary particles that may intrude in the way of their upward zig-zag Brownian path. Such exceedingly small bubbles can quickly ascend hundreds and even thousands of feet in a manner not available to large gas bubbles or to individual gas molecules. Large bubbles contain too large a surface area to be able to demonstrate kinetic or Brownian oscilation while individual molecules of dissolved gas possess insufficient difference of density for gravitational displacement even though they are jostled mightily by the Brownian effect.

The vertical ascent of colloidal gas bubbles to form hydrocarbon microseepage anomalies also explains the "chromatographic effect" of higher and higher concentrations of higher and higher molecular weight hydrocarbons with decreasing distance to hydrocarbon deposits, as observed by Horvitz (1939), Rosaire (1940), Kartsev et al. (1959), and Sokolov et al. (1971). This migration mechanism also explains why hydrocarbon microseeps are composed predominantly of the C_1 to C_4 hydrocarbons. Crude oil-methane solubility data of Price et al. (1983) establish that hydrocarbon gas can dissolve and transport significant amounts of higher molecular weight hydrocarbons. As temperature and especially pressure decrease, however, both the amount (Fig. 1) and the molecular weight (Fig. 2) of the hydrocarbons carried in the gas phase significantly decrease. Microbubbles of hydrocarbon gas migrating vertically from an oil deposit will carry dissolved higher molecular weight (C_{6+}) hydrocarbons. As such bubbles ascend the sedimentary column above the oil deposit, they encounter lower pressures and temperatures. This causes C_{6+} hydrocarbons to exsolve progressively from the gas phase and be left behind in the sedimentary column over the oil deposit in precisely the observed pattern (i.e., increasing concentrations of higher molecular weight hydrocarbons with decreasing distance from the hydrocarbon deposit). A direct result of this migration mechanism is that surface anomalies from true hydrocarbon microseepage in onshore sedimentary basins cannot possibly involve significant concentrations of C_{6+} hydrocarbons. Such hydrocarbons exsolve from vertically migrating hydrocarbon gas microbubbles long before these bubbles reach the earth's surface. To test this hypothesis, a natural gas atmosphere was established in a closed container over hexane at 22° to 25°C and one atmosphere. After the system had equilibrated, the hexane concentration in the natural gas atmosphere was measured by gas chromatography. Within the limits of the analysis, this concentration was no different from that due only to hexane partial pressure. The latter was determined by measuring the hexane concentration in the atmosphere over hexane in another vessel containing only hexane and air without natural gas under the same conditions. The results of this experiment support earlier investigations and confirm that microseepage by migration of hydrocarbon gas microbubbles can involve only insignificant amounts of C_{6+} hydrocarbons *at the earth's surface.* Onshore surface anomalies involving C_{6+} hydrocarbons are due to other migration mechanisms (effusion/macroseepage, diffusion, vertical water transport, or basinal uplift erosion). Because of the high pressures generated by significant sea-water columns, however, offshore hydrocarbon microseepage anomalies should involve C_{6+} hydrocarbons.

An unanswered question relating to any proposed migration mechanism for hydrocarbon microseepage is the ascent of hydrocarbon gases through thick shale sections with near zero vertical permeability. All shales do have some degree of lateral permeability. However, in shales which have undergone even moderate lithification and compaction, 10% to 90% of the shale-pore throats are reduced to sizes too small for the larger complex molecules of petroleum to fit through (Momper, 1978, p. B-38). Thus, migration of gas microbubbles through shales appears to be difficult laterally and impossible vertically. Migration of single C_1 to C_5 hydrocarbon molecules is possible laterally, but not vertically, through a shale.

Some investigators, including Hoffman (1939), Fash (1940, p. 1446–1447, in Rosaire et al., 1940), Horvitz (1950), Davidson (1963), and Duchscherer (1981A) have dismissed the problem of zero vertical shale permeability with the observation that diffusion of noble gases and hydrogen has been documented through the atomic lattice of metals and glass (T'sai and Hogness, 1932; Urry, 1932; Alty, 1933; Burton et al., 1933; Barrer, 1934, Braaten and Clark, 1935; Taylor and Rast, 1938; and Weyl and Pincus, 1938). This allows the conclusion that nothing is truly impermeable and therefore hydrocarbon gases can vertically pass through shales, which certainly have more permeability, even in the vertical direction, than glass and/or metal. Duchscherer (1981A, p. 203) has even called for the diffusion of hydrocarbon gases through the crystal lattices of minerals. There is a flaw in the above reasoning. The molecular diameters of hydrogen and the noble gases used in the above experiments are *much* smaller than those of the hydrocarbon gases. No documentation could be found of hydrocarbon gases diffusing through metals, glass, mineral lattices, or vertically through shale. The relatively large size of the C_1 to C_5 hydrocarbon molecules would prevent such diffusion. As Dickey and Hunt (1972, p. 159) noted, "It has been shown repeatedly that hydrocarbons do not diffuse upward from an oil accumulation through shale in measurable amounts."

The very existence of surface hydrocarbon anomalies, and the speed with which these anomalies appear and disappear, indicate that whatever the migration mechanism, it must take place through a pervasive microfracture system above hydrocarbon deposits. Published research (Price et al., 1984, p. 113–334, 121) and the author's unpublished research suggest that when hot oil moves through cooler sediments along a major secondary migration route, or into a trap, the oil will leave a distinct thermal imprint in the overlying (and to some extent underlying) sediments. These thermal imprints can be observed by examining the commonly employed inorganic or organic maturation indices. Such thermal imprints have been observed in the Williston, Powder River, Denver, and Anadarko Basins. The heat necessary for such imprints would also cause microfracturing in sediments above either major oil deposits or crude oil secondary migration routes. It is also possible that all lithified sedimentary rocks have a pervasive microfracture system emplaced in them simply from regional tectonism or "Earth stress." Park Jones observed that the earth's crust consists of a near infinite number of "cryptocracks" as small as one or two molecules wide (Davidson, 1963, p. 96).

Meakin (1971, p. 128), discussing the migration mechanism of hydrocarbon microseepage, noted:

> Unfortunately, research on the migration of hydrocarbon gases appears to be limited to the U.S.S.R.... Stklyanin et al. (1968), in a detailed paper on the conditions, parameters, and means for the ascending migration of natural gas, determined vertical microfractures to be the main avenue of hydrocarbon migration in the Voskresenski petroleum field.

In the author's opinion, the ascent of microbubbles of natural gas through microfracture systems above hydrocarbon deposits is the principal migration mechanism responsible for C_1 to C_5 hydrocarbon microseepage and the resulting surface hydrocarbon anomalies. It is the only proposed mechanism in agreement with observed data. The unequivocal existence of such microfracture systems, however, has yet to be demonstrated.

BACTERIAL INFLUENCE

Bacteria

The one element of hydrocarbon microseepage most commonly overlooked, considering the importance of the role it plays, is the interaction of bacteria with vertically migrating hyrdocarbons. This interaction not only accounts for the fundamental basis of most surface geochemical exploration methods, but also explains many of the mysteries of hydrocarbon microseepage, including "halo" anomalies, "disappearing anomalies," chimneys, radiation lows, iron/manganese anomalies, paraffin dirt, and many more. Yet it is the one element which largely has been ignored in Western investigations and thought patterns concerning hydrocarbon microseepage. Davis (1967, p. 117) expressed a degree of amazement for the disregard given to bacterial processes by Western investigators:

> Microbial activity in soils, and the effect it might have upon hydrocarbon gases or their "oxidized or polymerized products" was not discussed by Rosaire (1939) or his associate, Horvitz (1939). Rosaire did allude to the microbial production of methane in soils and to the biological formation of ethylene, as well. Ethane, propane and butane in soils were particularly significant because they were not known to be produced ordinarily by biological means (Horvitz, 1939; Rosaire, 1939)... In general, workers active in the field of gas surveying including McDermott (1939) and Rosaire et al. (1940) did not discuss microbial activity as a possible factor affecting gas surveys or other geochemical investigations from the standpoint of modifying or destroying the index hydrocarbons.

Kartsev et al. (1959, p. 302), discussing the oxidation of microseeping hydrocarbons by bacteria, noted:

> The actual magnitude of processes of oxidation of hydrocarbon gases now taking place and their practical significance in oil and gas prospecting were not immediately apparent. It must be noted that even today these phenomena are not adequately studied.

Kartsev et al. (1959, p. 306–315) discussed the various forms of bacteria and related organisms known to oxidize hydrocarbons. All fall under the broad classification of heterotrophic organisms, or those which subsist mainly on preformed organic matter. There are three broad types of bacteria: aerobic, facultative anaerobic, and obligate anaerobic. Aerobic bacteria depend on the presence of free oxygen for their life processes. Facultative anaerobes can exist either in the presence or in the absence of oxygen. Obligate anaerobes cannot tolerate even the lowest concentrations of free oxygen. In addition to the many types of aerobic bacteria which oxidize hydrocarbons, important anaerobes also exist (e.g., denitrifying bacteria and especially sulfate-reducing bacteria). All bacteria produce carbon dioxide and organic acids from hydrocarbon oxidation. In addition, the sulfate reducers produce hydrogen sulphide and the denitrifiers produce free nitrogen and nitrous oxide. As Davis (1956, p. 1444) noted, not only bacteria, but also fungi and actinomycetes readily oxidize hydrocarbons in soils. In discussing microbes capable of hydrocarbon oxidation, McKenna and Kallio (1965, p. 183) noted:

> Codification of the numerous microbial forms capable of degrading or assimilating hydrocarbons has not been attempted on several grounds. Many reports carry descriptions of new isolates that are, to be kind, meager. Bacteria, yeasts, and fungi possessing the capability are legion, new members of the group are continually being found, and such a list would be relegated to senicide before it appeared in these pages.

Davis (1967, p. 228) discussed the characteristics of actinomycetes, which have traits of both bacteria and fungi. These microbes (1) heavily populate soils to depths greater than three feet, (2) have remarkable oxidative versatility, (3) can readily adapt to the consumption of hydrocarbons, and (4) are spore formers. Many microbes have remarkable adaptive ability.

At the earth's surface and into the soil layer, the main microbial activity is carried out by aerobic bacteria, fungi, and actinomycetes. Other bacteria, however, are active over a staggering range of conditions. The basic requirements for bacterial activity are moisture, nutrients, some form of oxygen, whether free or reduced, and a food source (Davis, 1967). Zobell (1946, p. 3, 7) noted that a nitrogen source is also required, and that dispersion of hydrocarbons, as would be present in a microseepage situation, greatly promotes bacterial activity.

> By dispersing hydrocarbons adsorbed on the surfaces of inert solids, the microbial attack on nearly all kinds of hydrocarbons tested is greatly increased. Certain hydrocarbons which seemed to be invulnerable to microbial attack were quite rapidly decomposed when dispersed throughout the medium on the surfaces of inert solids. (Zobell, 1946, p. 9–10)

Davis (1967, p. 121) noted that although the ordinary view is that soil microbial activity is curtailed in the winter months because of low temperatures, Samtsevich (1955) found that soil microbial activity did not slow down at temperatures above 0°C with sufficient soil moisture. Kartsev et al. (1959, p. 305) noted: "The metabolism of microorganisms can proceed in a relatively wide range of temperatures between a few degrees below zero and 85–90°C above zero." Oxidation of hydrocarbons at the earth's surface apparently is controlled to a greater extent by soil moisture than by soil temperature. Low soil moisture contents can significantly reduce, or even eliminate, soil microbial activity (Soli, 1954, p. 2557; Kartsev et al., 1959, p. 319; and Davis, 1967, p. 121).

Bacterial oxidation of vertically migrating hydrocarbons, and bacterial activity in general, is most pronounced at the soil surface layer (Kartsev, 1959, p. 303; Sokolov et al., 1959, p. 620). Zobell (1946, p. 23–25) gave detailed documentation of the universal occurrence of hydrocarbon-oxidizing bacteria under various surface conditions, including those of surface soils. "Söhngen (1913) found from 50,000 to 200,000 paraffin-oxidizing bacteria per gram of garden soil and up to 3,000 per ml of ditch water. A marked increase in the abundance of microorganisms in soil treated with paraffin was observed by Jensen (1931)." (Zobell, 1946, p. 24). Both Zobell (1946) and Davis (1967) have noted that an introduction of hydrocarbons into almost any environment invariably leads to pronounced bacterial response. For example, Haas et al. (1941) found a bacterial population of 49,000,000 bacteria per ml of crude oil-soaked soil in a Texas crude oil tank battery. Zobell (1946, p. 25) noted:

> There is ample evidence from field and laboratory observations that crude oil is attacked by soil microorganisms. The rapid disappearance of oil from waterways, from soil around refineries, leaking pipe-lines or oil wells, and from polluted beaches is believed to be due largely to the activity of hydrocarbon-oxidizing microorganisms.

Zobell (1946, p. 33) also discussed marked bacterial responses around leaking natural gas lines. Although bacterial activity is most pronounced in surface soils, it can occur at all depths above a leaking hydrocarbon deposit. Kartsev et al. (1959, p. 302–305) have noted that different hydrocarbon-oxidizing bacteria are found at all depths of the sediment column above hydrocarbon deposits, and in the edge waters of oil deposits. For example, Telegina et al. (1963) found hydrocarbon-oxidizing bacteria associated with an oil field at a depth of 2,680 m (8,793 ft) and Rybakova (1957) isolated denitrifying and sulfate-reducing bacteria from crude oils at 4,200 m (13,780 ft) in the Apsheron Peninsula.

Crude oil degradation by bacterial attack is an almost universally documented process which has resulted in the loss of huge volumes of crude oil (Davis, 1967; Winters and Williams, 1969; Evans et al., 1971; Bailey et al., 1973; Deroo et al., 1974; Connan et al, 1975, Phillipi, 1977; Tissot and Welte, 1978; Hunt, 1979; and Price, 1980).

As Kartsev et al. (1959, p. 305) noted:

> Owing to their small dimensions, microbes easily penetrate the finest cracks, pores, and fissures, and are carried for great distances by dust particles. Because of their small dimensions, microbes have an enormous surface of contact with the surrounding medium. Therefore, encounters of microbes with various molecules occur quite frequently, even when the latter are very highly dispersed in the medium. This permits a number of microorganisms easily to utilize gaseous substances as a source of nourishment.

Thus, bacteria would be likely to permeate microfracture networks above hydrocarbon deposits, and by so doing would oxidize hydrocarbons migrating toward the earth's surface. Some types of bacteria form spores enabling them to "hibernate" at the onset of unfavorable growth conditions. Kartsev et al. (1959, p. 308) noted that "spored" and active bacteria are almost ubiquitous in sediments at considerable depths, and also, . . . there can be no doubt that the various forms of microorganisms found in deep-seated sedimentary layers of the earth's crust subsist in these layers for long geological periods." Zobell (1946, p. 4) also noted that bacteria can be recovered from ancient sediments and stimulated to growth. The grand result of all this is a series of environmental niches from the soil layer to hydrocarbon deposits where different microbial forms may exist, all of which utilize hydrocarbons as a food source.

The effects of microbial oxidation of vertically migrating hydrocarbons are staggering in relation to surface geochemical exploration methods. The most obvious result of such oxidation is a decrease in the concentration of soil-air hydrocarbons, hydrocarbons dissolved in sediment pore waters, and soil-sorbed hydrocarbons, especially in the surface or near-surface soil layers where the most intense microbial activity takes place. Kartsev et al. (1959, p. 303) appropriately referred to a "biological filter" through which microseeping (or any vertically migrating) hydrocarbons must pass in order to establish a surface anomaly or signal. As those authors noted, quantification of this signal may take any one of three measurement approaches, alone or in any combination: (1) the intensity of bacterial activity for bacteria specific to C_{2+} hydrocarbons (microbial prospecting), (2) the concentration of soil-air hydrocarbons, or (3) the concentration of soil-sorbed hydrocarbons. All three analyses simultaneously carried out in Soviet studies have shown that unless hydrocarbon microseepage is intense, microbial anomalies will not correspond to soil-sorbed or soil-air hydrocarbon anomalies (Kartsev et al., 1959, p. 334, 338, 340). In areas where intense bacterial activity occurs, hydrocarbon microseepage anomalies cannot be established by soil-sorbed or soil-air hydrocarbon analysis even in areas of significant hydrocarbon microseepage. In discussing surface microbial activity, Kartsev et al. (1959, p. 319) noted: "As a result of this, hydrocarbon gases migrating from the depths may be completely oxidized in the presence of a sufficiently developed bacterial filter." Davis (1967, p. 118) noted:

> The utilization of gaseous hydrocarbons in soils is obviously an important factor to consider with respect to soil-gas surveys and this activity becomes all the more important when the concentration of gases being analyzed is in the order of parts per billion or a few parts per million of the interstitial soil gas (Sokolov, 1936; Horvitz, 1939; McDermott, 1939). The question might arise as to why there should be hydrocarbon gases detectable in soils particularly since hydrocarbon-oxidizing microorganisms are so prevalent.

Davis (1967, p. 119) described experiments carried out in glass tubes wherein 50–90 ppm of ethane at low flow rates was passed through soil containing bacteria and actinomycetes and no ethane could be detected in the effluent gas at a sensitivity limit of 5 ppb. Zobell (1946, p. 37) noted:

> Nearly any method of prospecting by chemical assay or "soil analysis" which ignores the effect of bacteria on the hydrocarbon content or other properties of soil can be expected to give anomalous results under certain conditions. Hydrocarbons may be decomposed or altered by microorganisms in the soil almost as fast as the hydrocarbons enter the biosphere from subterranean deposits. The rapidity with which hydrocarbons are decomposed in soil may vary seasonally with the waste content of the soil, its oxygen content, abundance of organic matter, nitrogen content, temperature, growth phase of the organisms and other factors which influence the abundance, kinds and activity of microorganisms in soil.

Surface bacterial activity thus not only has the capability of obliterating a hydrocarbon microseepage anomaly as measured by analysis for soil-air or soil-sorbed hydrocarbons (Zobell, 1946, p. 37; Kartsev et al., 1959, p. 302, 319; and Davis, 1969, p. 140, 155), but it also can cause seasonal variations in the anomaly intensity (Zobell, 1946, p. 37; Kartsev et al., 1959, p. 302; and Davis, 1969, p. 141, 154). Figure 3 shows soil-sorbed hydrocarbon gas concentration as a function of time for the Krasnodar region in the Soviet Union. In this region, times which correspond to low hydrocarbon concentrations are times of heavy rains during which maximum microbial activity would take place. During the dry summer months microbial activity decreases and hydrocarbon concentrations increase.

Paraffin Dirt

Surface microbial activity is responsible for paraffin dirt, a waxy-appearing soil associated with gas seepages and thought to be a powerful indicator of hydrocarbon deposits. As W. E. Pratt in Milner (1925, p. 82) noted, "I feel it to be established that 'paraffin Earth' is found in soils in which aluminum- and iron-bearing clays are present in large proportion in the vicinity of seepages of gas." Also see Barton's (1927) discussion. Although paraffin dirt was most readily recognized in the onshore Gulf Coast province, it also was reported in other areas with tropical to temperate climates, including Colombia, Romania, and Burma (Milner, 1925). Davis (1952, 1967) carried out the most definitive published research on paraffin dirt, both in the laboratory and on natural occurrences. In nature, paraffin dirt is limited to areas of gas microseepages or macroseepages (Fig. 4) and has a high content of nitrogen and organic carbon and a low content of lipids and saturated hydrocarbons. Radioactive carbon analysis of the organic matter in a paraffin-dirt bed (Davis, 1967, p. 102) showed that only 15% of the carbon was "modern" carbon 14, indicating that the carbon of the paraffin-dirt bed was predominatly composed of "old" carbon (>5,700 years) derived from microseeping hydrocarbons. Microscopic examination of the organic matter in paraffin dirt revealed heavy concentrations of cell walls of fungi, yeasts, actinomycetes, and bacteria. Davis (1967) was able to manufacture a laboratory paraffin dirt indistinguishable from natural samples by encouraging bacterial growth by passing hydrocarbon gas through moist soils. He concluded that paraffin dirt was an accumulation of dead bodies and cell walls from microbial activity on microseeping natural gas, a view in total agreement with that of Kartsev et al. (1959, p. 215):

> "Paraffin dirt" occurs widely in the region of the Gulf Coast. It occurs in the top layer of the soil as yellow-brown, amorphous masses consisting mainly of carbohydrates. (Its composition does not correspond at all to its name.) There is no doubt that this substance is formed from hydrocarbon gases. It is found only where there are gas seeps, and the size of the dirt deposits is directly proportional to the size of the seeps.
>
> Consequently, the simplest hydrocarbon gases may be converted to complex organic substances. This is apparently the result of bacterial activity.

Apparently the necessary requirements for the formation of paraffin dirt are an adequate rate of gas microseepage to support an abundant microbial population, and adequate year-around soil moisture. Thus, paraffin dirt could form in any area with a temperate-to-tropical climate without a defined dry season. The fact that paraffin dirt originates from microbial activity on microseeping hydrocarbons has an interesting ramification. Probably because it is so inappropriately named, "paraffin dirt" in the past was sometimes associated with petroleum macroseepage, even though it is only an indirect consequence of hydrocarbon microseepage. Thus, oil exploration on the basis of paraffin dirt is (or was) an indirect geochemical detection technique, one responsible for the discovery of vast amounts of oil in the Gulf Coast province alone. Solely on the basis of paraffin dirt, surface geochemical exploration has resulted in the discovery of much more oil than it has been given credit for.

Chimneys

"Chimneys" are rock columns above hydrocarbon deposits thought to be modified by vertically migrating hydrocarbons, or by some other unknown association due to a hydrocarbon deposit. Because of these modifications, the rocks above hydrocarbon deposits are thought to take on different lithological properties compared to surrounding country rocks (not over hydrocarbon deposits) at equivalent depths. Hence a "chimney" of altered rock is created above a hydrocarbon deposit. These lithological differences can be gauged by different physical/chemical measurements. Although Pirson (1964C) apparently was the first to use the term "chimney," the concept can be traced farther back. Pirson (1960, 1963, 1964A, B, C, D, 1969) developed his hypothesis of chimneys based on radiation lows observed over oil fields (Fig. 5) which he sometimes could detect as much as 7,000 ft (2.13 km) above a deposit (Sunniland, Florida, Pirson, 1964C). The effect was also found in surface sediments (Pirson, 1964D). Pirson (1963, 1964C, p. 71) found that gamma-ray log decreases of up to 50% (compared to gamma-ray readings in the same shale off the deposit) were best developed over hydrocarbon deposits in organic-rich black shales (such as the Lower Mississippian/Upper Devonian Woodford Shale and equivalents).

Pirson (1969, p. 165) documented the effect in over thirty oil fields in France, Canada, and the United States, and developed an elaborate, though, in the author's opinion, incorrect scenario to explain his observations. This scenario (Pirson, 1969, p. 161–164) is based on an assumed geologically long-continued flow vertically through sediments to the earth's surface, of heavily mineralized deep-basinal compaction waters. Because of temperature and pressure decreases suffered by the upward movement of such water, minerals were assumed to precipitate in the sediments through which the waters passed. During the earliest stages of such flow, the waters carried "solubilizers" (Baker, 1959) which caused the solution and transport of crude oil to traps where the solubilizers and crude oil were filtered from the waters, resulting in an oil deposit. Eventually the supply of "solubilizers" from deep-basin sediments was used up, which terminated oil migration and accumulation. The flow of mineralized waters, however, was postulated to continue, such that mineralization of sediments along the path of the migrating water continued. Pirson (1969) recognized that such water flow would have to be diverted around oil fields. As the sediments above oil fields would no longer be subjected to mineralization, these sediments over hydrocarbon deposits would have lower concentrations of the elements causing gamma-ray radiation (Pirson, 1969).

Pirson's (1969) theory can be dismissed on four points:

1. The black organic-rich shales which Pirson used (his "optimum significant stratigraphic level" or OOSL) have far above average concentrations of radioactive elements from depositional conditions. Such shales clearly do not become mineralized with radioactive elements during postdepositional burial.
2. Baker's (1959) solubilization hypothesis of primary petroleum migration has been shown to be impossible (Price, 1976, and McAuliffe, 1978).
3. The vertical flow of water through impermeable sediments is hydrodynamically impossible.
4. There is no evidence of such vertical water flow which, if it were occurring, would be quite evident. (The documented flow of deep-basinal compaction waters up faults and fractures is not evidence supporting Pirson's 1969 concept.)

Pirson (1969) rejected Alekseyev's (1959) hypothesis that radiation lows over oil fields were due to a *leaching* of radioactive elements from shales. Although Pirson's (1969) postulated cause of chimneys of radiation lows over hydrocarbon deposits is considered to be in error, it is not disputed that radiation lows sometimes do exist. Such radiation lows (including surface radiation lows) have also been reported by Lundberg (1952, 1956), Merritt (1952), Alparone (1963), MacElvain (1963), Tilloy and Monchaux (1965), Alekseyev and Gottikh (1966), Kilmar (1983), and Curry (1984). Armstrong and Heemstra (1973) provided literature reviewing the subject.

On the other hand, a straightforward, one-to-one correlation of radiation lows over hydrocarbon deposits can hardly be taken as a given. In a detailed radiation survey over the Sutherland Field, Greenwood County, Kansas, Heemstra et al., (1979) found no correlation between hydrocarbon production and surface radiometry. They did, however, find a correlation between surface radiometry and topography, with low radiation readings corresponding to bedrock outcrops along ridges and in depressions. High radiation readings were recorded in areas with deep rich topsoil. Potassium-40 in clays is responsible for most soil radiation. If the outcropping rocks over the Sutherland Field were carbonates or sandstones, radiation lows over bedrock would be expected. Heemstra et al. (1979) encountered other problems in this and other radiation surveys they carried out. They found that truck radiation surveys did not correlate to airborne radiation surveys run over the same area. They also found that repetitive truck radiation measurements at the same locality showed large enough diurnal variations to be mistaken for anomalies. Heemstra et al. (1979, p. 80) concluded:

> Statements of fact in the literature indicate a relationship exists between petroleum at depth and observable gamma radiation anomalies at the surface and that this relationship can be correlated to light hydrocarbons, gas anomalies at the surface. From studies on the field data made thus far, however, neither of these stated facts can be substantiated. Although the presence of anomalies in the level of natural radiation and hydrocarbon gases at the Earth's surface was measured and observed in this study, their uniqueness to an oil reservoir has not been proven.

Horvitz (1950, p. 939) alluded to the existence of chimneys before Pirson's research: "During the migration of hydrocarbon gases from a petroleum deposit, the formations penetrated are believed to be modified chemically." However, Rosaire (1938) first described and documented chimneys, although he did not actually use the term. Rosaire (1938) noted that in processing geophysical data, a tacit assumption was made that lateral-lithologic variations of stratigraphically equivalent rocks were minimal. He believed, however, that sediments over some oil fields had abnormally high velocities because these sediments were more highly indurated than sediments over areas away from oil fields. Rosaire (1938, p. 101) believed that this induration was due to a greater degree of cementation or mineralization in sediments over hydrocarbon deposits [a direct contradiction to Pirson's (1960) doubtful hypothesis], and noted that induration of sediments over oil fields was apparent to rotary drillers of the early 1900s, a situation they described as "dome digging."

Rosaire (1938) pointed out that developmental drilling of some Gulf Coast oil fields showed that geophysical predictions of the structural relief of these fields were greatly exaggerated, and that the fields were actually in lower relief structures than originally thought, and some were in stratigraphic traps. There is no manner of gauging the amount of hydrocarbons discovered in low-relief structures and stratrigraphic traps because of misinterpretation of seismic data due to the lateral-lithologic inhomogeneity of equivalent-stratigraphic units (chimneys) created by bacterial oxidation of vertically migrating hydrocarbons. Whatever volume of reserves were discovered because of this induration feature, an indirect geochemical detection technique would have been responsible for the discovery (as in the case of paraffin dirt) of a much greater volume of hydrocarbon reserves than has been previously acknowledged.

The creation of mineralized chimneys over hydrocarbon deposits has largely been attributed to hydrocarbons being "chemically reduced compounds." By this thinking, as hydrocarbons migrate into or through an environment, they "oxidize" to create a reducing environment, which leads to substantial change in the mineralogy of the host sediments (Lundberg, 1952; Donovan, 1974, p. 432; Pirson, 1969, p. 179; Donovan and Dalziel, 1977, p. 1; Heemstra et al., 1979, p. 2; Duchscherer, 1980, p. 194, 1981B; and Sivenas and Beales, 1984). However, this is an erroneous hypothesis. *Hydrocarbons are kinetically stable* under all microseepage conditions, and in microbially sterile environments hydrocarbons will show absolutely no tendency to oxidation. Nature, however, is not microbially sterile and *many microbial forms oxidize hydrocarbons in utilizing them as a food source* in hydrocarbon reservoirs, in the sediment columns over reservoirs through which vertically migrating hydrocarbons pass, and in soil layers. Microbial hydrocarbon oxidation consumes either free oxygen (O_2) or chemically complexed oxygen (SO_4^{-2} or NO_3^{-2}, for example). In any case, the redox potential (eH) of the system is significantly affected. Perhaps the main product of microbial hydrocarbon oxidation is CO_2, which associates with water to form carbonate-biocarbonate species which drastically change the pH of the system. In addition to CO_2, sulfate-reducing bacteria also produce H_2S, which also significantly affects the pH of the system. Zobell (1946, p. 12) noted that bacterial activity also produces organic acids. Such pH/eH changes result in totally new mineral-stability fields. Some minerals become unstable and are dissolved and mobilized, and others are precipitated from solution. [Ed.'s note: see Matthews, this volume.] The rock column in which these acidic species are generated acts as a buffer to neutralize the acidic solutions. Such a neutralization (pH increase) takes place by the solution of different mineral species. The most apparent mineralogical change from the significant increase in the (neutralized) CO_2 concentration is the precipitation of various carbonates, particularly calcite. Davis (1967, p. 161) made a most relevant point concerning oxidation of crude oil by sulfate-reducing bacteria:

> There is evidence that the activity of sulfate-reducing bacteria in subsurface waters can affect the physical characteristics of the formation (Ashirov and Sazonova, 1962). Lithological examination of certain oil-bearing carbonate reservoirs have revealed that rock in the areas of oil-water contact has been made impermeable due to deposition of calcite (crystalline calcium carbonate). In at least one of these oil accumulations of the Ural-Volga area the pressure of the oil deposit dropped 900 lb./sq. inch relative to the water pressure without water entering the wells, the water being sealed off. Ashirov and Sazonova experimentally showed that sulfate-reducing bacteria in the presence of petroleum can reduce calcium sulfate to sulfide with deposition of calcium carbonate crystals in the cultures. The prismatic crystals formed (particle size, 0.01 mm or less) appear to be similar to the secondary calcite in the reservoir rock. Abd-El-Malek and Rizk (1963) recently reported the development of alkalinity and precipitation of carbonate due to bacterial sulfate reduction in water-logged soils.

The most intense microbial activity takes place in surface soils. When compared to microbial activity in soil layers, bacterial activity at greater depths in the proposed microfracture chimney above hydrocarbon deposits clearly would be greatly diminished. However, as Davis (1967, p. 244) noted, even the lowest levels of microbial activity have a cumulative effect over geologic time. The observations of Kartsev et al. (1959, p. 305) concerning the ability of bacteria to "... penetrate the finest cracks, pores, and fissures . . . ," has relevance here.

Bacterial oxidation of vertically migrating hydrocarbons also causes precipitation of calcite and other carbonates in surface and near-surface soils, in addition to the calcite precipitation in the sediment column above hydrocarbon deposits. Kartsev et al. (1959, p. 294–295) discussed "soil carbonate-siallite" surveys carried out in Russia. Such surveys are based on enrichments of calcite, silica, and alumina over leaking hydrocarbon deposits (Fig. 6). The high CO_2 concentrations from microbial hydrocarbon oxidation of vertically migrating hydrocarbons not only causes significant calcite precipitation, but, as Kartsev et al. (1959, p. 295) noted, the acidic solutions from the high CO_2 concentrations also alter or decompose clays (from the neutralization of these solutions) which results in increased concentrations of silica and alumina. Such clay alteration is most probably responsible for the radiation lows or halos reported over oil fields. Clay decomposition or alteration would liberate potassium from clays, including potassium-40, which, as Pirson (1969, p. 164), Heemstra et al. (1979) and Kilmar (1983) noted, is by far the major contributor to soil radioactivity. In addition, other radioactive elements would be freed and mobilized from clay lattices from the neutralization of carbonic acid. The shape of a radioactive anomaly would depend on the degree of mobilization of the radioactive elements. If these elements were transported entirely from the system, a "hole," or low, in the normal background readings would result. If the elements were transported to the edges of the vertical projection of the hydrocarbon deposit to the earth's surface, where they precipitated from solution, a "halo," or "doughnut," of high values surrounding the low values over the surface projection of the hydrocarbon deposit would result.

Pirson (1960, p. 95) reported that Lasky (1950) developed an indirect geochemical detection technique based on the "thermal alteration" of surface clay minerals from underlying hydrocarbon deposits. However, Lasky (1950) gave no details of the various surface geochemical exploration techniques he discussed. It is possible that Pirson's (1960) statement was based on personal knowledge of Lasky's clay-alteration techniques. Whatever the case, it is also possible that Lasky's (1950) technique was based on changes in clays due to the carbonic acid resulting from microbial oxidation of vertically migrating hydrocarbons.

The surface geochemical exploration techniques of Horvitz (1945, 1950, 1954, 1957, 1965, 1969, 1981) and Duchscherer (1980, 1981A, 1981B,

1983) are based on a significant development of "acid soluble soil salts" (mainly calcite) from CO_2 produced by the microbial oxidation of microseeping hydrocarbons. Different investigators have noted a strong correlation between acid-soluble soil salts and "sorbed" hydrocarbons (Hoffman, 1939; Horvitz, 1950, 1954, 1972; Colombo, 1959; Debnam, 1969; Poll, 1975; Devine and Sears, 1975, 1977; and Philp and Crisp, 1982). Heating soil samples to liberate "sorbed" hydrocarbons without acid treatment results in determinations of much lower hydrocarbon concentrations than if the samples are treated with acid and heated (Horvitz, 1950, 1954; Colombo, 1959; and Debnam, 1969). Horvitz (1954, p. 1207–1208), discussing the results of two different soil surveys over the Hastings Field, Brazoria County, Texas, one with acid treatment of samples and one with only heat treatment, noted:

> This survey indicates the importance of the acid treatment in the extraction procedure. In an earlier survey (Horvitz, 1939) of the Hastings Field conducted in 1937, prior to the introduction of this improvement, values of a much lower magnitude were obtained, and although an anomaly developed, a much less distinct pattern resulted. Apparently heating of samples without the acid treatment liberates only the loosely held hydrocarbons.

Horvitz (1950, 1969) passingly referred to hydrocarbons occluded, rather than sorbed, in the soil. From this and other evidence it is clear that the bulk of the hydrocarbons that Horvitz measures are not "sorbed" in or on the soil at all, but instead have been incorporated into calcite, and perhaps other minerals, which have been precipitated because of CO_2 buildup from the microbial oxidation of microseeping hydrocarbons. Such hydrocarbon occlusions are analogous to the fluid inclusions in ore minerals.

Duchscherer (1981A, p. 208) described his analytical technique as preheating soil samples in an oxygen atmosphere at 500°C to destroy organic matter and desorb loosely bound CO_2. The samples are then heated to 600°C, in a nitrogen atmosphere, which results in a pulse of CO_2 (which Duchscherer has termed "ΔC") being "generated" from the carbonates found in soil layers above oil deposits. As Duchscherer (1981A) noted, this CO_2 is not due to thermal decomposition of any known carbonate; and although the origin of this CO_2 is unknown, he noted that it is consistently found in soil carbonates above hydrocarbon deposits and is an excellent indicator of hydrocarbon deposits. Duchscherer's "ΔC" carbon dioxide may in part result from decrepitation of carbonates due to the thermal expansion of fluid inclusions or CO_2 otherwise occluded in soil carbonates. It is also possible that low undetected concentrations of hydrocarbon gases are also present in Duchscherer's "ΔC" CO_2. These are the hydrocarbons that are measured by Horvitz.

Geochemical anomalies (including soil-carbonate content, radiation lows, soil-mineral alteration, and occluded CO_2 in soil carbonate) caused by abnormal amounts of CO_2 *are nonspecific for thermogenic hydrocarbon deposits.* Abnormal CO_2 concentrations can result from factors other than microbial oxidation of microseeping hydrocarbons. Such factors could be geothermal, hydrothermal (ore mineralization), igneous, or metamorphic activity (especially the interaction of any of these with limestones or dolomites), or microseepage from a CO_2 deposit. In petroleum basins it is probable that nearly all such anomalies are caused by hydrocarbon deposits. This would not, however, be true in other geological provinces.

Other surface geochemical methods, resulting from calcite precipitation due to abnormal CO_2 concentrations in soils, are specific for hydrocarbon deposits. For example, because Horvitz measures the concentrations of C_2 to C_5 hydrocarbons, his method *is* specific for thermogenic hydrocarbon deposits. Donovan (1974) and co-workers (Donovan and Dalziel, 1977; Donovan et al., 1974, 1975, 1979A, 1981; and Henry and Donovan, 1978) studied cementation of near-surface rocks (mainly sandstones) by carbonates and the development of magnetite over oil fields (Donovan and Dalziel, 1977; Donovan et al., 1979A, 1979B, 1981; Henry and Donovan, 1978; and Dalziel and Donovan, 1984). Donovan and Dalziel (1977) explained the precipitation of such carbonate cements as the result of the "oxidation" of "microseeping" hydrocarbons resulting in the production of CO_2 which reacts with water to form the carbonate ion, which in turn causes the precipitation of various carbonate minerals. Anomalous C^{13}/C^{12} ratios in carbonate cements were correctly attributed to the fact that the carbon in the carbonate ions originated from vertically migrating hydrocarbons. The carbonate cements studied by Donovan and his co-workers, as well as crude oil and thermogenic natural gas, are depleted in the C^{13} isotope (Donovan and Dalziel, 1977). The precipitation of calcite in soils is not specific in itself for hydrocarbon deposits; however, the anomalous carbon isotope ratios which Donovan and his co-workers observed in these cements are specific for hydrocarbon deposits. Most of Donovan's and his co-workers' investigations were carried out with outcropping sandstones. As Kartsev et al. (1959, p. 319) noted, the possibility of microbial activity greatly increases with an increase in porosity.

Donovan and Dalziel (1977, p. 1) explained anomalous distributions of iron and manganese in surface sediments over hydrocarbon deposits as due to the fact that hydrocarbons (and associated H_2S) are reduced compounds and, "the passage of hydrocarbons and associated compounds such as hydrogen sulphide through surface rocks causes a reducing environment . . ." They proposed that this environment reduced iron and manganese to lower valence states, which resulted in the mobilization and removal of the elements, or precipitation of magnetite or maghemite, or co-precipitation of iron and/or manganese with calcite in carbonate cements over hydrocarbon deposits. Although H_2S is generated with petroleum in source rocks, the ubiquitous association of H_2S with crude oil is clearly due to sulfate-reducing bacteria oxidizing petroleum hydrocarbons after accumulation (Kartsev et al., 1959, p. 36, 310–311; and Davis, 1967, p. 157–194). The creation of the reducing environment called for by Donovan and Dalziel (1977) is clearly from microbial oxidation of vertically migrating hydrocarbons which depletes both free (O_2) and complexed (SO_4^{-2}, NO_3^{-2}, etc.) forms of oxygen. Magnetic anomalies from such reducing environments would indicate a hydrocarbon deposit only if the anomalies result from microbial oxidation of microseeping hydrocarbons. It is probable, however, that such anomalies can also result from microbial oxidation of vertically migrating hydrocarbons not directly related to a thermogenic hydrocarbon deposit. Therefore, some of these anomalies may be "false anomalies." Magnetite formation, or other iron-manganese anomalies, form the basis for a number of different electrical-geophysical exploration techniques, including, but not limited to, aeromagnetic detection (Donovan et al., 1974A, 1974B) and induced polarization (Sternberg and Oehler, 1984).

Biogenic Methane

Such methods as carbon isotopic measurement of carbonate cements, aeromagnetic detection, induced polarization, etc., are dependent on the interaction of microbes with vertically migrating hydrocarbons, and therefore can indicate a hydrocarbon deposit. Such a hydrocarbon deposit, however, is not necessarily a thermogenic hydrocarbon deposit. In a certain sense, therefore,it can be a type of false anomaly. Rice and Claypool (1981), in a definitive paper, discussed the causes of and constraints on the origins of biogenic methane, a byproduct of the attack of organic matter by obligate-anaerobic bacteria during shallow sediment burial and diagenesis. Although most biogenic methane is lost at the sediment-seawater interface during burial, enough is retained and later mobilized such that significant commercial accumulations can result. Such accumulations have been reported worldwide in many different petroleum basins and in many different types of traps (Rice, 1980; Rice and Claypool, 1981; Gautier, 1981; and Mattavelli et al., 1983), suggesting that biogenic methane accumulations can occur any place in a sedimentary basin in any type of trap. The contribution of biogenic methane to the world's natural gas reserves is much greater than generally recognized. Rice and Claypool (1981, p. 23) noted: "Gas of predominantly biogenic origin has made major contributions to the world's reserves (more than 20%). We predict that biogenic gas will make an even greater contribution to world energy needs in the future." Rice and Claypool (1981) are probably conservative in their estimate. For example, from analyses of 90% of the gas fields in the Po Valley Basin, Italy, Mattavelli et al. (1983) concluded that 80% of the gas in place was biogenic, 10% was thermogenic, and 10% was mixed. Mattavelli et al. (1983) documented biogenic gas fields in the basin as deep as 15,000 ft (4,500 m). Rice and Claypool (1981) estimated Soviet NW Siberian reserves alone at 372 TCF. Clearly, biogenic gas is more widespread and plays a greater energy role than is generally recognized.

Biogenic methane is a very "dry" natural gas with C_1/C_1-C_5 ratios always greater than 0.99 (Rice and Claypool, 1981). For example, the biogenic gas deposits of the Po Valley range from 99.40% to 99.97% CH_4 (Mattavelli et al., 1983, Table 1). Further, the bulk of the C_{2+} fraction is usually ethane. Deposits of biogenic methane, like all hydrocarbon deposits, are subject to leakage, which causes problems for certain indirect geochemical detection methods. Rice and Claypool (1981, p. 11) noted: "In the marine-shelf environment, the deposition of discontinous and relatively impermeable silt and sand enveloped by organic-rich mud and clay provides extremely favorable conditions for both in situ generation and entrapment of biogenic methane." Gautier (1981) described biogenic methane deposits of the northern Great Plains (United States) as occurring in such siltstone "microlenses" with good porosity (30% to 40%) but poor permeability (3 to 30 md). Because of poor interconnection of the silt lenses, the overall reservoir permeability of such deposits is less than 0.4 md. These (and other) reservoir characteristics make such accumulations next to impossible to recognize by electric log response, and fracture treatment is required for commercial production.

Microseepage from biogenic gas accumulations results in various anomalies such as magnetic, soil carbonate (including calcite-occluded CO_2), and radiation lows. Such anomalies would be identical to anomalies created by microseepage from a thermogenic hydrocarbon deposit. Drilling such an anomaly would fail to result in a commercial oil discovery. Also, because of the poor reservoir characteristics of some biogenic methane accumulations and the need for fracture treatment for production, even the existence of the biogenic methane might be missed. "Tight" biogenic methane accumulations probably have been responsible for a number of past surface geochemical exploration failures. Direct geochemical detection techniques which measure C_2-C_5 hydrocarbons are not subject to such a problem because of the lack of C_2-C_5 hydrocarbons in biogenic methane. It is possible that false anomalies caused by biogenic methane are partly responsible for the lower success rate reported by Duchscherer (1981B) for his "first class" anomalies, 38/170, or 22%, versus 23/39, or 59%, for Horvitz's (1954) "first class" anomalies.

Stahl and Faber (1981) described an offshore Santa Barbara Basin geochemical survey where two hundred sediment samples were taken over a 400 km^2 (249 mi^2) area and analyzed for sorbed methane. Isotopic analysis differentiated biogenic methane (being produced temporally) from thermogenic methane leaking from known oil fields. Thermogenic methane "anomalies" had concentrations up to 200 ppb versus biogenic methane concentrations of up to 50,000 ppb, or 250 times greater. The limited utility of methane, without expensive and difficult isotopic analyses, as an indicator of hydrocarbon deposits is clearly demonstrated by these data. [Ed.'s note: see Horvitz, this volume.]

Surface geochemical exploration techniques have significant potential as exploration tools in the search for oil in stratigraphic traps on the stable shelves of petroleum basins, especially maturely explored basins. Unfortunately, biogenic methane deposits appear to be abundant in these same areas (Rice and Claypool, 1981; Gautier, 1981; and Rice, 1984), and most often the deposits are in "tight" reservoirs. Explorationists using surface geochemical exploration methods should always be aware of the possibility of false anomalies caused by biogenic methane deposits on the normally oil-bearing stable shelves of sedimentary basins.

Halos

Before discussing the cause of halo anomalies found with soil-occluded hydrocarbon (and other) analyses, the association of the anomaly shapes found with each direct geochemical detection method will be reviewed. Such associations set the limiting conditions on halo formation. Kartsev et al. (1959, p. 101–106, 332–333) and Sokolov (1970) discussed the various anomaly shapes resulting from the vertical migration of leaking hydrocarbons. Such shapes are: apical (focal), annular (halo), and linear. Apical or focal anomalies are high readings directly over a hydrocarbon deposit and may be either continuous in the form of one large patch, or discontinuous (spotty) in the form of unconnected patches covering a given area. Halo anomalies have doughnut forms with a ring of high values surrounding the surface trace of a hydrocarbon deposit, and lower central and background values. Halos may demonstrate continuous doughnut forms, but more often they are discontinuous. Halo and apical anomalies apparently result mainly from hydrocarbon microseepage. Linear anomalies are found over surface traces of faults, fractures, or faulted-fractured, structurally intense anticlines. These linear anomalies involve C_{6+} hydrocarbons and result from effusion or vertical movement of deep-basinal water carrying significant amounts of dissolved hydrocarbons. Once such hydrocarbons reach the near-surface environment, they are subject to dispersion through the diffusive action of ground waters. An anomaly resulting from such a dispersion would be diffuse and difficult to distinguish from an apical anomaly.

As many investigators have noted, halo anomalies are the most common form found with soil-occluded hydrocarbon analysis (Horvitz, 1950, p. 948–949; 1954, p. 1207; 1957; 1969; 1980, p. 245; 1981; Soli, 1957, p. 134; and Siegel, 1974, p. 235–237). The anomaly shape associated with soil-sorbed hydrocarbon analysis is variable. Kartsev et al. (1959, p. 132) stated that soil-sorbed hydrocarbon anomalies take the same form as soil-air hydrocarbon anomalies (discontinuous or continuous apical or halo anomalies). Early Western work, carried out before the advent of acid treatment of soils (for soil-occluded hydrocarbon analysis), reported mainly halo anomalies (Rosaire, 1938, 1940; Horvitz, 1939; McDermott, 1940; Pirson, 1940; and Rosaire et al., 1940), although apical anomalies were reported to a lesser extent (Rosaire, 1938, Fig. 6; see Fig. 10, this paper). By this paper's model, both apical and halo anomalies could be present anywhere with soil-sorbed hydrocarbon analysis; however, in areas of high soil humidity (such as the Gulf Coast, where most early Western work was carried out) halos would predominate. Kartsev et al. (1959, p. 101–105) reported that discontinuous or continuous apical or halo anomalies are found with soil-air hydrocarbon analysis, although only apical anomalies were observed in the earliest investigations (Laubmeyer, 1933; and Sokolov, 1936). A number of investigators have stated that soil-air hydrocarbon analysis gives only, or mainly, halo anomalies (Horvitz, 1939, p. 215; 1950, p. 942; Pirson, 1940, p. 1471; 1941, p. 32; 1969; Soli, 1957, p. 134; Baijal, 1962; and Jones and Drozd, 1983, p. 949). By this paper's model, soil-air hydrocarbon analysis could result in either halo or apical anomalies. Microbiologic prospecting always yields apical anomalies. The rare halo anomalies observed with this method were associated with large faults (Kartsev et al., 1959). As Siegel (1974, p 239) noted:

> It is noteworthy that several other factors that have been investigated as potential guides in the search for hydrocarbons also present anomalies in halolike forms (radioactivity, electrical conductance, and/or resistivity, and other), except for microbial density, which presents a full target as an anomaly.

Of all the hypotheses advanced to explain the celebrated halo anomaly found with soil-occluded hydrocarbon analysis (and other surface geochemical exploration methods), the most popular (Fig. 7) is that the sediments (caprock) over the hydrocarbon deposit have become plugged as a result of mineralization. This forms an indurated seal, forcing hydrocarbon microseepage to occur only at the edges of the deposit where such mineralization has not taken place (Rosaire, 1940, p. 1427). Two lines of evidence show that this theory cannot be correct. First, if this were the case, analyses for soil-air and soil-sorbed hydrocarbons, and microbiological surveys would not yield apical anomalies, but would only yield halo anomalies. The existence of apical anomalies are incontestable evidence that hydrocarbon microseepage is occurring directly over the hydrocarbon deposit and not solely on the edges of the deposit. The second line of evidence against the "plugging theory" is in the form of hydrocarbon-gas logging, wherein hydrocarbon-gas concentrations are measured as a function of depth down either exploration wells or deep-reconnaissance bore holes such as the Soviets use. In bore holes directly (apically) over hydrocarbon deposits (Fig. 9), hydrocarbon-gas concentrations continually increase and reach high values just over the hydrocarbon-bearing zone irrespective of the hydrocarbon anomaly shape at the earth's surface (Rosaire, 1940; Horvitz, 1950, p. 959; Sokolov et al., 1959; Stroganov, 1969; Sokolov, 1970, p. 546; and Thompson, 1979, p. 668). In fact, Stroganov (1969) specifically carried out hydrocarbon-gas logging investigations of surface halo anomalies, and uniformly found high hydrocarbon concentrations below such anomalies. These observations show that halo-anomaly formation cannot be due to caprock plugging, but instead must be due to a near-surface effect.

In the author's opinion, Soli (1957) correctly explained the cause of halo anomalies observed with soil-air and soil-sorbed hydrocarbon analysis. This explanation has largely gone ignored except for a reference by Davis (1967, p. 212–214), and MacElvain's (1963A, 1963B) discussion supporting Soli's hypothesis. Soli (1957) hypothesized that intense bacterial activity in the uppermost soil layers destroyed most of the vertically migrating hydrocarbons over the center of the surface trace of a hydrocarbon deposit. As one moved laterally off the center of the deposit, the concentration of vertically migrating hydrocarbons would decrease to values too low to support viable bacterial growth but still higher than the background values (Fig. 9). Kartsev (1959, p. 319) noted that a minimal threshold concentration of surface hydrocarbons must be exceeded before a viable microbial population can develop, which supports Soli's (1957) hypothesis. Soli's (1957) explanation accounts for halos reported for both soil-sorbed and soil-air hydrocarbon analyses, but does not fully explain the halos found with soil-occluded hydrocarbon analysis.

The origins of halos reported for soil-occluded hydrocarbon analysis, Duchscherer's (1980, 1981A, 1981B, 1983) ΔC technique, the soil-carbonate halos of Kartsev et al. (1959, Fig. 127; Fig. 6, this paper), and the "mineralization" halos of McDermott (1940), are all interrelated. In the environment above a hydrocarbon deposit, only a limited amount of Ca^{+2} ions are available to precipitate as calcite from the CO_2 concentration buildup from microbial oxidation of vertically migrating hydrocarbons. After precipitation of this limited amount of Ca^{+2} ions, the system will be depleted in Ca^{+2} ions. The only possibility for further calcite precipitation lies in replenishment of Ca^{+2} ions to the system. This occurs either through physical transport by sediment waters or by diffusion of Ca^{+2} ions due to the concentration gradient established by the original calcite precipitation. In any case, such Ca^{+2} ions moving toward the system will encounter elevated concentrations of CO_2, HCO_3^-, and $CO_3^=$ diffusing away from the apical area of intense microbial activity. Calcite would thus be precipitated at this ionic meeting at the edges of the apical area before the new Ca^{+2} ions had a chance to migrate to the center of the apical area, thereby creating the various halo anomalies associated with this precipitated calcite.

Soil-occluded hydrocarbon analysis should almost always yield halo anomalies by the above model. However, soil-sorbed hydrocarbon analysis and soil-air hydrocarbon analysis could yield either halo or apical anomalies, depending on the intensity of microbial activity. In fact, because microbial activity can be seasonal, soil-sorbed hydrocarbon analysis or soil-air hydrocarbon analysis could at one time of the year yield a halo anomaly and at another time yield an apical anomaly. Also, a strong anomaly at one time of the year could entirely disappear at another time of the year. Philp and Crisp (1982, p. 5–6) described such an instance in soil-air hydrocarbon surveys they carried out in the Cooper basin, Australia:

> The results of this preliminary study were encouraging, and anomalously high concentrations of C_2 to C_4 hydrocarbons were detected in an area adjacent to a fault previously determined by a seismic survey. The frustrations of this method of prospecting were made abundantly clear one year later, when another survey was performed in the same region and the anomaly could no longer be detected.

Seasonal variation of both anomaly shape and intensity (Fig. 10) was also observed by Rosaire (1938) in soil-sorbed ethane surveys run over the Hastings Oil Field, Brazoria County, Texas, in May and September, 1937, where in spite of increased field development, anomaly intensity increased, and anomaly shape changed from halo to apical. The acid extraction treatment, incidentally, was not used as early as 1937. By coincidence, Horvitz (1969) reported a disappearing anomaly at Hastings due to reservoir depletion from production. Kartsev (1959, p. 83–86, 120, 302) discussed seasonal fluctuations of both soil-sorbed and soil-air hydrocarbons and noted that the concentration of such hydrocarbons vary throughout the year and sometimes vanish entirely. Such seasonal variations and/or disappearances of soil-sorbed or soil-air hydrocarbons is in all probability due to seasonal variation in the intensity of the microbial filter.

DIRECT GEOCHEMICAL DETECTION METHODS—PROS AND CONS

Because of the possibility of false anomalies with some surface geochemical exploration methods, only direct geochemical detection methods which measure C_2-C_5 hydrocarbon concentrations (unassociated with C_{6+} hydrocarbons), either relatively or absolutely, are specific for thermogenic hydrocarbon deposits. Such methods are: soil analysis for occluded and/or sorbed hydrocarbons, soil-air analysis, microbiologic analysis, and integrative absorption.

Depending on the climatological environment, all five methods may show large seasonal variations, and/or not be applicable at any time in a given area.

The first three methods require the drilling of deep bores (*at least* 6 ft, 1.8 m), which results in long field times and high field expenses. Further, the need for a truck-mounted auger drastically decreases field mobility. For the first four methods, sampling and sample analyses can be time consuming, difficult, and expensive. Other problems also exist individually for all five methods.

Soil-Air Hydrocarbons

The low concentrations of C_2-C_5 hydrocarbons in soil-air make soil-air sampling and analysis both difficult and open to large errors. Diurnal breathing of soil, and especially climatological variations (wind, rain, heat, cold, snow, barometric pressure, and humidity) which make soil "breathe," put *severe* limitations on soil-air surveying. Large regional surveys cannot be carried out under equal meteorological or diurnal conditions, and data from different sample localities cannot be directly compared to each other. See Kartsev et al. (1959, p. 72–73) for a discussion of soil "breathing." Intense bacterial activity can obliterate all soil-air hydrocarbons. Kartsev et al. (1959, p. 85–87) discussed other problems with soil-air analysis. The lack of published results in the Western literature on soil-air analysis suggests that it is not an easily applicable technique. Although Baijal's (1962) research is sometimes referred to as a successful example of soil-air analysis, this is not the case as he incorrectly contoured his data (Fig. 11). Kartsev et al. (1959, p. 64–106) provided an excellent discussion of soil-air surveying, and on page 106 ended with the statement: "In conclusion it should be noted that the gas survey can be perfected much further and is in need of serious investigations of a theoretical character."

Soil-Sorbed Hydrocarbons

Soil-sorbed hydrocarbon analysis is also beset with problems. A detailed discussion of sorption (adsorption, absorption, and chemisorption) is certainly outside the scope of this paper. To this day, sorption is still one of the more or less "black art" areas of physical chemistry. Sorption under controlled laboratory and/or industrial applications is quite complex. In nature, where additional variables are present, defining sorption becomes impossible. Kartsev et al. (1959, p. 109–115) discussed sorption in soils. For more rigid physical chemical treatments, see McBain, 1932; Prutton and Maron, 1951, p. 226–239; Bikerman, 1958; Crockford and Knight, 1959, p. 382–392; Barrow, 1961, p. 753–775; Moore, 1962, p. 740–753; and Vold and Vold, 1983, p. 71–114). The exact mechanism or mechanisms of sorption and (more importantly) desorption during analysis, of microseeping hydrocarbons has not been defined and probably never will be. Soil is highly complex and its hydrocarbon sorption capacity can be highly variable even over limited lateral distances (Sanderson, 1940; Bronston, 1947; Kartsev et al., 1959, p. 112–114; Sokolov et al., 1971; Horvitz, 1980; and Klusman and Voorhees, 1983). For example, Kartsev et al. (1959, p. 114) noted that "soil, rich in organic matter, absorbs nearly 50 times as much methane as clay that has no trace of organic matter." Kartsev et al. (1959, p. 117–121) also noted that as a result of inhomogeneity of sorption characteristics in soils, even sample splits from the same core can have much different sorbed-hydrocarbon concentrations. Devine and Sears (1975) reported poor repeatability of twenty-eight duplicate samples taken 1 m (0.30 ft) apart in their study.

Sanderson (1940) discussed factors which could affect hydrocarbon sorption in soils, including the surface, physical shape, and chemical composition of the sorbing particles, as well as the amounts and nature of materials previously sorbed onto the particles, such as H_2O, CO_2, and mineral ions; not to mention the various soil gases generated by plant and microbial activity. All these entities will compete for active soil sorption sites with microseeping hydrocarbons. It has been stated that variations in soil sorption capacities can be avoided by sieving soil and working only with finer sieve sizes, which purportedly normalizes the soil (Pirson, 1940; and Horvitz, 1980). However, Pirson (1940, p. 1446) noted:

> Measurements, reported to the writer by C. F. Bowden (personal communication), made on samples of clays of the same crystalline composition, but of varying surface chemical nature, indicate that one clay having a surface at least 10 times smaller than another retained twice as much ethane, per equal weight of clay

sample, when brought in intimate contact with a sample of air containing a given proportion of ethane. This is a direct contradiction with the expected results and gives ample proof that the mechanism of retention of hydrocarbons by soil depends more on the chemical nature of soils than on their degree of subdivision, grain size, exposed grain surface, and other physical characteristics. From this example, it appears that the particle size plays a secondary role in retaining the gases of leakage in the soil and that correction methods based solely on particle size will yield nondependable results.

Another concern with soil-sorbed hydrocarbon analysis is the probable microbial attack, in moist samples, of the sorbed hydrocarbons after the sample has been taken. This possibility has also been noted by Rosaire et al. (1940. p. 1461) and Soli (1957) and could be avoided by freezing samples or by the use of bactericide in the samples after they were taken. Smith and Ellis (1963, p. 1900) found a marked bacterial oxidation of hydrocarbons in 200 g soil samples they took in sealed tin cans in their study:

A series of tests showed that a rapid decrease in saturated hydrocarbon gases occurred in some soil samples when stored at 20°–35°C. Tests showed this activity was stopped by cooling to 0°C or pasteurizing by heating to more than 60°C.

This issue would not be present in analysis for soil-occluded hydrocarbons, as the hydrocarbons are "sealed" in protective carbonates. The concentration of lightly sorbed hydrocarbons, like soil-air hydrocarbons, are most likely highly influenced by factors (diurnal, seasonal, and climatological) which cause soil to breathe. In other words, such surveys may not be reproducible. Contrary to the claims of some investigators (Rosaire, 1940), sorbed hydrocarbons, unlike occluded hydrocarbons, are not a signal which has necessarily been integrated over long time periods or even over geologic time.

Perhaps the biggest problem with soil-sorbed hydrocarbon surveys is in *quantitative* desorption of hydrocarbons during analysis. Kartsev et al. (1959, p. 122–131) discussed the analytical techniques employed in the Soviet Union for analysis of soil-sorbed hydrocarbons. As of 1954, the original publication date in Russian of Kartsev et al. (1959), Soviet investigators were not using acid treatment to liberate occluded hydrocarbons. The Soviet analytical techniques are based on thermal or vacuum desorption or combinations of the two. Kartsev et al. (1959) discussed the impossibility of assuring that all hydrocarbons sorbed in soils have been released during heating (a point previously addressed by Sanderson, 1940, p. 287).

Desorption temperatures must be kept below 200°C, as temperatures above 200°C result in the generation of measurable amounts of C_{2+} hydrocarbon gases from indigenous soil organic matter (unpublished hydrous-pyrolysis data, L. C. Price). As Kartsev et al. (1959, p. 122) noted: "Overheating of the rock during gas extraction leads to the formation of new gases, previously not present in the rock. Therefore, it was thought that heating above 100°C in order to evolve sorbed hydrocarbon gases was undesirable." This author's experiments have shown that even if soil samples with sorbed hydrocarbons are heated for periods of ten to twenty-four hours, to the point that no more gas is evolved (even under vacuum), the addition of another chemical to the system will result in the release of more hydrocarbon gas. The added chemical, dichloromethane, competes effectively for active sites in the soil, thus displacing hydrocarbon gas not evolved earlier. (For an analogous situation with activated charcoal, see McBain, 1932, p. 92–95.) Quite simply, quantitative desorption of soil-sorbed hydrocarbons is not possible using only mild heating (less than 200°C) even under vacuum. All one can do is keep analytical conditions constant, and hope that lithological variation between samples is not so great as to affect significantly the outcome of the survey, an unlikely situation.

Soil-Occluded Hydrocarbons

Many of the problems associated with soil-sorbed hydrocarbon analysis are removed by acid treatment in the analysis for soil-occluded hydrocarbons. Acid treatment of soil with complete dissolution of all "acid-soluble salts" (hereafter referred to as carbonates) results in quantitative liberation of occluded hydrocarbons. In addition, the acid ions would likely compete for active clay sorption sites, perhaps also resulting in quantitative or near quantitative liberation of sorbed hydrocarbons. In some geologic situations, the amount of occluded hydrocarbons could be much greater than the amount of sorbed hydrocarbons. In such cases it may not matter if acid treatment did not quantitatively liberate sorbed hydrocarbons. Soil-occluded hydrocarbon surveys do not show the seasonal variation (resulting from variable intensity of microbial activity) which soil-air and soil-sorbed hydrocarbon surveys show, as soil carbonates seal and protect occluded hydrocarbons from microbial destruction (L. Horvitz, personal communication, April 17, 1985). In cases, however, where the concentration of soil-sorbed hydrocarbons was greater than or approximated that of the soil-occluded hydrocarbons, quantitative desorbtion of soil-sorbed hydrocarbons during occluded hydrocarbon analysis could become important. In such cases, seasonal variations would be possible with soil-occluded surveys. Research into the relative amounts of sorbed versus occluded hydrocarbons in different geologic and geographic-climatological environments would be apropriate.

By far the greatest amount of data on soil-occluded hydrocarbon surveys have come from the Florida to Texas Gulf Coast. The published anomalies from this area are impressive; however, attempted applications of soil-occluded hydrocarbon analysis in other geographic areas have demonstrated limitations of the technique. McCrossan et al. (1972) carried out an extensive program over 540 mi² (1,399 km²), taking 4,561 samples of glacial soil at a depth of 5 ft (1.5 m) from the oil-productive Olds-Caroline area of western Alberta. Analysis for occluded hydrocarbons was carried out by acid treatment of the soil samples under vacuum. The authors found no correlation between proven oil production and surface hydrocarbon anomalies. Subsequent research led them to conclude that most of the measured hydrocarbons were syngenetic, originating from carbonate glacial rock fragments eroded from the foothills of the Canadian Rockies to the west. (Before uplift and erosion, the sediments of that area of western Canada had achieved a high organic-maturation rank, with in situ generation of large amounts of hydrocarbon gases.) The authors concluded that direct geochemical detection techniques for soil hydrocarbons were not applicable in areas of Canada covered by glacial till.

Kartsev et al. (1959, p. 122, 103) briefly alluded to problems caused by syngenetic gas. It is widely believed that substantial amounts of hydrocarbon gases are only formed at high maturation ranks. By this belief, syngenetic hydrocarbon gases would only be a problem where carbonate-bearing rock (including carbonate-bearing shales) of mature organic rank were present in the soil. Research demonstrates, however, that significant generation of C_1 to C_7 hydrocarbons takes place at relatively low maturation rank (Price, 1986).

Another example was given by Heemstra et al. (1979) of the confusion and/or errors which can be caused by syngenetic hydrocarbon gases in carbonate-bearing rock, even of low organic-maturation rank, during soil-occluded hydrocarbon analysis by acid treatment. Heemestra et al. designed research to test the validity of surface geochemical exploration by drilling exploratory wells based on surface geochemical anomalies. They performed soil-occluded hydrocarbon analysis ("gas logging") on the cuttings chips from such a well drilled to 2,000 ft (610 m) in Greenwood County, Kansas [see Horvitz (1980, p. 260–262) for a detailed discussion of this technique]. The rocks in that area of the Anadarko basin are organically immature. The hydrocarbon gas logging analyses which Heemstra et al. (1979) carried out showed a strong correlation between high concentrations of "absorbed" hydrocarbon gases with limestones. Heemstra et al. (1979) were at a loss to explain this correlation. That correlation was most probably due to acid treatment dissolving organically immature limestones and freeing low concentrations of syngenetic hydrocarbon gases, which overwhelmed sorbed-occluded hydrocarbon gases from microseepage. In order to avoid any problems of syngenetic gases in analysis for soil-occluded hydrocarbons, it may be necessary to run surveys only in reworked, weathered Quaternary soils.

Debnam (1969) ran soil-occluded hydrocarbon survey traverses across six Canadian oil fields: Gobles, Colchester, and Morpeth in southwestern Ontario; and Innisfail, Pincher Creek, and Joffre Viking in Alberta. Samples were taken from 9 ft (2.4 m) and acid extracted with both 0.05N HCl and 2N HCl. A moderate anomaly (hydrocarbons normalized to acid soluble material) was obtained at Gobles, with the 0.05N HCl acid extraction. However, the anomaly disappeared with 2N HCl acid treatment. The Colchester field showed no anomaly. At Innisfail, Debnam (1969, p. 20) called for "... a pronounced hydrocarbon anomaly, probably of the 'halo' type ..." In this author's opinion no anomaly was evident at Innisfail, from Debnam's (1969) data. Debnam (1969) found no anomaly at Pincher

Creek and did not discuss the data for the other two fields most probably because of poor results. Debnam (1969, p. 1) concluded:

> Geochemical anomalies appear to be associated with all the fields studied with anomaly intensity roughly proportional to productivity. Each of the "halo" and "solid" types of anomalies can be recognized.
>
> The writer concludes that the geochemical method can be a valuable addition to the usual geological and geophysical exploration methods if used under favourable circumstances.

Debnam's (1969) conclusions are unjustified considering: (1) the anomalies changed as a function of acid strength, (2) there was the alternate citing of halo or apical anomalies for soil-occluded hydrocarbon analysis, and (3) there was the lack of sharp anomalies of either type in most of the data. If anything, Debnam (1969) succeeded in demonstrating the inapplicability of soil-occluded hydrocarbon analysis in his study areas.

Devine and Sears (1975, 1977) carried out a study in the Cooper basin, Australia, centered around the Della gas field in the Strzelecki desert, a region which receives less than 100 mm (4 in) rain per year. Their study was characterized by difficult analyses and an "extremely noisy range" of low hydrocarbon concentrations. In the first phase of their study, they took 379 samples from 5 m (16.4 ft) for soil-occluded hydrocarbon analysis. Of these samples, only eight had hydrocarbon concentrations greater than 14 ppb, and only two greater than 20 ppb. The authors carried out a rigorous statistical analysis (not usually the case in surface geochemical exploration methods) that showed soil-occluded hydrocarbon concentrations exhibited, at best, a moderately poor correlation with the Della gas field. The correlation of soil-occluded hydrocarbons to the field was improved by normalizing hydrocarbon concentrations to the percentage carbonate ("acid soluble material") for each sample. This resulted in a vague, discontinuous apical anomaly.

The studies of Debnam (1969), McCrossan et al. (1972) and Devine and Sears (1975, 1977) document that soil-occluded hydrocarbon analysis may not be universally applicable. The technique appears to be of limited use in true deserts, and situations where noise from syngenetic hydrocarbons in soils may overwhelm signals generated from hydrocarbon microseepage. However, when applied in favorable geographical/climatological conditions, proper use of soil-occluded hydrocarbon analysis appears to give very favorable results. The technique clearly is superior to soil-air and soil-sorbed hydrocarbon analyses.

Microbiologic Surveys

Microbiologic surveys attempt to measure surface and near-surface microbial activity caused specifically by microseeping hydrocarbons. The Soviet investigator Mogilevskii (1938, 1940) is clearly the father of this technique. Although the Soviets appear to have had marked success with this method, little Western work has been published, the most authoritative of which is Davis (1967, 1969). Impressive exploration results from microbial surveys have been published. Kartsev et al. (1959, p. 333, 340) discussed a number of microbial-survey case histories. Mogilevskii (1953) claimed an 80% predictive success rate for microbial surveys, but Subbota (1954) maintained not all of these predictions were made on the basis of microbial surveying alone. Sealey (1974A) claimed an 82% prediction success, and with "improved techniques" and 85.7% prediciton success using microbial surveys. However, in the true spirit of Western surface geochemical exploration, Sealey (1974A, 1974B) disclosed absolutely no information on his methods. If 80% predictive capabilities for microbial surveys are realities, it is difficult to understand why the method has not been more extensively employed in Western petroleum exploration.

Kartsev et al. (1959, p. 319) discussed the four main types of microbial surveys carried out in the Soviet Union: (1) microbial-soil surveys, (2) microbial surveys of cuttings chips from exploration or reconnaissance wells, (3) microbial surveys on core samples ("biologging"), and (4) groundwater microbial surveys. Although as of 1954 microbial groundwater surveying was the most widely utilized method in the Soviet Union, only microbial-soil surveys are discussed here. Contradictions exist in the published literature regarding microbial-soil surveying. Mogilevskii (1940) adopted the use of soil surveying for methane-oxidizing bacteria. Western investigators have roundly rejected the use of methane-oxidizing bacteria, because of the non-specificity of methane for hydrocarbon deposits. Mogilevskii attempted to avoid this problem by deep sampling [at least 1.5 m (4.9 ft), and preferably much deeper] to avoid the zone where cellulose-decomposing bacteria (which produce biogenic methane) were active. He also ran concurrent analyses for cellulose-decomposing bacteria, as significant populations of methane-oxidizing bacteria in the absence of cellulose-decomposing bacteria would imply high methane microseepage from hydrocarbon deposits. However, contradictions even exist concerning the utility of bacteria specific for C_{2+} hydrocarbons. Kartsev et al. (1959, p. 314) noted:

> However, in the examination *en masse* of ground samples from known oil-bearing regions, ethane-oxidizing bacteria are found relatively rarely. It may be that these bacteria require special conditions for development. For these reasons ethane-oxidizing bacteria are not now used as indicating organisms in oil-prospecting work.

Thus, as Davis (1969, p. 144) noted, the obvious choice of Soviet investigators carrying out microbial surveying is the use of propane-oxidizing bacteria (*Mycobacterium rubrum* var. *propanicum*). The Soviet research in this area largely has been carried out by E. N. Bokova and S. I. Kuznetsova. Contrary to the Soviet experience, Davis (1967, 1969) reported no problem in using ethane-oxidizing bacteria (*Mycobacterium perrugosum* var. *ethanicum*) in his extensive research on microbial surveying. However, on the basis of his research using ethane-oxidizing bacteria, Davis (1969, p. 154) reached far different conclusions regarding hydrocarbon microseepage than most other investigators in the field using many different surface geochemical exploration methods dependent on hydrocarbon microseepage:

> Thus, we have learned that natural gas seepage over oil accumulations is far from invariable, both from an areal and a frequency standpoint. We do not believe that a "blanket emanation" of petroleum gas in soils over petroleum accumulations often exists. In fact, evidence for microseeps suggests that the surface areal diameter of such seeps probably is of the order of 100 ft or less.

Davis also expressed general negative conclusions about surface geochemical exploration in Meakin (1971, p. 128). In light of Davis's conclusions, the Soviet opinion of the utility of ethane-oxidizing bacteria in microbial soil surveying would appear valid. On the other hand, Miller (1976) reported spectacular results for microbial soil surveys using ethane-oxidizing bacteria over producing oil fields (Glenburn, North Dakota; Recluse and Belle Creek, Wyoming; and Hyde, Colorado). However, Miller's (1976) results can be explained by a most disturbing conclusion drawn by Davis (1969, p. 153–154):

> Our methods for detecting and enumerating hydrocarbon-oxidizing bacteria were developed by using soil samples from oilfields, and the samples were taken at depths which ordinarily precluded surface contamination by hydrocarbons. However, gas seepage from oil well boreholes can permeate soils overlying oilfields. The invariability of finding bacteria adapted to the oxidation of ethane over established fields, in contrast to a marked variability in their presence in soils over newly productive areas, strongly suggests that gas seepage often occurs over an oil accumulation only after the oil reservoir is punctured by wells.

Davis's (1969) conclusions in this matter, if correct, would have staggering consequences for surface geochemical exploration, as clearly the best published examples of geochemical anomalies through the years have come from surveys over producing fields. It is likely that there may be at least some validity to Davis's (1969) point.

There is no uniform agreement on the correct sampling depth for microbial-soil surveys. Strawinski (1955) recommended a depth of 0.61 m (2 ft). Kartsev et al. (1959) recommended minimum depths of 2 to 3 m (6.6 to 9.8 ft). Most of the surveys they discuss apparently were carried out over the 2 to 4 m (6.6 to 13.2 ft) range, although some were carried out as deep as 15 to 30 m (49.2 to 98.4 ft). Völz and Schwartz (1958, 1962) called for a depth of 1.5 to 2.5 m (4.9 to 8.2 ft). Davis (1967) advised depths of a least 1 to 1.5 ft (0.30 to 0.46 m), although he believed depths greater than 3 ft (0.92 m) are better. Apparently he carried out much of his research at depths of 1 to 15 ft (0.30 to 4.6 m). Miller (1976) advised sampling depths of 2 to 3 ft (0.61 to 0.92 m). It is likely that sampling depth is one of the critical controlling parameters of microbial-soil surveying, and different sampling depths in different studies could lead to divergent results and conclusions. The most intense microbial activity clearly takes place in the uppermost soil layers, and it is here that one would suppose the largest bacterial populations would develop in response to microseeping hydrocarbons. In would seem that intense bacterial activity at depths in excess of 2 m (6.6 ft) would be limited by atmospheric exchange with the soil-air to replenish the necessary oxygen.

In addition to the apparent ill-defined nature of microbial soil surveying, serious sampling and analytical problems also exist. Three analytical techniques have been generally utilized: manometric (gas-uptake), visual, and microscopic. With the first two analytical techniques, a given amount of nutrient solution is added to a fixed amount of soil in a suitable small reaction vessel, resulting in a slurry. In a closed system, as Davis (1969, p. 148) noted, bacterial activity results in a decrease in system gas pressure as more gas is utilized than produced. For example, with ethane, Davis (1969) wrote the reaction as:

$$C_2H_6 + 2.5\ O_2 \rightarrow CO_2 + (CH_2O) + 2H_2O,$$

where (CH_2O) is bacterial cell substance. Such a decrease in system gas pressure, which is directly related to the intensity of bacterial activity, can be monitored with a manometer attached to the reaction vessel. For example, over a given time interval, a significant population of ethane oxidizers in a sample will result in a much greater pressure decrease in a system where ethane is the food source, than in the same system with a small original population of ethane oxidizers. The visual analytical technique involves an estimate of the volume of a distinctive bacterial film which forms in a reaction vessel from bacterial activity on a given food source. This volume is calculated by multiplication of the estimated film thickness by the estimated area of the vessel that the film covers. As of 1954, this was the analytical technique exclusively employed by the Soviets (Kartsev et al., 1959). The microscopic (or plating) analytical technique consists of spreading small amounts of soil as thinly as possible on mineral agar plates and allowing bacterial colonies to form in an air and ethane (or other hydrocarbon) atmosphere. After a sufficient time, the colonies are microscopically examined and classified. Davis (1967, p. 234–245) discussed other unsatisfactory analytical techniques he attempted and other less widely employed analytical techniques.

Common problems to all three analytical methods are: sampling, laboratory adaptation to food sources by different microbes, and incubation times. Incubation times for measurable reactions apparently run fifteen to thirty days, thus preventing quick analyses. As Davis (1967, p. 220–221) noted, sampling presents the biggest problem, as soil within the same sample can vary considerably in texture and moisture content; and indigenous hydrocarbon oxidizers are not uniformly distributed throughout soil, but exist as isolated colonies. Attempts to disperse evenly such colonies by homogenization by drying the soil followed by grinding or sieving results in a promotion of actinomycetes, which can flourish under dry conditions. During the subsequent incubation period, actinomycetes then adapt to the hydrocarbon gas food source, thus negating the manometric analysis. Reproductibility in sample analysis is also of prime concern, as Davis (1967, p. 221) stated:

> Actually, the preferred method of treating soil to be assayed for ethane-oxidizing bateria involves a minimum of handling and analysis as soon after collection as possible. Lack of reproducibility in the assay of a given sample can be compensated by averaging the results of replicate analyses of the soil sample.

In large areal surveys, significant time is spent in the field before the samples are returned to a laboratory for preparation and incubation. Thus, samples taken at the first part of the survey will likely suffer a pronounced decrease in bacterial activity, compared to samples taken at the end of the survey. Such samples from the same survey would not be directly comparable.

Davis (1967, p. 227) raised a very serious question concerning microbial adaptation to C_2-C_5 hydrocarbon consumption during laboratory incubation. The whole purpose of microbial-soil surveying is to measure microbial populations which specifically utilize C_2-C_5 hydrocarbons from microseepage or which have adapted to the use of C_2-C_5 hydrocarbons in the field. Strawinski and Tortorvich (1955) described a case wherein methane-oxidizing bacteria from a hydrocarbon microseep had adapted to the utilization of ethane, propane, and butane; however, the same bacteria laterally off the seep could only utilize methane. Davis (1967, p. 227) has noted:

> Microorganisms exist, however, which can adapt rapidly by means of their versatile enzymic systems to utilize a large variety of compounds. These organisms will therefore utilize ethane in the laboratory without having previously been associated with it in the field.

Davis (1967) noted that actinomycetes and molds are very adaptive in the laboratory; however, because they are morphologically distinct, they can be easily identified under the microscope. Such adaptive capability renders manometric (gas-uptake) analytical results untrustworthy, especially considering the long required incubation times. It also requires microscopic examination of the incubated samples by personnel trained in microbiology. This significantly increases analysis time and expense.

There are obviously major problems in microbial soil surveying. Microbial oxidation of microseeping hydrocarbons clearly plays a pivotal role in all direct geochemical detection methods (and surface geochemical exploration in general), yet no easily employed, inexpensive, and trustworthy method is available to measure anomalous surface microbial activity due to hydrocarbon microseepage. Davis (1967, 1969) clearly raised disturbing questions concerning microbial-soil surveying and surface geochemical exploration in general. Davis's questions are perhaps meaningless, since they could well be due to Davis's exclusive employment of ethane-oxidizing bacteria in all his investigations, a tool the Soviets clearly found unsatisfactory. Davis's (1967, p. 207; 1969, p. 144) cursory dismissal of the contradiction between his and the Soviet findings as a result of the "possible" lack of pure ethane in the Soviet Union is unsatisfactory and uncharacteristic of the rigorous approach that the Soviets have taken regarding surface geochemical exploration.

Integrative Absorption

Integrative absorption is the use of absorbents (charcoal, silica gel, etc) or any other "collectors," in shallow (1 to 2 ft, 0.3 to 0.6 m) surface holes to measure the relative flux of microseeping hydrocarbons over areas of interest. The technique has not been widely used in the past. Pirson (1946) was the first to report its use, but Horvitz (1950, p. 942–943) gave a more complete description of Pirson's (1946) technique. Pirson buried sorbent tubes in 5 cm (2 in.) diameter holes, 0.6 m (2 ft) deep for twenty-four hours. The tubes were removed and desorbed in the laboratory, and the amount of desorbed ethane was "quantitatively" measured. Pirson (1946, p. 312) reported precisions of ±10% for low leakage areas and ±100% in high leakage areas, and also reported only mixed success. Pirson (1946) found it necessary to apply a number of different "correction factors" at each sample locality, including topographic, ground-water flow, climatic, seasonal atmospheric, soil nature (particularly soil permeability), and "deep geologic effects." Pirson (1946, p. 320) concluded that the results of nine years of research did not suggest that the "emanometric" technique showed promise of being any more effective than other surface geologic exploration techniques in use at the time.

Several points concerning Pirson's (1946) investigations warrant comment. His specific approach, as well as the absence of sophisticated techniques for hydrocarbon gas analysis (e.g. gas chromatography), negated some of the possible advantages of integrative absorption. The need for so many "correction factors" inserts an unwelcome degree of subjectivity into the method which, in turn, introduces a high degree of uncertainty into the data. The pre-1950 analytical techniques make it very possible that Pirson (1946) was not always measuring microseeping hydrocarbon gases but rather indigenous surface hydrocarbon and nonhydrocarbon gases. Pirson assumed excessively high hydrocarbon gas microseepage rates and consequently left his "emanometers" buried for only twenty-four hours. This obviated the principal advantage of the integrative absorption technique which, compared to soil-air and soil-sorbed hydrocarbon surveys, requires an "integrative" time period long enough to overcome diurnal, climatological, barometric, or even seasonal effects.

Kartsev et al. (1959, p 65, 75–82) discussed the technique of drawing soil air through sorbent tubes to concentrate hydrocarbon gases, which greatly increased the signal generated by soil-air hydrocarbons. That method, however, is not equivalent to integrative absorption. As Kartsev et al. (1959, p. 82) noted, "In practical use this method has encountered a number of difficulties of a procedural character and has not as yet been used as extensively as it deserves." Heemstra et al. (1979) described a rudimentary technique similar to integrative absorption which they carried out at the Sutherland Field, Greenwood County, Kansas. A soil-air sample was taken from a 10 cm (4 in) hole and analyzed for C_1 to C_4 hydrocarbons using a field gas chromatograph. In addition, a large paper soft-drink cup was inverted and sealed with dirt in the hole. After twenty-four hours, personnel returned to the sample sites and withdrew air samples from the inverted cups by inserting syringe needles through the tops of the cups (actually their inverted bottoms). These samples were also analyzed for C_1 to C_4 hydrocarbons by gas chromatography, and the difference between the two analyses was taken as the hydrocarbon flux at each site for twenty-four hours.

Understandably, a number of problems were encountered with this technique. The cups adsorbed hydrocarbons, were made soft and pliable by water and heat, and were crushed by livestock in the area. In addition, "heavier hydrocarbons," which caused analytical problems, were present in the soil-air samples. The "cup" survey which Heemstra et al. (1979) carried out showed an inverse correlation to the results of an occluded-hydrocarbon soil survey taken from 2.7 m (9 ft) holes over the same area. It did, however, show correlation with an "aerial hydrocarbon survey" run by Geophysical Systems Corporation, Dallas [see Thompson (1981) for description of this method].

PETREX personnel (Hickey et al., 1983A, 1983B; and Klusman and Voorhees, 1983) reported an integrative-absorption technique combining pyrolysis with mass spectrometry. Ferro-magnetic wires, whose ends are coated with activated carbon bonded to the wire by inorganic cement, are placed in holes 6 to 10 in. (15.2 to 25.4 cm) deep and covered by aluminum cans which are then covered by 2 to 4 in. (5 to 10 cm) of dirt. The assembly is left in the ground for at least a week, during which time vertically migrating hydrocarbons are sorbed by the charcoal. After retrieval from the field, the wire is heated to the curie point of the metal. This causes desorption of the sorbed hydrocarbons which are analyzed using mass spectrometry and a multivariant computerized statistical package, "ARTHUR." Published examples of the use of this technique are limited to areas of petroleum production, or possible petroleum production, associated with faulting or fracturing (Hickey et al., 1983B, p. 1342) where upward migration of hydrocarbon by effusion or, more likely, where transport by deep-basinal compaction waters is occurring.

Klusman and Voorhees (1983) noted major advantages to using integrative-absorption techniques to measure microseeping hydrocarbons. By negating the need for deep (2 m to 600 m, 6.6 to 1,969 ft) bore holes, the number of samples which can be taken per unit time is dramatically increased, even considering that two field trips are necessary with integrative-absorption techniques. For example, Klusman and Voorhees (1983) discussed a study which they carried out in the Green River basin at the Patrick Draw Oil Field which duplicated an earlier study carried out by Richers et al. (1982). Whereas the Richers et al. (1982) survey required 120 field man days and power equipment for augering, Klusman and Voorhees (1983) carried out the same survey in six field man days with hand tools. Decreasing field time by a factor of twenty greatly decreases field expenses. The use of hand tools with integrative-absorption techniques, instead of truck-mounted power augers, greatly increases field mobility, allowing more detailed surveys to be carried out. Another major advantage with integrative-absorption techniques, compared to soil-air and soil-sorbed hydrocarbon surveys, lies in the integrative aspect itself. Barometric, climatological, diurnal, and even seasonal variations in microseepage hydrocarbon concentrations apparently can be negated by leaving the collectors buried for sufficient time. Depending on the analytical method used, integrative-absorption analyses can be much quicker and cheaper than soil-air, soil-occluded, or soil-sorbed hydrocarbon analyses.

There are problems with the technique of integrative absorption The principal problem arises from its strongest attribute, the ability to carry out surveys at soil depths of 6 to 24 in. (15 to 60 cm). Unfortunately, this is also the realm of most intense microbial and plant activity. As Kartsev et al. (1959, p. 315) noted, "The zone with the greatest bacterial population and variety of morphological and physiological groups of bacteria is the soil zone." Kartsev (1959, p. 65) discussed early (1945) Soviet research which revealed that nitrous oxide (N_2O), from the soil-microbial processes, was the main component of upper layer soil gas:

> At that time the first experimental work on gas surveying by accumulation of hydrocarbons on silica gel was done. E. K. Gerling and B. M. Nakashidze then proved conclusively that the incombustible component of the heavy fraction consists of nitrous oxide and often constitutes 80–95% of the entire gas. Since nitrous oxide is mainly formed in the uppermost soil layer, it was decided to resort to the gas survey only for combustible gases.

This early Soviet research also demonstrated that the thermogenic hydrocarbon component of soil gas is usually less than 1%, most of which is methane. In discussing this Soviet research, Davis (1967, p. 126) critically noted:

> Nitrous oxide which has the same molecular weight as propane can be a problem in soil-gas analysis. It is practically impossible to remove it prior to analysis. In mass-spectral analysis nitrous oxide is ionized to nitric oxide (NO) which has the same molecular weight as ethane.

Nitrous oxide will be taken up by absorbents, and although not a problem in gas chromatographic analysis using flame ionization detection, N_2O could possibly lead to errors in mass-spectrometric analysis, as Davis (1967) noted. Besides N_2O, other products of microbial and plant life processes are quite capable of interfering with integrative absorption. In discussing early Soviet work, Kartsev et al. (1959, p. 73–75) noted:

> Investigations with the TG-5A apparatus showed that the methane fraction contains, besides methane, carbon monoxide and volatile organic substances of complex composition . . . The heavy hydrocarbon fraction also consists of certain complex volatile organic substances which are absorbed by activated charcoal. Volatile organic compounds contained in the soil may be evolved by plants and in the bacterial decomposition of the remains of organic matter.
>
> N. G. Kholodnyi ascertained that various organs of plants liberate complex volatile organic compounds into the atmosphere and that these (phytogenic) substances are a constant component of the atmosphere at the earth's surface wherever there is plant cover; the odors of flowers, barks, and leaves are due to phytogenic substances in the atmosphere. Further investigations by N. G. Kholodnyi led to the conclusion that gaseous organic substances contained in the soil air are formed in part under the influence of microorganisms in dead organic remains and in part in the tissues of living plants and animals for which the soil is the natural habitat. The author considers that the gaseous organic substances may consist of methane and volatile esters, alcohols, and acids.
>
> Special investigations conducted in 1946–1948 by M. I. Subbota showed that volatile organic substances are liberated not only by the soil, but also by plant roots. It is of interest to note that the liberation of volatile organic substances proceeds at different intensities in different seasons. The maximum quantity of emanation is observed during the period of vigorous growth of the plants and, after that, during the time of their flowering. The unique seasonal character of the liberation of organic substances is well known to specialists who study ester-bearing plants.
>
> It is evident from this that soil gas contains a great variety of combustible gases formed within the soil layer itself: methane, volatile organic substances, carbon monoxide, and others. In order to avoid these soil gases, it is necessary to go to greater depths below the soil or to construct apparatus that will determine hydrocarbon gases not by the products of combustion, but directly. For correct interpretation of the results of gas surveying, it is necessary to employ heavy hydrocarbons and methane for the most part.

Zobell (1946, p. 33, 41) also noted that plants and bacteria produce unsaturated gaseous hydrocarbons in the upper soil layers.

Little Western research has been carried out on the indigenous organic compounds existing in the upper soil layers, as such compounds pose no problem for soil-air, soil-occluded, or soil-sorbed hydrocarbon analyses whose samples are from depths where microbial and plant activity is greatly reduced. Therefore, the possible interference of indigenous organic compounds with those methods is minimal. However, the possible degree of interference of such compounds with integrative-absorption techniques is high. Because of the integrative aspect of the technique, even small concentrations of these compounds could be summed up over time to give possibly large interfering or misleading signals. An example of such confusing signals was given by Pirson (1946), who observed that his integrative-absorption technique led to "artificial anomalies" which were often of a greater magnitude than the significant measurements, and which in turn led to a number of dry holes being drilled. Pirson (1946) found that topographic highs resulted in low fluxes of hydrocarbon microseepage, whereas higher than normal fluxes were associated with areas where the water table was closer than normal to the earth's surface, such as parallel to (dry or running) stream beds or "ground-water mounds." Such areas of high moisture content are optimum for both plant and microbial life and the associated organic compounds which they produce. Although Pirson (1946) believed he was measuring only ethane, because of the pre-1950 rudimentary organic analytical techniques it is quite likely that he was measuring minor amounts of thermogenic hydrocarbon gases overshadowed by large concentrations of plant and/or microbially produced volatile organics.

Another example of the confusion which can be caused by plant and/or microbially produced organics is the exchange between Smith and Ellis (1963) and Horvitz (1972). Smith and Ellis (1963) contended that data they presented (Fig. 12) showed that saturated C_2-C_5 hydrocarbon gases could be produced by plants, and thus analysis of soils for C_2-C_5 hydrocarbons was not necessarily specific for thermogenic hydrocarbon deposits. Horvitz (1972) effectively replied to and negated the points made by Smith and Ellis (1963) in regard to occluded hydrocarbon soil analysis. However, the compounds which Smith and Ellis (1963) discussed could be a significant problem for integrative-absorption techniques. Heemstra et al. (1979, p. 53) discussed soil-air analytical problems with their "cup integration" in surface holes, problems which were not encountered in their analysis of soils for occluded hydrocarbons. Those problems were likely due to organic compounds indigenous to the soil they worked with.

Volatile organic compounds (as well as CO_2 and N_2O) generated by plants and/or microbes also could "poison" collectors by taking up active sorbent sites and reducing sorbent capacity to take up microseeping thermogenic hydrocarbons. This would depend on the surface area of the collector, the time of burial, and the relative intensity of plant and/or microbial activity. Another possible problem is bacterial attack of sorbed hydrocarbons in moist collectors after they are taken from the ground.

Klusman and Voorhees (1983, p. 2) believe that computerized pyrolysis-mass spectometric analysis is superior to gas-chromatography as an analytical tool for integrative absorption.

> Gas chromatography has some inherent disadvantages as a highly automated system. The ability to separate hydrocarbons changes with column characteristics resulting in possible changes in retention times as well as relative peak positions. This makes the system difficult to computerize. Separation of hydrocarbons heavier than about C_6 usually requires a long anlysis time and the need of an additional analytical device, such as a mass spectrometer or ft-infrared spectrometer.

This author is in disagreement with Klusman and Voorhees (1983). Perhaps for the apparent application of their technique (the identification of hydrocarbon anomalies caused by either the effusion of C_{6+} hydrocarbons, or the migration of C_{6+} hydrocarbon-bearing deep-basin waters up faults and fractures) C_{6+} hydrocarbon analysis is important. However, the C_1 to C_5 hydrocarbons make up all or most of *microseeping hydrocarbons* and gas chromatographic analysis of C_1 to C_5 hydrocarbons can be accomplished in two to four minutes. (Gas chromatographic analysis of hydrocarbons up to C_{18} can be carried out in, at most, fifteen minutes with oven-temperature programming.) As the C_{6+} hydrocarbons are not usually present in significant concentrations with true hydrocarbon microseepage, analysis for C_{6+} hydrocarbons is clearly unnecessary in most cases. Thousands of analyses using gas-solid gas chromatography columns have been carried out by the author and no change in peak retention time for C_1 to C_5 hydrocarbons have been observed which would cause problems for bench computers. Hundreds of analyses carried out using gas-liquid gas chromatography have not resulted in the problems cited by Klusman and Voorhees, 1983 (David Osborne, Exploration Research Laboratories, Salt Lake City, personal communication). Horvitz (1969, p. 207) incorporated gas chromatography into his occluded hydrocarbon soil analysis method after having investigated mass spectrometry, as well as several other methods, and found them unsuitable for his needs. In discussing soil-gas analysis by gas-chromatographic, as opposed to mass-spectrometric, methods Davis (1969, p. 154) stated:

> Gas-chromatographic methods are the most recent and the best methods of soil-gas analysis, because the gases are readily identified and quantitatively measured. The methods are relatively simple, sensitive and inexpensive.

Gas chromatographic analysis offers distinct advantages over pyrolysis-mass spectrometric analysis. The two most important advantages are: (1) analysis cost, and (2) the possibility for multiple analyses with gas chromatography as opposed to a "one shot" pyrolysis-mass sprectrometric analysis. The latter advantage allows both a check on accuracy/reproducibility as well as the co-injection of a spike of a known internal standard of C_1-C_5 hydrocarbons, together with the unknown, for the purpose of establishing a positive identification of sample peaks in the unknown. This is not possible with pyrolysis-mass spectrometric analysis, and the distinct possibility of confusing mass-spectral ions from organic and inorganic compounds indigenous to soils (Davis, 1967, p. 126) makes positive compound identification a first priority in integrative-absorption techniques. Certainly Horvitz's (1972) resolution of the questions raised by Smith and Ellis (1963) would not have been possible using only mass spectrometric analysis. Analysis cost per sample for computerized mass spectrometry with multivariant statistical analysis is much greater than that of gas chromatographic analysis using a desktop computer. There is little advantage in slashing field costs with integrative-absorption techniques only to burden the method with unnecessarily high analytical costs.

There are, no doubt, a host of other yet unrecognized problems and disadvantages with integrative absorption. Given past experience in surface geochemical exploration, such problems will surface if the technique is more widely employed. Also, like all direct geochemical detection methods, the interaction of integrative absorption with seasonal development, and the subsequent decline of the microbial filter, must always be kept in mind. Historically, the uppermost soil layer has always been the most difficult niche in which to carry out direct geochemical detection. As Sokolov et al. (1959, p. 670) noted:

> For most methods, and especially for gas methods, the layers at the very surface are unfavorable for the detection of anomalies, owing to intensive gas exchange between these layers and the atmosphere, and to the intensive biochemical processes occurring in them.

DISCUSSION

Hydrocarbon-Microseepage Fluxes

Different investigators have called for high microseepage hydrocarbon rates (fluxes) when considered over geologic time. Other investigators have subscribed to hydrocarbon microseepage mechanisms which, a priori, involve very high hydrocarbon fluxes. Rosaire (1938, p. 110) called for an oil "microseepage" rate at the Hastings Field of 10,000 barrels per day. On the basis of both his "geodynamic" and "emanometric" methods, (Pirson, 1941, 1946) has given different estimates of hydrocarbon fluxes. For example, Pirson (1941) called for an ethane flux of 23 mm^3/24 hours/ft^2 (or 8,000 ft^3/year/mi^2) at the Music Mountain Gas Field in northern Pennsylvania.

Duchscherer (1981B, p. 320) called for fluxes of "several hundred cubic feet of gas being lost to the atmosphere per square yard of surface area per year." When projected over geologic time, such fluxes result in the loss of staggering amounts of hydrocarbons (Table 1).

If one assumes a lower Miocene oil emplacement at Hastings, 84 trillion bbls of oil would have been lost there due to microseepage since accumulation. A mid-Pennsylvanian gas emplacement at Music Mountain would have resulted in a loss of 480 trillion ft^3 of gas from that pool. Clearly the "microseepage" rates of Table 1 are geologically untenable. Even fluxes millions of times less would be untenable. Two assumptions are possible with regard to the "reservoir" of hydrocarbons supporting hydrocarbon microseepage. It may be assumed that hydrocarbon generation, primary and secondary migration, and accumulation are processes which constantly occur in sedimentary basins, and hence the hydrocarbon pool is constantly being replenished. Or it may be assumed that these processes occur only during short periods of geologic time. This author's research would indicate that the latter scenario is usually the case.

A detailed discussion of the origin and evolution through geologic time of petroleum basins, hydrocarbon origin and migration, and the accumulation of commercial hydrocarbon deposits is certainly outside the scope of this paper. Thus, only a brief and undocumented review will be given here of what the author believes is the usual scenario based on as yet unpublished research. Hydrocarbon generation and primary migration are heat-driven processes resulting from either increasing burial or an increasing geothermal gradient, or a combination thereof. The cessation of burial or a decreasing geothermal gradient causes a termination of both hydrocarbon generation and migration. Also, a steeply decreasing geothermal gradient, coupled with or due to basinal uplift and erosion, results in a marked cooling and contraction of a sedimentary basin and its pore waters. Such pore-water cooling results in two events: (1) the precipitation of a substantial amount of silica-alumina minerals (mainly clays), and other minerals, in sediment pores, and (2) the creation of less than hydrostatic fluid-pressure gradients in most sediments of the basin (this is

TABLE 1

Hydrocarbon Microseepage Rates as Calculated by Several Investigators

Source	Flux	Amount lost in one million years
Rosaire (1938)	10,000 bbls/day of oil	3.65 trillion bbls of oil
Pirson (1941)	23 mm^3/24 hrs/ft^2 of ethane	1.6 trillion ft^3 of gas
Duchscherer (1981B)	300 ft^3/year/yd^2 of HC gas	17.90 quadrillion ft^3 of gas

Assumptions: Pirson (1941)—Compositionally, ethane represents only 10 volume percent of the gas in the pool. Field size is 20 mi^2.
Duchscherer (1981B)—Field size is 20 mi^2. "Several" equals three.

especially prevalent in sediments on the stable shelves of such basins). The precipitation of clays in the porosity of coarse-grained sediments in such basins greatly restricts (but does not totally halt) basinal fluid flow from either meteoric water recharge or deep-basin compaction. This mineral precipitation also effectively helps to seal off hydrocarbon deposits throughout the basin, especially in stratigraphic and low-relief traps on the stable shelves of sedimentary basins. Without such a sealing and disruption of fluid flow, commercial hydrocarbon pools would be destroyed by water washing and diffusion (Price, 1980, p. 21–26). Petroleum basins, which have had such large decreases in geothermal gradients, may be termed "dead" basins. This distinguishes them from "live" basins, such as the coastal and offshore portions of the Gulf Coast basin, as well as some California coastal basins, where processes leading to the accumulation of commercial hydrocarbon deposits are still proceeding. Most interior cratonic American petroleum basins have fluid-pressure gradients less than hydrostatic, especially sediments and hydrocarbon deposits on the stable shelves of these basins. [See Dicky and Cox (1977) and Magara (1981) for examples and discussion.] Such basins have had substantial decreases from their maximal geothermal gradients. Therefore, it is unlikely that the hydrocarbon deposits of these basins can be "constantly replenished" to account for the very high hydrocarbon microseepage fluxes proposed by different investigators.

Mills and Wells (1919) first proposed gas-induced evaporation of ground water, a concept amplified by Nisle (1941). Other investigators (including Tripp, 1945; Donovan and Dalziel, 1977; and Duchscherer, 1981B, p. 320) have also called upon the mechanism to explain different observed features, including halos of any kind, radiometric anomalies, "soil salt" anomalies, and anomalous oxygen and/or carbon-isotopic compositions of soil carbonates. According to Mills and Wells (1919), natural gas vertically diffusing through the sediment water column by molecular solution exsolves near the earth's surface (as a result of pressure decrease) and forms a free-gas phase. Because it is undersaturated with respect to water, this free gas phase takes groundwater into solution, thereby evaporating groundwater, which causes precipitation of soil salts, isotopic fractionations, etc.

Gas-induced groundwater evaporation is most unlikely for three reasons: (1) As discussed, diffusion of hydrocarbon gas vertically through impermeable sediments in not possible, (2) the amounts of gas required are unrealistically large, and (3) hydrocarbon microseepage probably results in exactly opposite features as those proposed to occur with gas-induced groundwater evaporation. As to the second point, the solubility of water in a hydrocarbon gas phase, especially at temperatures below 100°C, is limited at best (Fig. 13). "Mineral salts" are present in groundwaters from the ppm to the 0.5% level, depending on the species. In cases of low concentrations of trace "salts" or metals, almost all the groundwater would have to be "evaporated" to cause precipitation of the mineral species. Mass balance calculations can be carried out to show that unrealistically large gas volumes would be required for precipitation of the reported amounts of various soil salts. The fact that most of the microseeping hydrocarbon gas will be destroyed by bacterial oxidation makes the amount of gas required even larger and more unrealistic. As proposed in this paper, the principal hydrocarbon microseepage mechanism is the vertical migration of microbubbles of natural gas. Thus, by the data of Figure 13, such a migration mechanism would transport deep-basinal water by such gas bubbles to the earth's surface where the water would exsolve from the gas phase due to pressure and temperature decline. This is the exact opposite result of the hypothesis of gas-induced groundwater evaporation. Whether or not this entrained water could in turn take up other elements (e.g., iodine) or salts which form surface anomalies (Kartsev et al., 1959, p. 293; Gallagher, 1984) is another matter not addressed here.

Hydrocarbon gas microseepage rates, by this paper's model, are probably millions and perhaps even hundreds of millions of times smaller than the fluxes called for by some investigators. Effusion rates, however, are another matter and result in a more rapid destruction of oil pools as compared to hydrocarbon microseepage. The change in sediment lithology over some oil pools in southern Oklahoma has been attributed to hydrocarbon microseepage by some investigators (including Seager, 1938, in the discussion section of Rosaire, 1938, p. 120; Donovan and Dalziel, 1977; and Duchscherer, 1981A). The southern Oklahoma portions of the Anadarko-Ardmore basins have some of the most pronounced structural intensity of petroleum basins worldwide, rivaling and/or surpassing that of the Ploesti district of Romania, and the Los Angeles-Ventura basins. The intense faulting and macrofracturing of southern Oklahoma make it probable that the various distinct anomalies observed over many of the fields there are not due to hydrocarbon microseepage but have been caused by effusion. These fields have been occasionally cited as models for hydrocarbon microseepage. The anomalies developed over these fields, however, are much stronger and more intensely evolved than normal. These fields do serve as a model, but a model wherein the effects of effusion overwhelm those of hydrocarbon microseepage. The structural intensity and magnitude of anomaly development over these fields make it remarkable that any oil is left in the fields at all, much less the large reserves they do have.

Kartsev et al. (1959, p. 67) and Sokolov et al. (1959, p. 669) both discussed cementation of fractures above oil deposits. Sokolov et al. (1959) noted that such fractures in Paleozoic rocks are more heavily cemented than fractures in Tertiary rocks in regions of similar structural intensity. Baijal (1962) stated that fractures above oil pools become mineralized and closed with time. These observations, and the earlier discussion of chimneys and lithologic reservoir alteration by bacteria, suggest that leaking hydrocarbon deposits may "heal" themselves over geologic time from mineral precipitation resulting from bacterial oxidation of leaking hydrocarbons. However, conjecture on the behavior of these systems over geologic time is highly speculative, especially considering that so little is known about even their temporal behavior.

Use of direct geochemical detection and other surface geochemical exploration methods in "live" petroleum basins may result in false anomalies, which would not be encountered in "dead" petroleum basins. There is a general but undocumented belief among petroleum organic rock geochemists that most of the hydrocarbons that migrate from source rocks never form commerical deposits. Instead, especially in structured basins, these hydrocarbons are lost to seepage (and presumably microseepage) as well as to other factors. If this belief is true, then such hydrocarbon effusions and microseepages might create false surface anomalies not necessarily caused by commercial hydrocarbon deposits.

Optimum Use of Direct Geochemical Detection

Since the inception of surface geochemical exploration, the West has attempted to use this approach as a front-line exploration tool. Few would argue with the conclusion that, at least in the West, surface geochemical exploration techniques have not yet achieved major success as primary exploration tools. Limited data are occasionally quoted to suggest that surface geochemical exploration might be more effective as a tool to check prospects generated by other exploration methods, rather than as a primary exploration tool. For example, Kuzmin (1983) noted that while 58% of the anomalies he worked with were proven productive on drilling, 79% of the prospects condemned by surface geochemical exploration were dry holes on drilling. However, such percentages for negative prediction are, in reality, unimpressive. If one assumes a new field wildcat success ratio of 1 in 9 to 1 in 11 for the last fifty years, by simply condemning all prospects and not drilling any wells at all, one could achieve a negative predictive success ratio of 89% to 91%. Thus, negative predictions for wildcat wells based on surface geochemical exploration would have to be in the range of 95% to 99% correct to be an effective "negative" exploration tool.

Perhaps the most powerful utilization of surface geochemical exloration would be as a tool to determine field boundaries subsequent to new field discoveries, or for extensions of older producing fields. The graveyard of dry and abandoned wells which surround established fields testify to the tremendous waste of developmental drilling. It is probable that had surface geochemical exploration methods been proven first as tools for development drilling (and for extensions of older producing fields) then the contribution of surface geochemical exploration methods to rank wildcat petroleum exploration may have been more obvious. This in turn might have led to the investment in basic research needed to delineate fully the controlling parameters for the full and proper use of surface geochemical exploration methods.

Different investigators have noted that hydrocarbon "microseepage" is most intense along fault and fracture zones, and that direct geochemical detection methods should be based on the identification of such fault zones (Weisman, 1979; Jones,

1982; Richers et al., 1982; Jones and Drozd, 1983; Hickey et al., 1983B; Klusman and Voorhees, 1983). "In general, the best application of the surface prospecting technique is in areas with tectonic activity" (Weisman, 1979, p. 388). Hickey et al. (1983B, p. 1342) noted:

> The appliction of integrative gas geochemistry combined with mass spectrometry and the use of pattern recognition procedures, has enabled rapid characterization of microseepages of gases along faults and fractures. This has been accomplished by incorporating the activated carbon/Curie point wire collector in gas geochemical surveys of faulted and fractured structures which serve as conduits to the subsurface . . . From these interpretations, an exploration model which utilizes fault leakage as a parameter will be discussed.

Jones (1982, p. 245), discussing hydrocarbon microseepage, noted that

> the upward migration of reservoired light hydrocarbons into near-surface soils represents a viable mechanism, allowing surface geochemical exploration techniques to be utilized for regional hydrocarbon evaluations.
>
> Geochemical investigations indicate that seep magnitudes depend on tectonic activity to aid migration along the fault and fracture avenues which appear to provide the major migration pathways.

There is no question that surface "microseepage" hydrocarbon concentrations are greatest over surface traces of faults and fractures. Sokolov et al. (1959, p. 672), commenting on the experience of the Soviets, noted:

> Many points where high hydrocarbon concentrations were observed were related to zones of disjunctive disturbances. In some cases it was possible thus to trace disturbance lines in solid rocks capped by drift. The shape of the gas anomalies was often determined by the disjunctive disturbances.

Finding a fault zone, however, is not equivalent to finding commercial hydrocarbons. If anything, a fault zone would be more likely to yield false anomalies. Even if a pool of commercial hydrocarbons were fault associated, the pool could be laterally displaced from the fault zone. Fault zones, in all probability, also yield anomalies not caused by or associated with commercial hydrocarbon deposits. The vertical movement of deep-basin waters up faults is well documented, even in "dead" petroleum basins (Price, 1976, p. 229–230; Roberts, 1981; Jones, 1981, 1984). Such waters, compared to surface waters, are enriched in dissolved hydrocarbons because of increased aqueous solubility of hydrocarbons at elevated burial temperatures (Buckley et al., 1958; Price, 1976, 1979, 1981A, 1981B; Price et al., 1981). Deep-basin waters moving up fault zones undergo temperature decrease, causing the exsolution of both oil and gas from these waters. The exsolved globules of gas and oil continue to migrate with the waters vertically up the fault zone, but now also as a consequence of buoyancy. The end result is an enrichment of hydrocarbons along the surface traces of fault zones, compared to normal low background values. This enrichment would appear as a hydrocarbon anomaly but would be caused solely by hydrocarbons dissolved in deep subsurface waters and not by a commercial hydrocarbon deposit.

The migration mechanism(s) responsible for fault-associated hydrocarbon anomalies obviously differ from the migration mechanism responsible for hydrocarbon microseepage over unfaulted hydrocarbon deposits. Fault anomalies could be caused by intermittent monophasic gas and/or oil migration from buoyancy or effusion, and by diffusion through and up the waters of the fault plane, as well as by the vertical transport of deep-basinal waters. Anomaly signal characteristics over fault zones would be different from signal characteristics of anomalies created by hydrocarbon microseepage from unfaulted hydrocarbon deposits. The latter signal is dominantly composed of C_1-C_5 hydrocarbons with little or no concentrations of C_{6+} hydrocarbons (Sokolov, 1970; Sokolov et al., 1971; and Philp and Crisp, 1982, p. 3). However, because of the upward transport of C_{6+} hydrocarbons, "fault anomalies" can have significant C_{6+} hydrocarbon concentrations. "Studies conducted in the Denver-Julesburg basin of Colorado, Green River basin of Wyoming, the hingeline of Utah, and the Las Animas arch of Colorado all produced fault-related samples where higher ($<C_{7+}$) molecular weight components were encountered" (Hickey et al., 1983B, p. 1342).

The intensity and characteristics of fault-related anomalies depend on many parameters, such that it will not always be possible to distinguish between faults associated with possible commercial hydrocarbon production and "nonproductive faults." Discounting effusion, signal intensity of fault anomalies depend on the volume of water transported up a fault and the original solution capacity of the water for hydrocarbons. Water volume depends on hydraulic potential and fault transmissivity. Solution capacity depends on the original pressure, salinity, gas content, and especially the temperature of the water (Price, 1976, 1979, 1981A; and Price et al., 1981). A vertically transmissive fault tapping high-temperature deep-basin compaction waters which have a high hydraulic potential would generate a much stronger signal or anomaly than would a "tighter" fault linked with a hydrocarbon pool at much shallower depths associated with formation waters of lower hydraulic potential and lower hydrocarbon solution capabilities. Hydrocarbons dissolved in fault-borne waters and originating from a hydrocarbon pool, either close to or removed from a fault, would not necessarily yield a "fingerprint" characteristic of a hydrocarbon pool (Price, 1976, 1979, 1981A; and Price et al., 1981). In addition, a hydrocarbon pool removed from a fault by some distance could generate a false anomaly at the surface trace of such a fault. An example would be active water washing of an oil pool by meteoric-water recharge (Price, 1980), with the waters transported some distance from the hydrocarbon pool and up a fault. Hunt (1979, p. 430) pointed out that eastern European organic geochemists at the Third International Geochemical Conference "concluded that areas with complicated tectonics involving faulting and thrusting give unfavorable results in surface prospecting because fault zones are preferential migration paths." Such an example was given (Fig. 14) in a soil-gas survey across the Ryckman Creek Field, Wyoming, by Jones and Drozd (1983). The "anomaly" over the Ryckman Creek Field is not well developed compared to adjacent false anomalies caused by faulting. As Jones and Drozd (1983, p. 947) noted: "The hydrocarbon seeps occur directly over the subcrop of the thrust faults, and thus gases would appear to migrate along the thrust plane and then vertically through the Tertiary section . . ."

Contrary to the opinons of some investigators, direct geochemical detection methods are not best suited to tectonically disturbed areas; in fact, this is the one area for which they are most ill suited. Fault zones do give much stronger hydrocarbon anomalies than those produced by microseepage hydrocarbon deposits in unfaulted, flat-lying sediments of the stable shelves of sedimentary basins (Sokolov, 1970, p. 545). If such strong signals over faults are not necessarily caused by commercial hydrocarbon deposits, then such signals lose much of their relevance to petroleum exploration. We are not in the business of discovering and producing faults. Seismic methods have proven reliable at locating these features. As Jones and Drozd (1983, p. 932) noted:

> Geochemical prospecting must be used with caution, and only in conjunction with geologic and geophysical tools, because the location and shape of many geochemical anomalies are governed more by the local tectonic structure of the region than by the position and shape of the actual deposit.

Fedynsky et al. (1975) also noted that faults can shift hydrocabon anomalies laterally from the hydrocarbon deposits causing them.

Potential for Finding Subtle Traps

If surface geochemical exploration is to make a significant contribution to oil exploration, it is to be in the search for oil pools in stratigraphic and low-relief traps on the stable shelves of maturely explored onshore petroleum basins. Hamm (1969, p. 13) noted:

> The industry has not yet developed a finding tool with high resolving power for the discovery of stratigraphic-trap fields. The principal finding methods being used today are subsurface geology based upon information from outcrops, wells, and various types of geophysical data. The batting average on stratigraphic-trap drilling is quite low. It is here that unconventional methods of oil exploration may make a major contribution.

The need for an effective exploration method for stratigraphic traps has also been noted by a myriad of other investigators (Davidson, 1963, p. 98; Jones and Drozd, 1983, p. 933; and Kuzmin, 1983, p. 1094). Acknowledging this need and the possible contribution which surface geochemical exploration methods could make almost borders on redundancy; however, the record here is not encouraging. In discussing direct geochemical detection on the stable shelves of petroleum basins, Kartsev et al. (1959, p. 106), noted:

> Sometimes a large oil formation is in no way indicated by gas survey results. This has been known to occur with formations characterized by undisturbed tectonics and the virtual absence of faults. This is most often observed in the case of platform formations.

The Soviet use of "platform" is equivalent to the stable-shelf area of flat-lying sediments in a basin past the hingeline of the basin. Sokolov et al. (1959, p. 574) noted:

> In the platform regions of the U.S.S.R., the effectivity of gas surveying with sampling from several depths was not high, though in individual, more disturbed districts anomalies were obtained at the sites of oil and gas fields (Ukhta, etc.) . . . On the Russian

platform in most cases no hydrocarbon gas anomalies were detected at all in the surface layers above oil deposits in Paleozoic sediments, this evidently being due to the low intensity of vertical gas migration from the deposits.

Sokolov et al. (1971) and Hunt (1981) also noted that direct geochemical detection methods are weakest in areas of flat-lying sediments. Of all the examples of direct geochemical detection anomalies published in Western trade and scientific journals, few are over stratigraphic traps.

Direct geochemical detection may make a significant contribution to oil exploration, besides developmental drilling and field extension, as the primary exploration tool for stratigraphic hydrocarbon accumulations. This contribution has not yet been made and the record is not encouraging.

Hypotheses

Several hypotheses may be advanced on the basis of the hydrocarbon microseepage model advocated in this paper. If surface and near-surface bacterial activity is the dominant controlling parameter of the surface concentration of microseeping hydrocarbons, then the concentration of soil-air and soil-sorbed hydrocarbons should be inversely proportional to microbial activity. Evidence supporting or contradicting this premise can only come from microbiologic, soil-gas and/or soil-sorbed hydrocarbon surveys simultaneously run over the same area. Kartsev et al. (1959, p. 333–340), after noting that the Soviets frequently carry out such surveys, discussed examples. The results of a soil-gas/microbiologic survey carried out in the northern Caucasus region, led Kartsev et al. (1959, p. 334) to conclude: "It should be noted that the greatest microconcentrations of gases are found in areas where hydrocarbon microflora is absent." (Also see Kartsev et al., 1959, Figure 148 with accompanying discussion.) Kartsev et al. (1959, p. 340) also noted:

When the influx of hydrocarbon gases is intense, complete or partial superposition of gas maxima on the microbiological ones is observed. With a feeble gas source, on the contrary, the zones of microbiological and gas anomalies may fail to coincide within the boundaries of a given gas field. Therefore, for correct interpretation of geochemical anomalies it is necessary to take into account not only the microbiological and gas indices separately, but also the specific way they are related in space. Only on the basis of these data can the actual output of hydrocarbon gases at any given depth be determined; this includes not only freely evolved gases, but also those absorbed by the bacterial filter.

An obvious converse corollary to the first premise is that microseepage hydrocarbon areas, which have weak or no microbiologic activity, would be expected to have significant concentrations of soil-sorbed or soil-air hydrocarbons.

A second premise, by this paper's model, is that the shape, intensity, and generation of occluded hydrocarbon soil anomalies is dependent on the presence of significant microbial activity in the rock column above an oil deposit. Occluded hydrocarbons clearly depend on the precipitation of calcite for occlusion, and such precipitation in turn clearly depends on the generation of high CO_2 concentrations from microbial oxidation of microseeping hydrocarbons. Further, with one exception, the shape of occluded hydrocarbon anomalies should always be continuous or discontinuous halos and not apical anomalies. The one exception would be microseepage from small hydrocarbon deposits, whose anomalies could appear as apical anomalies due to either sampling density (sample sites too far apart to distinguish fine details) or insufficient amounts of microbially generated CO_2 to overcome a greater incoming flux of Ca^{+2} ions. Horvitz (1954, p. 1207; 1980, p. 247) noted that small hydrocarbon deposits sometimes yield apical and not halo anomalies with occluded hydrocarbon soil surveys.

Bacterial oxidation of vertically migrating hydrocarbons is clearly the major unacknowledged controlling parameter of surface geochemical exploration. It is likely, however, that there are other unrecognized parameters which play lesser, yet significant roles. For example, Horvitz (1939, p. 217) noted that a sorbed hydrocarbon soil survey at the Ramsey Field, Payne County, Oklahoma, conducted in the winter over snow-covered ground, yielded much higher hydrocarbon concentrations than a later replicate survey conducted in the summer. Horvitz (1939) attributed the higher hydrocarbon concentrations to the snow cover producing a seal, which reduced the escape of soil hydrocarbons to the atmosphere. Kartsev et al. (1959, p. 121) also discussed the increase in soil-sorbed hydrocarbon gas concentrations as a consequence of the plugging of surface capillaries by water or ice. Conceivably, a rain storm could temporarily block soil porosity and also cause a brief concentration increase in soil-air and soil-sorbed hydrocarbons. Part of Horvitz's (1939) winter hydrocarbon concentration increase also would be expected to be the result of a decrease in microbial activity from low winter air temperatures, although the data of Samtsevich (1955), as discussed by Davis (1967, p. 121), seemingly contradict this hypothesis.

Correcting the Record

Dickey and Hunt (1972), Hunt (1979, 1981), and Neglia (1979) have taken negative positions with respect to surface geochemical exploration. Neglia (1979, p. 594) stated: "The basic concept of geochemical prospecting—the assumption that a superficial halo of dispersion of hydrocarbons must be present above every oil and gas accumulation—is not true." Unfortunately, Neglia (1979) gave no data or discussion supporting his conclusion. Dickey and Hunt (1972) concluded that it was unlikely surface geochemical exploration would ever prove useful for two reasons. The first was that "many surveys" by major oil company research laboratories had shown that no hydrocarbon diffusion aureole exists radially outward from effusive macroseeps. This is similar to a logic employed by many proponents of surface geochemical exploration, which is that if macroseeps exist, then a priori, microseeps must exist. For example, McDermott (1940, p. 860) noted: "In view of the prevalence of visible oil and gas seeps, that is, leakage from buried accumulations, it is only reasonable to expect a much greater occurrence of microscopic leakage." Debnam (1969, p. 1) made a similar statement. By this paper's model, such logic simply is not correct. If macroseeps and microseeps originate by two different and independent migration mechanisms, the existence of one does not guarantee, a priori, the existence of the other. The second reason given by Dickey and Hunt (1972) which would prevent any possible significant contribution by surface geochemical exploration was their valid contention that vertical diffusion of hydrocabons in not possible through impermeable shales, nor can diffusion possibly explain the shape of halo or any other anomalies. Just because hydrocarbon microseepage does not occur by diffusion, however, does not mean that it does not occur at all. The point that Dickey and Hunt (1972) make here has no bearing on the possible existence or nonexistence of hydrocarbon microseepage.

Concerning the Soviet effort, Dickey and Hunt (1972, p. 158) noted: "Surface geochemical methods have been used extensively in Russia with the same discouraging results as in the United States." This statement simply does not agree with the published literature. The Soviets have made a major commitment over the last fifty years to surface and near-surface geochemical prospecting, approaching the topic much more scientifically than has the West. Consequently, the Soviets have published by far the largest share of the meaningful research on hydrocarbon microseepage. Kartsev et al. (1959, p. 65), discussing the early Soviet commitment, noted: "In 1940 the specialized office Neftegazos'emka (Oil-gas survey) was organized for prospecting work; the scientific director was V. A. Sokolov. Every year Neftegazos'emka sent twenty-five to thirty production and research parties to various parts of the Soviet Union." In discussing the magnitude of the more recent Soviet commitment, Thompson (1981, p. 240–241), noted:

> While predominantly employed conventional methods of exploration continue to offer less than promising results in the United States, Soviet research institutes have spent years attempting to develop and refine geochemical techniques for locating oil and gas reservoirs. In fact, more research effort in this area has probably been expended in the Soviet Union than in any other nation. It has been reported (Sweet, 1966) that in the 1950s the Soviets maintained as many as three hundred geochemical crews in the field at one time. The reason for this intensive Soviet investigation is clear—success in developing direct and indirect detection methods could effect huge economic impact in exploration (after Davidson, 1967).

Soviet technology in many fields clearly lags behind Western, and especially American, technology. Yet the Soviets now produce more crude oil than any other nation on earth, and probably would produce even more if it were not for their lag in production technology. I find it too coincidental that the Soviets, who have made such concentrated efforts in surface geochemical exploration, have achieved such exploration successes in spite of a general overall lagging technology.

Sokolov et al. (1959, p. 674) noted: "Reviewing the data on gas surveying predictions we get the following results. Under favorable geological conditions the proportion of correct predictions is rather high—about 70 percent." Discussing the success of surface geochemical exploration, Kartsev et al. (1959, p. 65) noted:

> With the aid of the gas survey, several oil formations were independently discovered. Several tens of formations were discovered by a combination of prospecting methods, among which the gas survey was not the least important. The great majority of gas anomalies have not yet been tested by drilling. Gas surveying became firmly established in the general complex of oil prospecting work.

Kartsev et al. (1959, p. 333–340) described a number of fields discovered on the basis of surface geochemical exploration (microbiologic, soil-air, and soil analysis). Sokolov et al. (1979, p. 539) also discuss a number of case histories:

> According to the data of A. Ya. Krems, L.A. Anischenko and others, surface gas anomalies in the Komi region led to the discovery of the Kush-Koy, Djebol, Njamed, Nibel, Voy-Voj and other new petroleum and gas fields. The success of gas surveys on the platform structures of the Komi region is about 90 per cent.
>
> During a survey of seismic shot-holes in 1957 at the mouth of the Pechora river near Naryan-Mar, a marked heavy hydrocarbon (C_2-C_5) gas anomaly was discovered. This information was the basis for carrying out oil and gas exploration work in the Zapolyarye area of the European part of the U.S.S.R. Later, oil and gas pools in Permian carbonate and terrigenous sediments were discovered there in a zone called the Yuryahinsko-Kumzhinskii swell. This is one of the more recent rich oil-bearing provinces of the U.S.S.R.
>
> In the middle Volga region, according to the data of M. I. Zeidelson and others, the efficiency of surface gas surveys was about 70–80 per cent. Some nineteen fields were discovered, including Muchanovo, Kajemjakino and others.

Kuzmin (1983) discussed results of other Soviet geochemical surveys where of forty-five anomalies, twenty-six resulted in commercial pools for a 65% success ratio. Dickey and Hunt's (1972) appraisal of "discouraging Soviet results" is erroneous.

Hunt (1979, p. 425–433), in some detail, strongly condemns surface geochemical exploration. His discussion and rejection of diffusion as the migration mechanism for hydrocarbon microseepage is valid. Hunt (1979, p. 433) concluded that such migration had to occur via "erratic pathways" (presumably faults and fractures), and that because of such migration routes, "surface geochemical prospecting cannot outline oil or gas accumulations at depth except in rare cases." This statement and that of Neglia (1979, p. 594) simply do not agree with the published record, which clearly demonstrates that in many cases surface geochemical exploration distinctly outlines surface boundaries of known fields: Horvitz (1939, Fig. 2, Hastings, Friendswood; 1945, Fig. 1, 2, Heidelberg; 1954, Fig. 1, Hastings, Manvel; 1957, Fig. 1, 2, large area with eleven fields at junction of Wharton, Matagorda, and Jackson Counties, Texas, Fig. 3, Bonney, Fig. 5, 6, Little Beaver; 1965, Fig. 1, 2, unnamed gas field, Fig. 4–6, North League City; 1969, Fig. 4–6, Pierce Junction, Fig. 7, 8, Kemah; 1980, Fig. 6, Flomaton-Jay-Blackjack Creek; 1981, Fig. 1–3, Francitas, Fig. 4–8, Cognac, offshore field), McDermott (1940, Fig. 2, 3, Cedar Lake, Fig. 7, Coles Levee, Canal, Ten Section), Ransone (1947, Fig. 2, Hardy; 1969, Fig. 2, Sixto), Kartsev et al. (1959, numerous examples), and Duchscherer (1981B, Fig. 4, Hanson, Fig. 6, New Cypress, Fig. 8, Midland; 1982, Fig. 6, Ten Section, Fig. 8, Sleepy Hollow, Fig. 10, Conley). Rosaire et al. (1940, p. 1459) pointed out that halo patterns were also present at Ezzell, Eureka, Welsh, Turkey Creek, Reynolds, and Sugarland in the Gulf Coast; Coalinga, Northeast Coalinga, Greeley, Rio Bravo, Huntington Beach, Rosecrans, Dominguez, Montebello, and Signal Hill in California; Goldsmith and Bennet in the Permian basin; Ramsey, Stroud, Sac, and Fox in Oklahoma; and Hebron in Pennsylvania. In discussing the efforts of Horvitz Laboratories, Weisman (1979, p. 371), noted:

> Horvitz has examined tens of thousands of near-surface sediment samples from both onshore and offshore areas for the saturated hydrocarbons through pentane. Recognizable hydrocarbon distribution patterns were observed over known fields. Similar patterns were found in areas which have been examined but not yet tested for hydrocarbon production. Follow-up studies indicate a high percentage of anomalies in originally unexplored areas subsequently were found to be associated with petroleum accumulations.

Jones and Drozd (1983, p. 940), discussing the results of an extensive Gulf Oil Company soil-air analyses research program, ". . . show that both structural and stratigraphic traps produce positive or direct surface anomalies . . ." Johnson (1963, p. 98, then Exploration Research Director for Sinclair Research) noted that ". . . many soil-hydrocarbon maps are in the literature and indicate definite relationship between surface accumulation of hydrocarbons and oil fields."

Clearly, some detractors of surface geochemical exploration have overstated their case, which is understandable considering the general lack of Western rigorous scientific development of the topic. There is no doubt, however, that hydrocarbon microseepage occurs. Nor is there any doubt that it can result in anomalies that distinctly outline the surface traces of hydrocarbon deposits. There is no doubt that surface geochemical exploration can be, and has been, a powerful petroleum exploration tool. The Soviets appear to have had marked success using surface geochemical exploration techniques. A number of literature examples exist of American oil fields discovered solely on the basis of surface geochemical exploration: Stormont (1939, East Mathis); Horvitz (1945, Fig. 1, 2, West Heidelberg Field and Heidelberg Field definition after its wildcat discovery); Ransone (1947, Hardy; 1958, Sojourner; 1969, Sixto); Horvitz (1965, Fig. 1, 2, unnamed gas field; Fig. 4–6, North League City); and Denison (1983, p. 162, Stoney Point). Horvitz (1954) reported that of 109 anomalies revealed by geochemical surveys, 39 had been drilled, with 23 new fields discovered. Duchscherer (1982) reported 39 new field discoveries based on the drilling of 171 geochemical anomalies.

It is probable that a significant number of "geochemical confirmations" exist that will never be published. A large but unknown amount of oil has been discovered because of paraffin dirt and lithologic enhancement of low-relief structures in seismic surveys (Rosaire, 1938). As these are both indirect geochemical detection methods resulting from hydrocarbon microseepage, much more oil has been discovered by indirect geochemical detection methods than has been previously realized by both detractors and proponents of surface geochemical exploration. The real question confronting surface geochemical exploration is whether it can be used as a viable exploration tool for oil in stratigraphic and low-relief traps on the stable shelves of maturely explored American petroleum basins. If we continue to use fifty-year-old procedures, and if the controlling parameters of hydrocarbon microseepage continue to go unresearched, then the answer to this question is probably no.

CONCLUSIONS

1. In this paper (A) hydrocarbon microseepage refers to the migration of the C_1 to C_5 hydrocarbons from hydrocarbon deposits to the earth's surface. (B) Direct geochemical detection is defined as any surface geochemical exploration method based on the measurement of concentrations of C_2 to C_5 hydrocarbons or measurement of the bacterial response from such hydrocarbons. Such methods are soil-air, soil-occluded or soil-sorbed hydrocarbon analyses, microbiologic analysis, and integrative absorption. (C) Methane is not considered to be a unique indicator of thermogenic hydrocarbon deposits.

2. In this paper indirect geochemical detection is defined as any method based on the measurement of a physical parameter which is an indirect result of and caused by the vertical migration of hydrocarbons (including hydrocarbon microseepage) from hydrocarbon deposits. Such methods are radiometric, redox, bitumen analysis, soil salt, trace metal or element, geobotanical, soil mineralogy, hydrochemical, electrical (e.g., induced polarization, self-potential), and magnetic. Other indirect geochemical detection methods are methods unrelated to vertically migrating hydrocarbons, such as helium surveying.

3. Diffusion, effusion, and vertical water movement through hydrocarbon deposits and/or impermeable sediments cannot be the migration mechanism responsible for hydrocarbon microseepage. That mechanism is hypothesized as the vertical ascent of colloidal-size microbubbles of hydrocarbon gases through microfracture systems over hydrocarbon deposits. However, diffusion of hydrocarbons through sediments and microfracture systems, slowly over geologic time, is partly responsible for background surface hydrocarbon concentrations in petroleum-bearing sedimentary basins. Effusion, defined as the continuous or intermittent monophasic flow of oil and/or gas up fault and/or fracture systems, is repsonsible for macroseepage. The vertical transport of deep-basin waters, carrying dissolved hydrocarbons up faults and fractures, is responsible (at times in conjunction with effusion) for hydrocarbon anomalies involving C_{6+} hydrocarbons which are observed at the surface traces of such fault and fracture systems in sedimentary basins.

4. The vertical ascent of colloidal-size bubbles of natural gas (carrying dissolved C_{6+} hydrocarbons) through microfracture systems over hydrocarbon deposits is the only proposed hydrocarbon microseepage migration mechanism in agreement with observed data: (A) the speed with which hydrocarbon surface anomalies appear or disappear in

response to reservoir pressure changes; (B) the chromatographic effect of increasing amounts of C_{6+} hydrocarbons with decreasing distance from hydrocarbon deposits; (C) the fact that C_1 to C_5 (mainly C_1 to C_3) hydrocarbons predominantly compose microseeping hydrocarbons at the earth's surface; and (D) the strong gravitational component reflected by the sharp surface boundaries of anomalies above hydrocarbon deposits. The existence of microfracture systems above hydrocarbon deposits, as advocated in this paper, has not yet been clearly demonstrated.

5. Bacterial interaction with vertically migrating hydrocarbons (including microseeping hydrocarbons) is the most important controlling parameter in surface geochemical exploration. This interaction has largely been disregarded in the past.

6. Although the maximum microbial activity occurs in surface soils, bacteria oxidize hydrocarbons over a staggering range of conditions, including all sediment depths from a hydrocarbon deposit to the earth's surface.

7. Surface bacterial activity can totally obliterate a hydrocarbon anomaly caused by microseeping hydrocarbons, as measured by soil-air or soil-sorbed hydrocarbon analyses, or integrative absorption. Such activity can also cause large seasonal variations in such anomalies.

8. Paraffin dirt is the result of surface bacterial utilization of microseeping hydrocarbons and is therefore an indirect geochemical detection method which has been responsible for the discovery of large amounts of crude oil in the Gulf Coast region.

9. Hydrocarbons are kinetically stable under all microseepage conditions and do not, in and of themselves, create reducing environments. Such environments are clearly due to microbiologic oxidation of vertically migrating hydrocarbons (including microseeping hydrocarbons).

10. "Chimneys" are rock columns above leaking hydrocarbon deposits having different physical characteristics than stratigraphically equivalent rocks away from the hydrocarbon deposit. By this paper's model, chimneys are created by the interaction of various bacteria with vertically migrating hydrocarbons (including microseeping hydrocarbons). Significant amounts of CO_2, and to a lesser extent H_2S and organic acids, are produced by this interaction. These reactive chemical species significantly change the eH and pH of the environments in which the bacteria are active and also change mineral stability fields in rock columns above hydrocarbon deposits. These changes result in the solution and mobilization of some mineral species and the precipitation of other species.

11. A number of mineralogic changes have been observed in rock columns above hydrocarbon deposits. (A) Radioactivity decreases have been documented in shales and soils above hydrocarbon deposits compared to background values. These decreases, by this paper's model, are due to acid solutions created by bacterial oxidation of vertically migrating hydrocarbons causing clay breakdown (as these solutions are neutralized). This neutralization leaches potassium 40 from clays and/or other radioactive elements from organic-rich black shales. (B) The induration of sediments documented above hydrocarbon deposits is believed to be due to carbonate precipitation from the microbially produced CO_2 and from silica cementation resulting from clay breakdown to silica and alumina. This in turn is due to neutralization of the acidic solutions caused by high CO_2 concentrations. (C) Changes in iron/manganese-bearing mineral species from eH/pH alterations have been documented and are also believed to be the result of bacterial oxidation of vertically migrating hydrocarbons. The consequent iron/manganese mineralogic modifications result in anomalous magnetic and/or electric potential measurements compared to unaltered country rocks.

12. By this paper's model, intense apical microbial activity over leaking hydrocarbon deposits creates anomalous concentrations of CO_2 and other carbonate species, which diffuse radially outward to meet inward diffusing Ca^{+2} ions. This ionic meeting results in a halo precipitation of carbonate surrounding the apical area of highest microbial activity in the entire rock column above a hydrocarbon deposit. CO_2 and hydrocarbons are occluded in the calcite during its precipitation. Subsequent basinal uplift and erosion expose this calcite sheath at the surface. This proposed scenario results in the halo anomalies observed for (A) Duchscherer's ΔC technique, (B) analysis for soil-occluded C_2-C_5 hydrocarbons (Horvitz), and (C) soil-$CaCO_3$ ("soil salt") surveys.

13. By this paper's model (A) soil-air and soil-sorbed hydrocarbon anomalies can be either apical or halos, with halos predominating in humid climates. (B) Soil-occluded hydrocarbon anomalies should always be halos. (C) Microbial anomalies should always be apical.

14. Intense apical microbial oxidation of microseeping hydrocarbons results in the halo anomalies sometimes found with soil-air and soil-sorbed hydrocarbon analyses. By this paper's model (and in accordance with Soli's 1957 hypothesis), the higher hydrocarbon concentrations of the ring portion of the halo anomaly represent microseeping hydrocarbon concentrations above background levels but too low to sustain intense microbial activity.

15. Conclusive evidence from nature demonstrates that caprock "plugging" above hydrocarbon deposits is not repsonsible for halo anomalies due to the sealing of the apical area of the deposit by cementation. This evidence is in the form of hydrocarbon gas-logging surveys run over fields with halo anomalies, and the fact that soil-air and soil-sorbed hydrocarbon surveys sometimes give apical anomalies, and almost universally, microbial surveys give apical anomalies.

16. Geochemical anomalies caused by abnormal CO_2 concentrations in near-surface soils and/or ground waters are nonspecific for thermogenic hydrocarbon deposits, since such anomalies can be caused by factors other than microbial oxidation of microseeping hydrocarbons. Such anomalies are soil-carbonate contents, radiation lows, and soil mineral alteration.

17. Hydrocarbon microseepage or leakage from biogenic methane deposits, including "tight" and/or noncommercial biogenic methane deposits, can create many of the same geochemical anomalies created by hydrocarbon microseepage from thermogenic hydrocarbon deposits (including magnetic and soil-carbonate anomalies). Therefore, only direct geochemical detection methods which measure C_2-C_5 hydrocarbons can be considered specific for thermogenic hydrocarbon deposits. "False anomalies" from microseeping biogenic methane deposits likely have been responsible for many geochemical failures, especially on the stable shelves of sedimentary basins.

18. Direct geochemical detection surveys specific for thermogenic hydrocarbon deposits (soil analyses for occluded or sorbed hydrocarbons, soil-air analysis, microbial analysis, and integrative absorption) all can show large seasonal variations in certain geographic-climatological environments. In different geographic-climatological environments, certain direct geochemical detection methods may not be applicable. Excluding integrative absorption, all methods require deep (at least 6 ft, 1.8 m) surface boreholes, which results in long field times, high field expenses, and low field mobility. Also, sampling for these methods can be difficult and/or expensive; and sample preparation and analysis can be time consuming, difficult, and/or expensive.

19. Soil-air analysis is difficult because of very low concentrations of C_2-C_5 hydrocarbons in soil air. Because of soil breathing, caused by a number of different factors, soil-air analysis for the most part is not reproducible and data from large regional surveys are not comparable.

20. Analysis for soil-sorbed hydrocarbons is undependable because of (A) inhomogeneous sorption characteristics of soils over short lateral distances, which results in poor sample reproducibility, (B) concentration changes caused by seasonal, diurnal, or climatological factors, and (C) nonquantitative desorption of soil-sorbed hydrocarbons during thermal/vacuum analyses.

21. Analysis for soil-occluded hydrocarbons by acid treatment of soils is clearly superior to soil-air and soil-sorbed hydrocarbon analyses, although problems also exist with this technique. (A) The ratio of sorbed to occluded hydrocarbons could affect the analysis. (B) The method is not applicable in certain cases; for example, where the soil contains unweathered carbonates bearing rock fragments with significant contents of syngenetic hydrocarbon gases.

22. The proper use of microbial soil surveying apparently continues to be ill defined, and problems exist despite very impressive prediction rates by some investigators. An analytical technique has not been developed that is quick, easily employed, trustworthy, and inexpensive. Serious contradictions exist between different studies. Considering the pivotal role which microbial oxidation of microseeping hydrocarbons (and vertically migrating hydrocarbons in general) plays in direct geochemical detection methods (and surface geochemical exploration methods in general), resolution of these contradictions and problems would be appropriate.

23. Integrative absorption is the use of absorbents (activated charcoal, silica gel, etc.), or other "collectors," in shallow (1–2 ft, 0.3 to 0.6 m) surface holes to measure the flux of microseeping hydrocarbons over a given area. Major advantages have been claimed with integrative absorption compared to other direct geochemical detection methods, such advantages as (A) greatly decreased field time and field expenses, (B) greatly increased field mobility, (C) the negation of barometric, climato-

logical, diurnal, and even seasonal variations by leaving collectors buried for sufficient time periods, and (D) significant possible decreases in sample-analysis expense and difficulty. Convincing evidence supporting these claims has not been published, and the method also has drawbacks and limitations. (A) Early attempts at integrative absorption were failures because of incorrect utilization of the technique, and insufficient analytical capabilities for microconcentrations of hydrocarbon gases. (B) The sparse application of the method has not yet revealed all its shortcomings. (C) The principal limitation of integrative absorption is its application in the upper soil zone, which is the realm of the most intense plant and microbial activity. This activity produces, or has the potential to produce, different organic and inorganic compounds, which can lead to significant analytical problems and/or errors.

24. Direct geochemical detection methods, contrary to previously expressed opinions, are ill suited as exploration tools in tectonically disturbed areas because of the possibility of false anomalies over surface traces of faults and fractures from (A) the discharge of deep basinal waters carrying dissolved hydrocarbons up faults, and (B) displacement of anomalies away from the surface trace of a field by faulting associated with the field.

25. The present optimum use of direct geochemical detection methods is in field definition after a wildcat discovery and in extension of older producing fields.

26. The greatest possible contribution of direct geochemical detection to petroleum exploration is in the search for fields in stratigraphic and low-relief traps, invisible to seismic surveying, in flat-lying sediments on the stable shelves of petroleum basins. This is an area, however, where the previous record of direct geochemical detection has not been encouraging.

27. The high hydrocarbon microseepage rates called for by some investigators are geologically untenable.

28. The creation of surface anomalies by gas-induced groundwater evaporation is unlikely. (A) The amounts of required gas are unrealistically large. (B) A hydrocarbon microseepage mechanism based on diffusion is not possible. (C) The hydrocarbon microseepage mechanism favored in this paper produces the exact opposite results of gas-induced groundwater evaporation.

29. It is possible, but undocumented, that some microseeping and/or effusing hydrocarbon deposits "seal" themselves as the result of mineral precipitation in macrofractures and microfractures from bacterial oxidation of vertically migrating hydrocarbons.

30. Ongoing primary and/or secondary migration in "live" petroleum basins (e.g., the Gulf Coast) may create false surface anomalies. There is, however, no documentation of this hypothesis.

31. Hydrocarbon microseepage does exist and can create surface anomalies which outline hydrocarbon pools and therefore can be used as a viable exploration tool.

32. A myriad of proposals for definitive research on hydrocarbon microseepage and surface geochemical exploration methods could be made. The most important step that could possibly be taken at this point, however, are high-quality translations of the post-1960 definitive Soviet publications, particularly translations of books and long journal review articles. The efforts put into the Kartsev et al. (1959) publication by the translators, P. A. Witherspoon and W. D. Romey of the University of California at Berkeley and by Standard Oil Company of California (now Chevron) which financed the translation, is to be highly applauded.

ACKNOWLEDGMENTS

Gratitude is extended to Pete Groth, Leo Horvitz, and Al Roberts for reviewing the manuscript.

REFERENCES

Abd-El-Malek, Y., and Rizk, 1963, Bacterial sulfate reduction and the development of alkalinity: Journal of Applied Bacteriology, v. 26, p. 7–26 (from Davis, 1967).

Alekseyev, F. A., 1959, Radiometric method of oil and gas exploration: Iadernaia Geofizkia: Moscow, Gostoptekhizdat, p. 3–26. Abstract, *in* Geophysical Abstracts, no. 183, item 525, (Oct.–Dec. 1960), p. 607.

———, and Gottikh, R. P., 1966, Mechanism of origin of radiometric anomalies over oil deposits: International Geologic Review, v. 8, no. 10, p. 1157–1171.

Alparone, N., 1963, Subsurface anomalous radioactivity in the environment of oil fields: Austin, University of Texas, M.S. thesis, 80 p.

Alty, T., 1933, Diffusion of monoatomic gases through fused silica: Philosophical Magazine, v. 15, p. 1035–1048.

American Geological Institute, 1976, Dictionary of geological terms: New York, Anchor Books, 472 p.

Armstrong, F. E., and Heemstra, R. J., 1973, Radiation halos and hydrocarbon reservoirs: A review: U.S. Bureau Mines Information Circular 8579, 52 p.

Ashirov, K. B., and Sazonova, I. V., 1962, Biogenic sealing of oil deposits in carbonate reservoirs: Mikrobiologiya, v. 31, p. 680–683.

Baijal, S. K., 1962, Geo-chemical survey of the Hilbig Oil Field, Bastrop County, Texas: Austin, University of Texas, M.S. thesis, 42 p.

Bailey, N. J. L., Jobson, A. M., and Rogers, M. A., 1973, Bacterial degradation of crude oil: Comparison of field and experimental data: Chemical Geology, v. 11, p. 203–221.

Baker, E. G., 1959, Origin and migration of oil: Science, v. 129, p. 871–874.

Barrer, R. M. 1934, The mechanism of activated diffusion through silica glass: Journal of the Chemical Society. p. 378–386.

Barrow, G. M., 1961, Physical chemistry: New York, McGraw-Hill, 694 p.

Barton, D. C., 1927, The indications of the oil fields in the Mid-Continent and Gulf Coastal plain of the United States: Journal of the Institute of Petroleum Technology, v. 13, p. 333–339.

Bernard, B., 1981, Use of sediment gas anomalies in surface prospecting: American Association of Petroleum Geologists Bulletin, v. 65, p. 900.

Bikerman, J. J., [1958], Surface chemistry, (2d ed.): New York, Academic Press, 501 p.

Boyle, R. W., and Garrett, R. G., 1970, Geochemical prospecting: A review of its status and future: Earth Science Reviews, v. 6, p. 51–75.

Braaten, E. O., and Clark, G. F., 1935, The diffusion of helium through fused silica: Journal of the American Chemical Society, v. 57, p. 2714–2717.

Bronston, A., 1974, Geochemical survey data forecast major field: World Oil, November, p. 112–116.

Buckley, S. E., Hocott, C. R., and Taggart, M. S., 1958, Distribution of dissolved hydrocarbons in subsurface waters, *in* Weeks, L. G., ed., Habitat of oil: A Symposium: American Association of Petroleum Geologists, Tulsa, Oklahoma, p. 850–882.

Burton, E. F., Braaten, E. O., and Wilhelm, J. O., 1933, Diffusion of helium through quartz: Relation to temperature: Canadian Journal of Research, v. 8, p. 463–467.

Chapman, R. E., 1982, Effects of oil and gas accumulation on water movement: American Association of Petroleum Geologists Bulletin, v. 66, p. 368–378.

Coleman, D. D., Meents, W. F., Liu, Chao-Li, and Keough, R. A., 1977, Isotopic identification of leakage gas from underground storage reservoirs: A progress report: Illinois State Geological Survey, Illinois Petroleum, no. 111, 10 p.

Colombo, U., 1959, In discussion section, p. 682–687, of Sokolov, V. A., Alexeyev, F. A., Bars, E. A., Geodekyan, A. A., Mogilevsky, G. A., Yurovsky, Y. M., and Yasenev, B. P., 1959, Investigations into direct oil detection methods: Fifth World Petroleum Congress, sec. I, Proceedings, Geology and Geophysics, p. 667–687.

Connan, J., Le Tran, K., and van der Weide, B., 1975, Alteration of petroleum in reservoirs: Ninth World Petroleum Congress, Proceedings, Applied Science Publishers, v. 2, p. 171–178.

Crockford, H. D., and Knight, S. B., 1959, Fundamentals of physical chemistry: New York, John Wiley & Sons, 463 p.

Curry, W. H., 1984, Evaluation of surface gamma radiation surveys for petroleum exploration in the deep Powder River Basin, Wyoming, *in* Davidson, M. J., and Gottlieb, B. M., eds., Unconventional methods in exploration for petroleum and natural gas III: Dallas, Southern Methodist Unviersity Press, p. 25–39.

Dalziel, M. C., and Donovan, T. J., 1984, Correlations of suspected petroleum-generated biogeochemical and aeromagnetic anomalies, Bell Creek Oil Field, Montana, *in* Davidson, M. J., and Gottlieb, B. M., eds., Unconventional methods in exploration for petroleum and natural gas III: Dallas, Southern Methodist University Press, p. 59–69.

Davidson, M. J., 1963, Geochemistry can help find oil if properly used: World Oil, July, p. 94–100.

———, 1967, Petroleum geochemical exploration from space: A radical new technology: Unpublished National Aero Space Administration (NASA) manuscript.

———, 1981, The vertical migration mechanism in surface geochemical prospecting for petroleum: American Chemical Society Annual Meeting, Atlanta, Georgia, April 1, 1981, (unpublished talk).

———, 1982, Toward a general theory of vertical migration: Oil and Gas Journal, June 21, p. 288–300.

———, 1984, State of the art and direction of unconventional oil and gas exploration, *in* Davidson, M. J., and Gottlieb, B. M. eds., Unconventional methods in exploration for petroleum and natural gas III: Dallas, Southern Methodist University Press, p. 4–8.

Davis, J. B., 1952, Studies on soil samples from a "paraffin dirt" bed: American Association of Petroleum Geologists Bulletin, v. 36, p. 2186–2188.

———, 1956, Microbial decomposition of hydrocarbons: Industrial and Engineering Chemisty, v 48, no. 9, p. 1444–1448.

———, 1967, Petroleum microbiology: New York, Elsevier, 604 p.

———, 1969, Microbiology in petroleum exploration, *in* Heroy, W. B., ed., Unconventional methods in exploration for petroleum and natural gas: Dallas, Southern Methodist University Press, p. 139–158.

Debnam, A. H., 1969, Geochemical prospecting for petroleum and natural gas in Canada: Geological Survey of Canada Bulletin 177, 26 p.

Denison, D., 1983, Geochemistry exploration techniques being used to help pin down hydrocarbon prospects in Michigan: Michigan's Oil and Gas News, December 23, 1983, p. 161–164.

Deroo, G., Tissot, B., McCrossan, R. G., and Der, F., 1974, Geochemistry of the heavy oils of Alberta, *in* Oil sands—Fuel of the future: Canadian Society Petroleum Geologists Memoir 3, p. 148–167, 184–189.

Devine, S. B., and Sears, H. W., 1975, An experiment in soil geochemical prospecting for petroleum, Della Gas Field, Cooper basin: Australian Petroleum Exploration Association Journal, v. 15, p. 103–110.

———, 1977, Soil hydrocarbon geochemistry: A potential petroleum exploration tool in the Cooper basin, Australia: Journal of Geochemical Exploration, v. 8, p. 397–414.

Dickey, P. A., and Hunt, J. M., 1972, Geochemical and hydrogeologic methods of prospecting for stratigraphic traps, *in* King, R. E., ed., Stratigraphic oil fields classification, exploration methods, and case histories: American Association of Petroleum Geologists Memoir 16, p. 136–167.

———, and Cox, W. C., 1977, Oil and gas reservoirs with subnormal pressures: American Association of Petroleum Geologists Bulletin, v. 61, p. 2134–2142.

Donovan, T. J., 1974, Petroleum microseepage at Cement, Oklahoma: Evidence and mechanism: American Association of Petroleum Geologists Bulletin, v. 58, p. 429–446.

———, Friedman, I., and Gleason, J. D., 1974, Recognition of petroleum-bearing traps by unusual isotopic compositions of carbonate-cemented surface rocks: Geology, July, p. 351–354.

———, Noble, R. L., Friedman, I., and Gleason, J. D., 1975, A possible petroleum-related geochemical anomaly in surface rocks, Boulder and Weld Counties, Coloradao: U.S. Geological Survey Open-File Report 75-47, 11 p.

———, and Dalziel, M.C., 1977, Late diagenetic indicators of buried oil and gas: U.S. Geological survey Open-File Report, 77-817, 44 p.

———, Termain, P. A., Henry, M. E., 1979A, Late diagenetic indicators of buried oil and gas II: Direct detction experiment at Cement and Garza oil fields, Oklahoma and Texas, using enhanced LANDSAT I and II Images: U.S. Geological Survey Open-File Report 79-243, 45 p.

———, Forgey, R. L., Roberts, A. A., 1979B, Aeromagnetic detection of diagenetic magnetite over oil fields: Geological Society of America Bulletin, v. 63, p. 245–248.

———, Roberts, A., and Dalziel, M., 1981, Epigenetic zoning in surface and near surface rocks resulting from seepage induced redox gradients, Velma Oil Field, Oklahoma: American Association of Petroleum Geologists Bulletin, v. 65, p. 919.

Duchscherer, W., 1980, Geochemical methods of prospecting for hydrocarbons: Oil and Gas Journal, December 1, p. 194–208.

———, 1981A, Carbonates and isotope ratios from surface rocks: A geochemical guide to underlying petroleum accumulations, *in* Gottlieb, B. M., ed., Unconventional methods in exploration for petroleum and natural gas II: Dallas, Southern Methodist University Press, p. 201–218.

———, 1981B, Nongasometric geochemical prospecting for hydrocarbons with case histories: Oil and Gas Journal, October 19, p. 312–327.

———, 1982, Geochemical exploration for hydrocarbons: No new tricks, but an old dog: Oil and Gas Journal, July 5, p. 163–176.

Evans, C. R., Rogers., M. A., Drushel, H. V., and Koons, C. B., 1971, Evolution and alteration of petroleum in western Canada: Chemical Geology, v. 8, p. 147–170.

Fedynsky, V. V., Karus, E. V., Polshkov, M. K., 1975, Geophysical and geochemical surface surveys for detection of oil and gas pools: Ninth World Petroleum Congress, Proceedings, v. 3, p. 279–288.

Gallagher, A. V., 1984, Iodine: A pathfinder for petroleum deposits, *in* Davidson, M. J., and Gottlieb, B. M., eds, Unconventional methods in exploration for petroleum and natural gas III: Dallas, Southern Methodist University Press, p. 148–159.

Gautier, D. L., 1981, Lithology, reservoir properties, and burial history of a portion of Gammon Shale (Cretaceous), southwestern North Dakota: American Association of Petroleum Geologists Bulletin, v. 65, p. 1146–1159.

Gervirtz, J. L., Carey, B. D., and Blanco, S. R., 1982, Surface geochemical exploration in the North Sea: Journal of Petroleum Geology, v. 5, p. 208.

Haas, H. F., Yantzi, M. F., and Bushnell, L. D., 1941, Microbial utilization of hydrocarbons: Kansas Academy Sciences, Transactions, v. 44, p. 39–45.

Hamm, W. D., 1969, A challenge to the oil finder of the future, *in* Heroy, W. B., ed., Unconventional methods in exploration for petroleum and natural gas: Dallas, Southern Methodist University Press, p. 11–14.

Heemstra, R. J., Ray, R. M., Wesson, T. C., Abrams, J. R., and Moore, G. A., 1979, A critical laboratory and field evaluation of selected surface prospecting techniques for locating oil and natural gas: Department of Energy, Bartlesville Energy Technology Center, BETC/RI-78/18, 84 p.

Henry, M. E., and Donovan, T. J., 1978, Isotopic and chemical data from carbonate cements in surface rocks over and near four Oklahoma oil fields: U.S. Geological Survey Open-File Report 78-927, 17 p.

Heroy, W. B., Jr., 1982, Field demonstration of unconventional methods in exploration for petroleum and natural gas: Final report to the Department of Energy, DE-AC01-81RA50653, 14 p.

Hickey, J. C., Klusman, R. W., and Voorhees, K. J., 1983A, An integrative gas geochemical technique for surficial petroleum exploration: American Association of Petroleum Geologists Bulletin, v. 67, . 483.

———, 1983B, Fault leakage characterization of integrative gas geochemistry/mass spectrometry/pattern recognition procedures: American Association of Petroleum Geologists Bulletin, v 67, p. 1342.

Hoffman, M. G., 1939, An advance in exploration by soil analysis methods: Oil and Gas Journal, March 16, p. 23–24, 115.

Horvitz, L., 1939, On geochemical prospecting, I: Geophysics, v. 4, p. 210–228.

———, 1945, Recent developments in geochemical prospecting for petroleum: Geophysics, v. 10, p. 487–493.

———, 1950, Chemical methods, *in* Jakosky, J. J., ed., Exploration geophysics, (2d ed.): Los Angeles, Trija Publishing, p. 938–965.

———, 1954, Near-surface hydrocarbons and petroleum accumulation at depth: Mining Engineering, December, 1954, p. 1205–1209.

———, 1957, How geochemical analysis helps the geologist find oil: Oil and Gas Journal, no. 45, p. 234–242.

———, 1965, Case histories show results of drilling geochemical prospects: World Oil, November, p. 161–165.

———, 1969, Hydrocarbon geochemical prospecting after thirty years, *in* Heroy, W. B., ed., Unconventional methods in exploration for petroleum and natural gas: Dallas, Southern Methodist University Press, p. 205–218.

———, 1972, Vegetation and geochemical prospecting for petroleum: American Association of Petroleum Geologists Bulletin, v. 56, p. 925–940.

———, 1980, Near-surface evidence of hydrocarbon movement from depth, *in* Roberts, W. H., and Cordell, R. J., eds., Problems of petroleum migration: American Association of Petroleum Geologists Studies in Geology no. 10, p. 241–263.

———, 1981, Hydrocarbon geochemical prospecting after forty years, *in* Gottlieb, B. M., ed., Unconventional methods in exploration for petroleum and natural gas II: Dallas, Southern Methodist University Press, p. 83–95.

Hunt, J. M., 1979, Petroleum geochemistry and geology: San Francisco, Freeman and Company, 617 p.

———, 1981, Surface geochemical prospecting: Pro and con: American Association of Petroleum Geologists Bulletin, v. 65, p. 939.

Jensen, H. L., 1931, Contributions to our knowledge of the Actinomycetales II: The definition and subdivision of the genus Actinomycetales, with a preliminary account of Australian soil Actinomycetales: Linnean Society, New South Wales, Proceedings , v. 56, p. 345–370.

Jones, P. H., 1981, Hydrogeology of petroleum and natural gas, *in* Gottlieb, B. M., ed., Unconventional methods in exploration for petroleum and natural gas II: Dallas, Southern Methodist University Press, p. 50–63.

———, 1984, Deep water discharge: A mechanism for the vertical migration of oil and gas, *in* Davidson, M. J., and Gottlieb, B. M. eds., Unconventional methods in exploration for petroleum and natural gas III: Dallas, Southern Methodist University Press, p. 254–271.

Jones, V. T., 1982, Prediction of oil or gas potential by near-surface geochemistry: American Association of Petroleum Geologists Bulletin, v. 66, p. 245.

———, and Drozd, R. J., 1983, Predictions of oil or gas potential by near-surface geochemistry: American Association of Petroleum Geologists Bulletin, v. 67, p. 932–952.

Kartsev, A. A., Tabasaranskii, Z. A., Subbota, M. I., and Mogilevskii, G. A., 1959, Geochemical methods of prospecting and exploration for petroleum and natural gas: Los Angeles, University of California Press, 349 p. (English Translation edited by P. A. Witherspoon, and W. D. Romey. The Russian original was published in 1954.).

Kilmar, C., 1983, Radiation lows over productive areas seen as soil geochemical phenomenon: Oil and Gas Journal, July 25, p. 179–182.

Klusman, R., and Voorhees, K., 1983, A new development in petroleum exploration technology: Mines Magazine, reprint, v. 73, no. 3, March 1983, p. 1–4.

Kuzmin, Yu., Ya., 1983, Analysis of results and methods of increasing the effectiveness of geochemical exploration for oil and gas on the Russian plate: International Geological Review, September, 1983, p. 1089–1094.

Lasky, B. H. 1950, Surface minerals reflect subsurface structure in new exploration technique: World Oil, October, p. 73–81.

Laubmeyer, G., 1933, A new geophysical prospecting method, especially for deposits of hydrocarbons: Petroleum (London), v. 29, p. 1–4.

Lewis, W. B., McDermott, E. M. and Rosaire, E. E., 1940, Geochemical methods, *in* Jakosky, J. J., ed., Exploration Geophysics (1st ed.): London, Times Mirror Press, p. 653.

Lundberg, A. H., 1952, An attempt to interpret radioactive patterns from airborne readings: Geological Association of Canada, Proceedings, v. 5, p. 117–125.

———, 1956, What causes low radiation intensities over oil fields?: Oil and Gas Journal, April 30, p. 192–195.

MacElvain, R., 1963A, What do near-surface signs really mean in oil finding?: Oil and Gas Journal, February 18, p. 132–136.

———, 1963B, What do near-surface signs really mean in oil finding?—Part 2: Oil and Gas Journal, February 25, 139–146.

———, 1969, Mechanics of gaseous ascension through a sedimentary column, *in* Heroy, W. B., ed., Unconventional methods in exploration for petroleum and natural gas: Dallas, Southern Methodist University Press, p. 15–28.

Magara, K., 1981, Hydrodynamics—does it trap oil?: Journal of Petroleum Geology, v. 4, p. 177–186.

Mattavelli, L., Ricchiuto, T., Grignani, D., and Schoell, M., 1983, Geochemistry and habitat of natural gases in Po basin, northern Italy: American Association of Petroleum Geologists Bulletin, v. 67, p. 2239–2254.

McAuliffe, C. D., 1978, Role of solubility in migration of petroleum from source, *in* Roberts, W. H., and Cordell, R. J., eds., Physical and chemical constraints on petroleum migration: Tulsa, Oklahoma, American Association of Petroleum Geologists, p. C-1 to C-39.

McBain, J. W., 1932, The sorption of gases and vapours by solids: London, George Routledge and Sons, 577 p.

McCrossan, R. G., Ball, N. L., and Snowdon, L. R., 1972, An evaluation of surface geochemical prospecting for petroleum, Olds-Caroline area, Alberta: Geological Survey of Canada, Department of Energy, Mines and Resources, Paper 71-31, 101 p.

McDermott, E., 1939, Concentrations of hydrocarbons in the earth: Geophysics, v. 4, p. 195–209.

———, 1940, Geochemical exploration (soil analysis). With speculation about the genesis of oil, gas, and other mineral accumulations: American Association of Petroleum Geologists Bulletin, v. 24, p. 859–881.

McKenna, E. J., and Kallio, R. E., 1965, The biology of hydrocarbons: Annual Reviews of Microbiology, v. 19, p. 183–208.

Meakin, C. P., 1971, The significance of surface indications of petroleum: Journal of the Australian Petroleum Exploration Association, v. 11, pt. 1, p. 126–130.

Merritt, J. W., 1952, Radioactive oil survey techniques: World Oil, July, p. 78–82.

Miller, G. H., 1976, Microbial surveys help evaluate geological, geophysical prospects: Oil and Gas Journal, October 4, p. 192–202.

Mills, R. V. A., and Wells, R. R., 1919, The evaporation and concentration of waters associated with petroleum and natural gas: U.S. Geological Survey Bulletin, 693, 104 p.

Milner, H. B., 1925, "Paraffin dirt": Its nature, origin, mode of occurrence, and significance as an indication of petroleum: Mining Magazine, v. 32, p. 73–85.

Mogilevskii, G. A., 1938, Microbiological investigation in connection with gas surveying: Razvedka Nedr., v. 8, p. 59–68.

———, 1940, The bacterial method of prospecting for oil and natural gases: Razvedka Nedr., v. 12, p. 32–43.

Momper, J. A., 1978, Oil migration limitations suggested by geological and geochemical considerations, *in* Roberts, W. H., and Cordell, R. J., eds., Physical and chemical constraints on petroleum migration: American Association of

Petroleum Geologists Continuing Education Course Note Series 8, p. B-1 to B-60.
Moore, W. J., 1964, Physical chemistry (3d ed.): Englewood Cliffs, New Jersey, Prentice-Hall, 844 p.
Neglia, S., 1979, Migration of fluids in sedimentary basins: American Association of Petroleum Geologists Bulletin, v. 63, p. 573–597.
Nisle, R. G., 1941, Considerations on the vertical migration of gases: Geophysics, v. 6. p. 449–454.
Phillipi, G. T., 1977, On the depth, time, and mechanism of origin of the heavy-to-medium-gravity naphthenic crude oils: Geochimica et Cosmochimica Acta, v. 41, p. 33–52.
Philp, R. P., and Crisp, P. T., 1982, Surface geochemical prospecting methods used for oil and gas prospecting: A review: Journal of Geochemical Exploration, v. 17, p. 1–34.
Pirson, S. J., 1940, Critical survey of recent developments in geochemical prospecting: American Association of Petroleum Geologists Bulletin, v. 24, p. 1464–1474.
———, 1941, Measure of gas leakage applied to oil search: Oil and Gas Journal, February 20, p. 21 and 32.
———, 1946, Disturbing factors in geochemical prospecting: Geophysics, v. 11, p. 312–320.
———, 1960, How to make geochemical exploration succeed: World Oil, April, p. 93–96.
———, 1963, Projective well log interpretation: World Oil, October, p. 116–120.
———, 1964A, Projective well log interpretation, Part 3: World Oil, October, p. 180–182.
———, 1964B, Projective well log interpretation, Part 4: World Oil, November, p. 156–166.
———, 1964C, Projective well log interpretation, Part 1: World Oil, August, p. 68–72.
———, 1964D, Projective well log interpretation, Part 2: World Oil, September, p. 83–86.
———, 1969, Geological, Geophysical and chemical modifications of sediments in the environment of oil fields, *in* Heroy, W. B., ed., Unconventional Methods in exploration for petroleum and natural gas: Dallas, Southern Methodist University Press, p. 159–186.
Poll, J. J. K., 1975, Onshore Gippsland geochemical survey: A test case for Australia: Journal Australian Petroleum Exploration Association, v. 15, p. 93–101.
Price, L. C., 1976, Aqueous solubility of petroleum as applied to its origin and primary migration: American Association of Petroleum Geologists Bulletin, v. 60, p. 213–244.
———, 1979, Aqueous solubility of methane at elevated pressures and temperatures: American Association of Petroleum Geologists Bulletin, v. 63, p. 1527–1533.
———, 1980, Crude oil degradation as an explanation of the depth rule: Chemical Geology, v. 28, p. 1–30.
———, 1981A, Aqueous solubility of crude oil to 400°C and 2,000 bars pressure in the presence of gas: Journal of Petroleum Geology, v. 4, p. 195–223.
———, 1981B, Primary petroleum migration by molecular solution: Consideration of new data: Journal of Petroleum Geology, v. 4, p. 89–101.
———, 1986, Primary petroleum migration from shales with oxygen-rich organic matter: Paper submitted to Journal of Petroleum Geology.
———, Blount, C. W., Mac Gowan, D., and Wenger, L., 1981, Methane solubility in brines with application to the geopressured resource, *in* Bebout, D. G., and Bachman, A. L., eds., Geopressured-Geothermal Energy, U.S. Gulf Coast: Fifth Conference, Proceedings, p. 205–214.
———, Wenger, L. M., Ging, T., and Blount, C. W., 1983, Solubility of crude oil in methane as a function of pressure and temperature: Organic Geochemistry, v. 4, p. 201–221.
———, Ging, T., Daws, T., Love, A., Pawlewicz, M., and Anders, D., 1984, Organic metamorphism in the Mississippian-Devonian Bakken Shale, North Dakota portion of the Williston Basin, *in* Woodward, J., Meissner, F. F., and Clayton, J. L., eds., Hydrocarbon source rocks of the greater Rocky Mountain region: Rocky Mountain Association of Geologists, p. 83–133.
Prutton, C. F., and Maron, S. H., 1951, Fundamental principles of physical chemistry, 2d ed.: New York, Macmillan, 591 p.
Ransone, W. R., 1947, Geochemical history of the Hardy Oil Field, Jones County, Texas: Geophysics, v. 12, p. 384–392.
———, 1958, Case history of the Sojourner Oil Field, Haskell County, Texas: A geochemical confirmation: Geophysics, v. 23, p. 574–578.
———, 1969, Case history of the Sixto Oil Field, Starr County, Texas: A geochemical discovery, *in* Heroy, W. B., ed., Unconventional methods in exploration for petroleum and natural gas: Dallas, Southern Methodist University Press, p. 187–196.
Rice, D. D., 1980, Chemical and isotopic evidence of the origins of natural gases in Offshore Gulf of Mexico: Gulf Coast Association of Geological Societies, Transactions, v. 30, p. 203–213.
———, 1984, Relation of hydrocarbon occurrence to thermal maturity of organic matter in the Upper Cretaceous Niobrara Formation, Eastern Denver basin: Evidence of biogenic versus thermogenic origin of hydrocarbons, *in* Woodward, J., Meissner, F. F., Clayton, J. L., eds., Hydrocarbon source rocks of the greater Rocky Mountain Region: Rocky Mountain Association of Geologists, p. 365–368.
———, and Claypool, G. E., 1981, Generation, accumulation, and resource potential of biogenic gas: American Association of Petroleum Geologists Bulletin, v. 65, p. 5–25.
Richers, D. M., Reed, R., Horstman, K. C., Michels, G. D., Baker, R. N., Lundell, L., and Marrs, R. W., 1982, LANDSAT and soil-gas geochemical study of Patrick Draw Oil Field, Sweetwater County, Wyoming: American Association of Petroleum Geologists Bulletin, v. 66, p. 903–922.
Roberts, W. H., 1980, Design and function of oil and gas traps, *in* Roberts, W., H., and Cordell, R. J., eds., Problems of petroleum migration: American Association of Petroleum Geologists Studies in Geology, no. 10, p. 217–240.
———, 1981, Some uses of temperature data in exploration, *in* Gottlieb, B. M., ed., Unconventional methods in exploration for petroleum and natural gas II: Dallas, Southern Methodist University Press, p. 8–49.
Rosaire, E. E., 1938, Shallow stratigraphic variations over Gulf Coast structures: Geophysics, v. 3, p. 96–121.
———, 1939, The Handbook of Geochemical Prospecting: Houston, Gulf Publishing Company, 61 p.
———, 1940, Geochemical prospecting for petroleum, *in* Symposium on Geochemical Exploration: American Association of Petroleum Geologists Bulletin, v. 24, p. 1400–1433.
———, McDermott, E., and Fash, R. H., 1940, Discussion of

geochemical exploration (soil analysis): American Association of Petroleum Geologists Bulletin, v. 24, p. 1434–1463.

Rybakova, S. G., 1957, Microflora in crude oil from deep tertiary formations on the Apsheron Peninsula: Doklady Akademia Nauk SSSR, v. 115, p. 813–815.

Samtsevich, S. A., 1955, Seasonal aspects and periodicity in the development of soil microorganisms: Mikrobiologiya, v. 24, p. 615–625.

Sanderson, R. T., 1940, Some neglected aspects of chemical exploration: Geophysics, v. 5, p. 284–294.

Sealey, J. Q., 1974A, A geomicrobiological method of prospecting for petroleum, Part 2: Oil and Gas Journal, April 15, p. 98–102.

———, 1974B, A geomicrogiological method of prospecting for petroleum, Part l: Oil and Gas Journal, April 8, p. 142–146.

Siegel, F. R., 1974, Applied geochemistry, *in* Geochemical prospecting for hydrocarbons: New York, John Wiley and Sons, Chapter 9, p. 228–255.

Sivaborvon, Vichai, 1974, Re-study of hydrocarbon distribution around the Hilbig Oil Field, Bastrup County, Texas: Austin, University of Texas, Masters thesis, 79 p.

Sivenas, P., and Beales, F., 1984, Electrochemical aspects of hydrocarbon reservoirs and surface seeps: American Association of Petroleum Geologists Bulletin, v. 68, p. 528.

Smith, G. H., and Ellis, M. M., 1963, Chromatographic analyses of gases from soils and vegetation, related to geochemical prospecting for petroleum: American Association of Petroleum Geologists Bulletin, v. 45, p. 1897–1903.

Söhngen, N. L., 1913, Benzin, petroleum, parraffinöl und paraffin als kohlenstoff und energiequelle für mikroben, Zentr. Bakt. Parasitenk. Infek. II, v. 37, p. 595–609.

Sokolov, V. A., 1936, Gas Surveying: Moscow, Gostoptekhizdat, 269 p.

———, 1970, The theoretical foundations of geochemical prospecting for petroleum and natural gas and the tendencies of its development, *in* Geochemical Exploration, Third International Geochemical Exploration Symposium, Toronto, Proceedings, April 16–18, 1970: Canadian Institute of Mining and Metallurgy, special volume 11, p. 544–549.

———, Alexeyev, F. A., Bars, E. A., Geodekyan, A. A., Mogilevsky, G. A., Yurovsky, Y. M., and Yasenev, B. P., 1959, Investigations into direct oil detection methods, *in* Fifth World Petroleum Congress, Proceedings, Section I: Geology and Geophysics, p. 667–687.

———, Geodekyan, A. A., Grigoryev, G. G., Krems, A. YA., Stroganov, V. A., Zorkin, L. M., Zeidelson, M. I., and Vainbaum, S. JA., 1970, The new methods of gas surveys, gas investigations of wells and some practical results: Canadian Institute of Mining and Metallurgy, special volume 11, p. 538–543.

———, Geodekyan, A. A., Stroganov, V. A., and Petzenko, M. G., 1971, Theory and techniques involved in direct geochemical prospecting for oil and gas: Translation of paper presented at Prague Symposium, supplied by Martin Davidson.

Soli, G., 1954, Geomicrobiological prospecting: American Association of Petroleum Geologists Bulletin, v. 38, p. 2255–2558.

———, 1957, Microorganisms and geochemical methods of oil prospecting: American Association of Petroleum Geologists Bulletin, v. 41, p. 134–140.

Stahl, W., and Faber, E., 1981, Geochemistry of gaseous hydrocarbons adsorbed in shallow sediments offshore Santa Barbara, California: American Association of Petroleum Geologists Bulletin, v. 65, p. 997.

Sternberg, B. K., and Oehler, D. Z., 1984, Electrical methods for hydrocarbon exploration I. Induced polarization (INDEPTH) method, *in* Davidson, M. J., and Gottlieb, B. M., eds., Unconventional methods in exploration for petroleum and natural gas III: Dallas, Southern Methodist University Press, p. 202–230.

Stklyanin, Yu. I., Breeva, T. F., Vasil'eva, L. B., Domina, N. A., Kotsonis, A. N., Kondiatov, L. S., Antoshina, I. N., Zav'-yalova, L. M., Kushchenko, V. N., and Shlepova, A., 1968, Conditions, parameters and means for the ascending migration of natural gases from petroleum and gas deposits, and some theoretical models of the migration: Trudy vses. nauchno-issled. Inst. Yad. Geofiz. Geochim., v. 4, p. 58–131.

———, Subbota, M. I., 1954, Mogilevskii's book: Microbiological method of prospecting for gas and petroleum deposits: Neft. Khoz., v. 32, p. 94–96.

Stormont, D. H., 1939, Gulf Coast field is opened on soil survey information: Oil and Gas Journal, July 20, p. 28–29.

Strawinski, R. J., 1955, A microbiological method of prospecting for oil: World Oil, November, p. 104–115.

———, and Tortorvich, J. A., 1955, Preliminary studies of methane-oxidizing bacteria and their possible role in oil prospecting: Bacteriological Proceedings, v. 56, p. 27.

Stroganov, V. A., 1969, Some recommendations of the employment of geochemical prospecting: Petroleum Gas Geology Geophysics, no. 7.

Sweet, G. E., 1966, The history of geophysical prospecting, Los Angeles, p. 249–250. (from Thompson, 1981).

Taylor, N. W., and Rast, W., 1938, The diffusion of helium and hydrogen through Pyrex chemically resistant glassware: Journal of Chemical Physics, v. 6, p. 612–619.

Telegina, Z. P., Subbota, M. I., ad Nikitina, E. A., 1963, Regularities underlying the occurence of hydrocarbon oxidizing bacteria in the waters of the Izbaskent oil-gas bed: Mikrobiologiya, v. 32, p. 33–38.

Thompson, C. K., 1981, The airborne gas sensor and its role in exploration reconnaissance, *in* Gottlieb, B. M., ed., Unconventional methods in exploration for petroleum and natural gas II: Dallas, Southern Methodist University Press, p. 240–248.

Thompson, K. F., 1979, Light hydrocarbons in subsurface sediments: Geochimica et Cosmochimica Acta, v. 43, p. 657–672.

Tilloy, R., and Manchaux, A., 1965, Radioactivé et recherches de pérole, *in* Petroleum National Congress, Fifth Le Touquet France, 1965: Association Francaise des Techniciens du Pérole, Proceedings, p. 975–983.

Tissot, B. P., and Welte, D. H., 1978, Petroleum formation and occurrence: New York, Springer-Verlag, 538 p.

Tripp, R. M., 1945, Thermodynamics of a gas migrating vertically through the sedimentary column: Geophysics, v. 10, p. 229–237.

T'sai, L. S., and Hogness, T. R., 1932, Diffusion of gases through fused quartz: Journal of Physical Chemistry, v. 36, p. 2595–2600.

Urry, W. D., 1932, Rare Gases I: The permeability of various glasses to helium: Journal of the American Chemical Society, v. 54, p. 3887–3901.

Vold, R. D., and Vold, M. J., 1983, Colloid and interface chemistry: London, Addison-Wesley.

Völz, H., and Schwartz, W., 1958, Basis of the geomicrobiological prospecting method: Erdöl Kohle, v. 11, p. 69–72.

———, 1962, The regional distribution of hydrocarbon oxidizing bacteria in various soils: Erdöl Kohle, v. 15, p. 426–429.

Weisman, T. J., 1979, Developments in geochemistry and their contribution to hydrocarbon exploration, *in* Tenth World Petroleum Congress, Proceedings, v. 2, p. 369–386.

Weyl, W., and Pincus, A. G., 1938, Role of gases in glass IV. Miscellaneous Gases: Glass Industry, v. 19, p. 455–456.

Winters, J. C., and Williams, J. A., 1969, Microbiological alteration of crude oil in the reservoir: American Chemical Society, Division of Petroleum Chemistry, New York, September 7–12, p. E-22 to E-31.

Zobell, C. E., 1946, Action of microörganisms on hydrocarbons: Bacteriological Review, v. 10, p. 1–50.

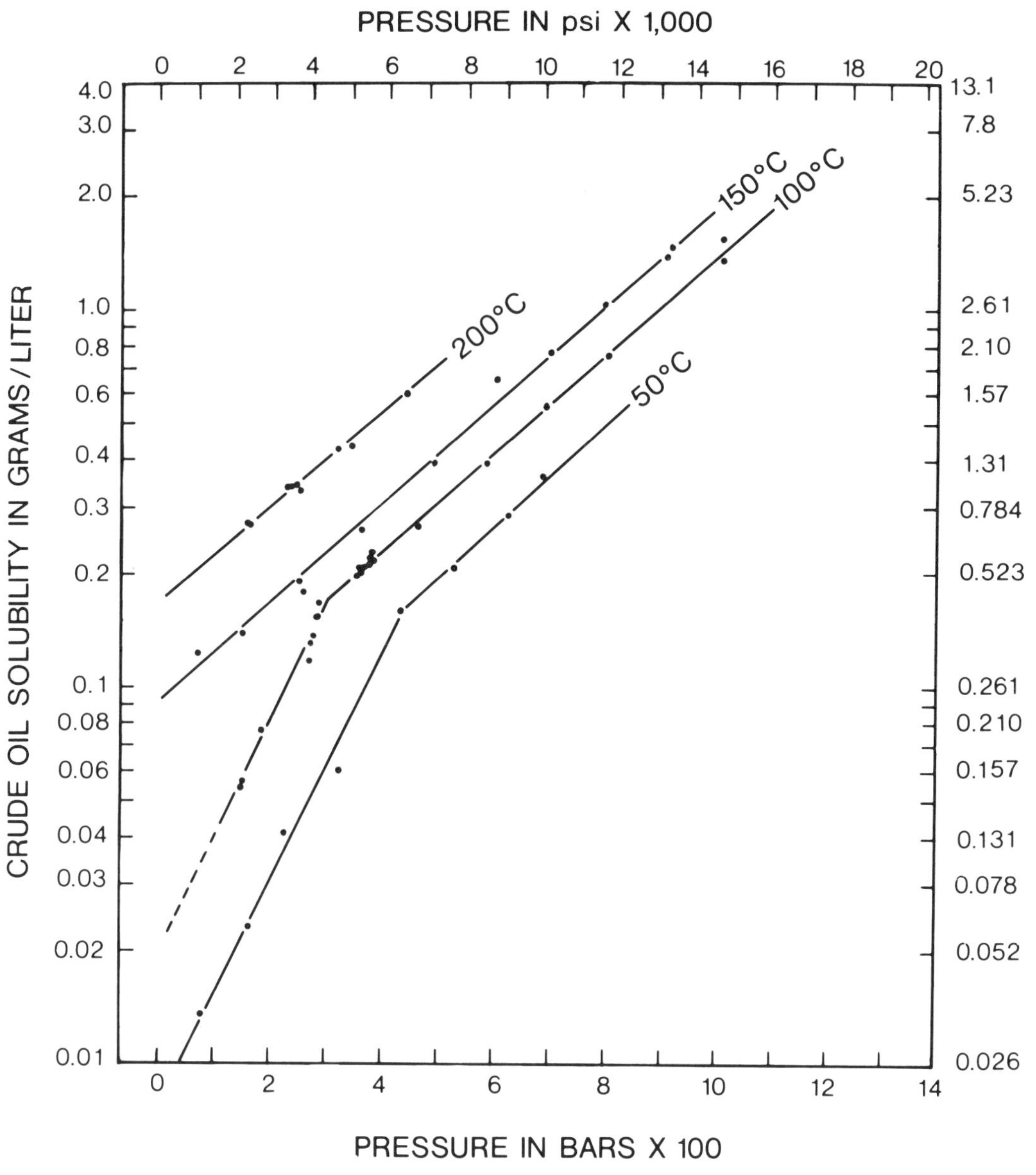

Fig. 1. Solubility of a whole crude oil in methane for different isotherms as a function of pressure with water present. Note that with decreases in pressure and temperature, the carrying capacity of methane for C_{5+} hydrocarbons decreases, especially at lower pressures and temperatures. After Price et al. (1983).

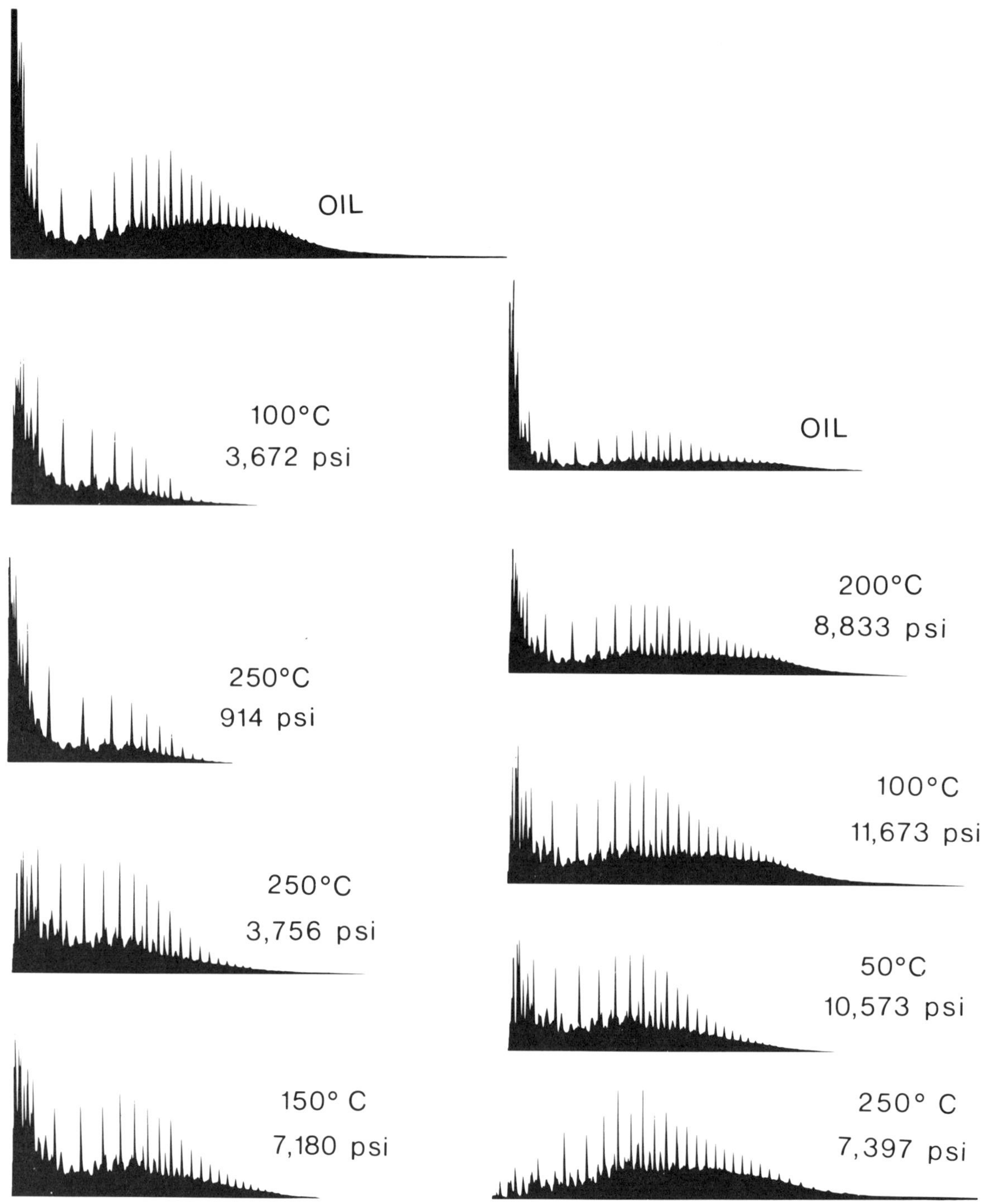

Fig. 2. Gas chromatograms of solute samples of crude oil showing distinct qualitative changes as a function of the pressure and temperature conditions at which the system was equilibrated. Note that with decreasing pressure and/or temperature, the molecular weight of the dissolved oil also decreases. After Price et al. (1983).

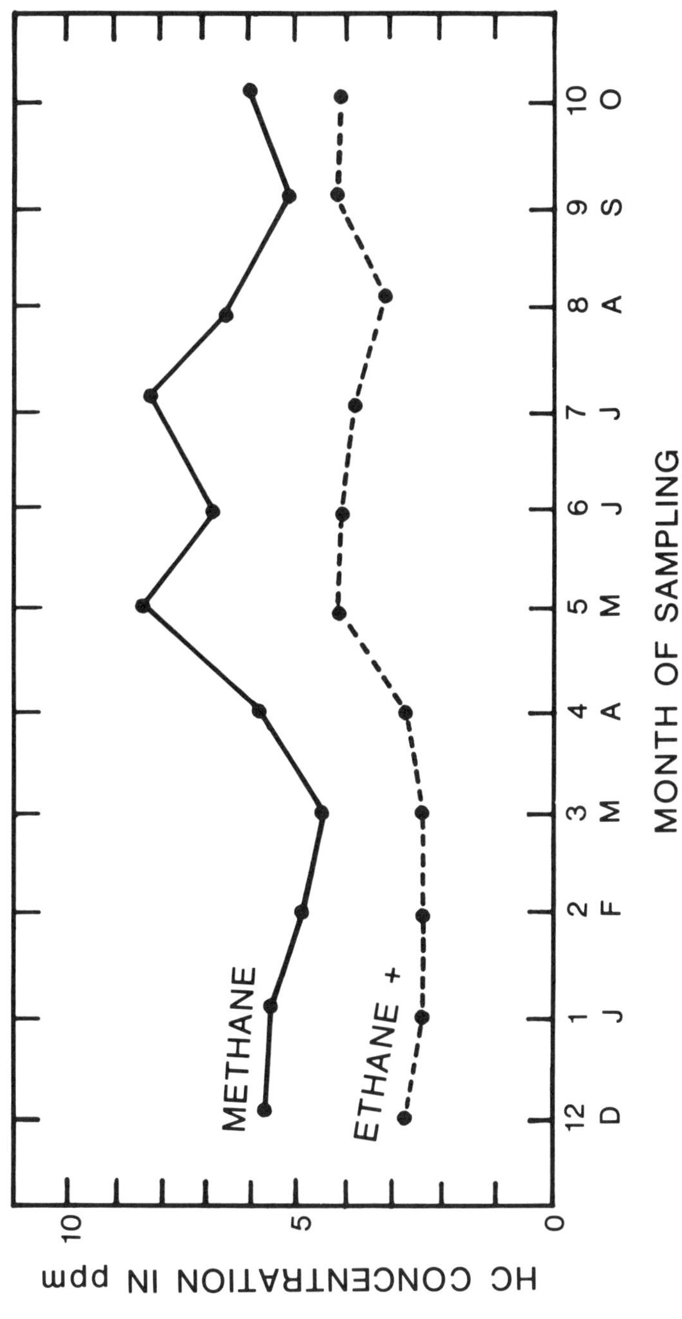

Fig. 3. Soil-sorbed hydrocarbon concentrations for the Krasnodar region, Soviet Union. After Kartsev et al. (1959, p. 84).

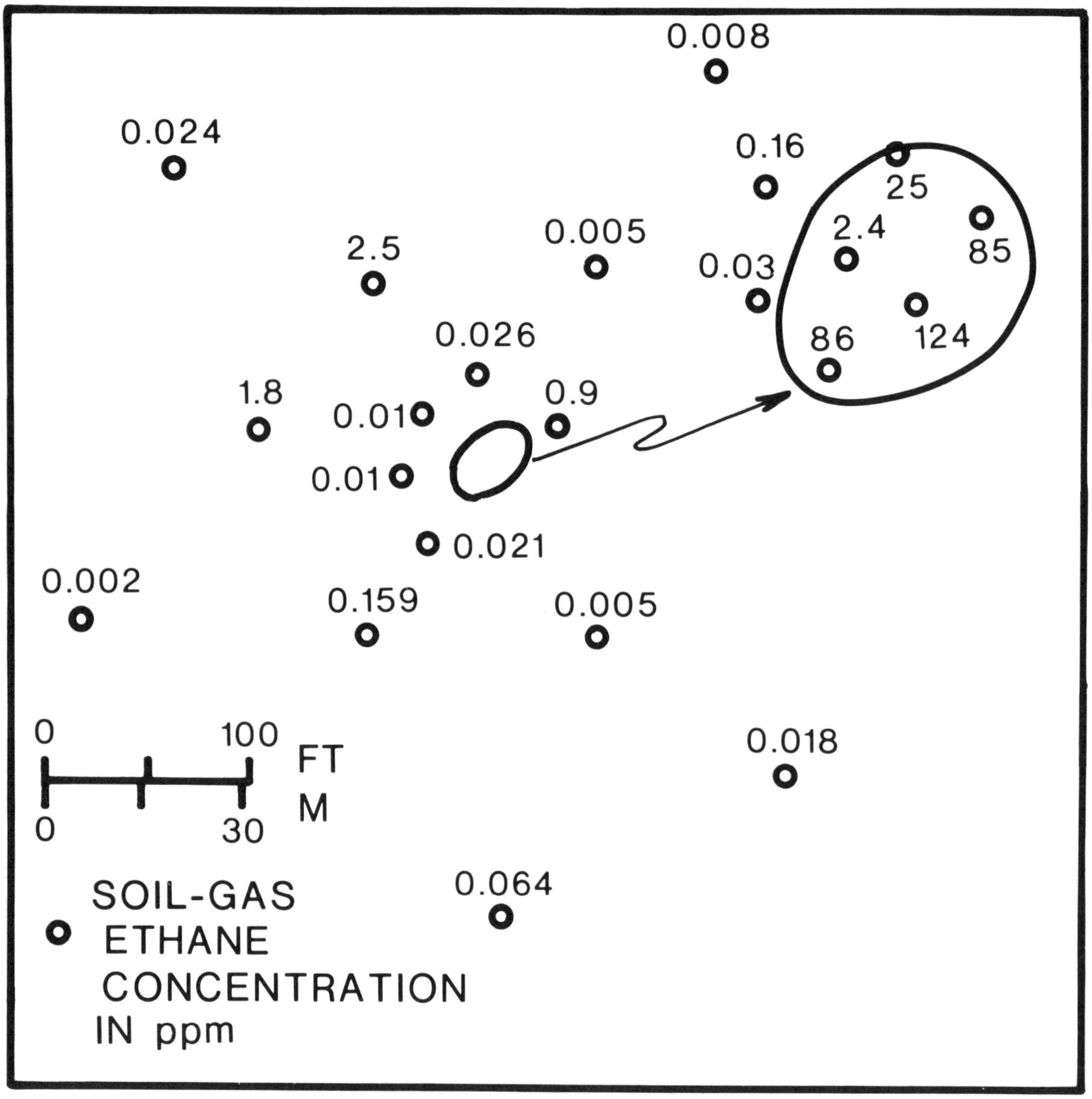

Fig. 4. Example of a paraffin dirt bed associated with hydrocarbon microseepage in Sabine County, Texas. Solid outline shows limits of paraffin dirt bed. Numbers are ethane soil-air concentrations in ppm. Note the association of higher ethane soil-air concentrations with the limits of the paraffin dirt bed. After Davis (1967, p. 102).

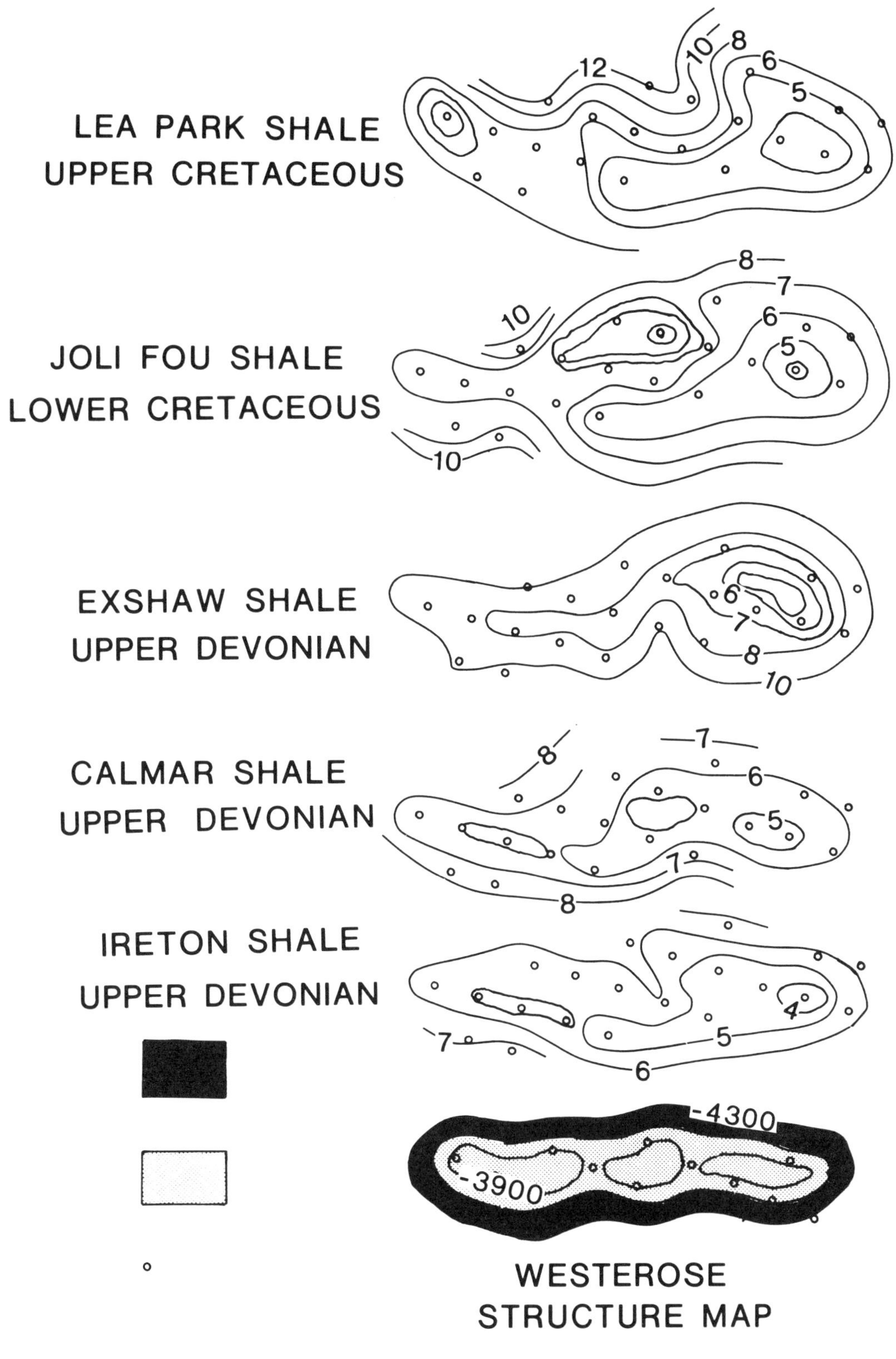

Fig. 5. Radioactivity in various shales overlying the Westerose Oil Field, Alberta, Canada, which produces from the Upper Devonian Leduc Formation. Radioactivity measurements as determined from well logs are given in "micro-grams Radium eq./ton units." After Pirson (1969, p. 176).

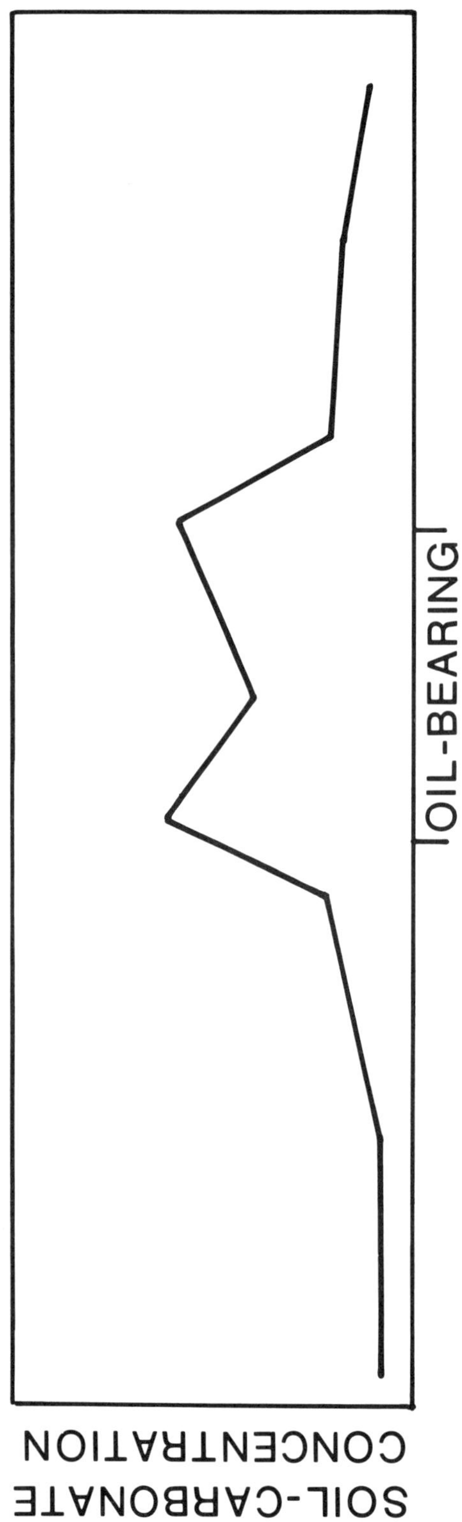

Fig. 6. Soil-carbonate traverse across an oil field. After Kartsev et al. (1959, p. 295).

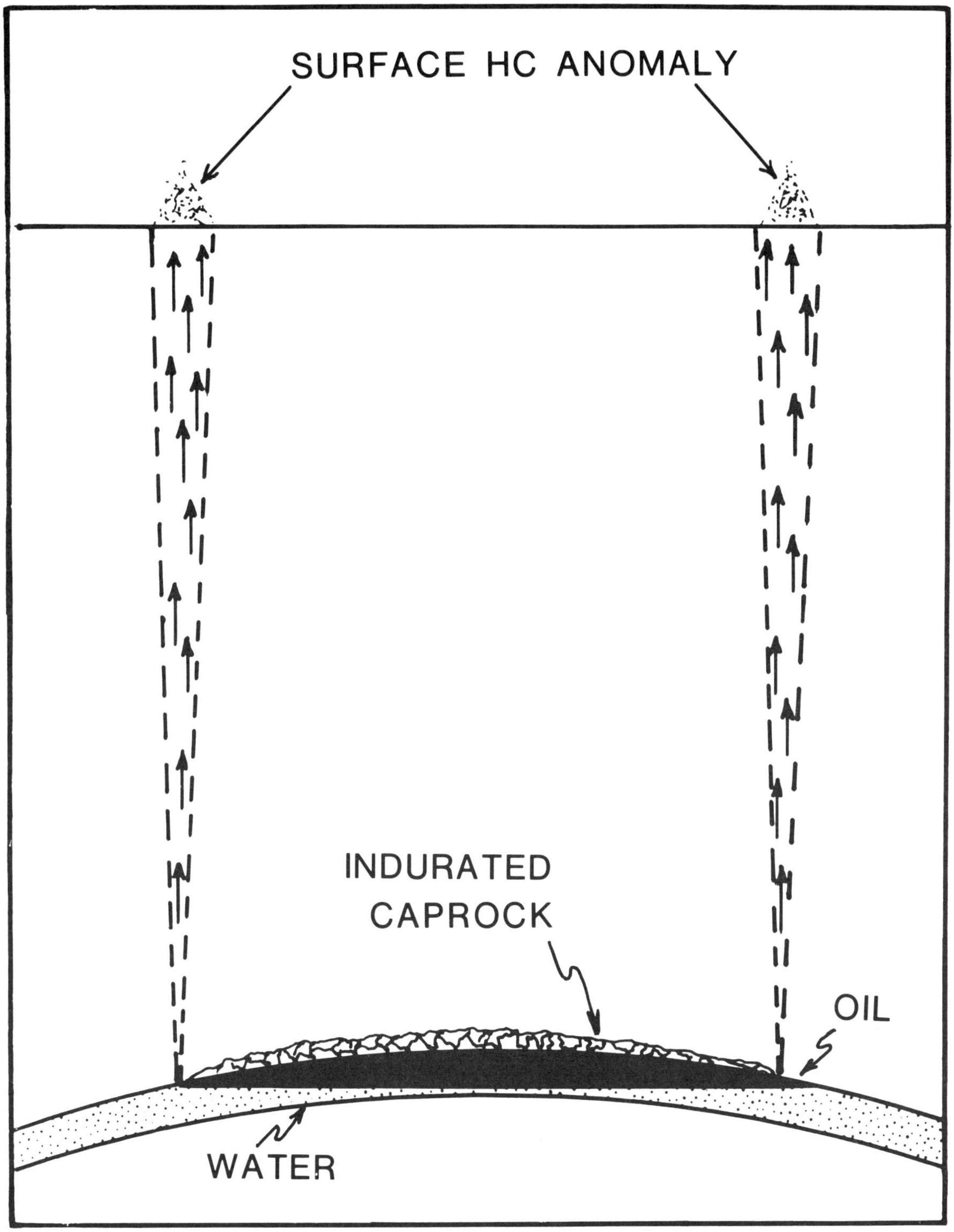

Fig. 7. Explanation of surface geochemical halos by caprock plugging.

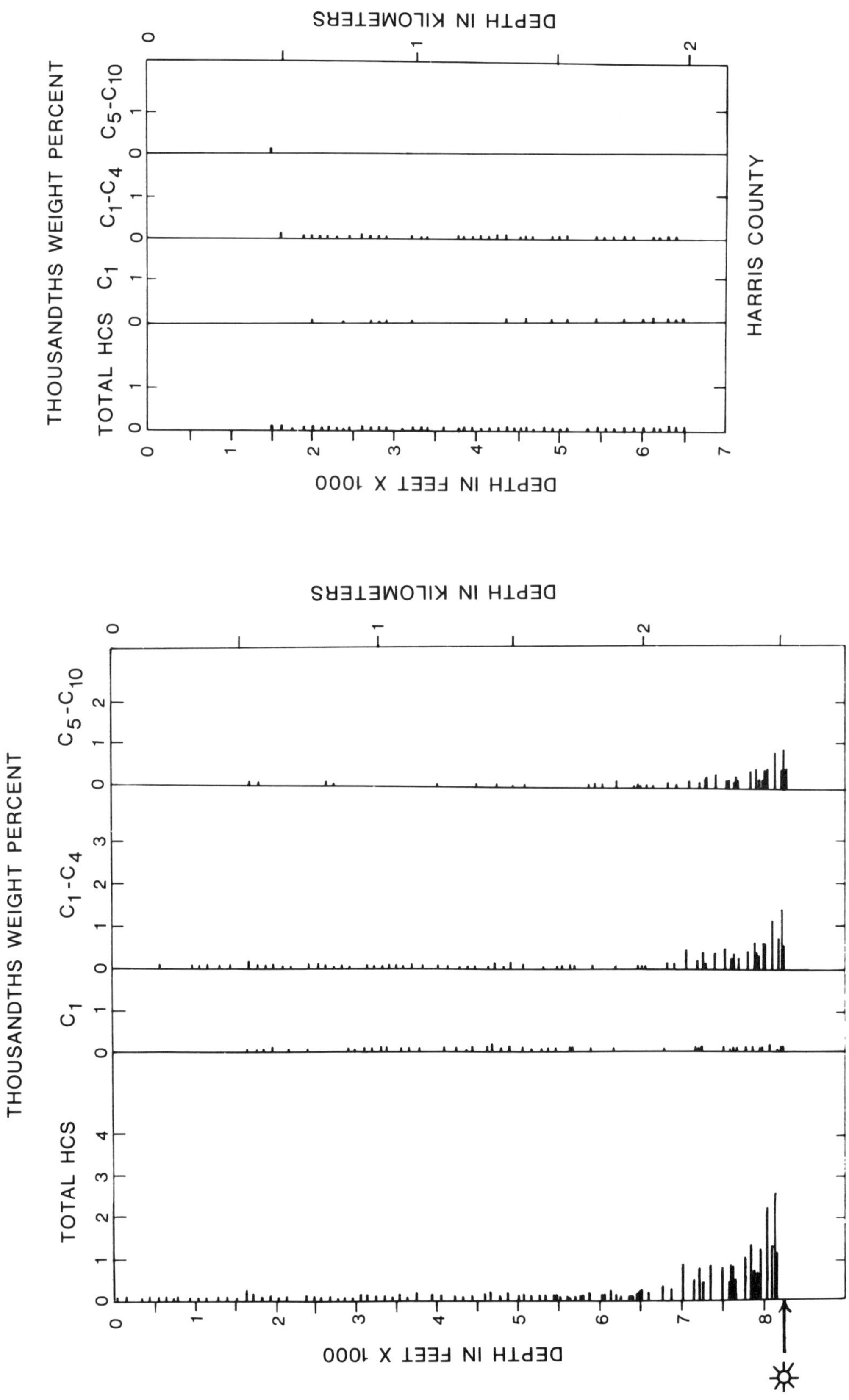

Fig. 8. Hydrocarbon gas logging examples for a new field wildcat discovery well near Rosenberg, Fort Bend County, Texas, and for a dry hole in neighboring Harris County, Texas. Note the increase in both total hydrocarbons and hydrocarbon molecular weight as the producing sand (8,246–8,254 ft, 2.513–2.516 km) is approached in the discovery well. After Rosaire (1940).

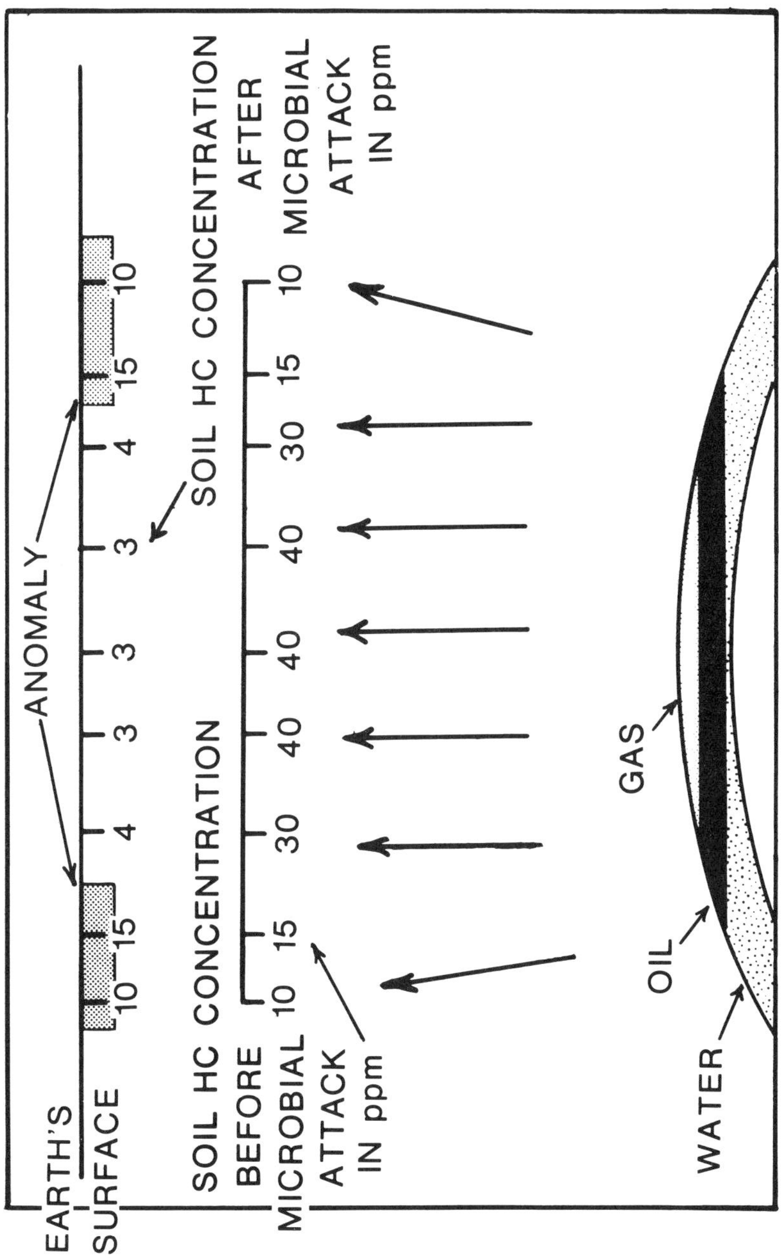

Fig. 9. Soli's (1957) explanation for soil-air and soil-sorbed hydrocarbon anomalies.

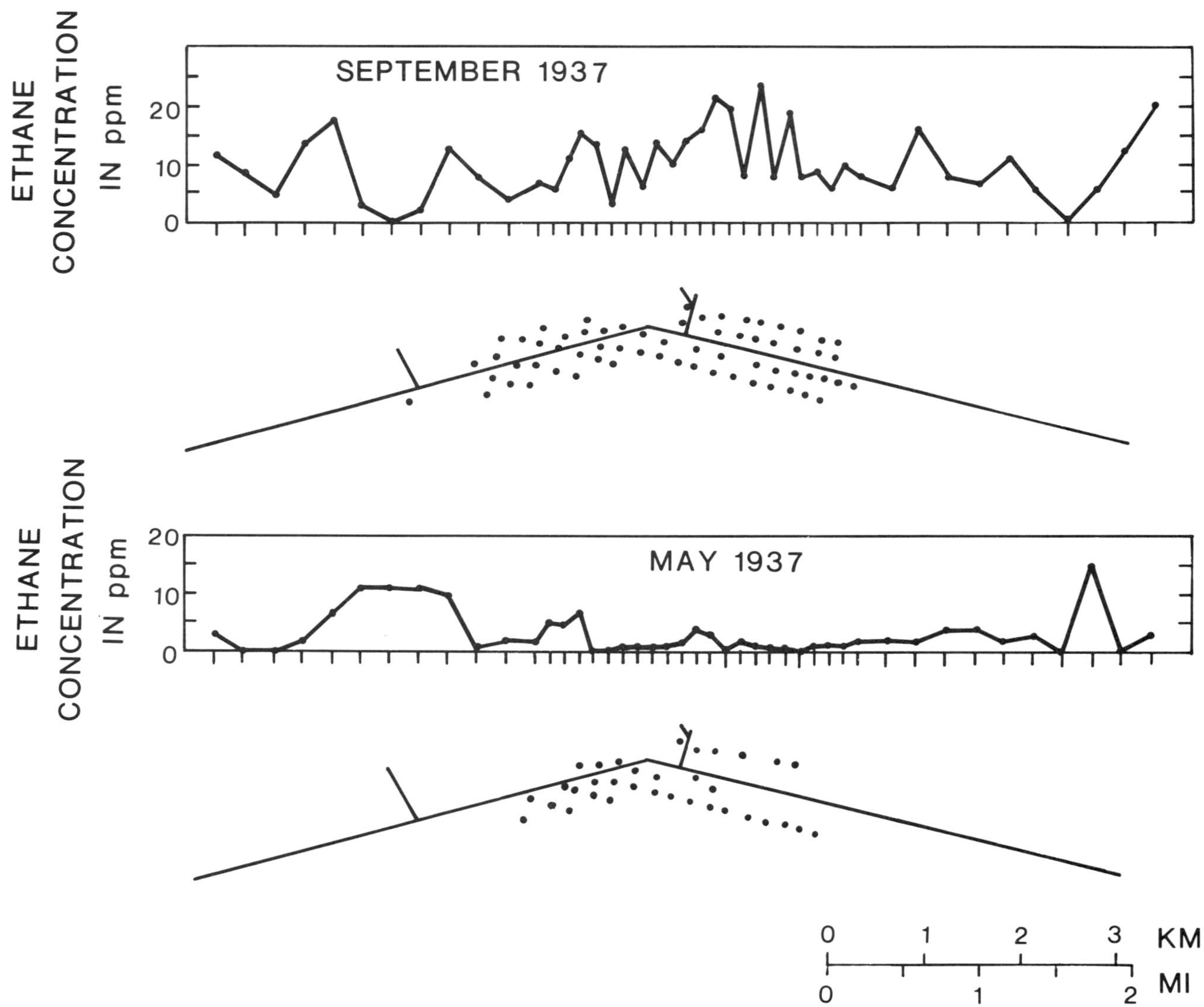

Fig. 10. Soil-sorbed ethane traverses run at different times (May and September, 1937) over the Hastings Oil Field, Brazoria County, Texas. After Rosaire (1938, p. 109).

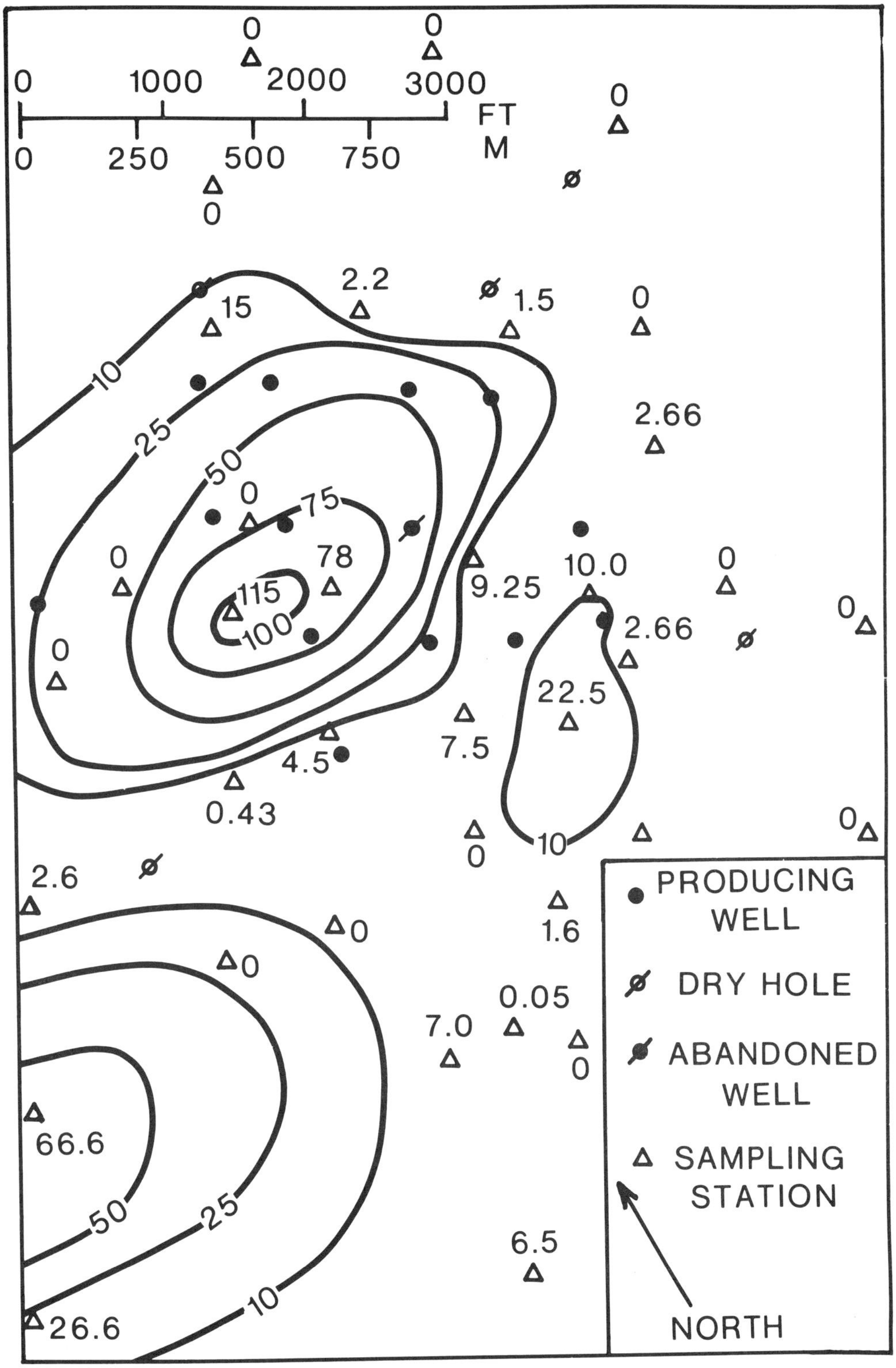

Fig. 11. Ethane soil-air survey of the Hilbig Oil Field, Bastrop County, Texas. Triangles show sampling stations; closed circles are oil-productive wells. After Baijal (1962), with Baijal's (1962) original iso-ethane contours.

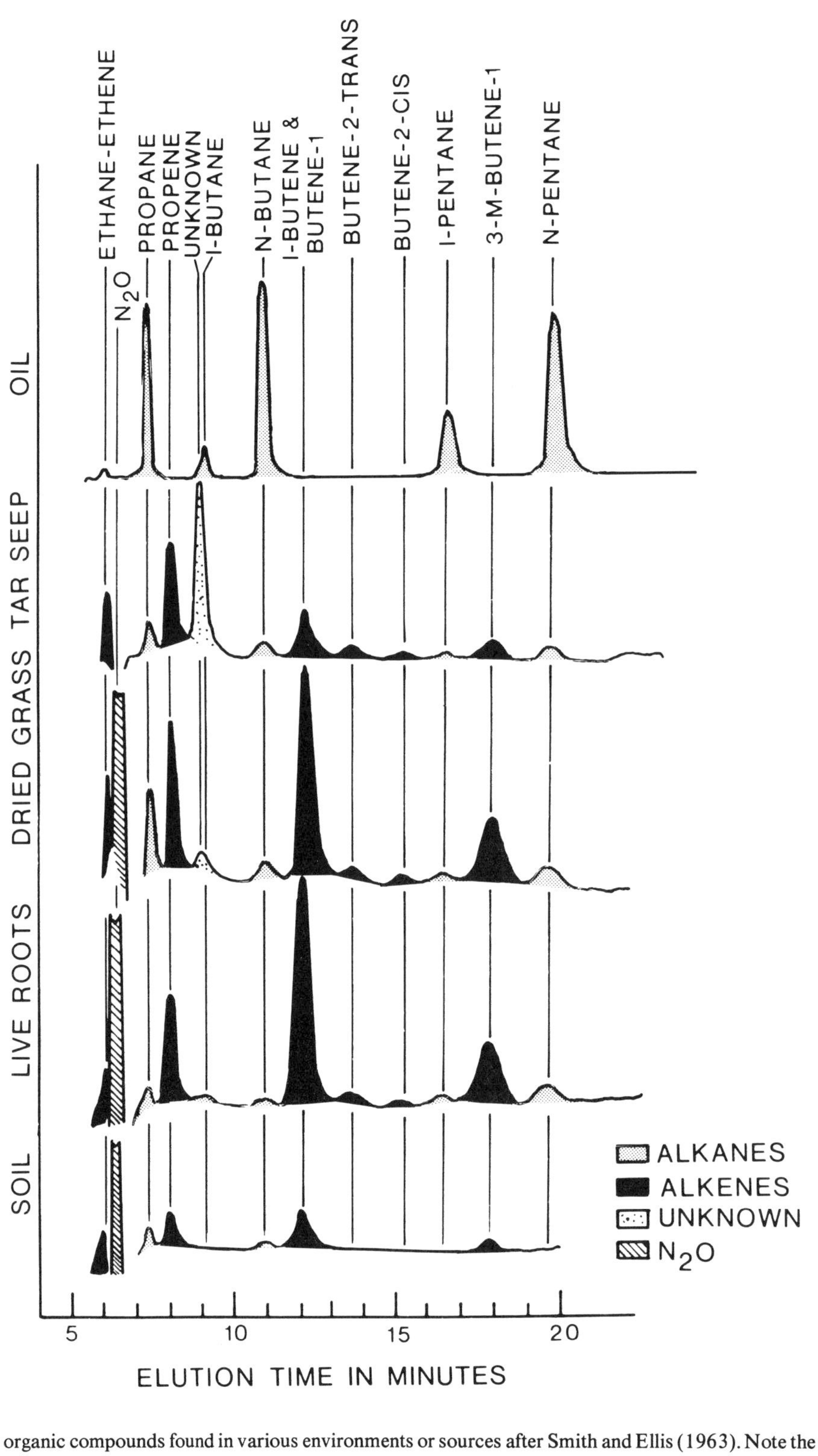

Fig. 12. Volatile organic compounds found in various environments or sources after Smith and Ellis (1963). Note the high component of nitrous oxide in roots, grass, and soil, as well as the fact that bacterial attack of a tar seep has resulted in a number of volatile organic compounds usually not found in crude oil. The manufacture of such compounds would be expected also in surface microbial oxidation of microseeping hydrocarbons.

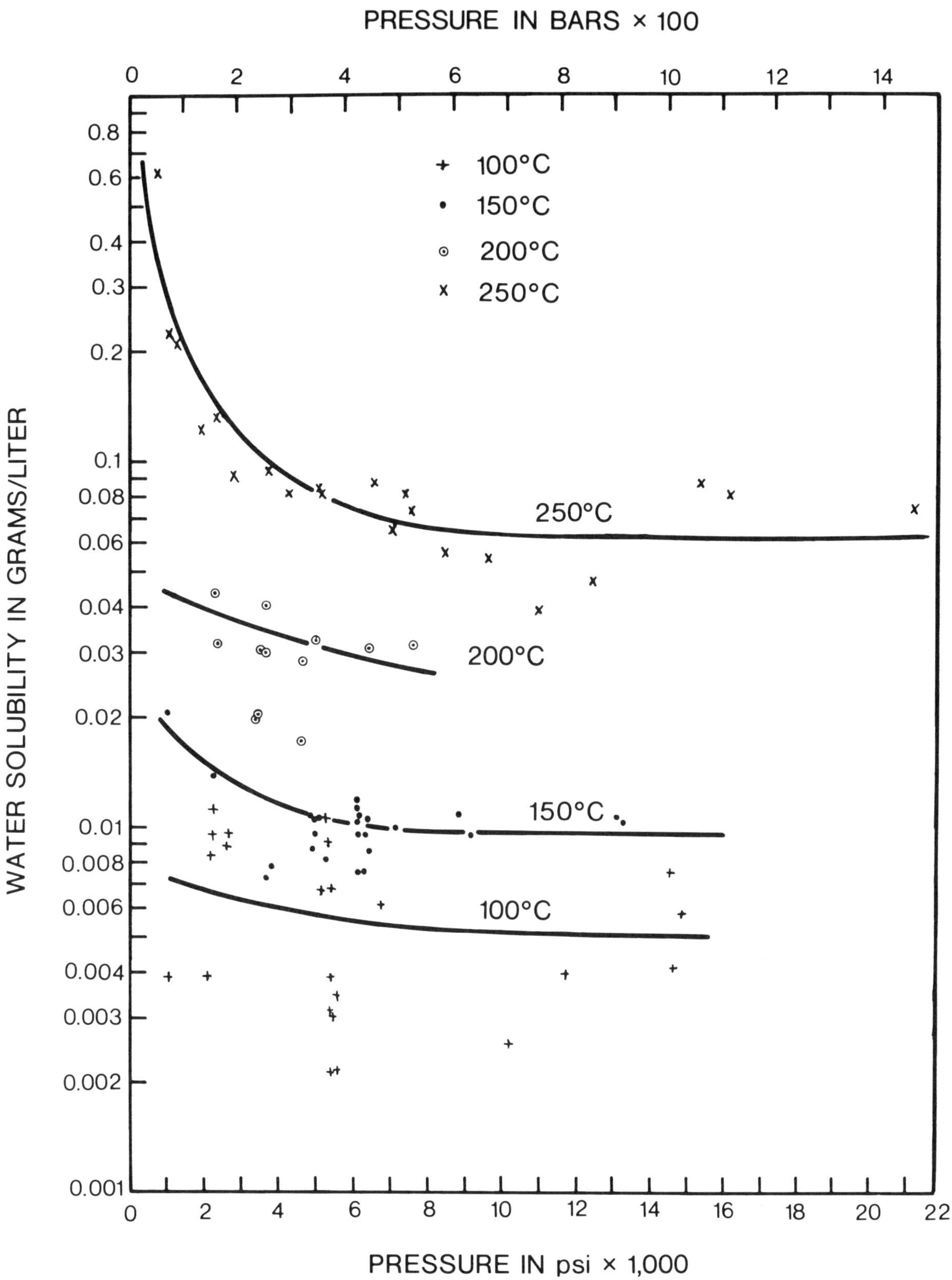

Fig. 13. Solubility of water in methane gas as a function of pressure and temperature. After Price et al. (1983).

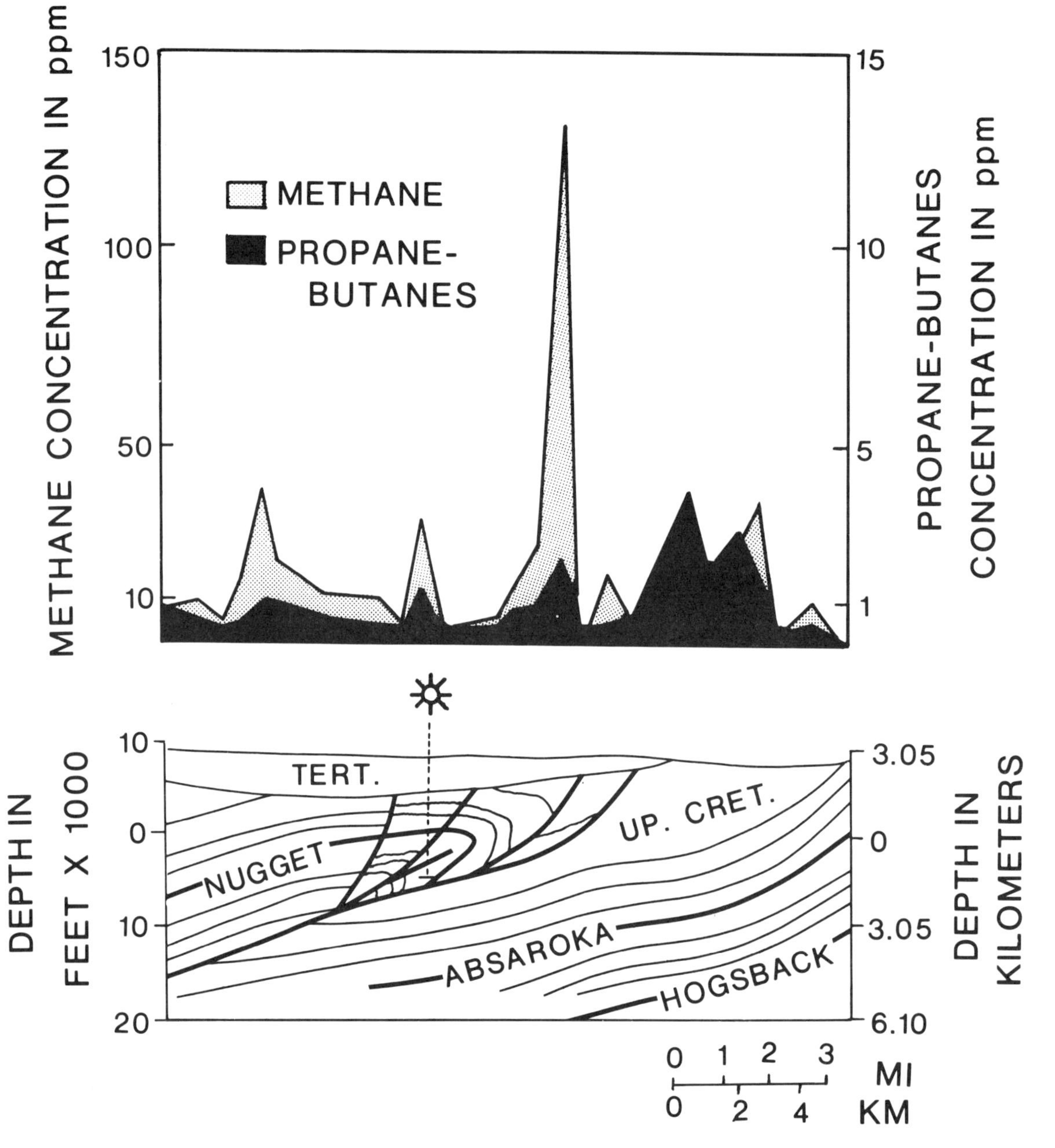

Fig. 14. Geology and results of a soil-gas survey over the Ryckman Creek Field, Wyoming. After Jones and Drozd (1983, p. 950). Note that the "anomaly" over the field is less intense than adjacent false anomalies. Nugget, Absaroka, and Hogsback, respectively, refer to those thrust faults.

LATEST ADVANCES IN THE APPLICATION OF REMOTE SENSING TO PETROLEUM EXPLORATION

Ken M. Morgan and David G. Koger

TEXAS CHRISTIAN UNIVERSITY

ABSTRACT

Remote Sensing is becoming widely accepted by energy companies as an exploration tool. Computer enhanced LANDSAT and radar imagery has proven useful in delineating structural and geochemical information often associated with petroleum exploration.

Recent advances in remote sensing technology were achieved with the launching of LANDSAT 4 and 5 equipped with the thematic mapper (TM). Improvements over earlier multispectral scanner (MSS) data include better spatial resolution, expanded spectral range, and three additional spectral bands.

Integrating LANDSAT MSS and TM data with geophysical information, production maps, and geochemical studies can lead to a better understanding of subsurface relationships important to exploration studies. This can be easily accomplished by use of computer systems adequately equipped for multilevel data integration. Examples of computer-oriented remote sensing geological investigations are presented in this study.

Interactive computer analysis of high resolution satellite imagery (TM) along with pertinent geologic information is providing a faster and more accurate means for evaluating the potential for exploration.

INTRODUCTION

Over the past few years, satellite remote sensing data have become increasingly important in the search for energy resources (Bailey et al., 1984). Many exploration companies now recognize the usefulness of satellite data for pinpointing areas that warrant further geologic investigation. Because of their accurate and broad coverage, satellite images can facilitate the worldwide search for mineral and petroleum deposits (Kinnucan 1982).

There are many types of sensor platforms currently being used in the industry. One of the most popular sensors is the multispectral scanner (MSS) on board the early LANDSAT satellites. LANDSAT MSS satellites have circled the earth since 1971 at an altitude of 920 km (570 mi) recording reflected radiation in two visible (band 4, 0.5–0.6 μm and band 5, 0.6–0.7 μm) as well as two near infrared (band 6, 0.7–0.8 μm and band 7, 0.8–11 μm) wavelength bands (Goetz and Rowan 1981).

Each band of MSS data is stored on computer-compatible tapes (CCTs) as digital data and can be used to produce LANDSAT imagery (79 m resolution). Computer processing of LANDSAT data can modify the image so that geometric, tonal, and spectral characteristics are enhanced. The synoptic view of large areas, spatially registered multispectral information, and repetitive imaging make LANDSAT MSS data especially useful as an exploration tool (Morgan et al., 1982).

Recognizing the popularity and potential of sensors like LANDSAT MSS, NASA recently launched (1982 and 1984) the LANDSAT D series. LANDSAT D differs from its predecessor in orbit, design, and remote sensing capabilities. These advanced satellites carry a payload that includes a higher resolution multispectral scanner (MSS) and a thematic mapper (TM). LANDSAT D (also called LANDSAT TM) is relatively new and, with its advanced sensor technology, may prove to be exceedingly valuable as a remote sensing tool for exploration over the next few years.

COMPARISON OF MSS AND TM DATA

Major improvements of TM over conventional MSS data are: (1) increased spatial resolution, (2) improved spectral separation, (3) expanded spectral range, and (4) collection of both reflected and emitted (thermal) information (Sabins, 1984). Table 1 shows a detailed comparison of these two sensor platforms. Note the improved resolution of TM over MSS (30 m vs. 79 m) and the increase in the number of spectral bands (four to seven). Table 2 gives the spectral location and general characteristics for each of the TM bands. Spectral TM bands, 2, 3, and 4 are similar to MSS bands 4, 5, 6, and 7 repectively. TM bands 1, 5, 6, and 7 were not available on previous LANDSAT systems.

Figure 1 is a comparison of the two image types. This is a split screen image over a portion of San

TABLE 1
COMPARISON OF LANDSAT MSS AND TM DATA

	LANDSAT MSS	LANDSAT TM
Altitude	920 km	705 km
MSS on board	Yes	Yes
TM on board	No	Yes
Spectral bands	4	7
Resolution (visible & IR)	79 m	30 m
Resolution (thermal)	–	120 m
Image coverage	34,225 km²	34,225 km²
Pixels per scene	29×10^6	266×10^6

Antonio, Texas, showing the higher resolution capability of LANDSAT TM. The two half-scenes are at approximately the same scale (1 in = 1 mi) and have the San Antonio International Airport in the center for comparison. An urban scene was chosen to illustrate the greater mapping detail of the TM data. Both scenes are multispectral composites of equivalent bands (i.e., TM 2, 3, and 4 and MSS 4, 5, and 7) shown here in black and white.

A potential drawback to the widespread use of TM data is that it costs up to six times more than MSS information. Another problem is that NASA has not produced LANDSAT TM tapes and images everywhere. TM data are currently limited to only about one-third of the United States. MSS data continue to have a very useful role as a reconnaissance exploration tool to be followed by more detailed TM analysis, if affordable and available.

MAPPING GEOLOGIC FEATURES

LANDSAT data have proven successful to date in the detection of diapirs, growth faults, and fracture patterns controlling hydrocarbon accumulation along the Gulf Coast (Morgan, 1984). Figure 2 is an example of MSS data showing a buried salt dome diapir in Duval County, Texas. A black and white 5/7 ratio image was generated on our LogE computer system to enhance the drainage pattern and tonal variation associated with this subsurface structure. In this case, a high resolution TM scene was not needed because the salt dome was accurately delineated using MSS data.

Another subsurface diapir currently being investigated is the "serpentine-plug." These features have significant potential for additional discoveries (Ewing and Caran, 1982). Criteria for recognition includes drainage patterns, geobotanical color anomalies, and radiating fault/fracture features. Although subtle in appearance, careful study of the surface expression of diapiric structures using remote sensing techniques has provided important information for planning or purchasing seismic lines across newly proposed structures.

Besides diapirs, growth faults are also very common along the Gulf Coast. Oil and gas deposits are often found on the down-thrown side (toward the

TABLE 2
SPECTRAL BANDS FOR LANDSAT TM

Bands	Spectral Range (μm)	Characteristics	MSS Equivalent
1	.45– .52	Some water penetration	None
2	.52– .60	Green wavelength for vegetation mapping	Band 4
3	.63– .69	Red wavelength for soils, geology, vegetation, and urban mapping	Band 5
4	.76– .90	Reflected IR for mapping drainage vegetation, and tonal patterns	Portions of Bands 6 & 7
5	1.55–1.75	Reflected IR, soil moisture, geobotanical, and vegetation mapping; penetrates thin clouds	None
6	2.08–2.35	Reflected IR; used with band 5 for hydrothermal alteration mapping; also, soils, geology, and tonal pattern identification	None
7	10.4–12.5	Emitted Thermal IR	None

Gulf). LANDSAT imagery can be useful for detecting growth faults if proper computer enhancement is used. Figure 3 is an MSS, black and white, band 5 image that has been contrast stretched. This is a simple computer technique that redistributes reflectance values for enhancement purposes (Sabins, 1978). The result is an image with enhanced tonal contrasts. Notice the dark tonal signatures trending NE/SW on the image. The areas of production are shown on the image and are numbered. This dark tonal pattern lies directly on top of a growth fault and appears to be related to both moisture content changes and potential vegetation stress from hydrocarbon leakage. Both of these factors can affect plant vigor and, therefore, the spectral reflectance of vegetation located along the fault trend. This type of analysis is currently being used to extend known faults or to map new ones. The geobotanical patterns produced from MSS data are useful for some fault/fracture systems. In many cases, however, a more detailed analysis is needed.

Figure 4 is a split image comparing the resolution of MSS versus TM for fault and fracture mapping where lineaments are much smaller and harder to map. This is a black and white multispectral composite covering portions of the Wichita Mountains and the frontal fault zone of the Anadarko Basin. Both scenes are at approximately the same scale (1 in = 1 mi). It is easy to see how the higher resolution TM data is superior to MSS in depicting the fault/fracture systems in this area. At this scale, it is difficult to trace-out faults on the MSS scene.

In many areas, fault/fractures may be controlling secondary porosity for shallow reservoirs. High density fracture zones often become potential sites for shallow drilling operations. Also, many companies follow up the remote sensing analysis with geophysical or geochemical techniques.

SUMMARY AND CONCLUSIONS

Many exploration companies now use LANDSAT in conjunction with existing data, or when other sources of information are inconclusive, when searching for geologic anomalies. Often the analysis is done on current lease holdings or as a quick evaluation of potential lease areas.

LANDSAT MSS data are well suited for these kinds of studies because of their relatively low cost and almost worldwide coverage. Computer enhancement of this data can be especially effective when higher resolution is not required.

The advanced LANDSAT TM satellite offers superior resolution and expanded spectral range for improved geologic analysis. The limitations include higher costs and somewhat limited availability of the data. As more TM data become available and their use more widespread, TM should prove to be an increasingly valuable remote sensing exploration tool.

In addition to the LANDSAT systems discussed here, other forms of remote sensing include thermal and radar imagery. These data are currently extremely limited in coverage and new data can be expensive to obtain. Novel operational uses of these and other remote sensing systems for exploration will become evident over the coming years.

REFERENCES

Bailey, G.B., Dwyer, J.L., Francica, J.R., and Feng, M.S., 1984, Update on the use of remote sensing in oil and gas exploration, *in* Davidson, M. J., and Gottlieb, B. M., eds., Unconventional methods in exploration for petroleum and natural gas III: Dallas, Southern Methodist University Press, p. 231–253.

Ewing, T.E., and Caran, S.C., 1982, Late Cretaceous volcanism in south and central Texas: Stratigraphic, structural, and seismic models, *in* Gulf Coast Association of Geological Societies Transactions, v. 32, p. 137–145.

Goetz, A F., and Rowan, L.C., 1981, Geologic remote sensing, *in* Science, v. 211, February, p. 781–791.

Kinnucan, P., 1982, Earth scanning satellites lead resource hunt, *in* High Technology, March/April, p. 53–60.

Morgan, K.M., Morris-Jones, D.R., and Koger, D.G., 1982, Applying LANDSAT data to oil and gas exploration along the Texas gulf coast: Oil and Gas Journal, September 27, p. 326–327.

——, 1984, Geologic application of LANDSAT data for exploration in Texas: San Antonio, American Society of Photogrammetry, Proceedings, Fall Convention, p. 575–580.

Sabins, F. F., Jr., 1978, Remote sensing: Principles and interpretation: San Francisco, W.H. Freeman and Company, 426 p.

——, 1984, Remote sensing laboratory manual: La Habra, California, Remote Sensing Enterprises, 253 p.

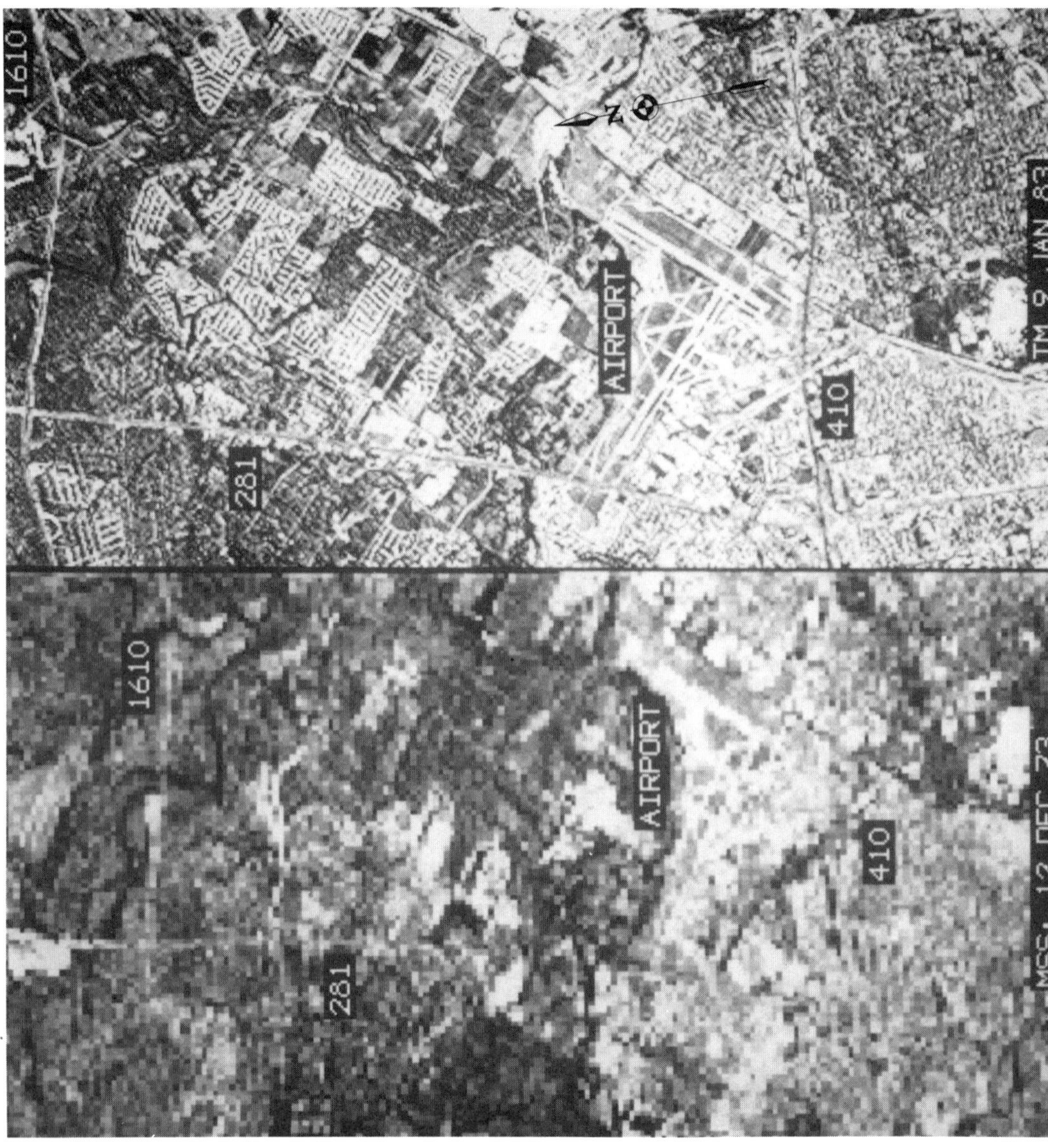

Fig. 1. Comparison of MSS (left) and TM (right) data over a portion of San Antonio, Texas (scale: 1 in = 1 mi).

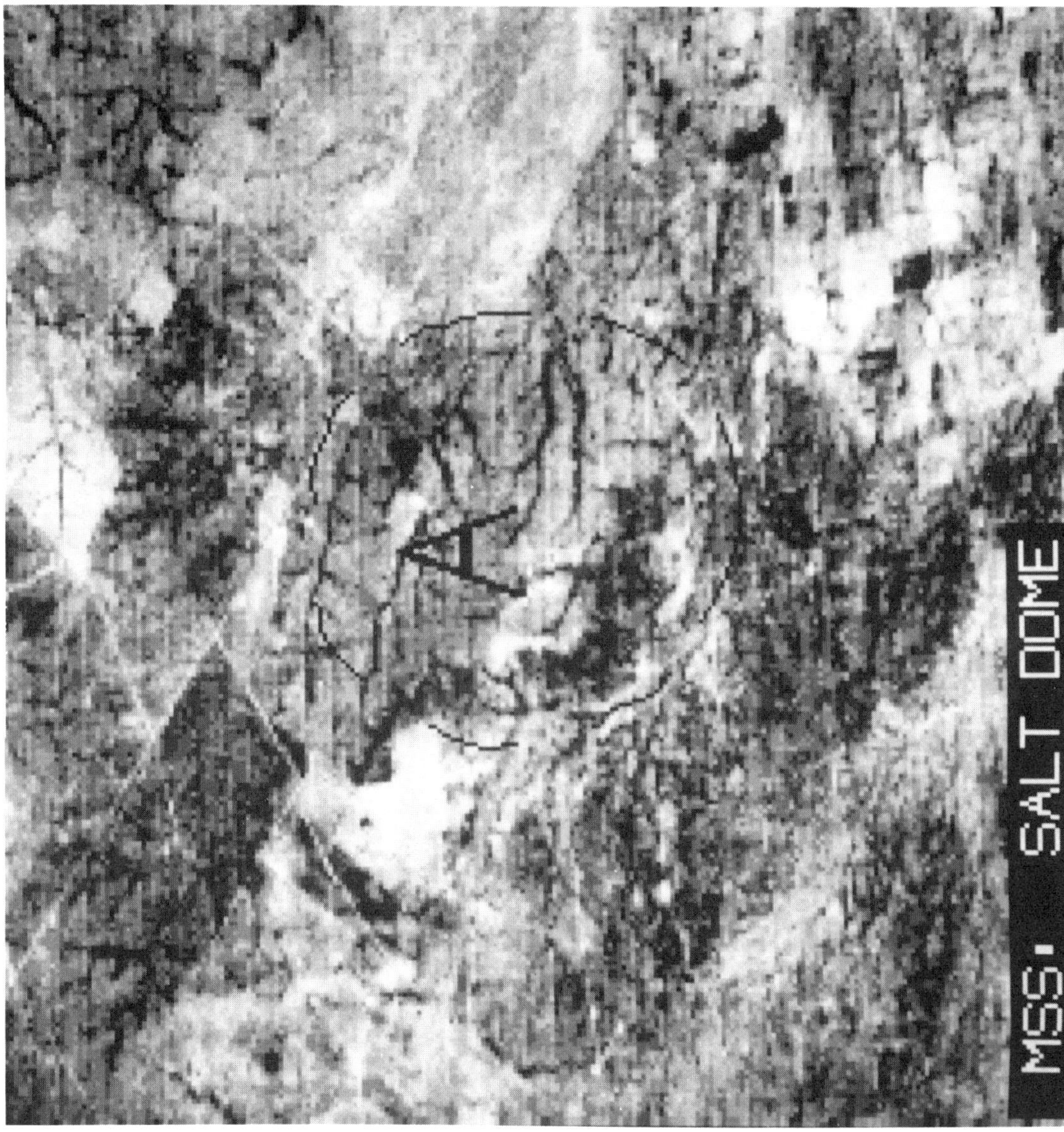

Fig. 2. MSS 5/7 ratio image over salt dome (A) in Duval County, Texas. Note radial and circular drainage pattern due to differential compaction over the subsurface feature (scale: 1 in = 1 mi).

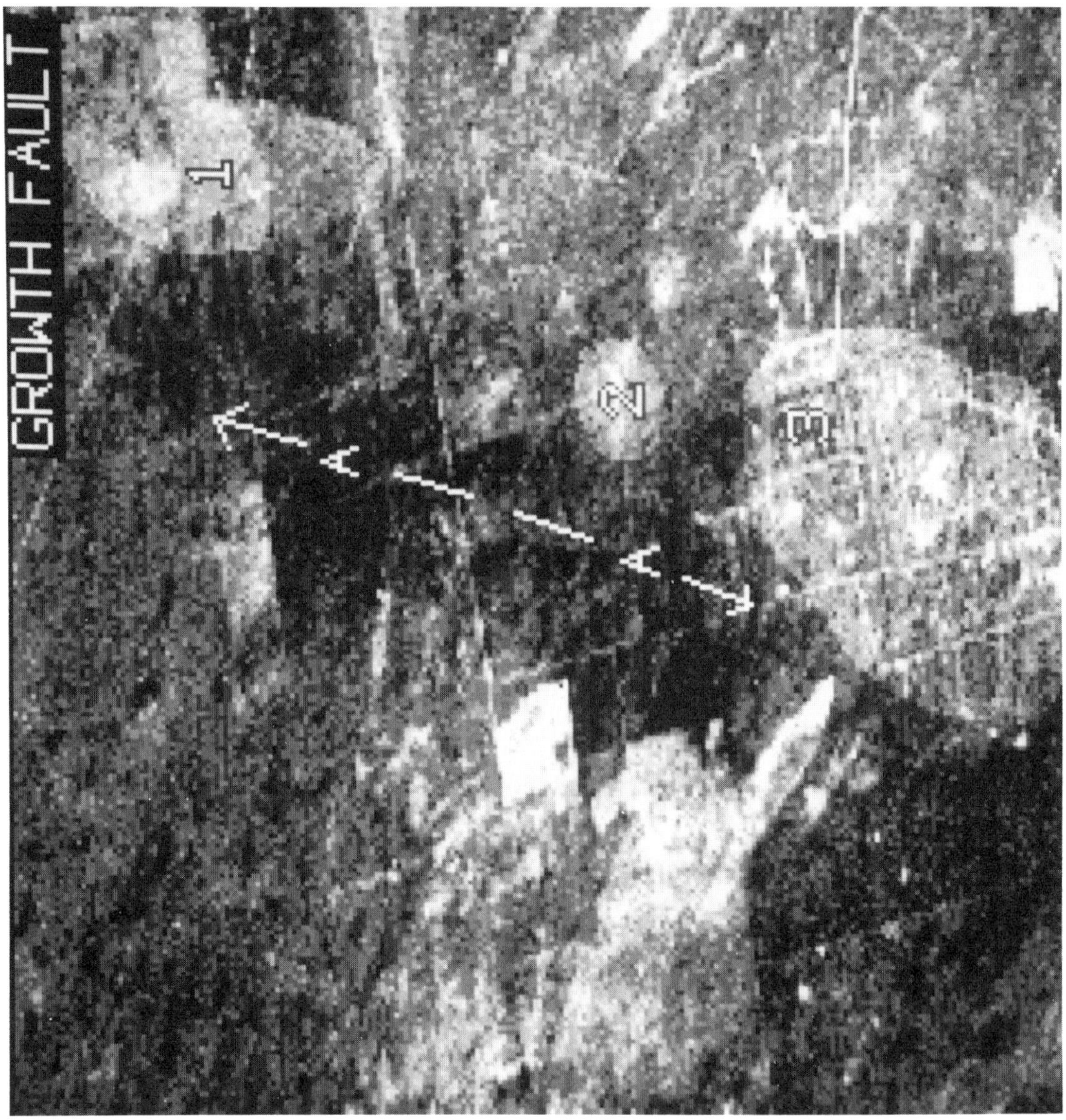

Fig. 3. MSS band 5 image showing tonal variation over growth fault (A) and production locations (1–3) (scale: 1 in = 1 mi).

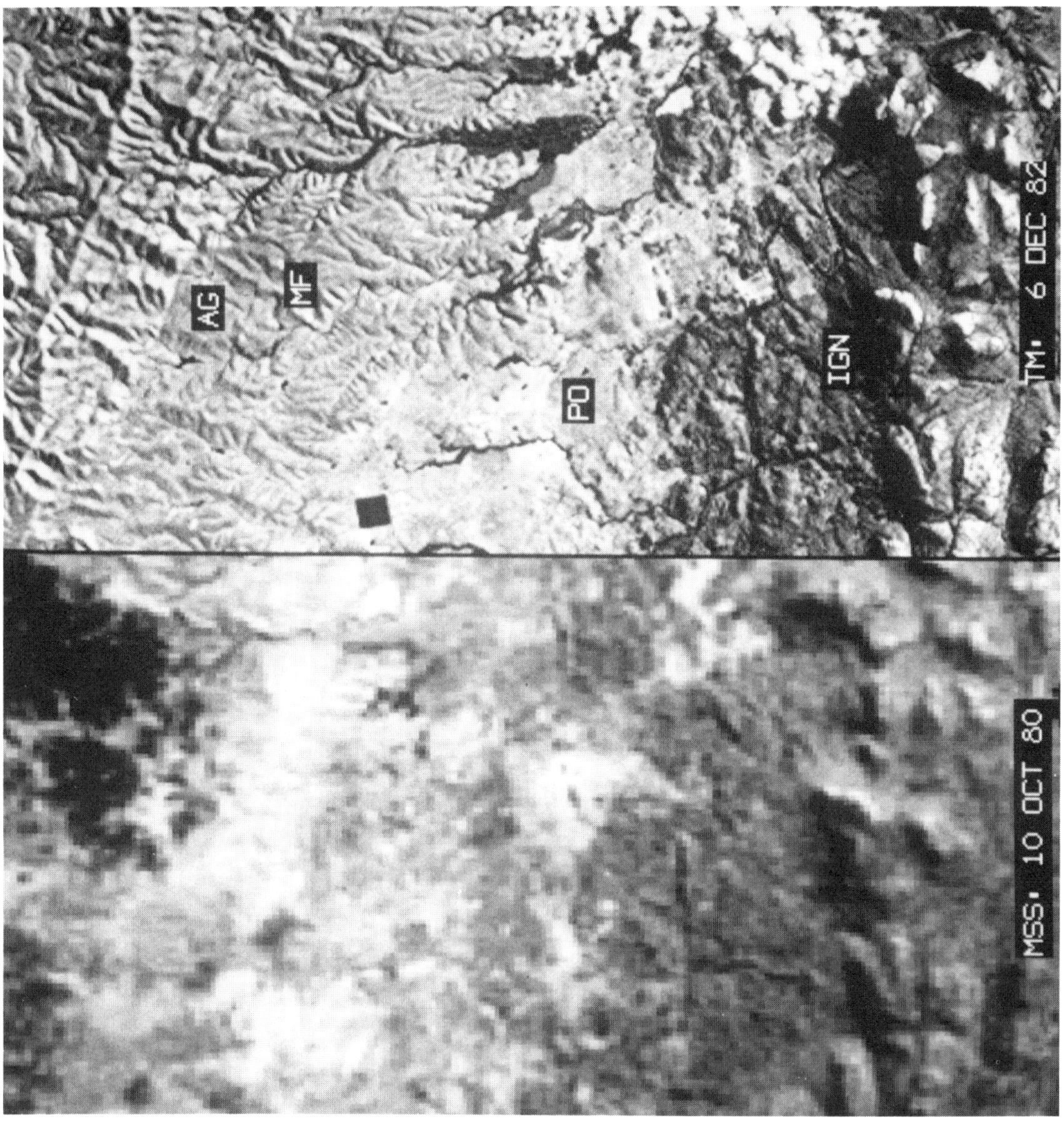

Fig. 4. Comparison of MSS (left) and TM (right) for mapping faults/fractures in a portion of Oklahoma (scale: 1 in = 1 mi). IGN represents the igneus Wichita Mountains, PO the Post Oak Formation outcrop, MF the Meers Fault, and AG the Arbuckle Group outcrop.

OPERATIONAL AND TECHNICAL RESULTS OF A MULTIWELL BOREHOLE GRAVITY SURVEY

George A. McCalpin

CONSULTING GEOPHYSICIST

ABSTRACT

A twelve-well borehole gravity meter survey was conducted for Sun Exploration and Production Company in August, 1983, in South Texas in temporarily abandoned wells which had been selected by the district geological staff. The twelve wells were logged in a period of nineteen days, including moving days and off-days. Average "hole time" for the surveys was slightly over eight hours, with rig-up and rig-down time adding an additional two to three hours. The time between station readings averaged about ten minutes for the 429 stations and repeats in the surveys.

One hundred eighty zones were examined, ranging in depth from 890 ft to 4,950 ft. A porosity, assuming 100% water saturation, was computed for each of these zones from the gravity-derived density and compared with the porosity determined from a compensated neutron log. On the basis of this comparison approximately 61% of the zones were interpreted as water sands, 28% as hard streaks, and 11% as prospective for either oil or gas.

INTRODUCTION

Since the appearance of the articles by Smith (1950) and Hammer (1950), other authors have made significant contributions to the theoretical, operational, and interpretational aspects of borehole gravity. Notable among these were the papers by McCulloh (1966, 1967), Jones (1972), Bradley (1974), Jageler (1975), and Rasmussen (1973, 1975). More recently the paper by Gournay and Maute (1982) treated specifically the single application of detection of shallow gas behind pipe in old wells. A further look at this type of survey, particularly with regard to the operational aspects to be anticipated, seems timely in view of the large number of old wells, now temporarily abandoned or soon to be permanently abandoned, especially along the Texas Gulf Coast.

PREPARATION FOR THE SURVEYS

Sun Exploration and Production Company decided in 1983 to log a number of wells in various fields in South Texas to assess remaining potentials for future development or for possible abandonment. Studies were conducted by the district geological staff and a recommended list was presented to the headquarters reserves development group. From these studies twelve wells in nine different fields were selected for the program.

To insure the greatest amount of efficiency in the actual logging operation, preparatory work on the well bores was begun some time in advance of the scheduled gravity surveys. Tubing was pulled and casing condition was checked to provide for the use of the 4 1/8 in OD borehole gravity meter tool (BHGM). The cost of this preliminary work varied from a few thousand dollars to several tens of thousands, depending upon conditions encountered.

Prior to the actual gravity logging, compensated neutron logs (CNL) were run through the targeted zones in each of the wells to give gamma ray and casing collar information for correlation with these same detectors which are included in the BHGM sonde, and also to give a neutron derived porosity which could be compared with the porosity derived from the bulk density calculated from the borehole gravity data.

OPERATION OF THE SURVEY

Although the BHGM can be operated on any seven-conductor cable, which can be supplied by a number of logging contractors, it was considered more efficient in this size survey to utilize the same mast truck and wireline truck throughout, rather than interface with different equipment as the survey progressed. Therefore, EDCON supplied these two pieces of equipment along with the borehole meters, and all were operated by EDCON personnel. The borehole gravity meters were the LaCoste & Romberg type which have temperature limitations of about 120°C, can tolerate hole deviations up to about 14° and have an outside diameter of 4 1/8 in. The total length of the sonde with gamma ray and casing collar locator is about 14 ft.

The first six wells were logged in a period of eleven days, which included some necessary equipment maintenance, swapping out a meter from Denver, and a crew move from McAllen to Alice. The last six wells were logged in an eight-day period, which included a move to Victoria and a scheduled "off" day. The operations within each of the wells also improved as the survey progressed.

Table 1 shows the program as it was preplanned from a study of the old electric logs and the newer compensated neutron logs. The individual zones were selected by the geological staff, and then individual station locations were chosen which would best investigate these zones, with thicker sands often being logged in several segments.

Table 2 is a time chart of the surveys and shows the typical time intervals required to do this type of program. It should be noted that one to two hours should be added at each end of the times shown to provide for rigging up and rigging down.

In Table 3 some of the information from Tables 1 and 2 has been rearranged and presented to show other aspects of the timing of this type survey. The second column, labeled "Correlation and Soak Time," shows the time between starting the tool into the hole and the reading time for the first station. This time interval includes the time necessary to run the tool to total survey depth at a normal speed of 100 ft/min. The "correlation" time is the time necessary to record a section of gamma ray curve and several casing collar locations with readily recognizable characteristics to allow the accurate positioning of the tool near the area to be logged. The "soak" time can be defined as the time necessary to allow the meter to reach stabilization with the downhole temperature, but it is also utilized by the operator to make some necessary sensitivity and reading parameter tests on the meter. The overall time is strongly dependent upon depth and its accompanying temperature, but can also be affected if correlation is difficult. Other columns on this chart show average times between stations or stations per hour after the logging operation begins.

A total of 429 stations were run in the complete survey with 363, or 85%, being new stations, and 66, or 15%, being repeats or reruns. The average repeat values in each of the wells varied from 0.9 microgals to 6.8 microgals, and for the total survey averaged 3.2 microgals. This level of repeatability will provide densities accurate to 0.02 gm/cc or less over the usual 10–15 ft measured intervals.

TABLE 1

PROGRAM PREPLAN

Well No.	Shallowest Depth	Deepest Depth	Zones	Stations
1	1519	4120	11	37
2	1015	4450	11	33
3	1655	4675	8	27
4	1013	4570	12	31
5	1022	3034	11	27
6	890	4030	9	32
7	1715	3175	13	38
8	748	4210	19	48
9	1446	3640	15	45
10	1790	4950	14	47
11	1015	4450	11	30
12	1712	3092	11	34

INTERPRETATION OF THE RESULTS

Based on the method of Gournay and Maute (1982), who used a cross-plot of neutron porosity versus BHGM density to characterize the measured zones, a variation of this approach was used in analyzing these survey results. A porosity was computed for each of the calculated densities from the borehole gravity data, assuming 100% water saturation. Since the calculated porosity from the CNL is dependent upon water saturation, the two porosities should agree in water sands. In gas sands the CNL porosity will be low due to the absence of water, while the computed BHGM porosity will be high because of the low bulk density.

The fundamental equation for porosity is

$$\phi = 100 \times \frac{\rho_g - \rho}{\rho_g - \rho_f}$$

where ϕ is the porosity expressed in percent, ρ_g is the grain density, ρ is the "in situ" bulk density of the zone being investigated, and ρ_f is the average density of the fluids contained in the pore space. Assuming 100% water saturation with a ρ_f of 1.0 gm/cc and a grain density of 2.65 gm/cc, the equation reduces to

$$\phi = 100 \times \frac{2.65 - \rho}{1.65}$$

TABLE 2
INTERVAL TIME CHART

Well No.	In Hole	First Station	Last Station	Out of Hole	Hole Time		Log Time	
					Hr	Min	Hr	Min
1	14:15	16:46	23:59	00:17	10	02	7	13
2	13:37	15:22	23:37	23:50	10	13	8	15
3	10:52	12:38	18:23	18:40	7	48	5	45
4	9:15	11:18	17:22	17:30	8	15	6	04
5	10:20	11:42	16:01	16:15	5	55	4	19
6	9:48	11:26	16:59	17:10	7	22	5	33
7	9:30	11:11	16:52	17:00	7	30	5	41
8	9:05	10:55	18:36	19:00	9	55	7	41
9	9:30	11:04	17:39	18:00	8	30	6	35
10	10:20	11:49	17:52	18:15	7	55	6	03
11	10:00	11:42	15:18	17:00	7	00	3	36
12	11:20	12:55	17:35	18:10	6	50	4	40

TABLE 3
SOAK TIMES AND AVERAGE READING TIMES

Well No.	Correlation & Soak Time		Number of Stations	Logging Time		Minutes Per Station	Stations Per Hour
	Hr	Min		Hr	Min		
1	2	31	37	7	13	11.7	5.1
2	1	45	33	8	15	15.0	4.0
3	1	46	27	5	45	12.8	4.7
4	2	03	31	6	04	11.7	5.1
5	1	22	27	4	19	9.5	6.3
6	1	38	32	5	33	10.4	5.8
7	1	41	38	5	41	9.0	6.7
8	1	50	48	7	41	9.6	6.3
9	1	34	45	6	35	8.8	6.8
10	1	29	47	6	03	7.7	7.8
11	1	42	30	3	36	7.2	8.3
12	1	35	34	4	40	8.2	7.3

and ρ, in this application, is the density of the formation as determined from the difference between two gravity readings. Solving this equation for a range of ρ values shows that for each .05 gm/cc of increase in density there is a corresponding decrease of 3% in porosity, ranging from 0% porosity at a density of 2.65 gm/cc to a 39% porosity at a density of 2.00 gm/cc.

Figure 1 shows a section of a CNL with porosity increasing to the left and scaling factor of 3% per division. From the foregoing discussion it follows that the gravity results can be plotted directly on the CNL either as porosity (as derived above) at the same scale as the neutron porosity, or as density with a scale starting at 2.65 gm/cc at the right edge and decreasing .05 gm/cc per scale division to the left.

Typically, then, in water sands the plotted porosity will agree quite closely with the neutron porosity, while in gas sands there will be a "cross-over" with the BHGM porosity being greater than the neutron porosity. In hard streaks the neutron porosity is depressed but usually not as much as the BHGM porosity. Examples of these occurrences are shown on Figure 2. Plotting directly on the CNL allows instantaneous recognition of anomalous zones, and aids in the scheduling of repeat observations while the survey is in progress.

After the completion of the surveys in all twelve wells, and the calculations of porosities as described above, 180 differences between the BHGM-derived porosities and the neutron porosities were available in the targeted sands. These differences (BHGM porosity minus CNL porosity) were tabulated and a plot of the differences versus frequency is shown in Figure 3. A three point running average was used to smooth this curve, which seems justified since the CNL porosities over the measured BHGM intervals were determined by visual estimation and could possibly have an error of 1% to 2%. It would be expected that a zero difference would occur in water sands, and that a positive difference would be present if the water were replaced by either oil or gas with densities that would be slightly or greatly less than the 1.00 gm/cc for water.

Several features of this plot are of interest. The "shoulder" on the curve around the −11 difference represents, for the most part, the hard streak section of the curve, as the densities associated with the great majority of the differences in this portion of the curve are in excess of 2.35 gm/cc and range up to 2.55 gm/cc. The portion of the curve which peaks at −4 difference is the "water" portion of the curve. It is noteworthy that this peak occurs not at zero, as would be expected theoretically, but in the negative difference area. This may be caused by shaliness of the sands, as neutron-derived porosities in shales generally exceed 40%, or it may be caused by some systematic miscalibration between the two different methods of determining porosities. The slight hump in the curve at the +4 to +6 difference may represent typical oil sands, where the density of the pore fluids is only slightly less than 1.0 gm/cc, while the hump near +12 is more typical of gas sands where the pore fluid density is considerably less than that of water. Obviously, the positive differences between these two humps and beyond are also prospective as indicators of different combinations of oil and gas saturations, but the fact that these two clusters of positive differences do occur indicates that all combinations of oil and gas saturation percents are not equally likely, at least not in this statistical sample.

SUMMARY AND CONCLUSIONS

Extended operations which are planned well in advance can result in very efficient BHGM surveys for gas and oil behind pipe in old wells. The accuracy of this type survey has been demonstrated in previous papers and reaffirmed by the repeatabilities in these surveys.

A method has been suggested that allows the plotting of the BHGM results directly on the CNL or similar logs which have a derived neutron porosity curve, to aid in the immediate identification of anomalous zones and to allow for timely survey adjustments, if needed or desired.

A plot of numerical differences between the BHGM-derived porosities and the neutron porosities for these surveys has shown that 28% of the targeted intervals can be classified as hard streaks on the basis of the porosity differences combined with a check of the actual density of the interval, 61% can be classified as possible water sands, and 11% show positive differences which can be interpreted as either oil, oil/gas, or gas saturations. The smaller positive differences should indicate oil saturation, or possibly pressure-depleted gas zones with some water encroachment, while the larger positive differences should indicate significant gas saturation.

Of the twenty zones which had been recommended in seven of the wells (five of the wells had no anomalous differences), twelve zones have not yet been tested. Seven of the twenty zones were found, after correlation with nearby wells, to have been already depleted by these other wells, and one zone with a +3 difference was tested unsuccessfully.

In spite of the rather incomplete testing program for these wells, the accuracy and repeatability of the data has been established in this and previous surveys, and the percentage distributions of the hard streaks, water sands, and possible hydrocarbon zones seems very reasonable for sands in known productive areas. The BHGM still appears to be a viable alternative either to extensive testing or to the final abandonment of wells without assessing the remaining potential.

REFERENCES

Bradley, J.W., 1974, The commercial application and interpretation of the borehole gravimeter: Symposium of the Houston Geophysical Society.

Gournay, L.S., and Maute, R.E., 1982, Detection of bypassed gas using borehole gravimeter and pulsed neutron capture logs: Log Analyst, May–June, p. 27–32.

Hammer, S., 1950, Density determinations by underground gravity measurements: Geophysics, v. 15, no. 4, p. 637–652.

Jageler, A.H., 1976, Improved hydrocarbon reservoir evaluation through use of borehole-gravimeter data: Journal of Petroleum Technology, June, p. 709–718.

Jones, B.R., 1972, The use of downhole gravity data in formation evaluation: Log Analyst, v. 13, no. 5, p. 15–21.

McCulloh, T.H., 1966, The promise of precise borehole gravimetry in petroleum exploration and exploitation: U.S. Geological Survey Circular 531, 12 p.

———, 1967, Mass properties of sedimentary rocks and gravimetric effects of petroleum and natural-gas reservoirs: U.S. Geological Survey Professional Paper 528-A, 50 p.

Rasmussen, N. F., 1973, Borehole gravity survey planning and operations: Society of Petroleum Well Log Analysts 14th Annual Logging Symposium, Transactions, Q1–Q28.

———, 1975, Borehole gravimeter finds bypassed oil, gas: Oil and Gas Journal, v.73, no.39, p. 100–104.

Smith, N. J., 1950, The case for gravity data from boreholes: Geophysics, v. 25, no. 4, p. 605–636.

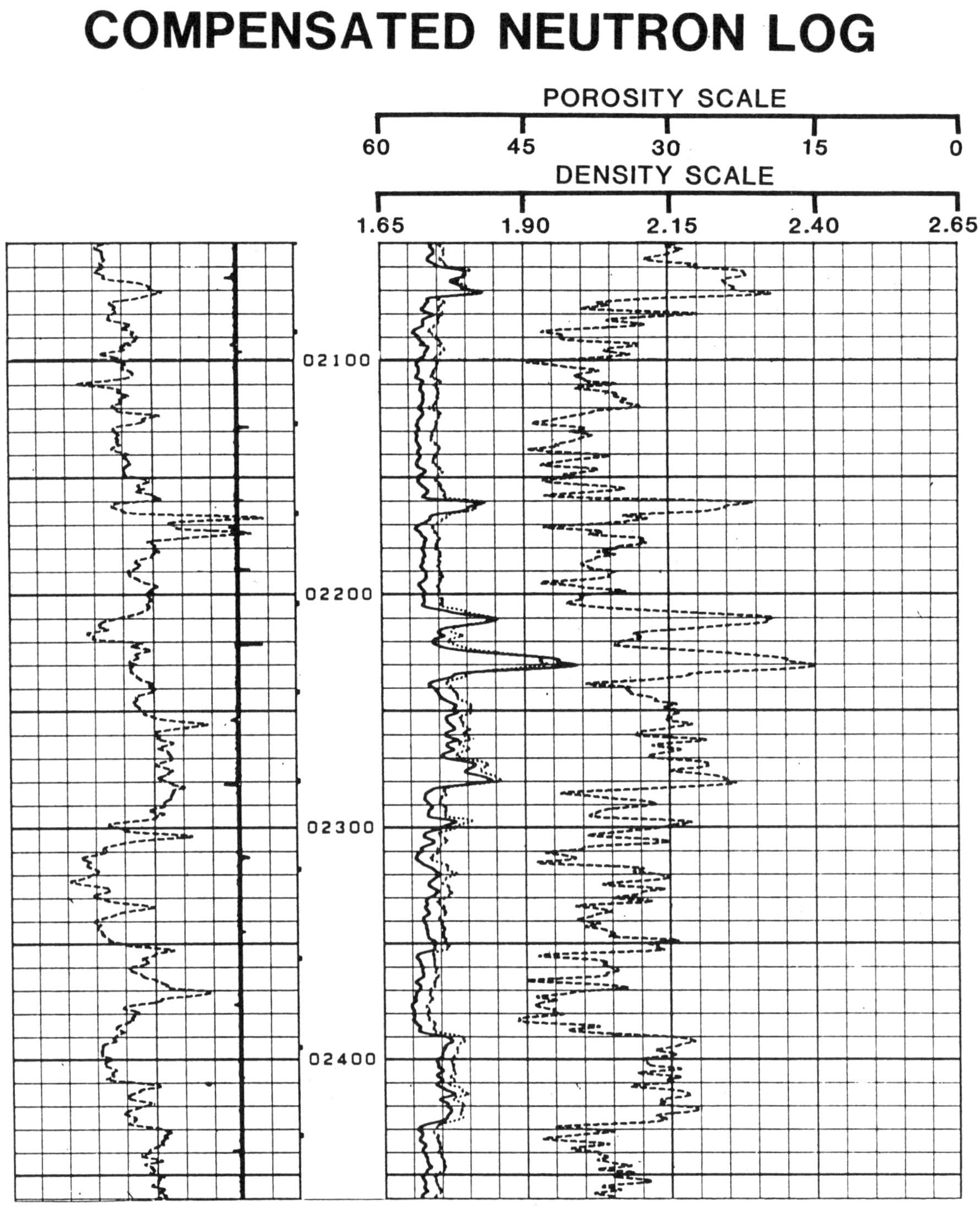

Fig. 1. A portion of a compensated neutron log showing a density scale corresponding to the porosity scale, assuming 100% water saturation in sandstones.

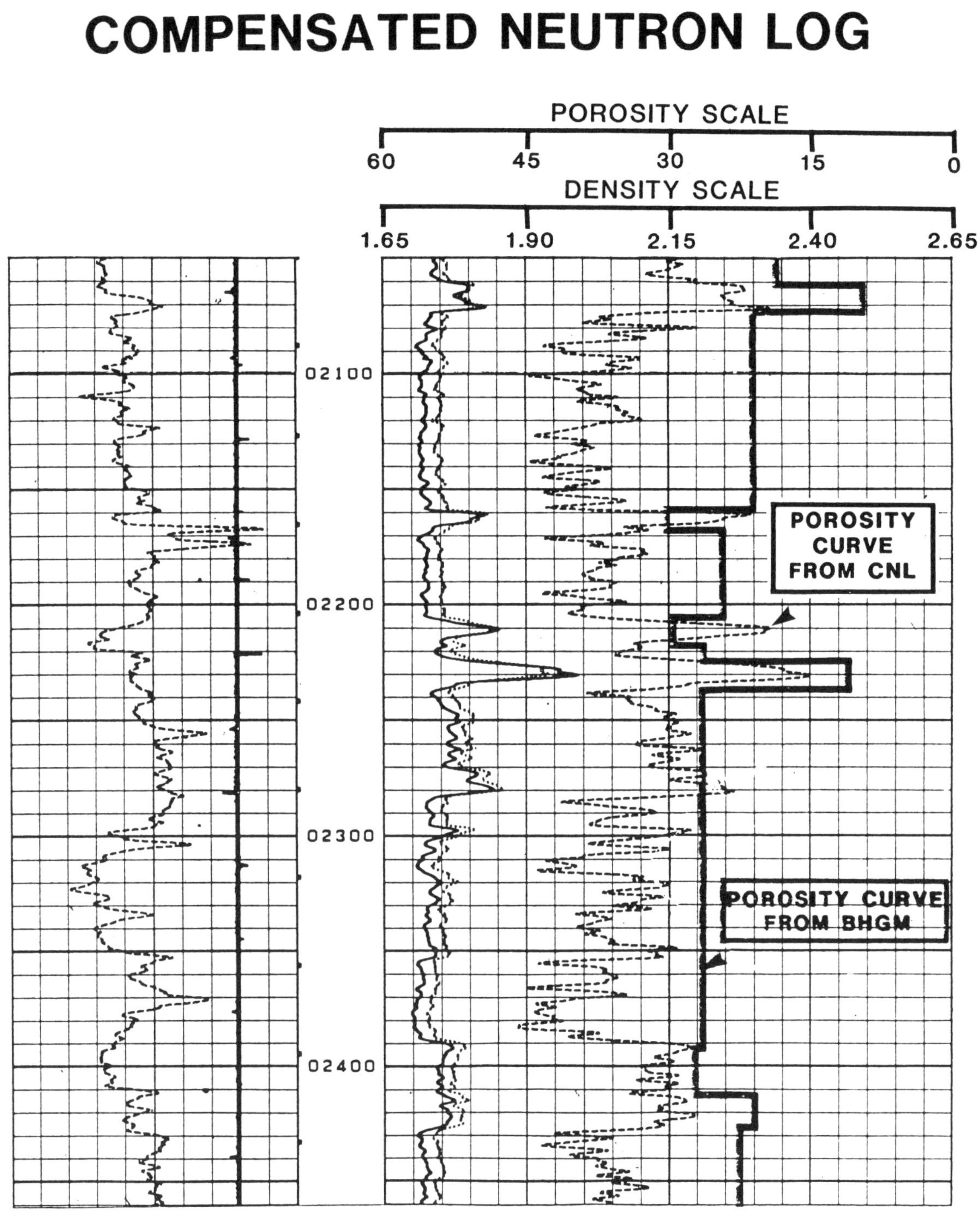

Fig. 2. A portion of a compensated neutron log showing the porosity/density curve from BHGM results plotted on the CNL to show comparisons with the neutron porosity. Crossovers indicate hydrocarbon saturation.

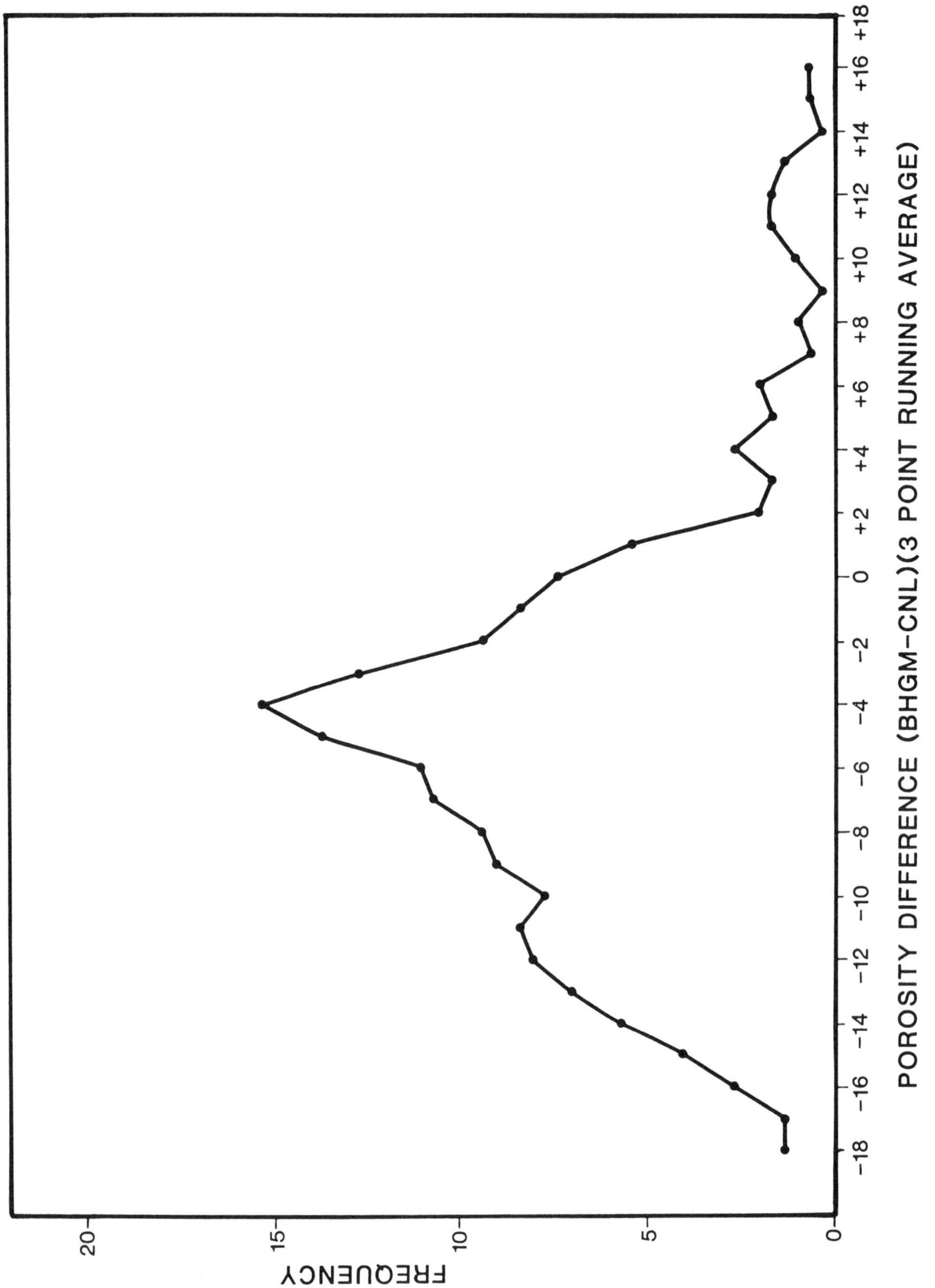

Fig. 3. Plot of number of occurrences of differences between BHGM porosity and CNL porosity, smoothed by a three-point running average.

USE OF HEAT FLOW/TEMPERATURE MEASUREMENTS, INCLUDING SHALLOW MEASUREMENTS, IN HYDROCARBON EXPLORATION

David D. Blackwell

DEPARTMENT OF GEOLOGICAL SCIENCES
SOUTHERN METHODIST UNIVERSITY, DALLAS, TEXAS

ABSTRACT

A rationale is described for the use of thermal techniques (temperature, temperature gradient, heat flow) in a hydrocarbon-prospect exploration package. Techniques for the collection, reduction, and interpretation of shallow and deep thermal data are described. The techniques discussed include measurement of horizontal and vertical variations in temperature, geothermal gradient, and heat flow (the product of measured temperature gradient and thermal conductivity). In most interpretations, heat flow is the appropriate property to use because the effects of rock property variations as well as superficial effects (such as topography) are included in its calculation. Thus anomalies not related to conditions associated with potential hydrocarbon traps may be most easily identified and dropped from further consideration. Thermal anomalies may be associated with hydrocarbon deposits in a trapping situation because of a variety of phenomena, including thermal refraction effects, thermal effects associated with vertical leakage from hydrocarbon traps, and local or regional fluid flow effects. Regional or local fluid flow systems may be associated with the migration of hydrocarbons, with the consequence that the zones of fluid flow with an upward component (and consequent high shallow heat flow) may be an indication of attractive regional or local hydrocarbon exploration targets. Thermal conductivity effects may be large if major geologic structures such as salt domes and shale diapirs are present. Thermal conductivity effects will be more subtle in most cases than fluid flow effects. Measurable thermal refraction anomalies, however, may be associated with hydrocarbon reservoirs. In addition, local maturation of regionally immature source rocks can be associated with local areas of high heat flow, such as geothermal systems. The latter effect may explain many of the hydrocarbon accumulations associated with active, high-temperature, rift basin tectonic settings, such as the Basin and Range province in Nevada.

INTRODUCTION

The use of thermal techniques remains an unconventional approach to hydrocarbon exploration, even as geochemical techniques become increasingly accepted. There are many different possible relationships of thermal phenomena to hydrocarbon accumulations. These relationships range from geochemically prospective associations (direct thermal anomalies associated with vertical or near vertical mass leakage from a reservoir), to the direct effects of oil and gas on thermal properties, such as thermal conductivity, to the use of thermal data to evaluate the thermal history and maturation stage of a package of rocks in a sedimentary basin. In addition, thermal techniques can be applied at all scales from basinwide analyses to individual reservoir definition. These multiple relationships increase the complexity of application of thermal techniques to hydrocarbon exploration. Combined with the typical lack of training in thermal techniques on the part of exploration geologists, and the often difficult collection of thermal data in hydrocarbon exploration settings, failure of the petroleum industry to use this geothermal hydrocarbon exploration tool, in any meaningful sense, is understandable.

There are few books or articles summarizing techniques of thermal exploration. Gretener (1981) has provided the most complete discussion of thermal techniques in a hydrocarbon exploration setting and has included an extensive set of references. Textbooks by Kappelmeyer and Haenel (1974) and Buntebart (1980) cover certain aspects of the application of thermal techniques to hydrocarbon exploration. Good examples of the results of application of such exploration techniques in the North American literature are few. In his rather complete summary of thermal associations with hydrocarbons, Roberts (1981) has been forced to rely heavily on examples from Russia. Throughout this paper, examples of exploration data from geothermal systems will be included. These data, derived from exploration for geothermal reservoirs, have taught us much about measuring and interpreting surface and subsurface thermal data in geothermal systems. These lessons and developed techniques, in many cases, can be applied to hydrocarbon exploration.

The quantities which are measured are temperature and thermal conductivity. Thermal conductivity is a measure of the ability of the rock to transfer heat.[1] The third quantity of interest, the heat

[1]Thermal conductivity units are mcal/cm sec °C (10^{-3} cal/cm² sec) in the cgs system while in the SI system the units are $Wm^{-1}K^{-1}$. Units of heat flow are μcal/cm²sec (10^{-6} cal/cm² sec) in the cgs system and mWm^{-2} in the SI system. Conversions between the two systems of units are 41.84 mWm^{-2} =

flow, may be calculated as a product of the other two as shown in equation (1).

$$Q_z = K \frac{dT}{dz} \quad (1)$$

In this equation Q_z is the vertical heat flow, K is thermal conductivity and dT/dz is the temperature- or geothermal gradient.

Equation (1) is written for vertical heat flow in a homogeneous or in a horizontally layered medium. In a horizontally layered medium with heat transfer by conduction only, the thermal conductivity and the temperature gradient would be inversely proportional so that the heat flow would not change with depth. In this case, over several lithologic layers, equation (1) can be rewritten as:

$$K_1 \left(\frac{dT}{dz}\right)_1 = K_2 \left(\frac{dT}{dz}\right)_2 = \cdots$$

$$= K_n \left(\frac{dT}{dz}\right)_n = Q_z \quad (2)$$

Measurements of "surface" heat flow are typically made in holes which are 30 to 150 m deep. The primary reason for this depth range is to avoid diurnal and annual effects and other kinds of thermal "noise" near the surface of the earth. Exploration experience in geothermal systems has shown that in some cases useful measurements may be made in holes as shallow as 1 to 3 m.

Because there are so many possible variations in thermal techniques and their applications, with consequent confusion as to which ones might be applicable in different situations, several techniques will be briefly discussed in this paper. These techniques include (a) collection and interpretation of temperature measurements in 1 to 3 m holes, (b) heat flow measurements in shallow holes (30 to 150 m deep), (c) use of bottom hole temperatures, and (d) applications of temperature gradients and vertical heat flow in deep holes. In addition, (a) the modelling of thermal data using numerical techniques to calculate the effects of thermal conductivity variations, (b) the use of mathematical continuation (equivalent to seismic migration) to outline subsurface temperature anomalies by calculating isotherms, and (c) analytical and numerical models of the thermal effects of subsurface fluid flow will be discussed.

Interpreting the thermal characteristics of particular sedimentary basins is important for the determination of the thermal maturation of organic material, and much attention has been paid to this subject in the past few years. The object of this paper, however, is the direct application of temperatures to hydrocarbon exploration. Determination of thermal histories and organic maturation of sedimentary basins will not be discussed.

TEMPERATURE MEASUREMENTS IN SEDIMENTARY BASINS

In the early phases of the petroleum industry a significant amount of attention was given to temperature studies in exploratory wells, and the development of accurate temperature logging devices has been discussed in earlier papers. Some of the studies remain the best data sets for illustration of the correlation of temperatures with structures and hydrocarbon accumulations available (Hawtof, 1930; Van Ostrand, 1934, 1937). In the twenties and thirties an extensive series of temperature measurements were made by personnel of the U.S. Geological Survey, and numerous papers by Van Ostrand and others were published. Much of this information has recently become available in open file form (Guffanti and Nathenson, 1981). The techniques in this early project included measurements using maximum reading mercury thermometers at several depths in holes that had reached thermal equilibrium.

It appears that the development of continuous downhole temperature logging tools more or less corresponded with the wide conversion from cable

---

1 μcal/cm^2 sec and .4184 Wm^{-1} K^{-1} = 1mcal/cm sec°C. The average global heat flow is 60 mWm^{-2} (1.5 μcal/cm^2 sec). Units of temperature gradient most commonly used are numerically the same in both the SI and cgs systems. The units respectively are mKm^{-1} and °C/km. In much of the exploration literature, temperature gradient is assigned to a confusing array of units. The most useful units are °F/100 ft although the units of ft/°F and °F/1,000 ft are sometimes encountered.

tool to rotary drilling techniques. Because most wireline logging is done immediately following completion of drilling, the temperatures are still disturbed by the massive circulation of surface fluid during rotary drilling. As a result, temperature logs taken at the time other wireline logs are run do not represent equilibrium conditions and are difficult to interpret. Cable tool holes, because of a lack of fluid circulation, would be more informative, but are generally no longer available. There are many studies which attempt to develop correction schemes to convert nonequilibrium temperatures into equilibrium temperature data. To this day, however, no detailed study of the recovery to thermal equilibrium of a deep hole has been published. Without such thorough studies, no truly valid inference can be made regarding the point (during recovery) when subtle correlations of temperature and lithology demonstrate the return to equilibrium data. The general low regard in which temperature logs are held has strongly discouraged an effective use of nonequilibrium logs; i.e., the location and characterization of porous zones in wells.

The technical ability to obtain accurate, precise temperature logs has been available for the last thirty to forty years. This technical ability, however, has all but been ignored by hydrocarbon exploration. Logging systems are available that can record temperatures with a precision of 0.001°C over a digitizing interval of a few centimeters (a temperature difference of 0.001°C corresponds to 1°C/km for a 1 m digitizing interval, 2°C/km for 0.5 m digitizing interval, and 4°/km for a 0.25 m digitizing interval). Logs are run with the tool going *down* the hole so that undisturbed temperatures are obtained. Logging velocities are typically 10 to 15 m/min, so that thermal equilibrium in the sensor is maintained (otherwise nonequilibrium corrections would have to be made). The limitation on the precision of temperature measurements is not associated with instrumentation (which could be improved at least one to two orders of magnitude), but with thermal noise in the well. Such noise exists even at thermal equilibrium and it is associated with small convection cells. Detailed studies of this convection phenomenon have been published by Gretener (1967), Diment (1967), and Sammel (1968). Blackwell and Spafford (1986) give an up-to-date discussion of the techniques and equipment used for temperature logging in continental heat flow measurements.

Measurements of the appropriate thermal properties of the rock (thermal conductivity) must be made in the laboratory because direct *in situ* techniques have not yet been developed. Typical techniques include divided bar measurements on core and cuttings, and the needle probe measurements on very soft materials and cuttings (see Gretener, 1981; Blackwell and Spafford, 1985), and the newer technique of a surface needle probe only now coming into vogue (Vacquier, 1985). Typical values of the thermal conductivity of sedimentary rocks are presented by Clark (1966) and Roy et al. (1981). Detailed thermal studies in holes in sedimentary basins, however, have shown the difficulty of obtaining *in situ* properties from laboratory measurements. Thermal conductivity values of shales in particular have been overestimated by a factor of two in the published literature compared to the apparent *in situ* values (Blackwell and Steele, 1986).

There have been a number of papers describing techniques utilizing well logs to obtain indirect *in situ* thermal conductivity values. These techniques have been reviewed in more detail by Blackwell and Steele (1986). A fully satisfactory and complete application of such techniques to the determination of thermal conductivity has yet to be described. While it is certainly possible to develop such techniques, at least for a limited suite of lithologies, general use will require extensive additional research.

Houbolt and Wells (1980) correlated the thermal properties of sedimentary rocks and P-wave travel times from reflection data. They argued that travel time would be related to the total thermal resistance (the inverse of thermal conductivity) of a sedimentary package. This technique may be useful empirically in some basins, but in the Gulf Coast, for example, sands and shales above 2 to 3 km are not readily identifiable by their velocity. Therefore, this technique has limited value.

The relationship of temperature to lithology in a conductive setting is illustrated in three high-precision equilibrium temperature gradient logs, as shown in Figure 1. The curves displayed are temperature gradient curves calculated from the difference of temperature over either a 0.5 or 1 m interval. This type of "gradient" log is used for interpretation because it displays the correlations with geology, anomalous zones, etc., better than a temperature-depth curve. The three logs shown in Figure 1 were made with two completely different sets of instrumentation, a conventional thermistor probe, and a probe containing a semiconductor sensor with an output of 1 mamp/°C. The logs were run at a speed of 12 m/min and digitally recorded at either 1.0 or 0.5 m depth intervals to a precision of 0.001°C. The lithology and formation names are given in the figure. The rocks are upper Cretaceous shales, marls, and limestones, with some sand (the Woodbine, Paluxy, and Trinity). The very close and reproducible correspondence of high temperature gradients with shale, and low geothermal gradients with sand and limestone is clear. Gradients are 40–50°C/km (2.2–2.7°F/100 ft), in the former units, and 15–20°C/km (0.8–1.1°F/100 ft) in the latter units. One interesting result is that the temperature gradient log shows the contact between the Woodbine sandstone and the Eagle Ford shale more clearly than does the total gamma-ray log often used to separate sand and shales.

The kind of detailed information available from an equilibrium temperature log may be compared to the information available from a set of bottom hole temperature (BHT) data often used in attempts to correlate thermal data with hydrocarbons. Figure 2 shows a comparison of bottom hole temperature data and precision temperature logs for eight sites in western Nebraska (Gosnold et al., 1981). Each log and its associated bottom hole temperature data (from holes within 10 km of the log) are shown by the same symbol. It is obvious that even with the correction of the bottom hole temperature for drilling disturbance (if it were possible) the information content would not compare to the temperature logs. For example, bottom hole temperature information is usually only available over a limited depth range. In a given area, most of the holes bottom and are logged at similar depths. Nonetheless, bottom hole temperature data are widely available and on occasion have been used effectively.

If temperature data are available as a function of depth in several wells, then vertical as well as horizontal variations can be explored. Thermal conductivity variations are generally the cause of vertical and lateral temperature variations (see Fig. 1). An extreme example (from eastern Kansas) is shown in Figure 3. The temperature gradient in the upper part of the hole (in Pennsylvanian shales and coals) is about 50°C/km (2.5°F/100 ft), while the temperature gradient in the lower part of the hole (in Mississippian and older carbonate rocks) is about 18°C/km (1°F/100 ft). This example illustrates the fallacy of using a single bottom hole temperature and the surface temperature to calculate a gradient, then using that gradient to calculate temperatures at other depths by extrapolation or interpolation. For accurate understanding of the subsurface thermal conditions, a knowledge of the thermal properties of the rocks, as well as temperature, is necessary.

In addition to the effects of thermal conductivity and conductive heat flow variations through a basin, the other major influence on temperatures (at least at certain locations in some sedimentary basins) is active fluid flow. The effects of fluid flow can be recognized on a local or regional basis. There are only a few documented examples of local temperature variations in sedimentary basins due to fluid flow, although the phenomenon is probably common. Gatenby (1980) and Harrison (1980) have documented such temperature variations in southern Louisiana and have associated these variations with the migration and trapping of hydrocarbons. More details of these studies may be found in the paper by Jones (1984) in an earlier volume of this series.

Much progress has been made in recent years in both understanding the characteristics of subsurface fluid flow and its association with temperature "anomalies." In this case "anomalies" reflect departures from expected "conductive" conditions, so conductive heat flow conditions must be understood before convective anomalies can be identified and correlated with hydrocabon accumulations. Techniques of interpreting thermal data in convective settings, which can be used for hydrocarbon exploration, are discussed in a following section.

The remainder of this paper will be a discussion of specific thermal techniques that can be used to evaluate temperature and heat variations in sedimentary basins, their possible associations with hydrocarbons, and ways in which hydrocarbon exploration techniques can be applied.

THERMAL TECHNIQUES

The techniques to be discussed are divided into two types: (a) those applied to measurements in shallow holes which are drilled specifically for "surface" exploration, such as shot holes for seismic studies or very shallow holes for geochemical measurements, and (b) those applied in deeper holes drilled directly as hydrocarbon exploratory wells.

The thermal techniques to be considered in order are: (a) temperature measurements in 1 to 30 m holes, (b) heat flow in shallow holes (30 to 150 m deep) (c) temperature studies in deep holes including bottom hole temperature comparison and (d) the use of temperature gradients and vertical heat flow measurements. Throughout this and the following section, many of the examples will be taken from the realm of geothermal exploration because of the many heat flow techniques that have been developed for application to geothermal exploration during the last ten years. A good understanding has been obtained of the way heat flow and geothermal gradient information can be used to determine subsurface temperature variations. In the hydrocarbon exploration setting, these techniques may be applied to identify temperature variations which might be associated with thermal property contrasts due to structural or stratigraphic effect, or associated with fluid flow related to regional or local hydraulic head differences.

There are techniques of thermal exploration that can be applied from the surface to any drillable depth. The objective of determining deep thermal conditions from "shallow" exploration results involves a determination of the minimum depth at which useful results can be obtained, always a thorny problem in thermal studies. The depth necessary is essentially a signal-to-noise problem. The signal is the increase in temperature with depth caused by the gradient (heat flow) difference, and the noise is caused by disturbances which would effect or mask this temperature increase.

Infrared Measurements and Measurements in 1 to 3 m Holes

In some ideally conductive situations, or situations where geothermal convection comes to very shallow depths, it may be possible to obtain useful exploration data in very shallow holes. The application of airborne infrared techniques to the location of temperature anomalies has been attempted many times. The general consensus is that the useful application of such a technique requires the existence of heat flow values of at least 2000 mWm^{-2}(50×10^{-6} cal/cm^2sec) or surface temperature differences of several degrees C. To work at this highest level of resolution, flight duplication, numerous extensive corrections, and ground truth observations are necessary (Del Grande, 1985). In general, the airborne infrared techniques are thus restricted to location of actual thermal manifestations (hotsprings, very shallow warm water, very shallow cold water, etc.). Of course, anomalies may be caused by shallow groundwater flow which is not representative of deep temperature or heat flow anomalies.

In holes 1 to 3 m deep, the primary effect on the temperature is the 2° to 10°C amplitude annual cycle. A minimum depth of 1 m is considered to be the shallow limit because any lesser depth would be confounded by the diurnal effect. Typical temperature gradients (20–100°C/km) will generate temperature differences of 0.04–0.2°C over a 2 m-depth range. Consequently, in holes of this depth, absolute measurement of geothermal gradient or heat flow is not attempted. This type of data are used to locate broad temperature anomalies in much the same way that holes in this depth range are used to locate geochemical anomalies.

In the interpretation of 1 to 3 m temperature data, many major temperature "noise" effects must be investigated. Such effects as microclimatic setting (albido, roughness, wind speed, slope orientation, elevation, and vegetation cover), thermal conductivity and diffusivity variations in time and space, the effects of rainfall, and very shallow groundwater movements must be considered. Temperature measurements must be made as nearly simultaneously as possible in order to avoid the effects directly associated with the annual cycle. Corrections are not simple or established for all of these effects.

The typical minimum noise associated with the technique is on the order of 1° to 2°C. The practical lower heat flow detection limit of the 1 to 3 m technique appears to be on the order of 400 mWm^{-2} ($10\mu cal/cm^2sec$) with a resolution of tens of mWm^{-2}. Based on empirical observations in some places, however, measurements in 1 to 3 m holes do outline thermal anomalies of even lower amplitude. In other studies, no relation between the shallow temperatures and deeper heat flow anomalies has been identified. An unresolved but important question affecting the possible usefulness of these techniques is the behavior of the earth's surface. Is it an isotherm or alternatively, are emissivity conditions of the surface sufficient to cause significant temperature variations at the earth's surface due to relatively small changes (on the order of tens of mWm^{-2}) in heat flow? There is no doubt that in some cases deep thermal conditions are reflected in shallow temperatures. An example is shown in Figure 4. Lange et al. (1982) have shown a close correspondence between 3 m temperatures and 30 m temperatures for the McCoy geothermal system in Nevada.

Detailed studies of 1 to 3 m temperature data have been discussed by Bowen et al. (1977), Olmsted (1977), and Leshack et al. (1979). Poley and Van Stevenick (1970) have described a significant study of temperatures in 1 to 3 m holes. They demonstrate a correlation of temperature anomalies on the order of 1° to 2°C with some fault zones and with buried salt domes. Zielinski and Bruchhauser (1983) presented results from a careful study of a series of 2 m holes over the San Sabastion Oil and Gas Field in Argentina. They located an anomaly of the order of 200 mWm^{-2} over the field using these data. They attributed the high heat flow to localized deep groundwater discharge.

Because of the possible limitations on the 1 to 3 m technique in a hydrocarbon exploration program, these types of studies might not be cost effective as a sole objective. A program of shallow temperature measurements in conjuction with geochemical studies would enhance the information content and interpretability of both sets of data.

Holes of 30 to 150 m Depth

Holes in the depth range 10 to 30 m are transitional between conditions from the 1 to 3 m holes on one hand and conditions from the 30 m holes on the other. The annual effect is small but significant, and very shallow groundwater flow can be a problem. If heat transfer is conductive and surface disturbances are small, however, significant anomalies can be identified. Detailed discussions of the annual temperature effect in this depth range have been discussed by Lovering and Goode (1963), Lovering and Morris (1965), Lachenbruch (1959), and Van Wijk (1966). In this depth range, the results obtained vary from data similar to shallow holes to data similar to deep hole results.

Many factors affect the subsurface temperatures and temperature gradient measured a each location. Extensive studies in the past twenty years have given us a good understanding of the mechanisms which affect subsurface temperatures. Corrections for (or means of avoidance of) many types of disturbances have been developed. Temperature gradient and heat flow measurements in holes in this depth range are what might be referred to as typical heat flow measurements. In most geologic terrains, depending on surface noise effects, a geothermal gradient can be measured with reasonable corrections and with thermal conductivity measurements which can be used to calculate within 5 to 20% the heat flow that is typical of the upper crust.

Once a hole is drilled, to make it ideal for heat flow/thermal measurements, it should be completed with small diameter observation tubing and grouted or backfilled to prevent water flow within the well as well as contamination of shallow aquifers. Backfilling may be accomplished with cuttings, heavy mud, or cement. After completion of temperature measurements the holes are abandoned by putting a cement plug in the tubing. The holes must be left for some period following drilling because the drilling effect itself causes a temperature disturbance. The temperature recovery from the thermal disturbance due to drilling has been discussed in may places in the literature, and some of these techniques are briefly summarized in the section on deep holes. Temperature measurements made in holes drilled by air, and at the bottom of the hole as diamond drilling proceeds, accurately reflect equilibrium temperatures. In these types of holes, subsequent loggings may not be necessary. Otherwise, from a practical point of view, it is useful to log each hole two or three different times following completion of drilling to make sure that equilibrium has been obtained. Short time data are also useful in recognizing effects that may be important but cannot be identified from a single equilibrium temperature log. Such data might recognize zones of fluid loss and anomalous recovery rates. Once the equilibrium temperatures are reached, temperatures remain stable over very long periods of time and below the depth of the annual wave, and there is no reason to expect variations with time unless some unusual effect has taken place.

In a variant of this technique, Sass et al. (1981) describe a procedure of driving a temperature and thermal conductivity measuring tool into soft sediments ahead of a drill bit. In this case measurements of heat flow and temperature gradient can be made without having to use tubing and grouting.

The first level of data analysis involves interpretation of the observed temperature-depth data in terms of the local geology, hydrology, and geothermal gradient. In areas where conductive heat flow dominates, temperature-depth curves of the type illustrated in Figures 1, 2, and 3 are typically observed. Conductive effects which might influence heat-flow values and for which corrections must be considered, include the effects of uplift and errosion, subsidence or deposition, climatic change, annual and diurnal temperature variation, lateral temperature differences associated with variations in surface conditions, and the effects of topography. The effects of uplift, erosion, deposition, subsidence, and climatic change, have been discussed extensively in the literature. *In most exploration situations these effects will not lead to significant errors.*

A common phenomena which does cause major variations in gradient, laterally and as a function of depth, is lateral variation in mean annual temperature at the surface of the earth. Typical amplitudes of temperature depth differences associated with climatic conditions at the surface of the earth are on the order of 2°C (see Poley and Van Stevenick, 1970). Typical plots of anomaly effect versus depth are shown in Figure 5. Corrections for these sorts of effects are discussed by Blackwell et al. (1980), and Lachenbruch (1957, Lauchenruch et al., 1962). It is clear from the figure that surface effects are generally relatively minor if holes as deep as 100 to 150 m are available. In this case the absolute magnitude of a gradient anomaly below 100 m is on the order of a few degrees C/km. This would not be true in the case of terrain effects due to very rugged topography.

If the holes cut different types of rock either within the same hole or between holes, then thermal conductivity must be determined so that heat flow values rather than just temperature gradients can be compared. As discussed in the first part of this paper, once the thermal conductivity has been measured and the temperature gradient calculated, the heat flow can be calculated. If thermal conductivity varies with depth, a reduction procedure devised by Bullard (1939) and discussed below can be used.

Once obtained, the heat flow values may be correlated with other types of exploration data to look for variations in heat flow associated with hydrocarbon accumulations, fluid flow, structures, basement depth variations, etc. An example of a geothermal gradient/heat flow anomaly due to fluid flow along a fault zone in a sedimentary basin, the western Snake River basin of Idaho and Oregon, is shown in Figure 6. The western Snake River basin is a large late Cenozoic basin filled with up to 4 km of volcanic and sedimentary rocks (Newton and Corcoran, 1963) and large potential gas reserves. The anomaly was identified by temperature measurements in 30 to 200 m holes drilled for uranium exploration. In these holes, lithology is essentially constant (siltstones) so that temperature gradient and heat flow variations have one-to-one correspondence. In geothermal exploration, attention has been paid to such local thermal anomalies as possible geothermal resources, and exploration of this area is in progress to determine its geothermal potential. The possible hydrocarbon implications of such phenomena will be discussed in a subsequent section.

Ocean Heat Flow Techniques

A specialized application of heat flow techniques is applicable to large ocean areas when hydrocarbon exploration takes place. Below the thermocline at a water depth of 30–300 mm, annual temperature variations are quite small and temperature gradient and heat flow measurements can be made using probes penetrating 2 to 20 m into the bottom muds. Extensive and detailed techniques and equipment have been developed to make heat flow measurements in the ocean basins. Some of the most recent equipment development is discussed by Hyndman et al. (1979). The techniques could be developed for application to shelf areas within or above the thermocline with a minimum amount of difficulty. With some information on annual temperature variations, heat flow measurements could be made routinely with shipboard equipment in conjuction with seismic reflection surveys, gravity surveys, etc. These measurements could provide additional exploration information and allow more direct evaluation of potential maturation conditions than can be obtained from modeling with no direct knowledge of the thermal regime.

A technique which might allow qualitative measurements of heat flow in oceans is the mapping of the base of the gas hydrate layer where the layer exists. In certain conditions, gas molecules are trapped inside an expanded lattice of water modules. The gas hydrates have a known phase diagram and distinctive physicial properties that cause the layers in many cases to show up clearly on shallow seismic data. Seismic determination of the base of the gas hydrate layer may allow relative determination of the temperature gradient (MacLeod, 1982).

Deep Exploration Hole Observations

The crudest, yet the most common form of thermal information available is represented by bottom hole temperature measurements from log headers. In areas where extensive drilling has been carried out, thousands of these measurements are available. Some of the quantitative difficulties with these measurements have been illustrated in a previous section. Since they are so abundant they could be a very valuable source of thermal information if they are usable on some statistical basis, or in some other way.

Most temperature data used in subsurface exploration are taken from well log headers. Because of several factors, including mainly fluid circulation associated with drilling and hole conditioning, bottom hole temperatures from well logs are lower than true formation temperatures. Thus, in order to calculate equilibrium temperature, a special procedure is involved. The Horner technique is one of the most common. Using the circulation time and the time elapsed until logging, a series of temperatures are used to estimate the equilibrium temperature. There are many other techinques used, including empirical techniques, variations in mathematical models, and consideration of additional information on (a) circulation times and (b) particular temperatures and rock properties. Essentially, each data set must be treated differently depending on the well conditions in which the data are obtained. Most papers dealing with use of bottom hole temperatures include discussions of equilibrium temperature techniques (Chapman et al., 1984).

Chapman et al. (1984) have also summarized results of the application of empirical corrections and argue that if temperatures are logged more than ten to twelve hours after the last circulation, then the majority of the temperatures are within $\pm 5\%$. Unfortunately, at depths of 3 to 4 km this difference may be on the order of $\pm 5°C$. Thus bottom hole temperatures can be used only if they extend over quite a large vertical range. They are not suitable for measurement of gradients over narrow vertical ranges unless large temperature anomalies are involved (on the order of several to 10°C).

One type of data which is the most suitable for interpretation is obtained from shut-in pressure measurements in gas reservoirs. In many cases temperatures are measured along with pressure measurements taken as part of the reservoir evaluation. These measurements are often taken at a long enough period of time after drilling to have reached equilibrium, and often are taken from nonflowing wells. Such data are not as commonly available in the United States as they are in some other countries, but where available, they can be used to great advantage in thermal interpretations.

In the interpretation of subsurface thermal data, where the temperature data are sparse, a techinque referred to as the thermal resistance method is one appropriate way to analyze the temperature data. Thermal resistance is a quotient of a certain thickness times the inverse of the characteristic thermal conductivity of that thickness of rock. In a horizontally layered earth, temperatures as a function of depth are related to the thermal resistance of the stratigraphic section according to the equation

$$T_D = T_o + Q \sum_{z=o}^{D} \left(\frac{\Delta z}{K}\right) \qquad (3)$$

where T_D is the temperature at depth, $z = D$, T_o is the surface temperature at $z = O$, Q is the surface heat flow, and a thermal resistance $(\Delta z/K)$ is summed for all rock units between the surface and depth D. This equation, in either integral or summation form, is commonly used in heat flow data reduction, because the heat flow can be calculated as the slope of a plot of temperature versus summed thermal resistance. Ideally, the thermal resistance would be summed individually for each well using conductivities and thicknesses for all rock units intersected (Carvalho et al., 1980; Chapman et al., 1984; Speece et al., 1985). Chapman et al. (1984) and Speece et al. (1985) have applied this technique by averaging over the stratigraphy of a basin. This process is useful if the basin stratigraphy and structure are sufficiently simple and well known so the contacts between rock units and the conductivity variations can be described by simple functions. In this way, thermal resistance as a function of depth can be calculated everywhere in the basin for use with the isolated (in the horizontal and vertical dimensions) temperatures. In this way, temperature variations can be associated with lateral variations in the stratigraphic or thermal conductivity package or associated with other effects, such as thermal refraction or fluid flow.

Bottom hole temperature information has been used to identify thermal anomalies associated with hydrocarbon migration in southern Louisiana, as described by Gatenby (1980) and by Harrison (1980). In addition, Speece et al. (1985) and Chapman et al. (1984), among others, have attempted to use the bottom hole temperature information to regionally characterize sedimentary basins (e.g., the Michigan and Uinta basins). The success of these applications is not clear because there are not many detailed thermal data availbale to cross check the results. A comparison of the results of the temperature gradient determination by Speece et al. (1985) at locations within the Michigan basin is shown in Figure 7. The calculated versus the observed curves have little detailed correlation. So there may be questions as to whether the copious bottom hole temperature data (even taken statistically) contain enough information to resolve important thermal characteristics of sedimentary basins.

Ball (1982) documented small temperature anomalies associated directly with oil fields using extensive bottom hole temperature data. The mechanism by which such anomalies are generated is not clear, but it could be any one of a number of mechanisms, including thermal conductivity variations (direct or indirect) related to the presence of hydrocarbons, or it could be small fluid flow variations.

Because of the limitations of bottom hole temperature data, a detailed understanding of the heat flow regime within an exploration area is important in order to apply thermal techinques most usefully in deep holes. It is critical to determine whether the regime is dominated by conductive effects or by a significant component of convective heat transfer. If one or more deep holes are available for detailed study, these holes can be used a calibration points for quantification of the heat flow regime and calibration of the bottom hole temperature data which may be used for the actual exploration investigations. A detailed equilibrium temperature log for the wells of the sort shown in Figures 1–3 must be available, and thermal conductivity measurements for samples encountered at various intervals in the well are also necessary. Thus, heat flow variations (or lack thereof) with depth in the drill hole can be determined in order to understand whether changes of gradient within the hole are asociated with lithology or with some other thermal effect, such as fluid flow or thermal refraction.

A lingering difficulty is one of exactly relating *in situ* properties to laboratory measurements. Not surprisingly, shales are particularly difficult to categorize. Most of the measurement techniques on samples of shale fail because typically only the most solid pieces of shale can be sampled, and measurements using a transient technique such as a needle probe are affected by the anisotropy of the shale. Blackwell and Steele (1986) pointed out that in a study of Paleozoic rocks in Kansas, laboratory measurements of thermal conductivity of shales based on the cutting technique were high by 50 to 100%. An examination of the literature indicated similar problems have frequently occurred in the past without being recognized. Since shale makes up such a significant component of the section in most sedimentary basins, the ability to evaluate its properties is very important.

INTERPRETATION OF THERMAL DATA

During the past ten years, many types of interpretation techniques for thermal data have been developed. For the purpose of this paper, the techniques will be broken into two types: (a) those applied in conductive situations and (b) those applied when convective heat transport is responsible for part of the overall heat budget. In addition to the location of temperature or "heat flow anomalies" which might be related in some way to hydrocarbon associations, different types of thermal data can be quantitatively interpreted to determine geometries of bodies of contrasting thermal conductivity as well as the location and configuration of convective fluid flow confined to specific regions.

Interpretation Assuming Conductive Heat Flow

The use of thermal data to investigate the geometries and locations of structures, such as salt domes, shale diapirs, basement uplifts, and so forth, has been discussed in the literature. In particular, the effect of salt on surrounding temperatures has been modeled numerically by Selig and Wallick (1966), Obrien and Lerche (1984), and others. In his discussion of geothermics, Gretener (1981, pp. 88–97) includes a section with models of the effects of salt domes and basement uplifts. Hawtorf (1930) demonstrated the presence of elevated temperatures above salt domes in Texas many years ago. Shale diapirs might be expected to show lower temperatures than normal because of the low thermal conductivity of shale. If the shale diapirs are active, however, there may be mass transport of heat by the updoming of the shale causing elevated temperatures to be associated with the shale diapirs as well as with salt diapirs.

Anticlines also might have higher or lower temperatures than the synclines associated with them, depending on the nature of the thermal conductivity contrasts between the two stratigraphic sections. Such effects, or possible fluid leakage through the anticlines, may be responsible for the elevated temperatures commonly observed in anticlinal regions, as documented in the United States by Van Ostrand (1937) and as illustrated for many areas in Russia by the data summarized by Roberts (1981). These examples make it clear that temperature variations should be associated with structurally related variations in the thermal conductivity.

The modeling techniques that are used in the investigation of the effects of conductivity contrasts on the subsurface temperature field are widely distributed and developed. Most of the techniques used are numerical (e.g., finite difference, finite element, etc.). If information is available from drilling or geophysical interpretation, it is possible to construct a geologic model and calculate its thermal effects using any number of possible numerical solution techniques.

In cases where fluid flow is confined, perhaps by impermeable shales on either side of an aquifer, by impermeable rocks on either side of a fracture or fault zone, or by self-sealed rocks *outside* a geothermal reservoir, the effect outside the paths of fluid flow can be modeled by conductive techniques (Brott et al., 1981). The temperature distribution at the confining boundary can be treated as a boundary condition for a conductive heat flow solution. In this case, conductive interpretation methods can be used to reconstruct the temperture along the zone of flow if the geometry of the flow region is known. As is often the case with geothermal systems, if the temperature of the reservoir is known, then the shape of the reservoir can be calculated.

Figure 8 shows a model which combines the two modelling approaches using conductive techniques discussed in this section. The basic structural setting is that of a graben filled with low thermal conductivity rocks and a bedrock horst terrain of higher thermal conductivity rocks. The structural displacement between the graben and the host is 1.5 km. A thermal conductivity contrast of a factor of two is assumed. Temperatures obtained by a numerical finite-difference solution of the conduction-only model are shown by the dashed curves. Where the dashed curves merge with the solid curves outside the immediate area of the fault contract, the conductive and convective solutions are the same. In the conductive case, the effect of the low thermal conductivity graben fill is to increase temperatures at a given depth compared to the adjoining horst. The geothermal gradient is approximately a factor of two higher in the graben than in the horst. If the thermal conductivity is taken into account, however, the heat flow values are virtually identical in both the horst and the graben blocks. Only in the vicinity of the fault is there refraction of heat so that slightly lower heat values are observed in the graben above the fault and slightly higher heat flow values are observed on the horst side of the fault. Variations in heat flow have an amplitude of about 50% of the assumed background value and are concentrated in the immediate vicinity of the fault.

In many cases, particularly in the Basin and Range provinces in the western United States, range/valley-bounding faults are either zones of fluid flow or they act as barriers to fluid flow. As a result, hot springs are associated with many of the faults. The geometry of the fluid flow may take many forms. For example, upflow may occur along fault intersections with the range-bounding fault. In such a setting the surface heat flow anomaly is not very large in size. An example of this type of anomaly is found at Leach hot springs in Nevada (Fig. 9).

In some cases, however, fluid flow may occur along much of surface of the fault zone so that larger size anomalies are generated. There may be little or no surface leakage, however, so that associated hot springs may be small or nonexistent. The Vail, Oregon, heat flow anomaly illustrated in Figure 6, is an example of such an anomaly. Since the fluid flow is confined to the fault zone, the temperature of the fluid along the fault can be treated as a boundary condition in a conductive solution. The model illustrated in Figure 8 is modified to match this situation by assuming that water with temperature of 100°C flows up the fault shown in Figure 8. The conductive solution outside the fault was calculated and is shown as the solid lines in Figure 8. Heat flow values associated with this fluid circulation are much higher than those associated with the conductivity contrast between the horst and the graben. The maximum heat flow is very sensitive to the actual upper depth to which the hot water rises, but the shape of the heat flow anomaly is related to the geometry of the fault zone and the thermal properties of the media. In fact, the shape can be used to obtain information on the geometry of the fault zone.

The geometry may be inferred by the technique of type-curve matching commonly used in geophysics. It may also be inferred directly using the mathematical technique of continuation of potential field data. This technique is equivalent to "downward" or "sourceward" continuation of gravity/magnetic data and to seismic migration. The concept of application of this technique to anomalies related to fluid flow has been discussed in detail by Brott et al. (1981), who also showed several applications of the technique. Other types of continuation approaches to the solution of the heat conduction equation have been discussed by Mareschal et al. (1985).

As illustrations of the power of this technique, two examples from the geothermal literature are shown in Figures 10 and 11. The heat flow anomaly at Roosevelt hot springs, Utah (see Ross et al., 1982, for a description of the area) is shown in Figure 10. The heat flow anomaly pattern based on measurements in holes 50 to 300 m deep was described by Ward et al. (1980). The setting of this geothermal system is at the edge of a Basin-Range horst block with the Milford Valley to the west and Mineral Range to the east. Since fluid flow so commonly occurs along Basin-Range bounding faults, it might be concluded that a range bounding fault between the Mineral Range and Milford Valley is the controling structure for the geothermal system. A drilling plan ought therefore to be designed to intersect this structure down dip. Figure 11 shows a superposition of the cross sections of the thermal anomalies from the Vail geothermal system in Oregon and the Roosevelt geothermal system in Utah. Also shown in the figure is a theoretical calculation of the heat flow anomaly for a dipping fault zone based on the type of model discussed in the previous paragraphs. It is quite clear that this dipping fault zone model provides a good fit to the heat flow anomaly observed at the Vail geothermal system. However, it is not a good model for the Roosevelt geothermal system because the anomaly is much broader than could be explained by a single fault. Deep drilling has shown that the Roosevelt geothermal system is in fact located within the granite bedrock of the range to the east of the range-bounding faults, a result which the heat flow data clearly supports. A geothermal power plant is now in operation in the Roosevelt area.

A final example is from the Marysville geothermal system in Montana (Fig. 12). This example was discussed by Brott et al. (1981). In this case the method of continuation has been applied to calculate the depth to the 95°C isotherm. This isotherm was chosen based on a deep well which found a temperature of 95°C as the reservoir temperature. The surface anomaly has a rather smooth shape. From downward continuation calculation of the depth of the 95°C isotherm, however, the controls on some parts of the reservoir became quite clear. The northern and eastern edges of the reservoir are bounded by a Mesozic granodiorite stock whose location was inferred based on interpretation of magnetic data. The salient at the south end of the surface heat flow anomaly collapses about an east/west fault which also marks the south edge of the granodiorite stock. Thus the continuation results suggest that the fault is associated with the fluid flow, a result not obvious from the surface heat flow map.

In neither the Roosevelt not the Marysville cases was the correct reservoir interpretation obvious from the surface heat flow data alone. However, interpretation using models and continuation techniques clearly elucidated the nature of, and controls on, the source of the heat flow anomaly. Heat flow anomalies should delineate hydrocarbon reservoirs associated with active motion of water and hydrocarbons. The techniques which have been developed and applied to geothermal systems can be used directly in some of these cases. These techniques have primarily been based on shallow exploration heat flow data in geothermal systems. However, if bottom hole or other temperature data are available, and information on thermal conductivity is also available, then the gradients can be calculated at some subsurface depth, and these gradients could be used for continuation of data at a greater depth to serve as a guide to deeper exploration.

Interpretation Using Convective Heat Flow

The interpretations are not quite so straightforward when dealing with temperature and heat flow anomalies associated with fluid convection through large bodies of rock where movements are slow but still significant enough to affect the thermal field. Many of the cases of association of hydrocarbon traps and migration with thermal anomalies have been assumed to be of this type.

On the scale of individual drill holes, if the direction of fluid flow is known, then an estimate of the velocity of fluid flow can be obtained from mathematical models discussed by Bredehoeff and Papadopulos (1965) (Sorey, 1970; Mansure and Reiter, 1979). In these cases, temperature-depth curves are observed in which temperature gradients vary with depth in a manner not consistent with the model referred to in Equation 2. Often in these cases the situation represents flow between two aquifers through an aquitard, or from an aquifer to the surface via an aquitard. The thermal technique is quite sensitive to the rate of flow, and detectable effects are present down to vertical flow rates of less than a centimeter per year. Some typical temperature depth curves obtained in situations dominated by slow vertical fluid flow are shown in Figure 13. These holes are from various aquifer systems in the western U.S. and show cases of both upflow and downflow.

Such a model might be used to investigate the origin of high gradients at the top of geopressure zones in the Gulf Coast. An upflow rate of approximately 2 cm/yr would be required to maintain a gradient of 45°C/km (2.5°F/100 ft) across the top of the zone compared to the background of 18–25°C/km (1.0–1.4°F/100 ft). A calculation of the volume of fluid required to maintain this flow rate indicates that the fluid from below the top of the geopressure zone could not maintain the gradient observed for a period of more than a few hundred thousand years at most. Therefore, it seems unlikely that convective effects are solely responsible for the high gradients, and the thermal properties of shale are primarily responsible for these changes in gradient (see Blackwell and Steele, 1986, for a more detailed discussion).

There is no guarantee that fluid flow, in general, in a sedimentary basin is vertical. Changes in heat flow as a function of depth in drill holes can be caused by horizontal as well as vertical fluid flow. It is expected that fluids may be in motion in most sedimentary basins. A complete rationale, however, for relating the directions and characteristics of fluid flow (which may vary systematically as a function of time and space) to basin evolution remains a fruitful field for investigation.

Regional groundwater flow in sedimentary basins has been documented in several different cases. In particular, Gosnold (1982) has shown results from the eastern part of the Denver basin which document the thermal effect of the long known (Darton, 1906; Bredehoeft et al., 1983; Back et al., 1983) west-to-east movement of water along various aquifers. There has been extensive discussion of heat flow variations in the Prairies Basin in Canada which may be associated with regional water flow, as discussed by Toth (1981) (Majorwicz and Jessop, 1981; Lam et al., 1984; Majorwicz et al., 1984). In addition, it has been suggested by Chapman et al. (1984) that apparent north-to-south heat flow variations in the Uinta Basin are associated with lateral fluid flow. Regional fluid flow in sedimentary basins has also been discussed extensively with respect to formation of mineral deposits (Cathles and Smith, 1983; Garven, 1985; Bethke, 1985).

Discussions by Roberts (1980) and Toth (1980) of the nature of oil traps emphasized the relationship of oil trapping and migration to fluid flow, and Roberts (1980) gives many references to other studies relating to such relations. Beginning with the simple analytical models of Domenico and Palciauskas (1973) there have been a number of papers dealing with theoretical models of large-scale fluid flow in sedimentary basins. More recently, Smith and Chapman (1983) presented some analytical models for the effects of fluid flow in regional confined aquifers. Gosnold (1984, 1985) presented a theoretical model and described temperature gradient data which related to the measurement of fluid flow in the basins of the Great Plains. Gosnold (1985) also presents examples of the correspondence of these fluid flow effects to heat transfer characteristics as a function of depth in drill holes.

Recognition of such effects allows a rationale for exploration in sedimentary basins which are not as well understood. The scenario for such applications has been developed by Zielinski and Bruchhauser

(1983) based on the example of the Magellan Basin in southern Argentina. There are a number of ways in which this information can be used. Measurements of anomalous heat flow in shallow holes could indicate the directions and magnitudes of fluid flow and thus the direction of movement of hydrocarbons. Measurements of heat flow versus depth would indicate where changes in heat flow occur with depth and pinpoint depths of fluid motion on a regional basis. It might also help to pinpoint the possible stratigraphic positions of reservoirs which might be filled by migration along these major paths of fluid flow.

The model shown in Figure 8 also has some interesting applications concerning the direct location of hydrocarbons. In an active tectonic setting with high heat flow and abundant fluid circulation, it may not be possible to characterize a large area in terms of a uniform stage of thermal maturity versus depth. A particularly good example of this setting is the western Basin and Range province in the United States. Source beds are abundant in this terrain and they range in age from Lower Paleozoic to Cenozoic. The structure in many areas is very complicated, and maturation, even within the same stratigraphic horizons, varies greatly from one part of the province to the other (Pool and Claypool, 1984). In many of the Cenozoic basins, young source beds may be undermature in general; however, if they are associated with geothermal systems, local maturity conditions may be much different. It is possible that local oil and gas maturation might be associated with local geothermal systems. If a trap is present, such local accumulations of oil or gas might be economically important. The well-known Bruffy oil seep in northeastern Nevada is closely associated with Bruffey hot springs. Duey (1983) has discussed some of these conditions and has presented a local geothermal gradient map for the Railroad Valley region of Nevada (Fig. 14). In the Railroad Valley, two of the three oil fields are associated with geothermal anomalies, one with a high heat flow anomaly and one with a low heat flow anomaly. Another prominent hot spring area does not, at least with the present state of knowledge, appear to be associated with a hydrocarbon accumulation. In active tectonic settings there may prove to be a closer relation between present (or fossil) geothermal fluid circulation and hydrocarbon accumulations than would have been realized before the full implication of the generation of hydrocarbons by thermal processes were understood. In these cases there may be a very direct relationship between geothermal exploration and hydrocarbon exploration. Examples presented in the previous sections, from geothermal systems, may be more appropriate than first realized.

REFERENCES

Back, W., Hanshaw, B. B., Plumer, L. N., Rahn, P. R., Rightmire, C. T., and Rubin, M., 1983, Progress and rate of dedolomitization: Mass Transfer and ^{14}C dating in a regional carbonate aquifer: Geological Society of America Bulletin, v. 94, p. 1415–1429.

Ball, S. M., 1982, Exploration application of temperatures recorded on log headings: An up-the-odds method of hydrocarbon-charged porosity prediction: American Association of Petroleum Geologists Bulletin, v. 66, p. 1108–1123.

Bethke, C., 1985, A numerical model of compaction-driven groundwater flow and heat transfer and its application to the paleohydrology of intercratonic sedimentary basins: Journal of Geophysicial Research, v. 90, p. 6817–1828.

Blackwell, D. D., and Chapman, D. S., 1977, Interpretation of geothermal gradient and heat flow data for Basin and Range geothermal systems, Geothermal Resources Council Transactions, v. 1, p. 19–20.

———, and Spafford, R. E., 1986, Experimental methods in continental heat flow, *in* Sammis, C. G., and Henyey, T. L., eds., Methods of experimental physics-geophysics: New York, Academic Press, in press.

———, and Steele, J. L., 1986, Thermal conductivity of sedimentary rocks—Measurement and significance: Society of Economic Petrologists and Mineralogists Special Publication, in press.

———, Steele, J. L., and Brott, C. A., 1980, The terrain effect on terrestrial heat flow: Journal of Geophysical Research, v. 85, p. 4757–4772.

———, Hull, D. A., Bowen, R. A., and Steele, J. L., 1978, Heat flow of Oregon: Oregon Department of Geology and Mineral Industries, Special Paper, v. 4, 42 p.

Bowen, R. G., Blackwell, D. D., and Hull, D. A., 1977, Geothermal exploration studies in Oregon: Oregon Department of Geology and Mineral Industries, Miscellaneous Paper, no. 19, 50 p.

Bredehoeff, J. D., and Papadopulos, I. S., 1965, Rates of vertical groundwater movement estimated from the earth's thermal profile: Water Resources Research, v. 1, p. 325–328.

———, Neuzil, C. E., and Milly, P. C. D., 1983, Regional flow in the Dakota aquifer: A study of the role of confining layers: United States Geological Survey, Water-Supply Paper 2237, 45 p.

Brott, C. A., Blackwell, D. D., and Morgan, P., 1981, Continuation of heat flow data: A method to construct isotherms in geothermal areas: Geophysics, v. 46, p. 1732–1744.

Bullard, E. C., 1939, Heat flow in South Africa: Royal Society London, Proceedings, series A, v. 173, p. 474–502.

Buntebart, G., 1980, Geothermics: An Introduction: Berlin, Springer-Verlag, 144 p.

Carvalho, H. da S., and Vacquier, V., 1977, Method for determining terrestrial heat flow in oil fields: Geophysics, v. 42, p. 584–593.

———, Purwoko, S., Thamrin, M., and Vacquier, V., Terrestrial heat flow in the Tertiary basin of central Sumatra: Tectonophysics, v. 69, p. 163–188.

Cathles, L. M., and Smith, A. D., 1983, Thermal constraints on the formation of Mississippi Valley-type lead-zinc deposits and their implications for episodic basin dewatering and deposit genesis: Economic Geology, v. 78, p. 983–1002.

Chapman, D. S., Keho, T. H., Bauer, M. S., and Picard, M. D., 1984, Heat flow in the Uinta basin determined from bottom hole temperture (BHT) data: Geophysics, v. 49, p. 453–466.

Clark, S. P., Jr., 1966, Thermal conductivity, *in* Clark, S. P., Jr., ed., Handbook of Physical Constants: Geological Society of America, Memoir, v. 97, 587 p.

Darton, N. H., 1906, Geology and groundwater resources of the Central Great Plains: U.S. Geological Survey, Professional Paper, v. 32, 433 p.

Del Grande, N. K., 1985, Airborne temperature survey maps of heat flow anomalies for exploration geology: Geothermal Resources Council Bulletin, v. 14, no. 3, p. 3–13.

Diment, W. H., 1967, Thermal regime of a large diameter borehole: Instability of the water column and comparison of air- and water-filled conditions: Geophysics, v. 32, p. 720–726.

Domenico, P. A, and Palciauskas, V. V., 1973, Theoretical analysis of forced convective heat transfer in regional groundwater flow: Geological Society of America Bulletin, v. 84, p. 3803–3814.

Duey, H. D., 1983, Oil generation and entrapment in Railroad Valley, Nye County, Nevada: Geothermal Resources Council, Special Report, v. 13, p. 199–206.

Garven, G., 1985, The role of regional fluid flow in the genesis of the Pine Point deposit, Western Canada Sedimentary Basin: Economic Geology, v. 80, p. 307–324.

Gatenby, G. M. 1980, Exploration ramifications of subsurface fluid migrations in the Lake Borgne-Valentine area of southeastern Louisiana: Gulf Coast Association of Geological Societies Transactions, v. 30, p. 91–104.

Geertsma, J., 1971, Finite-element analysis of shallow temperature anomalies: Geophysical Prospecting, v. 19, p. 652–681.

Gosnold, W. D., 1984, Heat flow and groundwater movement in the central great plains, p. 70–75, *in* Jorgensen, D. G., and Signor, D. C., eds., Geohydrology Dakota Aquifer Symposium, Proceedings: Worthington, Ohio, Water Well Journal Publishing Co., 213 p.

Gosnold, W. D., Jr., 1985, Heat flow and groundwater flow in the great plains of the United States: Journal of Geodynamics, v. 4, p. 247–264.

———, Eversoll, D. A., and Carlson, M. P., 1982, Three years of geothermal research in Nebraska, p. 142–157, *in* Ruscetta, C. A., ed., Geothermal direct heat program roundup technical conference, Proceedings: United States Department of Energy Report ID 12079-79, p. 727–738.

Gretener, P. E., 1967, On the thermal instability of large diameter wells: An observational report: Geophysics, v. 32, p. 727–738.

———, 1981, Geothermics: Using temperature in hydrocarbon exploration, Tulsa: Education Course Note Series no. 17: American Association of Petroleum Geologists, 170 p.

Guffanti, M., and Natheson, M., 1981, Temperature-depth data selected for drill holes in the United States obtained by using maximum thermometers: United States Geological Survey Open-File Report, p. 81–555.

Harrison, F. W., III, 1980, The role of pressure, temperature, salinity, lithology, and structure in hydrocarbon accumulation in Constance Bayou, Deep Lake, and Southeast Little Pecan Lake Fields, Cameron Parish, Louisiana: Gulf Coast Association of Geological Societies Transactions, v. 30, p. 113–129.

Hawtof, E. M., 1930, Results of deep well temperature measurements in Texas: American Petroleum Industry Production Bulletin, v. 205, p. 62–108.

Houbolt, J. J. H. C., and Wells, P. R. A., 1980, Estimation of heat flow in oil wells based on a relation between heat conductivity and sound velocity: Geologie en Minjnbouw, v. 59, p. 215–224.

Hyndman, D. D., Davis, E. E., and Wright, J. A., 1979, The measurement of marine geothermal heat flow by a multipenetration probe with digital acoustictelemetry and *in situ* thermal conductivity: Marine Geophysical Research, v. 4, p. 181–205.

Jones, P. H., 1984, Deep water discharge: A mechanism for the vertical migration of oil and gas, p. 254–271, *in* Davidson, M. J., and Gottlieb, B. M., eds., Unconventional methods in exploration for petroleum and natural gas III: Dallas, Southern Methodist University Press, 271 p.

Kappelmeyer, O., and Haenel, R., 1974, Geothermics with special references to applications: Berlin, Geopublication Associates, 238 p.

Keen, C. E., and Lewis, T., 1982, Measured radiogenic heat production in sediments from continental margin of eastern North America: Implications for petroleum generation: American Association of Petroleum Geologists Bulletin, v. 66, p. 1402–1407.

Lachenbruch, A. H. 1957, Three-dimensional heat conduction in permafrost beneath heated buildings: United States Geological Survey Bulletin, v. 1052-B, p. 51–67.

———, and Brewer, M. C., 1959, Dissipation of the temperature effect in drilling a well in Arctic Alaska: United States Geological Survey Bulletin, v. 1083-C, p. 73–109.

———, Brewer, M. C., Greene, G. W., and Marshall, B. V., 1962, Temperatures in permafrost, p. 791–803, *in* Temperature: Its measurement and control in science and industry, v. 3, no. 1: New York, Reinholt Publishing Corp., 848 p.

Lam, H. L., Jones, F. W., and Majorowicz, J. A., 1985, A statistical analysis of bottom-hole temperature data in southern Alberta: Geophysics, v. 50, p. 677–684.

Lange, A. L., Pilkington, H. D., and Deymonaz, J., 1982, Comparative studies of geothermal surveys in 3-meter and temperature-gradient holes: Geothermal Resources Council Transactions, v. 6, p. 133–136.

LeSchack, L. A., Lewis, J. E., Chang, D. C., Lewellen, R. I., and O'Hara, N. W., 1979, Rapid reconnaissance of geothermal prospects using shallow temperature surveys: United States Department of Energy Report for Contract EG-77-C-01-4021, 44 p.

Lovering, T. S., and Goode, H. D., 1963, Measuring geothermal gradients in drill holes less than 60 feet deep, East Tintic district, Utah: United States Geological Survey Bulletin, v. 1172, 48 p.

———, and Morris, H. T., 1965, Underground temperatures and heat flow in the East Tintic District, Utah: United States Geological Survey Professional Paper 504-F, 28 p.

MacLeod, M. K., 1982, Gas hydrates in ocean bottom sediments: American Association Petroleum Geologists Bulletin, v. 66, p. 2649–2662.

Majorowicz, J. A., and Jesop, A. M., 1981, Present heat flow and a preliminary paleogeothermal history of the Central Prairies Basin, Canada: Geothermics, v. 10, p. 81–93.

———, Jones, F. W., Lam, H. L., and Jessop, A. M., 1984, The variability of heat flow both regional and with depth in southern Alberta, Canada: Effects of groundwater flow?: Tectonophysics, v. 106, p. 1–29.

Mansure, A. J., and Reiter, M., 1979, A vertical groundwater movement correction for heat flow: Journal of Geophysical Research, v. 84, p. 3490–3496.

Mareschal, J. C., Cunningham, J. P., and Lowell, R. P., 1985, Downward continuation of heat flow data: Method and examples from the western United States: Geophysics, v. 50, p. 846–851.

McKenzie, D., 1981, The variation of temperature with time and hydrocarbon maturation in sedimentary basins formed by extension: Earth and Planetary Science Letters, v. 55, p. 87–98.

Newton, V. C., and Corcoran, R. E., 1963, Petroleum geology of the western Snake River basin, Oregon-Idaho: Oregon Department of Geological and Mineral Industries Oil and Gas Inv., no. 1, 67 p.

O'Brien, J. J., and Lerche, I., 1984, The influence of salt domes on paleotemperature distributions: Geophysics, v. 49, p. 2023–2043.

Olmsted, F. H., 1977, Use of temperature surveys at a depth of 1 meter in geothermal exploration in Nevada: United States Geological Survey Professional Paper, 1044-B.

Poley, J. Ph., and Van Stevenick, J., 1970, Geothermal prospecting: Delineation of shallow salt domes and surface faults by temperature measurements at a depth of approximately 2 meters: Geophysical Prospecting, v. 18, p. 666–700.

Poole, F. G., and Claypool, G. E., 1984, Petroleum source-rock potential and crude-oil correlation in the Great Basin, p. 179–229, *in* Woodward, J., Meissner, F. F., and Clayton, J. L., eds., Hydrocarbon source rocks of the greater Rocky Mountain Region: Rocky Mountain Geological Association, Denver.

Reiter, M. and Mansure, A. J., 1983, Geothermal studies in the San Juan Basin and Four Corners area of the Colorado plateau, I: Terrestrial heat-flow studies: Tectonophysics, v. 91, p. 233–251.

Roberts, W. H., III, 1980, Design and function of oil and gas traps, p. 217–240, *in* Roberts, W. H., III and Cordel, R. J., eds., Problems of petroleum migration: American Association of Petroleum Geologists, Studies in Geology, v. 10, 269 p.

———, 1981, Some uses of temperature data in petroleum exploration, *in* Gottlieb, B. M., ed., Unconventional methods in exploration for petroleum and natural gas II: Dallas, Southern Methodist University Press, p. 8–48.

Ross, H. P., Nielson, D. L., and Moore, J. N., 1982, Roosevelt hot springs geothermal system, case study, Utah: American Association of Petroleum Geologists Bulletin, v. 66, p. 879–902.

Roy, R. F., Beck, A. E., and Toyloukian, Y. S., 1981, Thermophysical properties of rocks, *in* Toyloukian, Y. S., Judd, W. R., and Roy, R. F., eds., Physical properties of rocks and minerals: New York, McGraw-Hill Cindus, v. II-2, p. 409–502.

Sammel, E., A., 1968, Convective flow and its effect of temperature logging in small diameter wells: Geophysics, v. 33, p. 1004–1012.

Sass, J. H., Kennelly, J. P., Jr., Wendt, W. E., Moses, T. H., Jr., and Ziagos, J. P., *In situ* determination of heat flow in unconsolidated sediments: Geophysics, v. 46, p. 76–83.

———, Ziagos, J. P., Wollenberg, H. A., Munroe, R. J., diSomme, D. E., and Lachenbruch, A. H., 1977, Applications of heat flow techniques to geothermal energy exploration, Leach hot springs area, Grass Valley, Nevada: United States Geological Survey Open File Report, p. 77–762, 125 p.

Sclater, J. C., and Christie, P. A. F., 1980, Continental stretching: An explanation of the post-mid-Cretaceous subsidence of the Central North Sea Basin: Journal of Geophysical Research, v. 85, p. 3711–3739.

Selig, F., and Wallick, G. C., 1966, Temperature distribution in salt domes and surounding sediments: Geophysics, v. 31, p. 346–361.

Smith, L., and Chapman, D. S., 1983, On the thermal effects of groundwater flow I., Regional scale systems: Journal of Geophysical Research, v. 88, p. 593–608.

Sorey, M. J., 1971, Measurement of vertical groundwater velocity from temperature profiles in wells: Water Resources Research, v. 7, p. 963–970.

Speece, M. A., Bowen, T. D., Folcik, J. L., and Pollack, H. N., 1985, Analysis of temperatures in sedimentary basins: The Michigan basin: Geophysics, v. 50, p. 1318–1334.

Toth, J., 1980, Cross-formational gravity-flow of groundwater: A mechanism of the transport and accumulation of petroleum (The generalized hydraulic theory of petroleum migration), p. 121–167, *in* Roberts, W. H., III, and Cordell, R. J., eds., Problems of petroleum migration: American Association of Petroleum Geologists, Studies in Geology, v. 10, 269 p.

Vacquier, V., 1985, The measurement of thermal conductivity of solids with a transient linear heat source on the plane surface of a poorly conducting body: Earth Planetary Science Letters, v. 74, p. 275–279.

Van Ostrand, C. E., 1934, Temperature gradients: Problems in petroleum geology: American Association of Petroleum Geologists, p. 989-1021.

———, 1937, On the estimation of temperatures at moderate depths in the crust of the earth: American Geophysical Union Transactions, p. 21–33.

Van Wijk, W. R., 1966, Physics of plant environment: Amsterdam, North Holland Publishing Co., 382 p.

Ward, S. H., 1978, A summary of the geology, geochemistry, and geophysics of the Roosevelt hot springs thermal area, Utah: Geophysics, v. 43, p. 1515–1542.

Zielinski, G. W., and Bruchhauser, P. M., 1983, Shallow temperatures and thermal regime in the hydrocarbon province of Tierra del Fuego: American Association of Petroleum Geologists Bulletin, v. 67, p. 166–177.

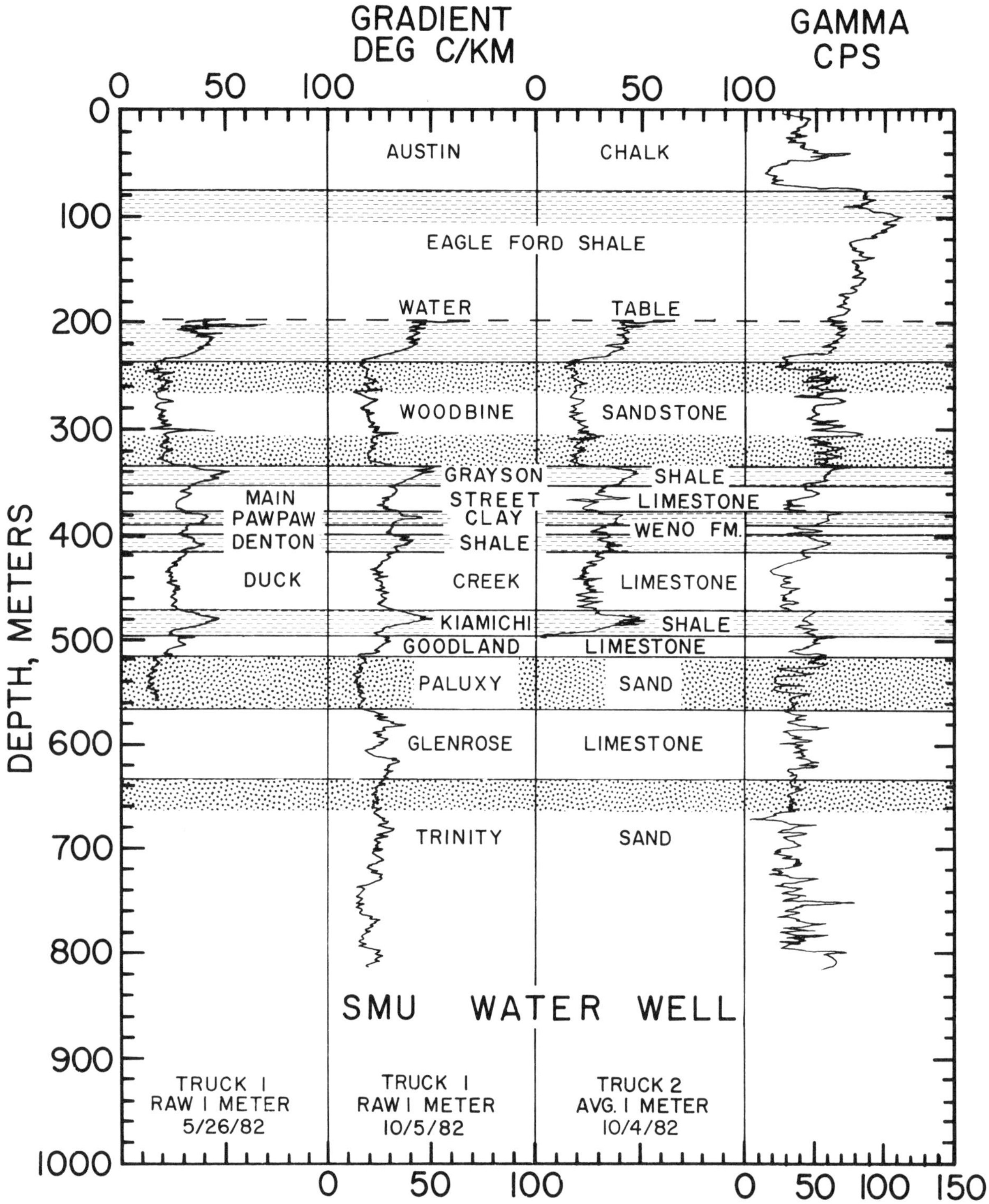

Fig. 1. Temperature gradient and total gamma count logs for a well on the Southern Methodist University campus. The section includes Upper and Lower Cretaceous rocks. Formation names are shown. Sand units are shown by dot pattern and shale units are shown by a line pattern. The remainder of the units are composed of marl or limestone. Temperatures were digitized at 1 or 0.5 m intervals to a precision of 0.001°C, and are not shown above the water table. Truck 1 logs were made with a conventional thermistor probe. Truck 2 log was made with a semiconductor sensor (see Blackwell and Spafford, 1986).

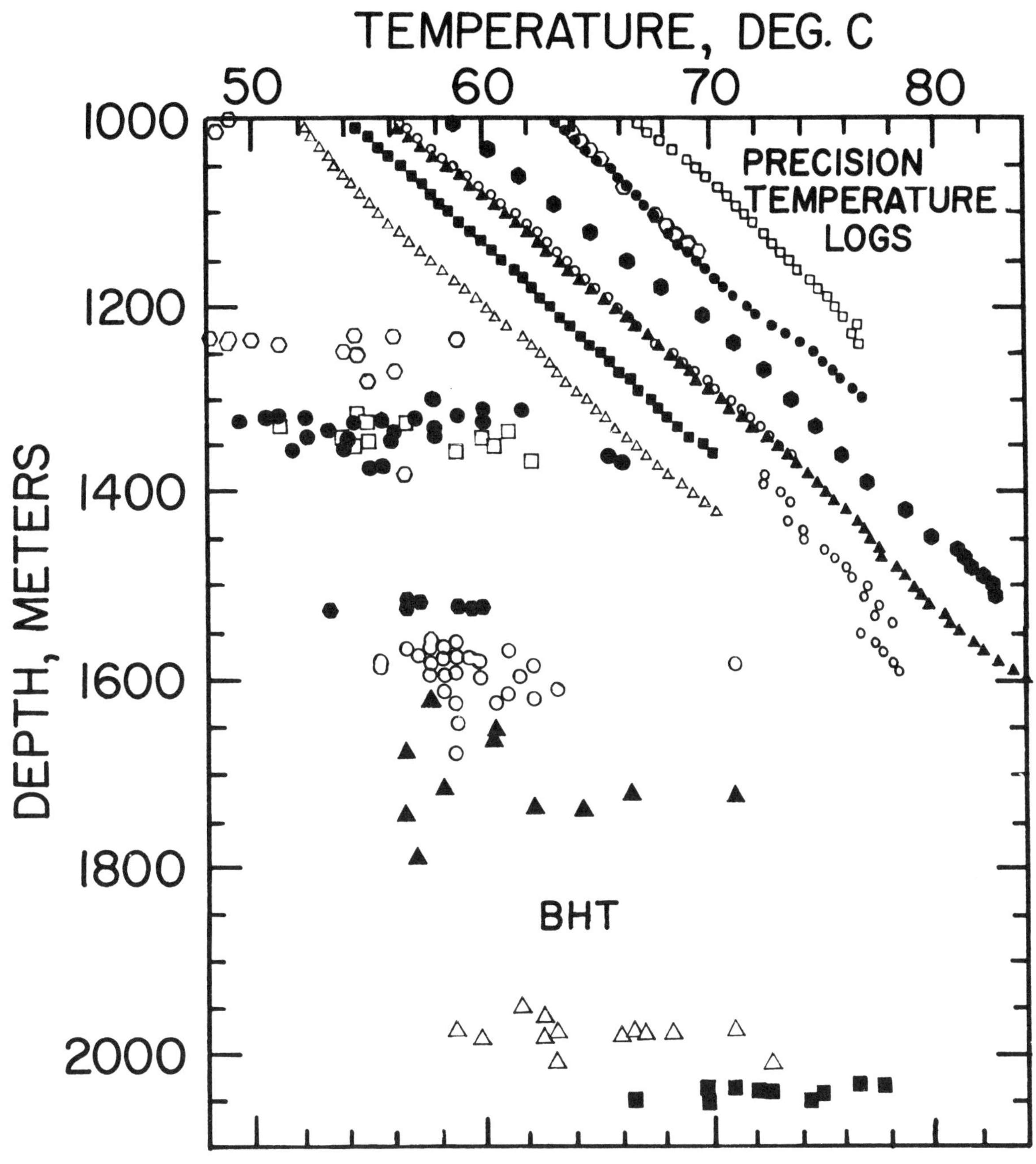

Fig. 2. Comparison of bottom hole temperatures and precision temperature logs for some holes in southwestern Nebraska (after Gosnold et al., 1982). BHT data within a 10 km radius of a particular deep well with a precision temperature log are shown by the same symbol used for plotting the temperature log for that well.

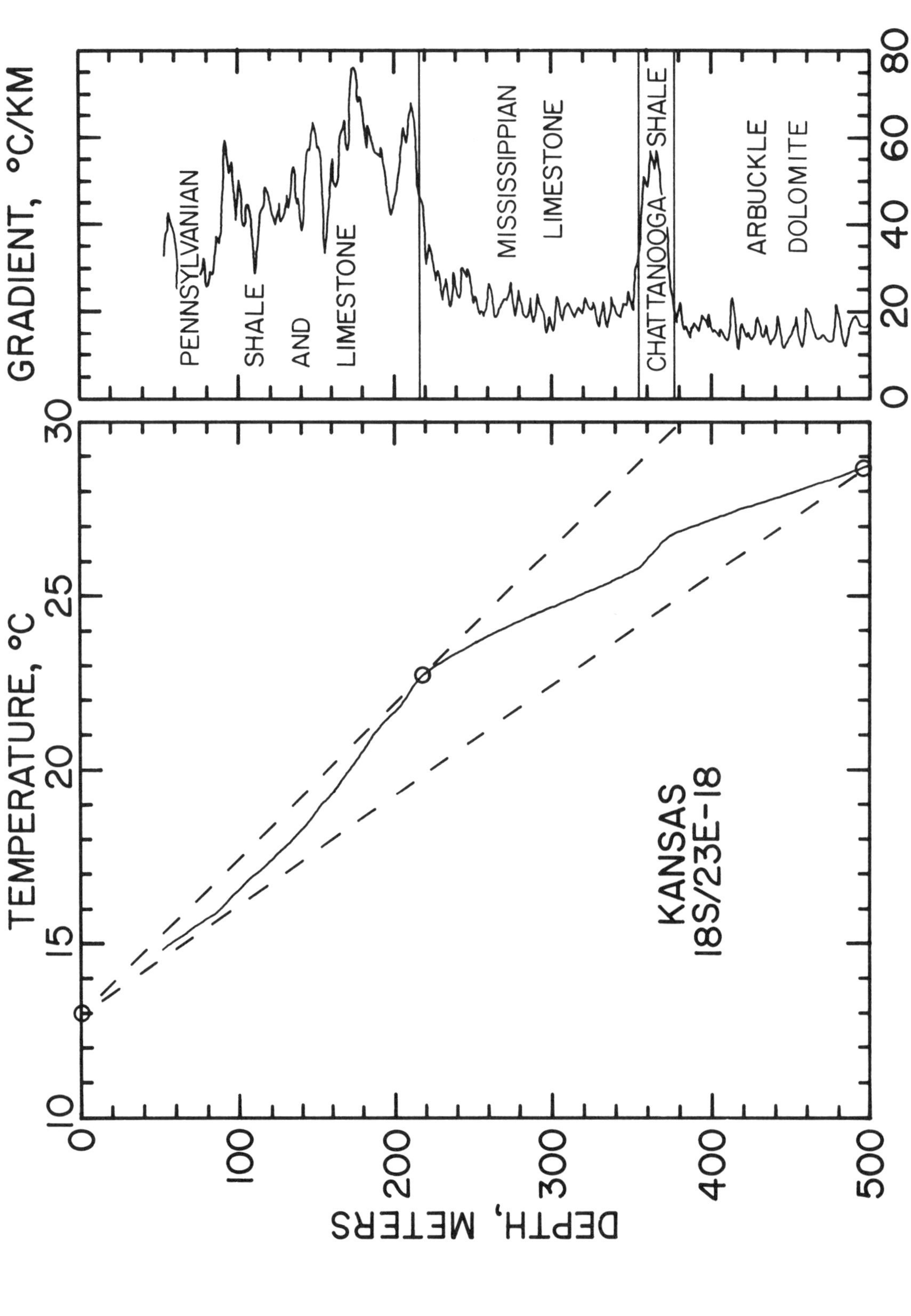

Fig. 3. Temperature-depth and temperature gradient-depth plots for a hole in eastern Kansas in T18S, R23E section 18 SE 1/4 of SW 1/4 of SE 1/4. The generalized stratigraphy encountered in the hole is shown. Dashed lines show the results that would be obtained by calculating an average gradient for the well using the mean annual surface temperature and a "bottom hole temperature" from 220 m depth and 460 m depth.

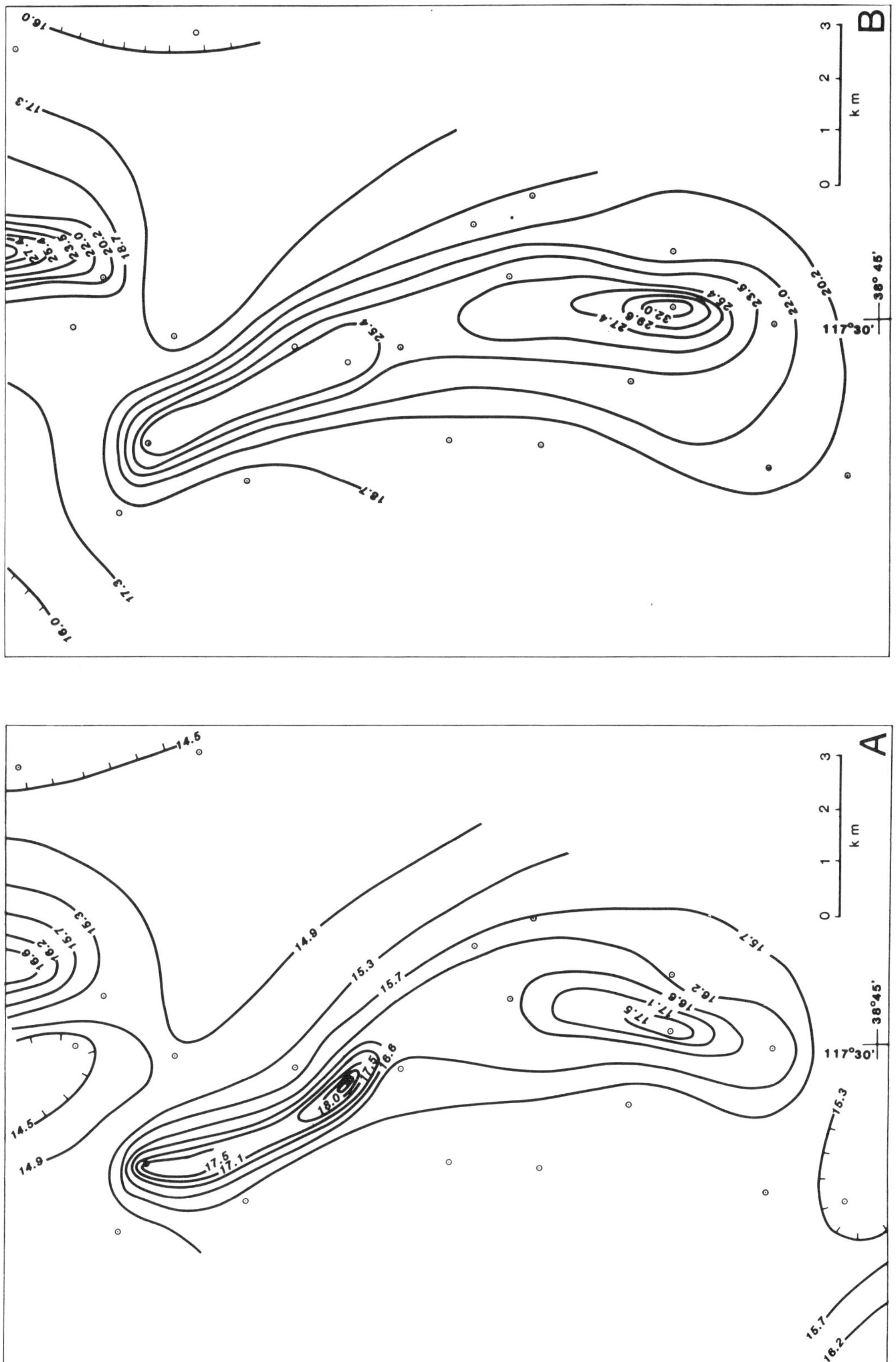

Fig. 4. Plan view of the thermal anomaly at McCoy, Nevada, July 11, 1978. Figure 4A shows temperatures at 3 m and Figure 4B shows temperatures at 30 m (both adjusted for altitude; after Lange et al., 1982, Figure 5).

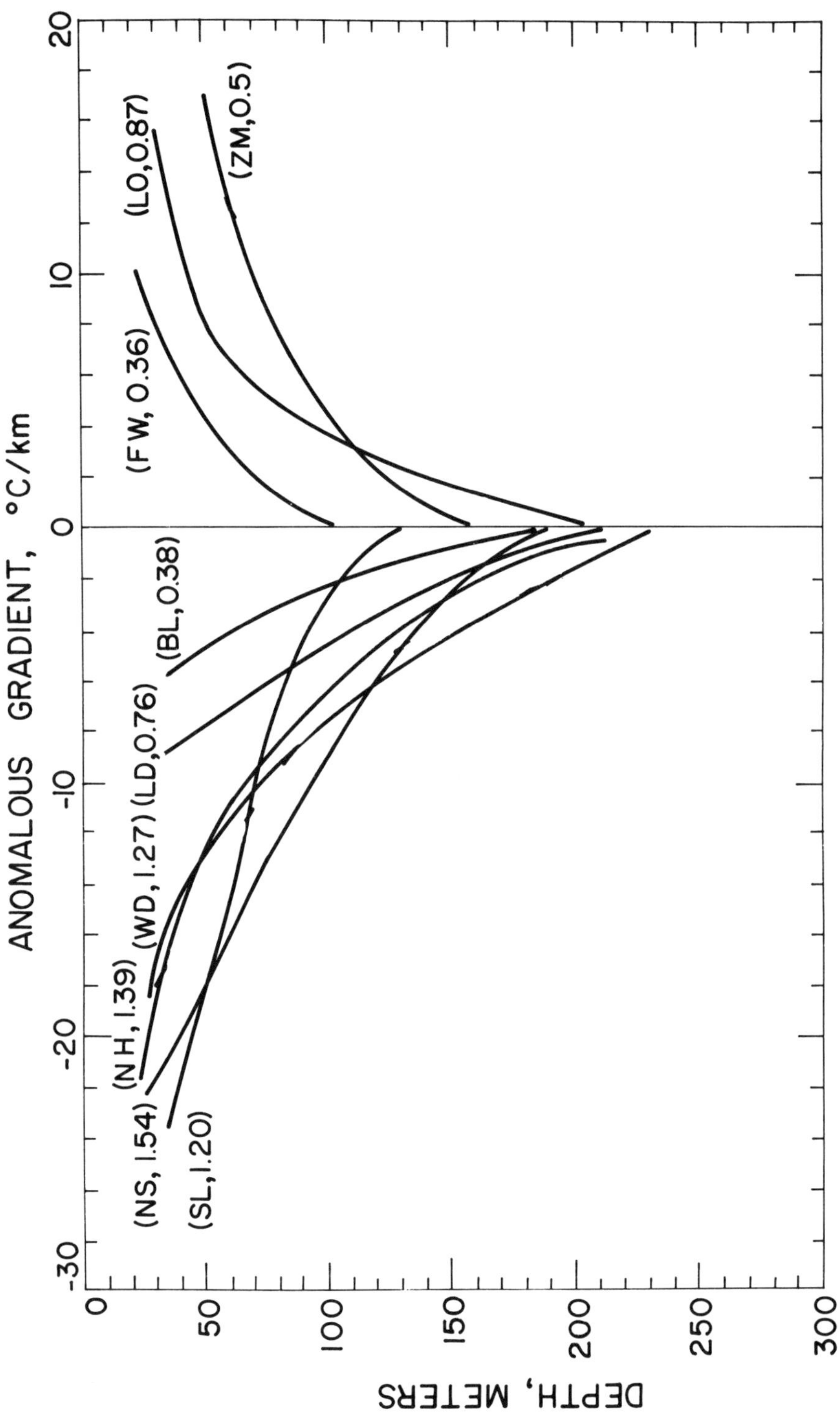

Fig. 5. Examples of smoothed "anomalous" gradient versus depth for a number of drill holes in New England and California. Characteristic amplitude of microclimatic "noise" is illustrated.

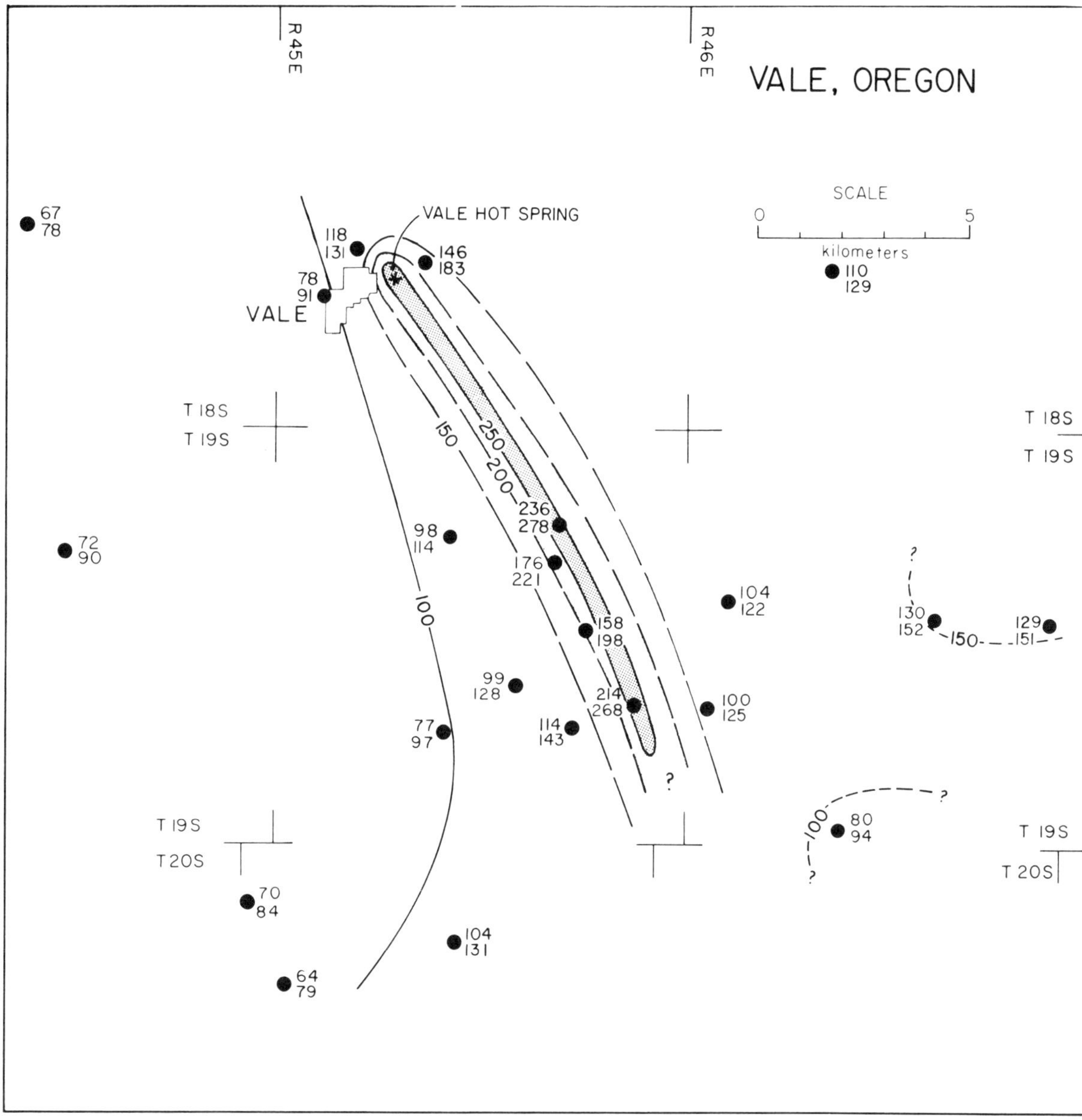

Fig. 6. Heat flow and geothermal gradients for the Vale-Cow Hollow geothermal anomaly near Vale, Oregon. The upper value associated with each hole is temperature gradient in °C/km. The lower value is the heat flow in mWm^{-2}. Heat flow is contoured at 50 mWm^{-2} intervals (after Blackwell et al., 1978, Figure 10).

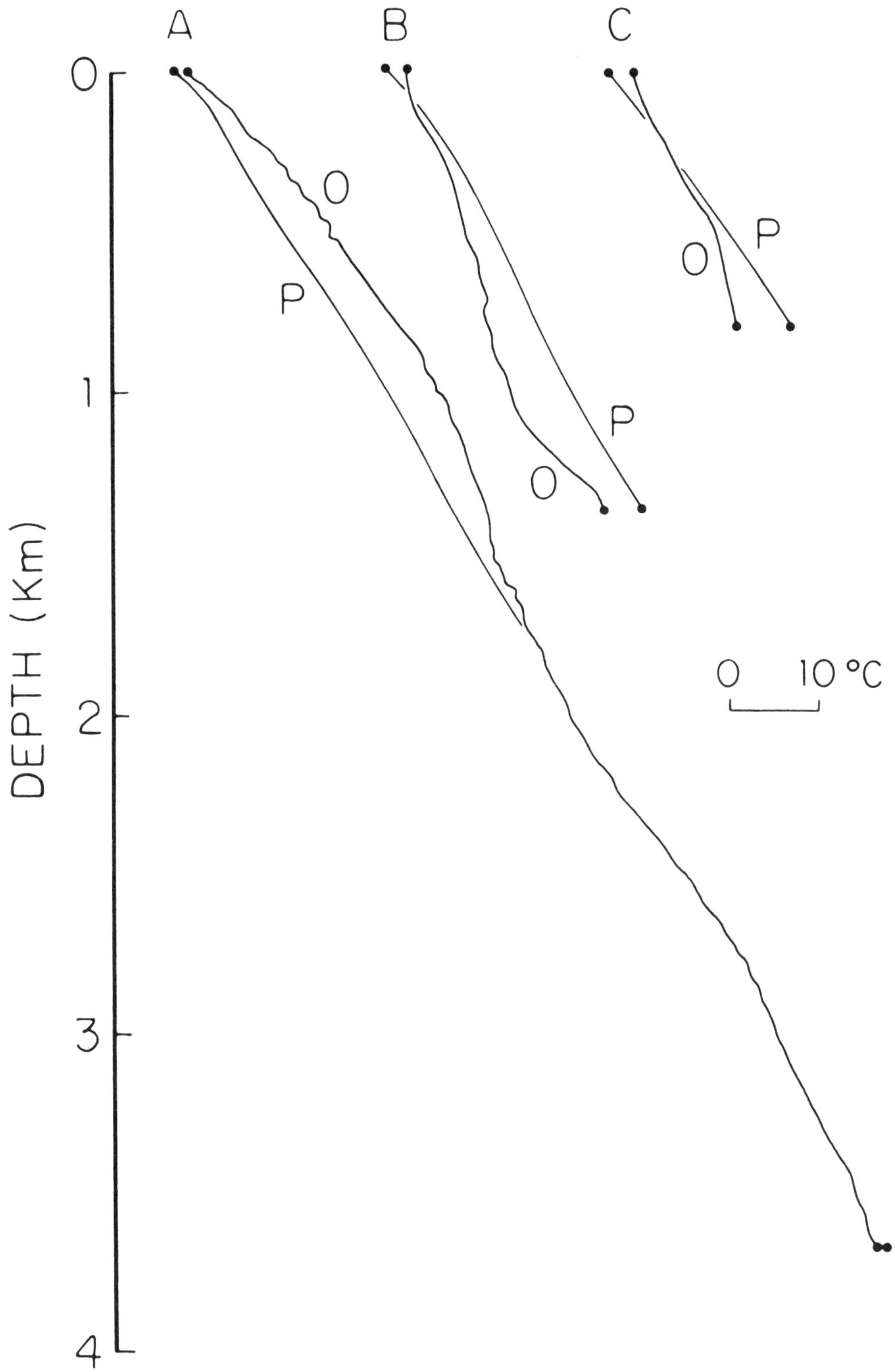

Fig. 7. Observed (O) paired with predicted temperature-depth curves (P) based on lithologically derived formation temperature gradients using large scale inversion of bottom hole temperature data (after Speece et al., 1985, Figure 6).

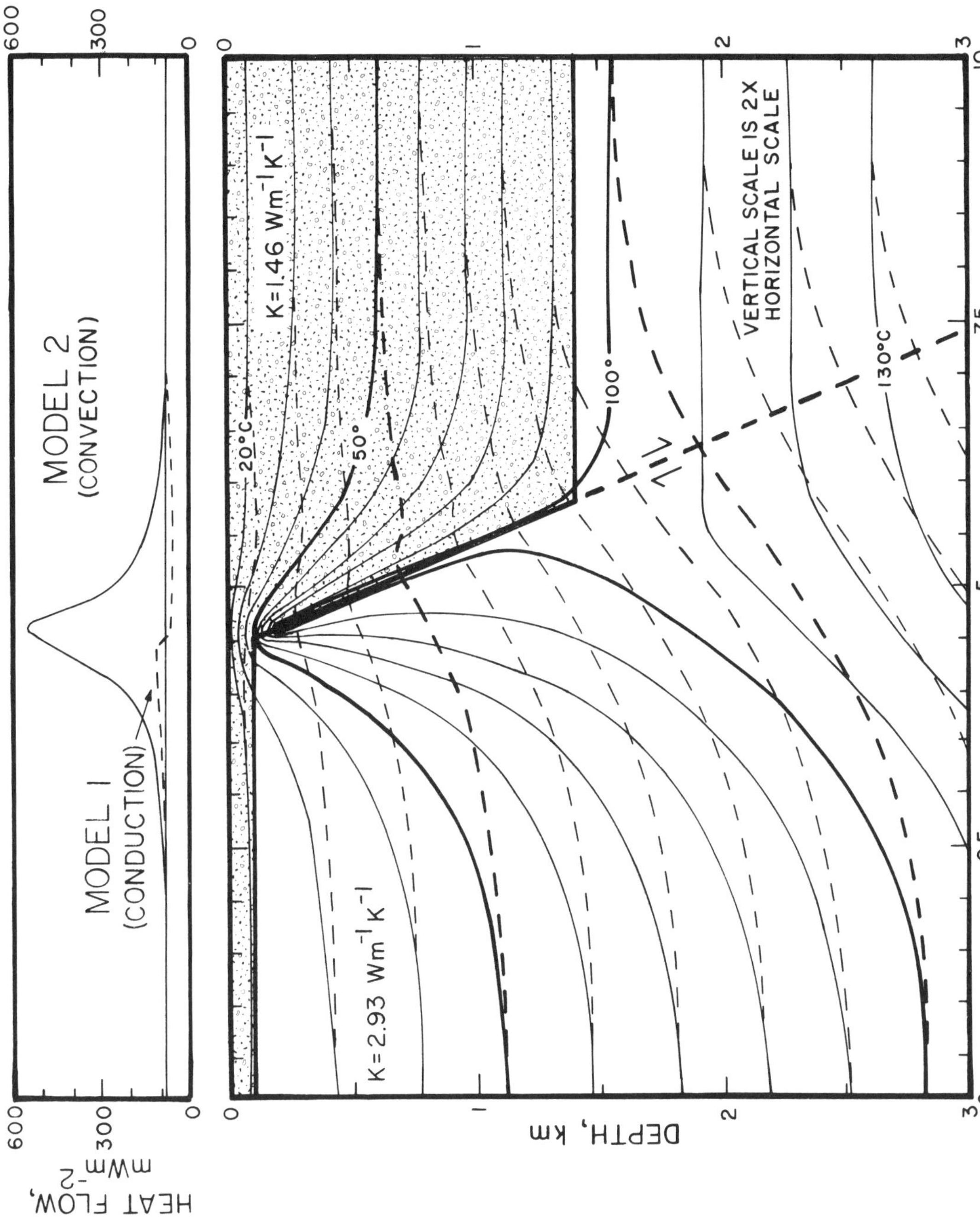

Fig. 8. Theoretical model for a low thermal conductivity graben and high thermal conductivity horst pair. The dashed lines show the theoretical temperature assuming conduction heat transfer only and the solid lines show temperatures for the convection/conduction model (100°C water flowing upward along the boundary fault). The figure is from Blackwell and Chapman (1977).

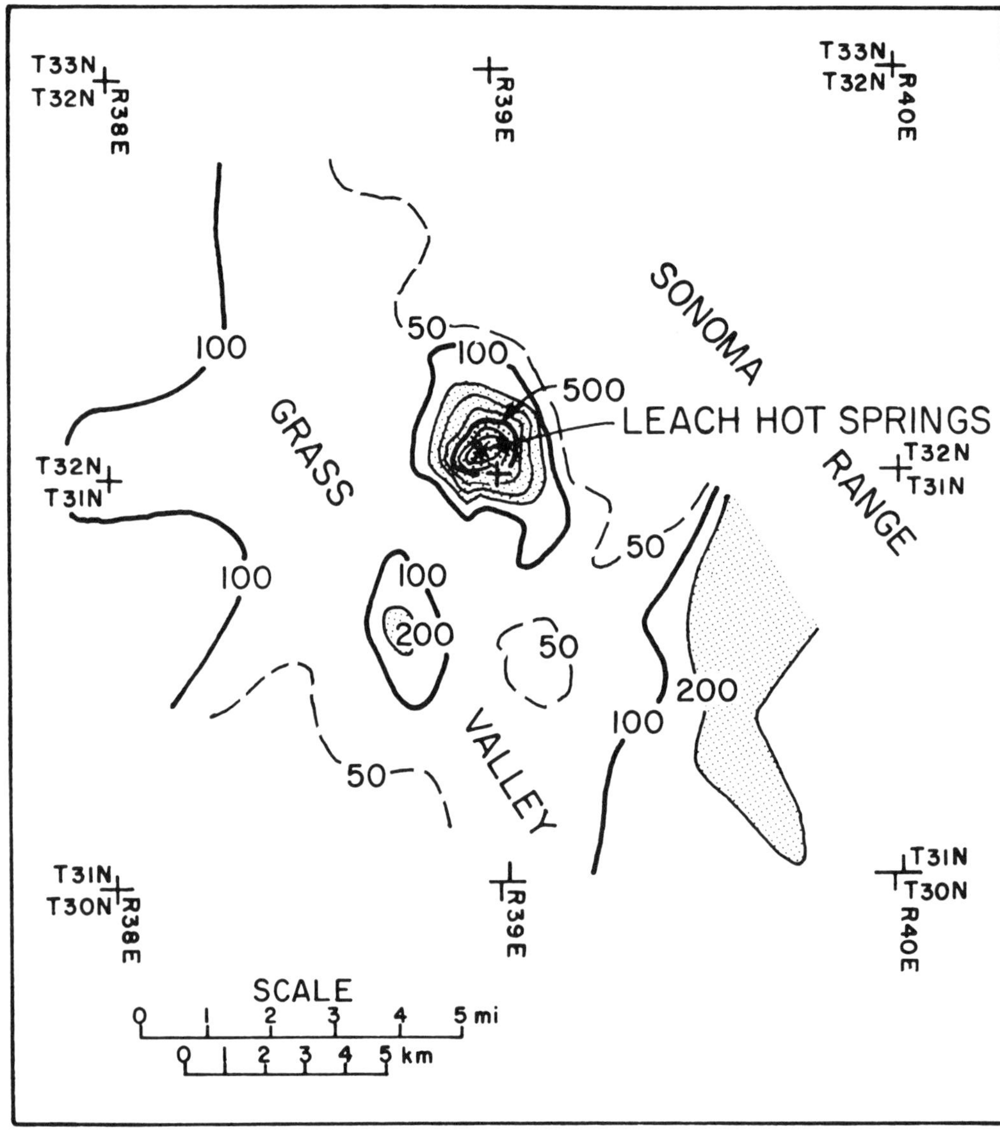

Fig. 9. Heat flow anomaly at Leach hot springs, Nevada. The contour interval is 100 mWm^{-2} with supplemental dashed contours at 50 mWm^{-2} (Sass et al., 1977).

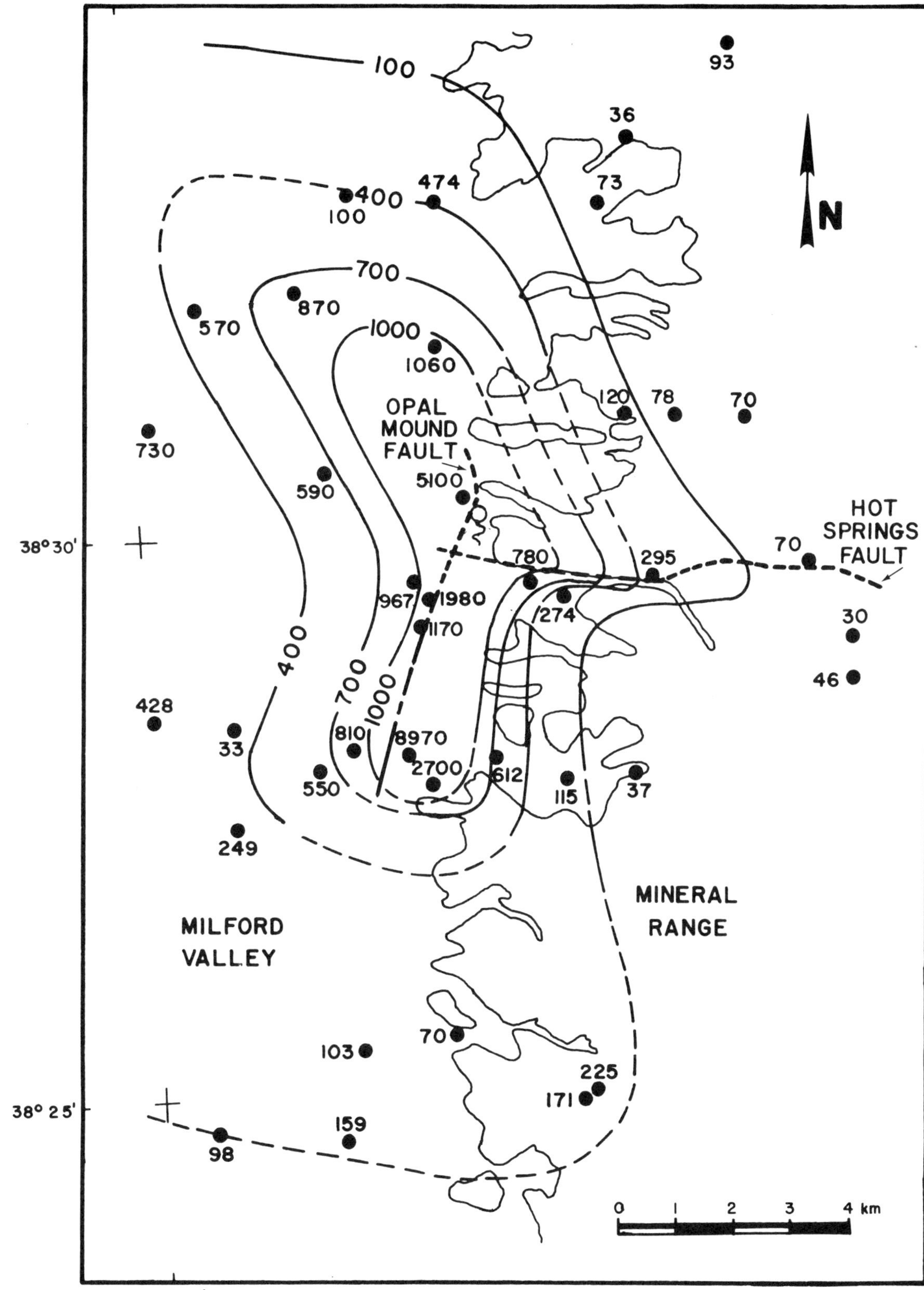

Fig. 10. Heat flow anomaly at Roosevelt hot springs, Utah. The heat flow values are in mWm^{-2} (Ward et al., 1980, Figure 8).

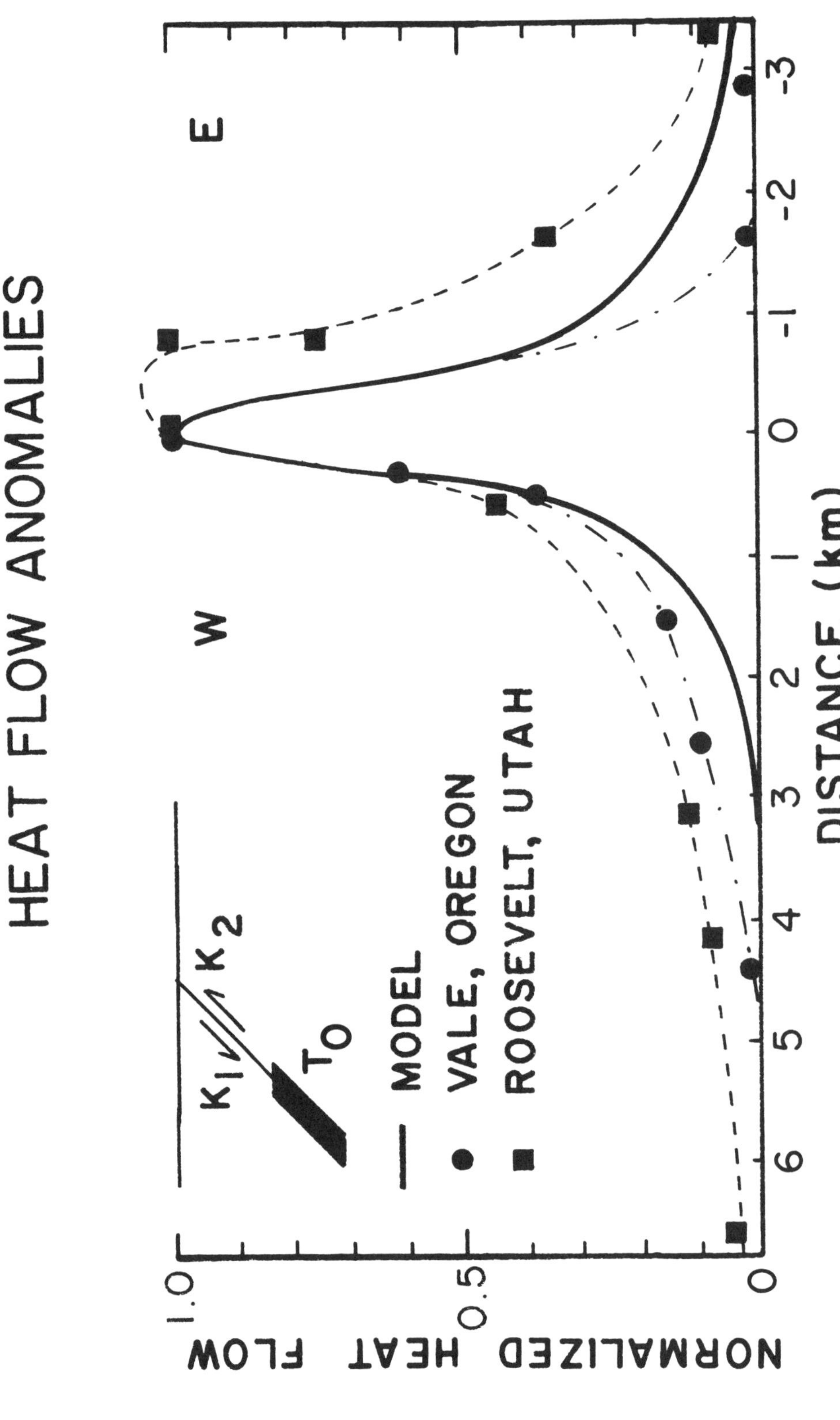

Fig. 11. Comparison of normalized heat flow cross sections of the Vale, Oregon and Roosevelt, Utah heat flow anomalies to theoretical heat flow anomalies due to a fault model.

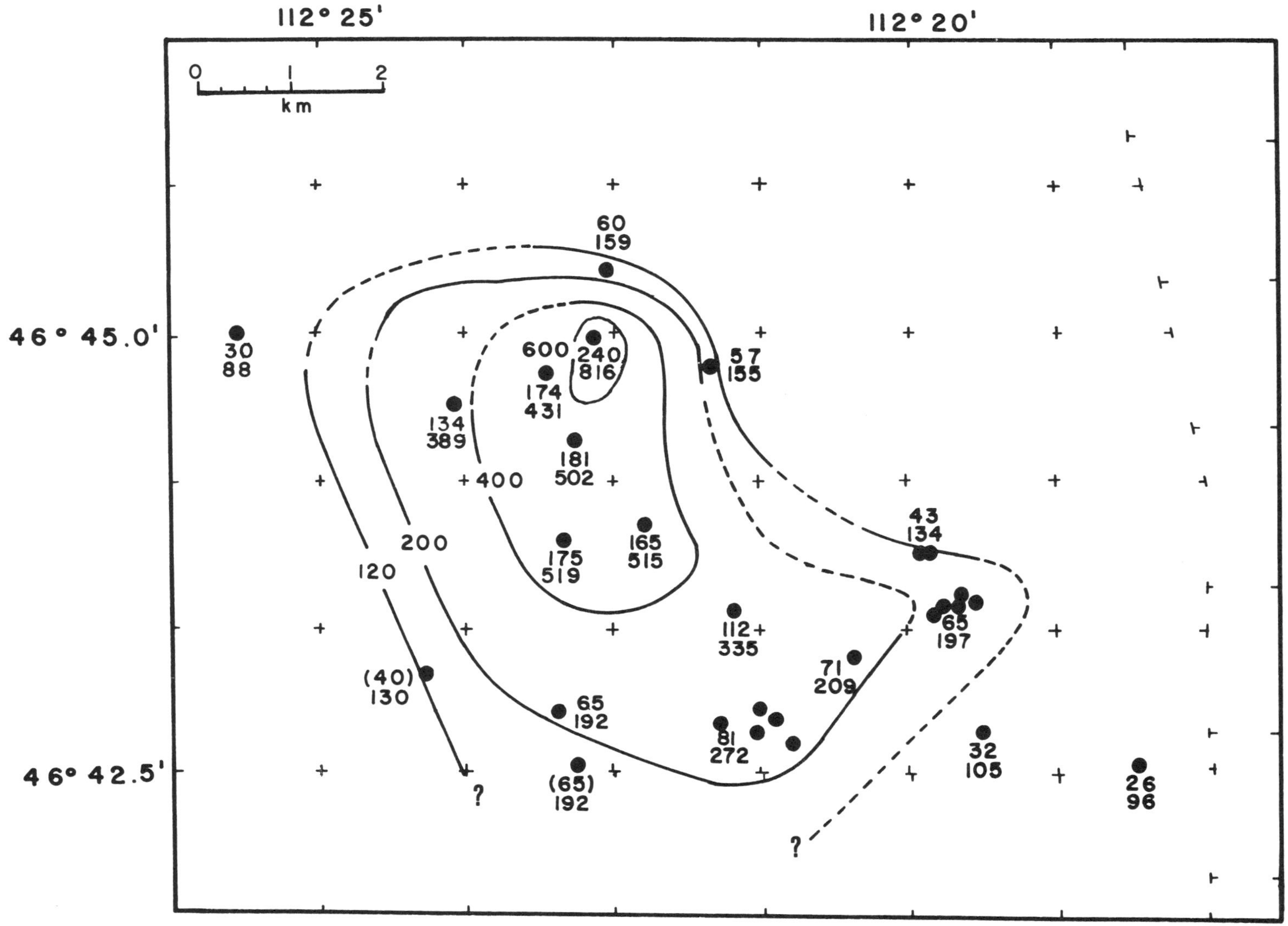

Fig. 12a. Heat flow anomaly at Marysville, Montana. Figure 12a is the heat flow anomaly. The heat flow contours are in mWm^{-2} (Brott et al., 1981). The upper number of each pair is the terrain corrected temperature gradient in °C/km. The lower value is the heat flow in mWm^{-2}. Gradient values shown in parentheses have been adjusted for thermal conductivity variation. Crosses are section corners.

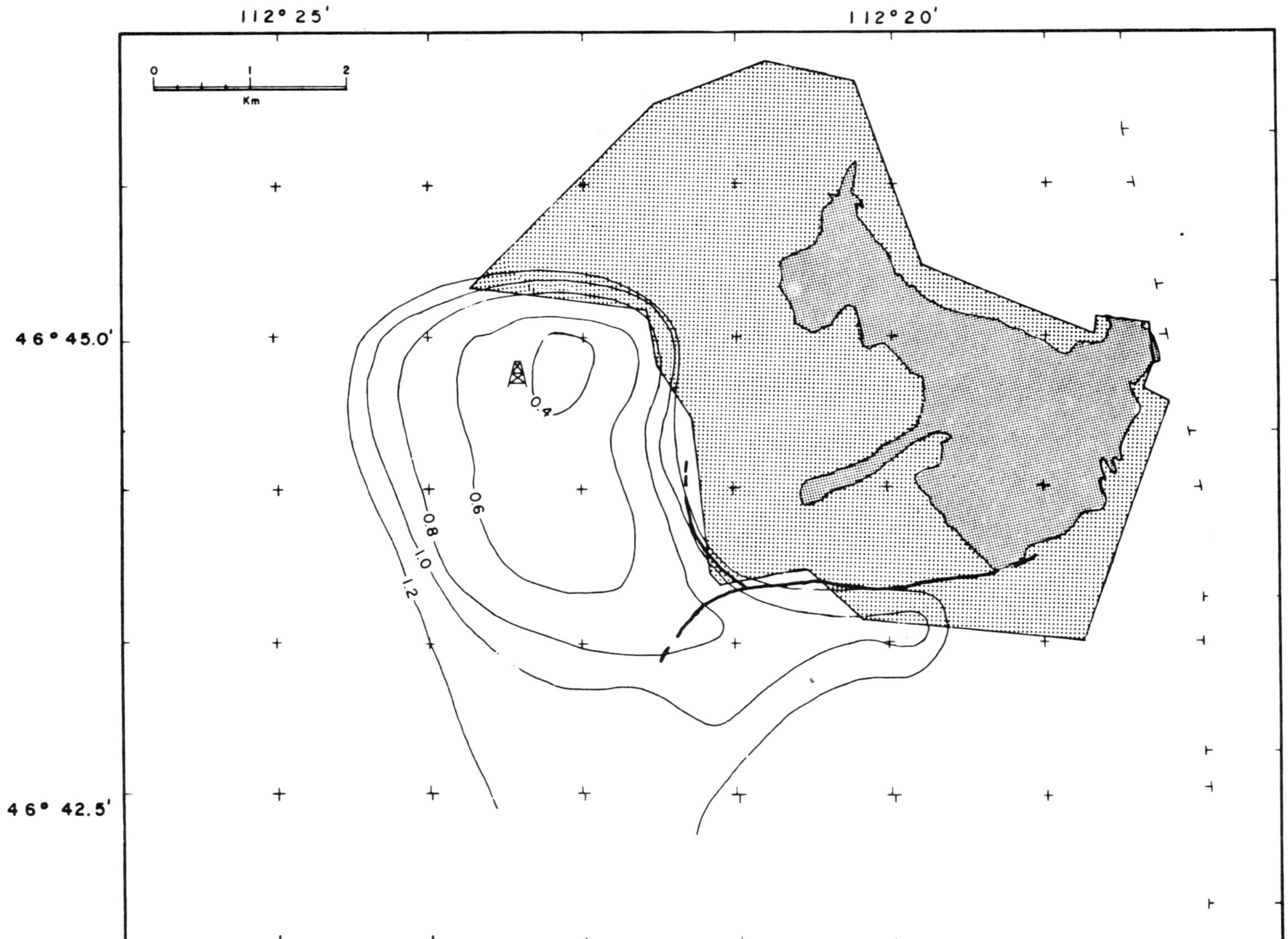

Figure 12b. The Marysville, Montana granodiorite stock and geothermal reservoir (assumed to coincide with the 95°C isotherm). The outcrop and maximum subcrop of the stock are shown as dark and light shaded regions, respectively. The heavy line is the surface trace of a major fault. Crosses are section corners (after Brott et al., 1981). Contours of 95°C isotherms are in kilometers below the surface.

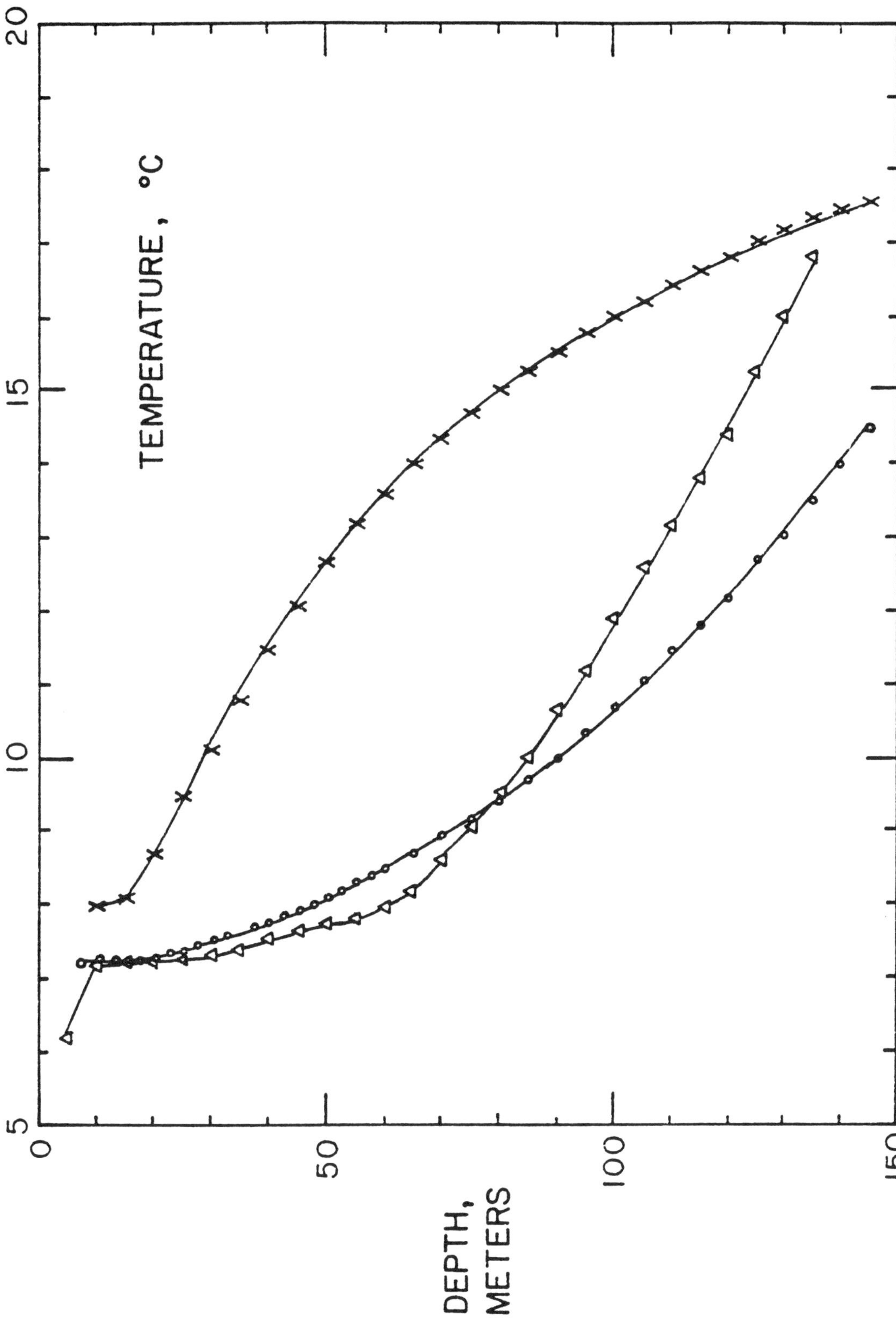

Fig. 13. Examples of large-scale aquifer motions on temperature-depth data. Thermal conductivity is approximately constant in each well. The curves indicated by the circles and triangles would represent typical conditions in a recharge location and the curve indicated by the Xs would represent typical conditions in a discharge location.

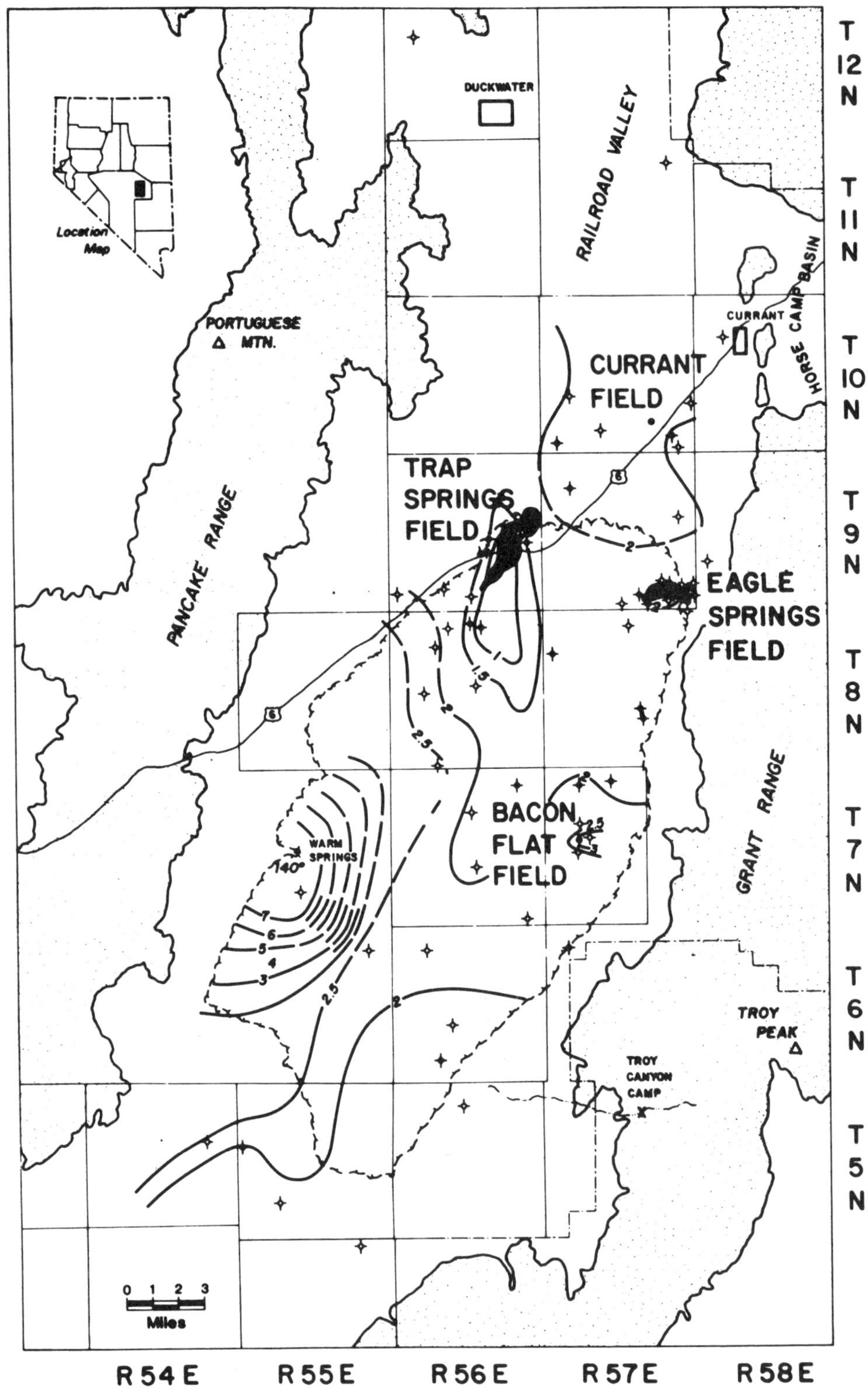

Fig. 14. Temperature gradient in Railroad Valley, Nevada (Duey, 1983). The contour interval is 0.5 or 1.0°F/100 ft (9 or 18°C/km) and the units are °F/100 ft. (18.2°C/km).